PROGRESS IN OPTICS
VOLUME XXXIII

EDITORIAL ADVISORY BOARD

G. S. AGARWAL,	*Hyderabad, India*
T. ASAKURA,	*Hokkaido, Japan*
C. COHEN-TANNOUDJI,	*Paris, France*
V. L. GINZBURG,	*Moscow, Russia*
F. GORI,	*Rome, Italy*
A. KUJAWSKI,	*Warsaw, Poland*
J. PEŘINA,	*Olomouc, Czech Republic*
R. M. SILLITTO,	*Edinburgh, Scotland*
H. WALTHER,	*Garching, Germany*

PROGRESS IN OPTICS

VOLUME XXXIII

EDITED BY

E. WOLF

University of Rochester, N.Y., U.S.A.

Contributors

M. J. BERAN, O. BRYNGDAHL, C. M. DE STERKE, V. I. KLYATSKIN,
A. LUKŠ, D. MALACARA, J. OZ-VOGT, V. PEŘINOVÁ,
T. SCHEERMESSER, J. E. SIPE, V. I. VLAD, F. WYROWSKI

1994

ELSEVIER
AMSTERDAM · LAUSANNE · NEW YORK · OXFORD · SHANNON · TOKYO

ELSEVIER SCIENCE B.V.
SARA BURGERHARTSTRAAT 25
P.O. BOX 211
1000 AE AMSTERDAM
THE NETHERLANDS

ISBN: 0 444 81839 1

© 1994 ELSEVIER SCIENCE B.V. All rights reserved.

No part of this publication may be reproduced, stored in a retrieval system, or transmitted, in any form or by any means, electronic, mechanical, photocopying, recording or otherwise, without the written permission of the publisher, Elsevier Science B.V., Copyright & Permissions Department, P.O. Box 521, 1000 AM Amsterdam, The Netherlands.

Special regulations for readers in the U.S.A.: This publication has been registered with the Copyright Clearance Center Inc. (CCC), Salem, Massachusetts. Information can be obtained from the CCC about conditions under which photocopies of parts of this publication may be made in the U.S.A.

All other copyright questions, including photocopying outside of the U.S.A., should be referred to the copyright owner, Elsevier Science B.V., unless otherwise specified.

No responsibility is assumed by the publisher for any injury and/or damage to persons or property as a matter of products liability, negligence or otherwise, or from any use or operation of any methods, products, instructions or ideas contained in the material herein.

PRINTED ON ACID-FREE PAPER

PRINTED IN THE NETHERLANDS

PREFACE

In this volume, six review articles dealing with topics of current research interest in optics and in related fields are presented. They have been contributed by several well known scientists.

The first article, by V. I. Klyatskin deals with the so-called imbedding method, which has found many useful applications in the study of wave propagation in random media. The method originated in the 1940's in connection with problems encountered in radiative transfer theory and has been usefully applied since that time to the analysis of many statistical wave-propagation problems arising, for example, in the study of ocean waves and in connection with light propagation in the atmosphere.

The second article, contributed by V. Peřinová and A. Lukš, presents a review of an interesting class of nonlinear optical phenomena which have their origin in the dependence of the complex dielectric constant of some media on the light intensity. These phenomena which include self-focusing, self-trapping and self-modulation, are finding many applications, for example in fiber optics devices, in signal processing and in computer technology.

The article by C. M. de Sterke and J. E. Sipe which follows is concerned with another topic of non-linear optics, namely gap solitons. Gap solitons are electromagnetic field structures which can exist in nonlinear media that have periodic variation in their linear optical properties, with periodicities of the order of the wavelength of light. The article presents both qualitative and quantitative descriptions of gap solitons and discusses some experimental schemes for their detection in the laboratory.

In the fourth article V. I. Vlad and D. Malacara describe methods for the determination of optical phase from phase-modulated images. These methods have found applications in plasma diagnostics, in connection with flow characterization and in the design of new optical instruments.

The fifth article by M. J. Beran and J. Oz-Vogt, reviews developments relating to imaging through turbulence in the atmosphere. It describes the current state of our understanding of this subject and discusses the most important methods that are presently being employed to compensate for image distortion caused by atmospheric turbulence.

The last article by O. Bryngdahl, T. Scheermesser and F. Wyrowski reviews developments in the field of digital half-toning. This new electronic technique has evolved from an old method, the half-tone technique, that converts a continuous tone image into a binary one.

We note that, just like its thirty-two predecessors, this volume brings into evidence the considerable progress that continues to be made both in optical theory and in its technological applications.

Emil Wolf

Department of Physics and Astronomy
University of Rochester
Rochester, New York 14627, *USA*

June 1994

CONTENTS

I. THE IMBEDDING METHOD IN STATISTICAL BOUNDARY-VALUE WAVE PROBLEMS

by V. I. Klyatskin (Moscow, Russia and Vladivostok, Russia)

§ 1. Introduction 3
§ 2. Multidimensional Stationary Problems 6
    2.1. Formulation of boundary-value wave problems 6
    2.2. Imbedding equations 9
    2.3. Statistical averaging 14
§ 3. Plane Waves in Layered Random Media 17
    3.1. Formulation of boundary-value wave problems 17
        3.1.1. Wave incidence on the medium layer 17
        3.1.2. Source of plane waves inside the medium 20
    3.2. Imbedding method in one-dimensional wave problems 23
        3.2.1. A wave incident on the layer 23
        3.2.2. Source location inside the layer of an inhomogeneous medium . . . 30
    3.3. Statistical description 32
        3.3.1. Coefficients of reflection and transmission 33
            3.3.1.1. Nondissipative medium 34
            3.3.1.2. Dissipative medium 40
        3.3.2. Source inside the layer of a medium 40
        3.3.3. Energy localization in the statistical problem 41
        3.3.4. Diffusive approximation 43
    3.4. Statistical description of wavefields 45
        3.4.1. Wave incidence on the layer of the medium 45
            3.4.1.1. Nondissipative medium 47
            3.4.1.2. Dissipative medium 54
        3.4.2. Source in infinite space 57
        3.4.3. Numerical modelling 59
    3.5. Eigenvalue statistics 63
        3.5.1. Common relationships 63
        3.5.2. Statistical averaging 64
        3.5.3. Numerical modelling 67
§ 4. Multidimensional Wave Problems in Layered Random Media . . . 69
    4.1. Nonstationary problems 69
        4.1.1. Formulation of boundary-wave problems 69
        4.1.2. Imbedding equations 71
        4.1.3. Statistical description 73
        4.1.4. Layer of finite thickness 78
    4.2. Point source inside a layered random medium 79
        4.2.1. Factorization of the wave equation 79

	4.2.2. Parabolic equation	81
	4.2.3. General case	84
ACKNOWLEDGEMENTS		86
APPENDIX		86
A.	STATISTICAL DESCRIPTION OF DYNAMICAL SYSTEMS	86
	A.1. The fokker-planck equation and its boundary conditions	86
	A.2. Caustics in random media	92
	A.2.1. General remarks	92
	A.2.2. Statistical description	94
B.	DYNAMICAL PROPERTIES OF THE WIENER AND LOGNORMAL RANDOM PROCESSES	98
	B.1. General remarks	98
	B.2. Majorant curves	103
	B.3. Statistics of a random area	105
	B.4. Isoprobable curve	107
	B.5. Some examples	108
	B.5.1. Stochastic parametric resonance	108
	B.5.2. Wave-beam propagation in a random parabolic waveguide	110
C.	FUNDAMENTAL SOLUTIONS OF WAVE PROBLEMS	115
D.	FACTORIZATION OF THE WAVE EQUATION IN A LAYERED MEDIUM	118
REFERENCES		121

II. QUANTUM STATISTICS OF DISSIPATIVE NONLINEAR OSCILLATORS
by V. PEŘINOVÁ AND A. LUKŠ (OLOMOUC, CZECH REPUBLIC)

§ 1.	INTRODUCTION	131
§ 2.	DISSIPATIVE THIRD-ORDER NONLINEAR OSCILLATOR	132
	2.1. Quantum dynamics	133
	2.2. Photon statistics	138
	2.3. Quantum coherence	139
	2.4. Squeezed states	142
	2.5. Phase properties	151
	2.6. Special initial states	155
	2.6.1. The coherent state	156
	2.6.2. The $SU(1,1)$ coherent state	165
	2.6.3. Gaussian mixed and pure states	167
	2.6.4. The $SU(1,1)$ two-photon coherent state	171
	2.6.5. The k-photon coherent state	171
	2.6.6. Squeezed and displaced number states	171
	2.7. Optically bistable two-photon medium	173
	2.8. Two-level atom in a Kerr medium	175
	2.9. Interaction of an electromagnetic field with a Kerr medium	178
§ 3.	HIGHER-ORDER NONLINEAR OSCILLATORS	184
§ 4.	COUPLED DISSIPATIVE THIRD-ORDER NONLINEAR OSCILLATORS	185
§ 5.	USE OF A KERR MEDIUM FOR GENERATION AND DETECTION OF NONCLASSICAL STATES	194
ACKNOWLEDGEMENT		198
REFERENCES		198

III. GAP SOLITONS
by C. Martijn de Sterke (Sydney, Australia) and J. E. Sipe (Toronto, Canada)

§ 1.	Introduction	205
§ 2.	Qualitative Description	208
§ 3.	Coupled-Mode Theory	212
§ 4.	Stationary Solutions	217
§ 5.	Solitary-Wave Solutions	225
§ 6.	Multiple Scales	230
§ 7.	Discussion of Analytic Results	237
§ 8.	Numerical Results	239
§ 9.	The Coupling Problem	244
§ 10.	Experimental Geometries	253
§ 11.	Outlook	256
	Acknowledgements	258
	References	259

IV. DIRECT SPATIAL RECONSTRUCTION OF OPTICAL PHASE FROM PHASE-MODULATED IMAGES
by Valentin I. Vlad and Daniel Malacara (León, Gto., México)

§ 1.	Introduction	263
§ 2.	Direct Spatial Reconstruction of Optical Phase (DSROP) from Fringe Patterns	265
	2.1. Phase demodulation in the space domain	266
	2.2. Phase demodulation in the spatial frequency (Fourier) domain	272
	2.3. Phase unwrapping and representation	278
§ 3.	Direct Spatial Reconstruction of Optical Phase (DSROP) from More General Phase-Modulated Images	281
	3.1. DSROP with a chirped radially symmetric pattern	282
	3.2. DSROP with a sheared periodic pattern	284
	3.3. DSROP with an arbitrary deterministic reference image	287
	3.4. DSROP with an arbitrary random reference image	291
§ 4.	Direct Spatial Reconstruction of Optical Phase (DSROP) for Time- and Space-Phase-Modulated Images	294
	4.1. Real-time method	294
	4.2. Double-exposure method	296
§ 5.	An Overview of DSROP Methods. Range of Measurement, Accuracy and Speed	298
	5.1. A systematic classification of DSROP methods	298
	5.2. The range of measurement, accuracy and speed of DSROP methods	299
§ 6.	Applications of the Methods of DSROP	302
	6.1. Optical testing	302
	6.2. Opto-mechanical testing and profilometry	303
	6.3. Optical diagnosis of inhomogeneous transparent materials	308
§ 7.	Concluding Remarks	311
	Acknowledgements	313
	References	313

V. IMAGING THROUGH TURBULENCE IN THE ATMOSPHERE
by Mark J. Beran and Jasmin Oz-Vogt (Ramat Aviv, Israel)

§ 1. Introduction 321
   1.1. Long exposures 322
   1.2. Isoplanicity 324
   1.3. Short exposures 326
      1.3.1. Adaptive optics 327
      1.3.2. Speckle interferometry and speckle masking 328
      1.3.3. Shift-and-add method 329
   1.4. General approach 330
§ 2. Propagation of Light through the Atmosphere 330
   2.1. Statistical properties of the index-of-refraction field in the atmosphere . . . 331
   2.2. Propagation of the coherence functions in the atmosphere 334
      2.2.1. Definition of the coherence functions 336
      2.2.2. Equations governing the coherence functions Γ_2 and Γ_4 337
      2.2.3. Solution of the equation governing the two-point coherence function 339
      2.2.4. Solution of the equation governing the four-point coherence function 342
   2.3. Two-point phase and log-amplitude coherence functions 349
      2.3.1. Use of Gaussian statistics 351
   2.4. Isoplanicity 352
      2.4.1. Horizontal propagation paths 355
   2.5. Summary of results of § 2 and their importance to the imaging problem . . 355
§ 3. Imaging of Objects in the Turbulent Atmosphere 358
   3.1. Basic imaging equations 358
      3.1.1. Long exposures 359
      3.1.2. Short exposures 363
   3.2. Long-exposure image-processing techniques 364
      3.2.1. Space-based systems imaging ground objects 365
      3.2.2. Horizontal paths within the atmosphere 366
      3.2.3. Inversion procedures 367
   3.3. Short-exposure image-processing techniques 370
      3.3.1. Adaptive optics 371
      3.3.2. Image sharpening 373
      3.3.3. Speckle interferometry 374
      3.3.4. Speckle masking (triple correlation theory) 378
      3.3.5. Intensity correlations and the three-point correlation function . . 380
      3.3.6. Shift-and-add method 382
   3.4. Concluding remarks 382
References 383

VI. DIGITAL HALFTONING: SYNTHESIS OF BINARY IMAGES
by Olof Bryngdahl, Thomas Scheermesser and Frank Wyrowski (Essen, Germany)

§ 1. Introduction and Trends 391
   1.1. Digital techniques on advance 391
   1.2. Narrowing the subject to image characterization 391
   1.3. The halftone technique 392
   1.4. Digital halftoning – a new branch of image synthesis 392

	1.5.	Three generations of electronic halftoning	394
§ 2.	THE QUANTIZATION PART OF AN IMAGE-PROCESSING SYSTEM	394
	2.1.	The model considered	394
	2.2.	Analysis of the model	398
§ 3.	SPECTRAL CHARACTERISTICS of BINARIZATION TECHNIQUES	400
	3.1.	Passive halftone techniques	401
		3.1.1. Binarization by constant threshold	401
		3.1.2. Binarization by carrier techniques	403
		3.1.3. Influence of sampling on passive techniques	405
	3.2.	Active direct halftone techniques	408
		3.2.1. Phase-modulated carrier procedure	409
		3.2.2. Error-diffusion procedure	409
	3.3.	Active iterative halftone techniques	412
		3.3.1. Definition of iterative techniques	413
		3.3.2. Iterative Fourier transform algorithm	413
		3.3.3. Simulated annealing algorithm	417
		3.3.4. Neural-network algorithms	417
	3.4.	Foundations for analysis of halftone techniques	419
§ 4.	ANALYSIS OF HALFTONE PROCEDURES WITH EXCESSIVE RESOLUTION	423
	4.1.	Transfer function and quality criteria	424
	4.2.	Carrier method	425
		4.2.1. Carrier with integer period ratio between carrier and raster	425
		4.2.2. Carrier with noninteger period ratio between carrier and raster	427
	4.3.	Error-diffusion method	431
	4.4.	Iterative Fourier transform method	436
	4.5.	Comparative remarks on quantization methods	438
	4.6.	Modification of residual detected noise	439
§ 5.	HALFTONE PROCEDURES WITH RESTRICTED RESOLUTION BELOW SAMPLING RASTER	442
	5.1.	Adaptation of noise spectrum to transfer function	443
	5.2.	Adjustment of halftone procedure to image spectrum	447
		5.2.1. Error-diffusion procedure	449
		5.2.2. Iterative procedure	450
		5.2.3. Control of phase in spectrum	453
§ 6.	HALFTONE RESOLUTION IN EXCESS OF SAMPLING RASTER	456
	6.1.	Perception of halftone dots	456
	6.2.	Formation of halftone dots	458
	6.3.	Overlap of halftone dots	458
	6.4.	Errors in position and shape of halftone dots	459
§ 7.	SYSTEM CONSIDERATIONS OF IMAGE QUANTIZATION GAIN IN IMPORTANCE	460
ACKNOWLEDGEMENTS	461
REFERENCES	461
AUTHOR INDEX	465
SUBJECT INDEX	475
CONTENTS OF PREVIOUS VOLUMES	479
CUMULATIVE INDEX	487

E. WOLF, PROGRESS IN OPTICS XXXIII
© 1994 ELSEVIER SCIENCE B.V.
ALL RIGHTS RESERVED

I

THE IMBEDDING METHOD IN STATISTICAL BOUNDARY-VALUE WAVE PROBLEMS

BY

V. I. KLYATSKIN

Russian Academy of Science,
Institute of Atmospheric Physics,
Pyzhevsky per. 3,
109017, Moscow, Russia,
and
Pacific Oceanological Institute,
Baltiiskay ul. 43,
690041, Vladivostok, Russia

CONTENTS

	PAGE
§ 1. INTRODUCTION	3
§ 2. MULTIDIMENSIONAL STATIONARY PROBLEMS	6
§ 3. PLANE WAVES IN LAYERED RANDOM MEDIA	17
§ 4. MULTIDIMENSIONAL WAVE PROBLEMS IN LAYERED RANDOM MEDIA	69
ACKNOWLEDGEMENTS	86
APPENDIX	86
REFERENCES	121

§ 1. Introduction

The spread of waves of different origin in natural media is currently under intensive study, in such fields as ocean acoustics, atmospheric optics, ionosphere, and plasma physics. In general, parameters of these media undergo large disturbances in space and time which are described mathematically by the boundary-value problem for the wave equation with a nonstationary time and nonuniform space refractive index. Since variations of the parameter are statistical, its boundary-value character creates difficulties of principle. In general, the solutions of such statistical problems are based on one of three approximate methods (see, e.g. Tatarskii [1971], Klyatskin [1975, 1980a, 1985], Apresian and Kravtsov [1983], Rytov, Kravtsov and Tatarskii [1987–1989]): (1) the use of a phenomenological linear theory of a radiative transfer, (2) the neglect of backwards scattering effects and passage to the small-angle approximation (parabolic equation), and (3) the reformulation of the problem in integral-equation terms. In analyzing statistical characteristics of this equation, one uses the Bourret approximation for the Dyson equation and the stairs approximation for the Bethe–Salpeter equation.

These approximate methods can only be substantiated for problems of a special type with the property of dynamical causality; that is, the solution is defined only by previous (in time or space) parameter values and does not depend on subsequent ones. Boundary-value problems are not of this kind. In such cases it is desirable to consider problems of an evolutional type with initial conditions. This method is indispensable for statistical problems and may be useful for the numerical analysis of deterministic ones. It enables construction of the imbedding method (also called invariant imbedding method in mathematical terminology) (Babkin, Klyatskin and Lyubavin [1980], Babkin and Klyatskin [1982a], Klyatskin [1985, 1986]). The primary equations for this method are obtained for many stationary, nonstationary, linear, and nonlinear boundary-value wave problems of different dimensions. In general, however, these equations are highly complicated (they may be nonlinear integrodifferential equations in finite-dimensional and, sometimes, in infinite-dimensional spaces). They are not investigated in practice, with the exception of problems of plane waves in layered media, which can be reduced to one-dimensional boundary-value problems, and permit total and systematic analysis (Klyatskin [1980a, 1985, 1986, 1991a, b, c], Babkin and Klyatskin [1982b]).

Note that natural media such as the oceans and the atmosphere can be regarded as layered in the first approximation. Thus plane-wave problems in layered media can be analyzed completely, allowing the primary peculiarities of the solution to be identified. This gives encouragement for considerable progress in the solution of multidimensional problems.

The idea of the invariant imbedding method was applied for the first time by Ambartsumian [1943, 1944, 1989] (Ambartsumian's principle of invariance) for solving some equations in linear radiative transport theory. Later it was mostly employed by mathematicians who wanted to reduce the boundary-value problems to those with initial conditions, since the latter are more convenient for numerical analysis. Several monographs (Casti and Calaba [1973], Kagiwada and Kalaba [1974], Bellman and Wing [1975]) introduced not only the physical aspects of the method but also computational possibilities associated with it.

It is interesting to note that these techniques which were first developed for solving some of the simplest equations in radiative transport theory can now be regarded as permitting the justification of linear radiative transport theory and also providing the means for its modification in cases where linear theory is no longer applicable.

Statistical characteristics of the wavefield in a random medium often differ considerably from wavefield behavior in separate realizations. In addition, in each specific realization of the process, some features of this process development can be observed that are absent in the statistical description. As an example of difference in principle, the existence in some realizations of the dynamic localization of the wavefield in layered random media can be observed, but in several cases the statistical energetic localization is absent (averaging is made over the total number of medium realizations) (Klyatskin and Saichev [1992]). Another example is an exponential divergence of rays on the average in a random medium in the geometric optics approximation (a relative diffusion of rays) (see, e.g. Klyatskin [1975, 1980a, 1985]), and at the same time the existence of caustics at finite distances with the probability equal to unity (Kulkarny and White [1982], White [1983], Zwilinger and White [1985], Klyatskin [1993]).

We shall examine the formulation of the main problems of waves in nonhomogeneous media, techniques for their solution, based on the imbedding method, and the primary fundamental features of statistical wave-problem solutions for waves in layered random media. General problems and statistical descriptions of dynamic systems are discussed in the appendixes to provide more exact detail and include additional material about the problems of wave localization in random media and the significance of boundary conditions for the Fokker–Planck equations.

The concept of localization originated in the physics of disordered systems (see, e.g. Anderson [1958], Lifshits, Gredescul and Pastur [1988], Anzygina, Pastur and Slusarev [1981]) described by the *Schrödinger equation* (time-independent or time-dependent) with a random potential, which is identical in form to the *Helmholtz equation* with a random refractive index. This coincidence, however, is an entirely superficial mathematical effect, derived from the formalism used for describing statistical phenomena in the cases considered. In the physics of disordered systems the fundamental entities are the self-averaging values, since they allow one to study the statistical properties of an object using a single realization that is sufficiently large (because of the absence of an ensemble of objects in general), and the main mathematical tool (characteristic of quantum-mechanical systems in principle) is a spectral expansion in the eigenfunctions of the corresponding boundary-value problem for the Schrödinger equation. The most important problems of wave propagation in randomly inhomogeneous media are considered to be based on the averaging over a collection of realizations of the parameters of the medium, and as mathematical tools one uses the standard classical theory of wave processes. Therefore, attempts to solve the problems of wave propagation in random media by the quantum-physics methods, as demonstrated by Freilikher and Gredescul [1990, 1992] and Gredescul and Freilikher [1988, 1990], are unsatisfactory. In addition, a major factor for these problems is wave absorption in the medium (even though arbitrarily small). In some cases the statistical properties are singular with respect to absorption. In these cases, by using the classical method of analytic continuation of the solution of the stationary problem into the complex plane of the parameter associated with the absorption, one can solve more complex problems, such as nonstationary problems or those concerning waves in layered media in three dimensions. At the same time in the approach based on the quantum-mechanical analogy, there is no dissipation *a priori*. Therefore, when solving these more complex problems, one has to start from scratch, ignoring the extensive existing information inherent in the solution of stationary problems. In addition, no mention has been made of situations where the limits of a vanishingly small absorption and passage to an infinite half-space (or an infinite space) simply do not commute. For these reasons the results obtained on the basis of quantum-mechanical analogy cannot be viewed with confidence.

This paper will examine an approach based on classical wave analysis. It is written primarily in mathematical terms, but as a specialist in theoretical and mathematical physics, I do not believe one can discuss such specific problems and effects or the significance of boundary conditions for wave problems and Fokker–Planck equations at a more visual level.

§ 2. Multidimensional Stationary Problems

2.1. FORMULATION OF BOUNDARY-VALUE WAVE PROBLEMS

Let a layer of inhomogeneous medium occupy a part of space $L_0 < x < L$. A point source is placed at the point with coordinates $(x_0, \boldsymbol{\rho}_0)$, where $\boldsymbol{\rho}$ are coordinates in the plane perpendicular to the x axis. Then the wavefield inside the layer is described by the equation for the Green function

$$\left\{ \frac{\partial^2}{\partial x^2} + \Delta_\rho + k^2[1 + \varepsilon(x, \boldsymbol{\rho})] \right\} G(x, \boldsymbol{\rho}; x_0, \boldsymbol{\rho}_0) = \delta(x - x_0)\delta(\boldsymbol{\rho} - \boldsymbol{\rho}_0), \quad (2.1)$$

where k is a wavenumber and $\varepsilon(x, \boldsymbol{\rho})$ is a deviation of the refractive index from unity. Let $\varepsilon(x, \boldsymbol{\rho}) = 0$ outside the layer. The wavefield is then described outside the layer by the Helmholtz equation

$$\left\{ \frac{\partial^2}{\partial x^2} + \Delta_\rho + k^2 \right\} G(x, \boldsymbol{\rho}; x_0, \boldsymbol{\rho}_0) = 0, \quad (2.1')$$

and at the layer boundary continuity conditions for the functions G and $\partial G/\partial x$ must be satisfied. Radiation conditions for $x \Rightarrow \pm \infty$ must also be satisfied for eq. (2.1').

Thus, the solution of eq. (2.1') with radiation conditions can be given in the following form:

$$G(x, \boldsymbol{\rho}; x_0, \boldsymbol{\rho}_0) = \begin{cases} \int d\boldsymbol{q}\, T_1(\boldsymbol{q}) e^{-i(k^2 - q^2)^{1/2}(x - L_0) + i\boldsymbol{q}\boldsymbol{\rho}} & (x \leqslant L_0), \\ \int d\boldsymbol{q}\, T_2(\boldsymbol{q}) e^{+i(k^2 - q^2)^{1/2}(x - L) + i\boldsymbol{q}\boldsymbol{\rho}} & (x \geqslant L). \end{cases} \quad (2.2)$$

Then, at the layer boundary $x = L_0$, the solution of the problem (2.1) is connected with the solution of eq. (2.1') by the formula

$$G(x, \boldsymbol{p}; x_0, \boldsymbol{p}_0) = \int d\boldsymbol{q}\, T_1(\boldsymbol{q}) e^{i\boldsymbol{q}\boldsymbol{p}}. \tag{2.3}$$

By analogy we obtain for the derivation $\partial G/\partial x|_{x=L_0}$

$$\frac{\partial}{\partial x} G(x, \boldsymbol{p}; x_0, \boldsymbol{p}_0)|_{x=L_0} = -i \int d\boldsymbol{q}\, (k^2 - q^2)^{1/2} T_1(\boldsymbol{q}) e^{i\boldsymbol{q}\boldsymbol{p}}$$

$$= -i(k^2 + \Delta_\rho)^{1/2} \int d\boldsymbol{q}\, T_1(\boldsymbol{q}) e^{i\boldsymbol{q}\boldsymbol{p}}. \tag{2.3'}$$

Hence, the boundary condition for eq. (2.1) at the boundary $x = L_0$ has the form

$$\left\{ \frac{\partial}{\partial x} + i(k^2 + \Delta_\rho)^{1/2} \right\} G(x, \boldsymbol{p}; x_0, \boldsymbol{p}_0)|_{x=L_0} = 0. \tag{2.4}$$

Similarly, the boundary condition at the boundary $x = L$ is

$$\left\{ \frac{\partial}{\partial x} - i(k^2 + \Delta_\rho)^{1/2} \right\} G(x, \boldsymbol{p}; x_0, \boldsymbol{p}_0)|_{x=L} = 0. \tag{2.4'}$$

Inside the layer the field G is continuous at all points, but the quantity $\partial G/\partial x$ has a jump at the point of the source location $x = x_0$:

$$G(x, \boldsymbol{p}; x_0, \boldsymbol{p}_0)|_{x=x_0+0} - G(x, \boldsymbol{p}; x_0, \boldsymbol{p}_0)|_{x=x_0-0} = \delta(\boldsymbol{p} - \boldsymbol{p}_0). \tag{2.5}$$

When inhomogeneities are absent ($\varepsilon = 0$), the Green function in free space

$$G(x, \boldsymbol{p}; x_0, \boldsymbol{p}_0)|_{\varepsilon=0} = g(x - x_0; \boldsymbol{p} - \boldsymbol{p}_0) = -\frac{1}{4\pi |\boldsymbol{r} - \boldsymbol{r}_0|} e^{ik|\boldsymbol{r} - \boldsymbol{r}_0|} \quad (\boldsymbol{r} = \{x, \boldsymbol{p}\})$$

is described by the integral form

$$g(x, \boldsymbol{p}) = \int d\boldsymbol{q}\, g(\boldsymbol{q}) e^{i(k^2 - q^2)^{1/2}|x| + i\boldsymbol{q}\boldsymbol{p}},$$

$$g(\boldsymbol{q}) = \frac{1}{8i\pi^2} \frac{1}{(k^2 - q^2)^{1/2}}. \tag{2.6}$$

When the observation point "x" and the point of the source location "x_0" are fixed (to be definite we shall assume that $x < x_0$), the function $g(x - x_0, \boldsymbol{p} - \boldsymbol{p}_0)$, being the function of the parameter x_0, satisfies for the equations of the first order

$$\frac{\partial}{\partial x_0} g(x - x_0; \boldsymbol{p} - \boldsymbol{p}_0) = i(k^2 + \Delta_\rho)^{1/2} g(x - x_0; \boldsymbol{p} - \boldsymbol{p}_0), \tag{2.7a}$$

$$\frac{\partial}{\partial x_0} g(x - x_0; \boldsymbol{\rho} - \boldsymbol{\rho}_0) = \mathrm{i}(k^2 + \Delta_{\rho_0})^{1/2} g(x - x_0; \boldsymbol{\rho} - \boldsymbol{\rho}_0), \tag{2.7b}$$

with the initial condition

$$g(0; \boldsymbol{\rho} - \boldsymbol{\rho}_0) = g(\boldsymbol{\rho} - \boldsymbol{\rho}_0) = -\frac{1}{4\pi|\boldsymbol{\rho} - \boldsymbol{\rho}_0|} e^{\mathrm{i}k|\boldsymbol{\rho} - \boldsymbol{\rho}_0|}. \tag{2.7'}$$

This formula expresses the factorization property of the Green function in free space (see Appendix C).

The equality

$$2\mathrm{i}(k^2 + \Delta_\rho)^{1/2} g(\boldsymbol{\rho}) = \delta(\boldsymbol{\rho}) \tag{2.8}$$

is the consequence of the integral representation (2.6), and there follows from (2.8)

$$\begin{aligned}(k^2 + \Delta_\rho)^{-1/2} \delta(\boldsymbol{\rho}) &= 2\mathrm{i}g(\boldsymbol{\rho}), \\ (k^2 + \Delta_\rho)^{1/2} \delta(\boldsymbol{\rho}) &= 2\mathrm{i}(k^2 + \Delta_\rho) g(\boldsymbol{\rho}).\end{aligned} \tag{2.8'}$$

The effect of the operator $(k^2 + \Delta_\rho)^{1/2}$ on an arbitrary function $F(\boldsymbol{\rho})$ can be written in the form of an integral operator

$$\begin{aligned}(k^2 + \Delta_\rho)^{1/2} F(\boldsymbol{\rho}) &= \int \mathrm{d}\boldsymbol{\rho}' \, (k^2 + \Delta_\rho)^{1/2} \delta(\boldsymbol{\rho} - \boldsymbol{\rho}') F(\boldsymbol{\rho}') \\ &\equiv \int \mathrm{d}\boldsymbol{\rho}' \, K(\boldsymbol{\rho} - \boldsymbol{\rho}') F(\boldsymbol{\rho}').\end{aligned} \tag{2.9}$$

Its kernel is defined by the second equality in eqs. (2.8')

$$K(\boldsymbol{\rho}) = (k^2 + \Delta_\rho)^{1/2} \delta(\boldsymbol{\rho}) = 2\mathrm{i}(k^2 + \Delta_\rho) g(\boldsymbol{\rho}), \tag{2.9'}$$

and the first equation of eqs. (2.8') determines the kernel of the inverse operator.

If the point source is placed at the layer boundary $x_0 = L$, the wavefield inside the layer $L_0 < x < L$ is described by the equation

$$\left\{ \frac{\partial^2}{\partial x^2} + \Delta_\rho + k^2[1 + \varepsilon(x, \boldsymbol{\rho})] \right\} G(x, \boldsymbol{\rho}; L, \boldsymbol{\rho}_0) = 0. \tag{2.10}$$

The condition (2.4) (at $x_0 = L$) remains as the boundary condition for the boundary $x = L_0$. For the boundary $x = L$ we have according to eq. (2.5) the equality

$$\frac{\partial}{\partial x} G(x, \boldsymbol{\rho}; x_0, \boldsymbol{\rho}_0)|_{x = L - 0} = \frac{\partial}{\partial x} G(x, \boldsymbol{\rho}; L - 0, \boldsymbol{\rho}_0)|_{x = L} - \delta(\boldsymbol{\rho} - \boldsymbol{\rho}_0),$$

which can be rewritten, taking into account eq. (2.4'), in the form

$$\left\{\frac{\partial}{\partial x} - i(k^2 + \Delta_\rho)^{1/2}\right\} G(x, \rho; L, \rho_0)|_{x=L} = -\delta(\rho - \rho_0). \tag{2.11}$$

The equality (2.11) defines the boundary condition for eq. (2.10) at the layer boundary $x = L$.

If a wave $\psi_0(x, \rho)$ is incident from the region $x > L$ on the layer of medium (in the negative direction of the x axis), at the boundary $x = L$ it creates a distribution of sources $f(\rho_0)$, such that

$$\psi_0(x, \rho) = \int d\rho_0 \, g(x - L; \rho - \rho_0) f(\rho_0). \tag{2.12}$$

According to eq. (2.8), the source distribution $f(\rho_0)$ will be described by the expression

$$f(\rho) = 2i(k^2 + \Delta_\rho)^{1/2} \psi_0(L, \rho). \tag{2.12'}$$

In this case the wavefield $\psi(x, \rho)$ inside the layer is connected with the solution of eq. (2.10) by the equality

$$\psi(x, \rho) = \int d\rho_0 \, G(x, \rho; L, \rho_0) f(\rho_0), \tag{2.13}$$

and, consequently, satisfies the boundary problem

$$\left\{\frac{\partial^2}{\partial x^2} + \Delta_\rho + k^2[1 + \varepsilon(x, \rho)]\right\} \psi(x, \rho) = 0,$$

$$\left\{\frac{\partial}{\partial x} + i(k^2 + \Delta_\rho)^{1/2}\right\} \psi(x, \rho)|_{x=L_0} = 0, \tag{2.14}$$

$$\left\{\frac{\partial}{\partial x} - i(k^2 + \Delta_\rho)^{1/2}\right\} \psi(x, \rho)|_{x=L} = -2i(k^2 + \Delta_\rho)^{1/2} \psi_0(L, \rho).$$

2.2. IMBEDDING EQUATIONS

We will begin by studying the Green function with a source at the point (L, ρ_0), when $x < L$, which is described by the boundary problem (2.10) and (2.11). Taking the solution of this problem as a function of the parameter L, we have the equality for it (Babkin, Klyatskin and Lyubavin [1980],

Klyatskin [1985, 1986]):

$$\left\{\frac{\partial}{\partial L} - \mathrm{i}(k^2 + \Delta_{\boldsymbol{p}_0})^{1/2}\right\} G(x, \boldsymbol{p}; L, \boldsymbol{p}_0)$$

$$= -k^2 \int \mathrm{d}\boldsymbol{p}_1 \, G(x, \boldsymbol{p}; L, \boldsymbol{p}_1)\varepsilon(L, \boldsymbol{p}_1) H(L; \boldsymbol{p}_1; \boldsymbol{p}_0). \tag{2.15}$$

Here $H(L; \boldsymbol{p}_1; \boldsymbol{p}_0) = G(L, \boldsymbol{p}_1; L, \boldsymbol{p}_0)$ denotes the field in the plane $x = L$, that is, the back-scattered wave. We can consider the equality (2.15) as an integrodifferential equation for G if we complete it by the initial condition for $L \Rightarrow x$,

$$G(x, \boldsymbol{p}; L, \boldsymbol{p}_0)|_{L=x} = H(x; \boldsymbol{p}; \boldsymbol{p}_0). \tag{2.15'}$$

For the function $H(L; \boldsymbol{p}, \boldsymbol{p}_0)$ we have the closed, nonlinear integrodifferential equation

$$\left\{\frac{\partial}{\partial L} - \mathrm{i}(k^2 + \Delta_{\boldsymbol{p}})^{1/2} - \mathrm{i}(k^2 + \Delta_{\boldsymbol{p}_0})^{1/2}\right\} H(L; \boldsymbol{p}; \boldsymbol{p}_0)$$

$$= -\delta(\boldsymbol{p} - \boldsymbol{p}_0) - k^2 \int \mathrm{d}\boldsymbol{p}_1 \, H(L; \boldsymbol{p}; \boldsymbol{p}_1)\varepsilon(L, \boldsymbol{p}_1) H(L; \boldsymbol{p}_1; \boldsymbol{p}_0). \tag{2.16}$$

The initial condition for it at $L = L_0$,

$$H(L_0; \boldsymbol{p}; \boldsymbol{p}_0) = g(\boldsymbol{p} - \boldsymbol{p}_0), \tag{2.16'}$$

corresponds with the source in free space.

By analogy, taking the Green function $G(x, \boldsymbol{p}; x_0, \boldsymbol{p}_0)$ as a function of the parameter L, ($G(x, \boldsymbol{p}; x_0, \boldsymbol{p}_0) = G(x, \boldsymbol{p}; x_0, \boldsymbol{p}_0; L)$), we obtain the equation

$$\frac{\partial}{\partial L} G(x, \boldsymbol{p}; x_0, \boldsymbol{p}_0; L) = -k^2 \int \mathrm{d}\boldsymbol{p}_1 \, G(x, \boldsymbol{p}; L, \boldsymbol{p}_1) G(x_0, \boldsymbol{p}_0; L, \boldsymbol{p}_1) \varepsilon(L, \boldsymbol{p}_1) \tag{2.17}$$

with the initial condition

$$G(x, \boldsymbol{p}; x_0, \boldsymbol{p}_0; L)|_{L=\max\{x, x_0\}} = \begin{cases} G(x, \boldsymbol{p}; x_0, \boldsymbol{p}_0), & x_0 \geqslant x; \\ G(x_0, \boldsymbol{p}_0; x, \boldsymbol{p}), & x_0 \leqslant x, \end{cases} \tag{2.17'}$$

which reflects the continuity condition with respect to the parameter L.

Thus, eqs. (2.15) to (2.17) form the closed system of imbedding equations for the given problem. The limit transition $L_0 \Rightarrow -\infty$, $L \Rightarrow +\infty$ corresponds with the problem solution for a point source within the inhomogeneous medium, which occupies the whole space.

Equation (2.17) can be integrated, and we see that the field of the point source, placed inside the medium, is connected in a simple way (by the quadrature) with the field appearing in the problem in which the source is located at the layer boundary.

Let us observe the case of the layered medium, in which $\varepsilon(L, \boldsymbol{\rho}) = \varepsilon(L)$. All functions G are then functions of $\boldsymbol{\rho} - \boldsymbol{\rho}_0$. Using the Fourier transformation, it is possible to pass from the integrodifferential equations system to the ordinary differential equations

$$\frac{\partial}{\partial L} G_q(x; x_0; L) = -(2\pi k)^2 \varepsilon(L) G_q(x; L) G_q(x_0; L), \tag{2.18a}$$

$$\left[\frac{\partial}{\partial L} - i(k^2 - q^2)^{1/2}\right] G_q(x; L) = -(2\pi k)^2 \varepsilon(L) G_q(x; L) H_q(L), \tag{2.18b}$$

$$\left[\frac{\partial}{\partial L} - 2i(k^2 - q^2)^{1/2}\right] H_q(L) = -\frac{1}{4\pi^2} - (2\pi k)^2 \varepsilon(L) H_q^2(L),$$

$$H_q(L_0) = g(q), \quad G_q(x; L)|_{L=x} = H_q(x), \tag{2.18c}$$

$$G_q = \frac{1}{(2\pi)^2} \int d\boldsymbol{\rho}\, G(\boldsymbol{\rho}) e^{-iq\rho}.$$

Equation (2.18b) describes the propagation of the plane wave, incident under an angle to the boundary $x = L$ with the amplitude

$$g(\boldsymbol{q}) = \frac{1}{8i\pi^2}(k^2 - q^2)^{-1/2}.$$

After normalizing the incident wave to the unity amplitude $\psi_0(x, \boldsymbol{\rho}) = \exp[-ik(x - L) + iq\rho]$, we have the system of equations

$$\frac{\partial}{\partial L} \tilde{G}_q(x; x_0; L) = \frac{ik^2}{2(k^2 - q^2)^{1/2}} \varepsilon(L) \tilde{G}_q(x; L) \tilde{G}_q(x_0; L),$$

$$\left[\frac{\partial}{\partial L} - i(k^2 - q^2)^{1/2}\right] \tilde{G}_q(x, L) = \frac{ik^2}{2(k^2 - q^2)^{1/2}} \varepsilon(L) \tilde{G}_q(x; L) \tilde{H}_q(L),$$

$$\left[\frac{\partial}{\partial L} - 2i(k^2 - q^2)^{1/2}\right] \tilde{H}_q(L) \tag{2.19}$$

$$= -2i(k^2 - q^2)^{1/2} + \frac{ik^2}{2(k^2 - q^2)^{1/2}} \varepsilon(L) \tilde{H}_q^2(L),$$

$$\tilde{H}_q(L_0) = 1.$$

The case of the normal incidence of the wave on the boundary corresponds to $\boldsymbol{q}=0$.

Note that the differential equation (2.15) can be rewritten as an integral equation

$$G(x, \boldsymbol{p}; L, \boldsymbol{p}_0) = \tilde{g}(x, \boldsymbol{p}; L, \boldsymbol{p}_0) - k^2 \int_x^L dx_1 \int d\boldsymbol{p}_1$$
$$\times G(x, \boldsymbol{p}; x_1, \boldsymbol{p}_1)\varepsilon(x_1, \boldsymbol{p}_1)\tilde{g}(x_1, \boldsymbol{p}_1; L, \boldsymbol{p}_0),$$

whence follows the integral equation for the wavefield $\psi(x, \boldsymbol{p}, L)$

$$\psi(x, \boldsymbol{p}; L) = \psi_0(x, \boldsymbol{p}; L) - k^2 \int_x^L dx_1 \int d\boldsymbol{p}_1$$
$$\times \tilde{g}(x, \boldsymbol{p}; x_1, \boldsymbol{p}_1)\varepsilon(x_1, \boldsymbol{p}_1)\psi(x_1, \boldsymbol{p}_1; L), \qquad (2.20)$$

where

$$\tilde{g}(x, \boldsymbol{p}; L, \boldsymbol{p}_0) = \exp\{i(k^2 + \Delta_{\boldsymbol{p}_0})^{1/2}(L-x)\}H(x; \boldsymbol{p}; \boldsymbol{p}_0),$$

$$\psi_0(x, \boldsymbol{p}; L) = \int d\boldsymbol{p}_0 \, \tilde{g}(x, \boldsymbol{p}; x_0, \boldsymbol{p}_0)f(\boldsymbol{p}_0),$$

and $f(\boldsymbol{p}_0)$ is the distribution of sources at the boundary $x = L$.

Equation (2.16) describes the field on the boundary $x = L$, that is, the back-scattered field. The back-scattered effect is essentially nonlinear, and is described by the last term in eq. (2.16). If we neglect this term, the solution of the remaining equation has the form $H(L; \boldsymbol{p}; \boldsymbol{p}_0) = g(\boldsymbol{p} - \boldsymbol{p}_0)$, which corresponds to the presence of the incident wave only in the plane $x = L$. In this case $\tilde{g}(x, \boldsymbol{p}; L, \boldsymbol{p}_0) = g(x - L; \boldsymbol{p} - \boldsymbol{p}_0)$, and eq. (2.20) takes the form of the causal integral equation

$$\psi(x, \boldsymbol{p}) = \psi_0(x, \boldsymbol{p}) - k^2 \int_x^L dx_1 \int d\boldsymbol{p}_1 \, g(x - x_1; \boldsymbol{p} - \boldsymbol{p}_1)\varepsilon(x_1, \boldsymbol{p}_1)\psi(x_1, \boldsymbol{p}_1),$$
$$(2.20')$$

which describes the spreading of the wave in the approximation, allowing, in general, the scattering of the wave under sufficiently large angles (but not exceeding $\pi/2$). It can be rewritten in the form of the operator equation. Differentiating eq. (2.20') with respect to x and using eq. (2.8), we have the equation (extended parabolic equation)

$$\left[\frac{\partial}{\partial x} + i(k^2 + \Delta_{\boldsymbol{p}})^{1/2}\right]\psi(x, \boldsymbol{p}) \quad -$$

$$= -i\frac{k^2}{2}(k^2 + \Delta_\rho)^{-1/2}\{\varepsilon(x,\rho)\psi(x,\rho)\}, \quad \psi(L,\rho) = \psi_0(L,\rho). \tag{2.21}$$

Malakhov and Saichev [1979] obtained this equation by another method. Other references related to the extended parabolic wave equation include Corones [1975], McDaniel [1975], Tappert [1977], and Hudson [1980].

For the case of small-angle scattering ($\Delta_\rho \ll k^2$), we obtain a parabolic equation

$$\frac{\partial}{\partial x}\psi(x,\rho) = -ik\psi - \frac{i}{2k}\Delta_\rho\psi - i\frac{k}{2}\varepsilon(x,\rho)\psi(x,\rho), \quad \psi(L,\rho) = \psi_0(\rho),$$

which for the function $u(x,\rho)$, defined by the equality

$$\psi(x,\rho) = u(x,\rho)e^{-ik(x-L)},$$

takes the form of the well-known parabolic equation (Tatarskii [1971], Klyatskin [1975, 1980a, 1985], Rytov, Kravtsov and Tatarskii [1987–1989])

$$\frac{\partial}{\partial x}u(x,\rho) = \frac{i}{2k}\Delta_\rho u + i\frac{k}{2}\varepsilon(x,\rho)u(x,\rho), \quad u(L,\rho) = u_0(\rho). \tag{2.22}$$

Note that the problems concerning the boundary conditions for eqs. (2.21) and (2.22) over ρ have not been discussed in the literature. It has been falsely assumed that the solutions of these equations tend to zero when $|\rho| \Rightarrow \infty$. It is true for small-angle approximation, but it cannot be valid for an extended parabolic equation. In fact, if there is a small attenuation in a medium, which can be introduced as an imaginary part of field $\varepsilon(x,\rho)$, that is,

$$\varepsilon(x,\rho) = \varepsilon_1(x,\rho) + i\gamma,$$

then the solution of eqs. (2.21) and (2.22) is proportional to the factor $e^{-\gamma x}$, which is not valid for large-angle scattering.

Let us introduce the Green function $G(x,\rho;x',\rho')$ for eq. (2.22), which satisfies eq. (2.22) with the initial condition

$$G(x,\rho;x',\rho')|_{x=x'} = \delta(\rho - \rho').$$

If inhomogeneities are absent (i.e. $\varepsilon = 0$), we have

$$G(x,\rho;x',\rho')|_{\varepsilon=0} = g_0(x-x';\rho-\rho') = \frac{k}{2i\pi(x-x')}\exp\left[\frac{ik(\rho-\rho')^2}{2(x-x')}\right]. \tag{2.23}$$

The wavefield $u(x,\rho)$ is the functional of the field $\varepsilon(x,\rho)$. The variational derivative of the field $u(x, \pmb{\rho})$ with respect to ε is expressed by the function G, using the equality (Klyatskin [1980a, 1985])

$$\frac{\delta}{\delta\varepsilon(x', \pmb{\rho}')} u(x, \pmb{\rho}) = i\frac{k}{2} G(x, \pmb{\rho}; x', \pmb{\rho}') u(x', \pmb{\rho}'). \qquad (2.24)$$

Fields in different planes are connected by the equality

$$u(x, \pmb{\rho}) = \int \mathrm{d}\pmb{\rho}' \, G(x, \pmb{\rho}; x', \pmb{\rho}') u(x', \pmb{\rho}'), \quad x \geqslant x'. \qquad (2.25)$$

The Green function satisfies the orthogonality condition (Klyatskin [1980a, 1985])

$$\int \mathrm{d}\pmb{\rho} \, G(x, \pmb{\rho}; x', \pmb{\rho}') G^*(x, \pmb{\rho}; x'', \pmb{\rho}'') = \delta(\pmb{\rho}' - \pmb{\rho}''),$$

and, hence, we obtain from eq. (2.25)

$$u(x', \pmb{\rho}') = \int \mathrm{d}\pmb{\rho} \, G^*(x, \pmb{\rho}; x', \pmb{\rho}') u(x, \pmb{\rho}), \quad x' \leqslant x. \qquad (2.26)$$

Using eq. (2.26), it is possible to rewrite eq. (2.24) in the form of equality

$$\frac{\delta}{\delta\varepsilon(x', \pmb{\rho}')} u(x, \pmb{\rho}) = i\frac{k}{2} G(x, \pmb{\rho}; x', \pmb{\rho}') \int \mathrm{d}\pmb{\rho}'' \, G^*(x, \pmb{\rho}''; x', \pmb{\rho}') u(x, \pmb{\rho}''), \qquad (2.27)$$

which expresses the variational derivative $u(x, \pmb{\rho})$ by the field $u(x, \pmb{\rho}')$ in the same plane.

2.3. STATISTICAL AVERAGING

When we consider ordinary methods of statistical analysis, in the general case, the boundary problems (2.1), (2.4) and (2.4') have no dynamic causality property, because the solution of the problem $G(x, \pmb{\rho}; x_0, \pmb{\rho}_0)$ for fixed values of x and x_0 is described by fluctuations $\varepsilon(x, \pmb{\rho})$ in the whole layer $L_0 < x, x_0 < L$. Equations of the imbedding method for $H(L; \pmb{\rho}; \pmb{\rho}_0)$, $G(x, \pmb{\rho}; L, \pmb{\rho}_0)$, $G(x, \pmb{\rho}; x_0, \pmb{\rho}_0; L)$, however, possess a dynamic causality property with respect to the parameter L, because the initial problem with the condition at $L = L_0$ was formulated for them. In the general case these equations were not studied. The analysis of eq. (2.16) for the backscattered field by Shevtsov [1981, 1982, 1983, 1985, 1987, 1989, 1990], Fortus and Shevtsov [1986] and Barabanenkov and Kryukov [1992] is an exception. In general, investigators are restricted by the analysis of the parabolic equation (2.22) (see, e.g. Tatarskii [1971], Klyatskin [1980a, 1985], Gulin [1984], Rytov, Kravtsov and Tatarskii [1987–1989], Manning [1993], Furutsu [1993]) which has a sufficiently large field of applicability in several cases. Qualitative results of back-scattering effects and results based on different approximation approaches are discussed in reviews by Kravtsov and Saichev [1982, 1985] and Barabanenkov, Kravtsov, Ozrin and Saichev [1991].

The primary method of statistical analysis of eq. (2.22) is the approximation of the delta-correlated field $\varepsilon(x, \boldsymbol{\rho})$ (see Appendix A), when its correlation function $B_\varepsilon(x, \boldsymbol{\rho}; x', \boldsymbol{\rho}') = \langle \varepsilon(x, \boldsymbol{\rho})\varepsilon(x', \boldsymbol{\rho}')\rangle$ can be approximated by an effective correlation function

$$B_\varepsilon^{\text{eff}}(x, \boldsymbol{\rho}; x', \boldsymbol{\rho}') = \delta(x - x')A(x; \boldsymbol{\rho}, \boldsymbol{\rho}'), \tag{2.28}$$

where

$$A(x; \boldsymbol{\rho}, \boldsymbol{\rho}') = \int_{-\infty}^{\infty} \mathrm{d}x' \, B_\varepsilon(x, \boldsymbol{\rho}; x', \boldsymbol{\rho}').$$

For homogeneous fluctuations $\varepsilon(x, \boldsymbol{\rho})$,

$$A(x; \boldsymbol{\rho}, \boldsymbol{\rho}') = A(\boldsymbol{\rho} - \boldsymbol{\rho}'). \tag{2.28'}$$

We suppose that the field $\varepsilon(x, \boldsymbol{\rho})$ is a Gaussian random field with zero mean value.

This approximation supposes that the longitudinal correlation radius is small compared with all parameters resulting from the process of propagation. The method for obtaining equations for statistical characteristics of the wavefield is well known (see, e.g. Tatarskii [1971], Klyatskin [1975, 1980a, 1985], Rytov, Kravtsov and Tatarskii [1987–1989]), and we shall not discuss this question further.

As an example, let us look at the mean field. Averaging eq. (2.22) over the ensemble of the function $\varepsilon(x, \boldsymbol{\rho})$ realization, we have

$$\left(2ik\frac{\partial}{\partial x} + \Delta_{\boldsymbol{\rho}}\right)\langle u(x, \boldsymbol{\rho})\rangle + k^2\langle \varepsilon(x, {}^{\boldsymbol{\rho}})u(x, \boldsymbol{\rho})\rangle = 0. \tag{2.29}$$

To find the last term in eq. (2.29) we shall use the Furutsu–Novikov formula

$$\langle \varepsilon(r)R[\varepsilon]\rangle = \int \mathrm{d}r' \, \langle \varepsilon(r)\varepsilon(r')\rangle \left\langle \frac{\delta}{\delta\varepsilon(x', \boldsymbol{\rho}')}R[\varepsilon]\right\rangle, \quad r = \{x, \boldsymbol{\rho}\},$$

which is valid for the Gaussian field $\varepsilon(r)$ and its arbitrary functional $R[\varepsilon]$.

Then, using eq. (2.24), we obtain

$$\left(2ik\frac{\partial}{\partial x}+\Delta_\rho\right)\langle u(x,\rho)\rangle$$

$$+i\frac{k^2}{2}\int_0^x dx'\int d\rho'\, B_\varepsilon(x,\rho;x',\rho')\langle G(x,\rho;x',\rho')u(x',\rho')\rangle = 0. \quad (2.30)$$

Substituting eq. (2.26) into eq. (2.30), it is possible to rewrite the latter in the form

$$\left(2ik\frac{\partial}{\partial x}+\Delta_\rho\right)\langle u(x,\rho)\rangle = -i\frac{k^3}{2}\int_0^x dx'\int d\rho'\, B_\varepsilon(x,\rho;x',\rho')\int d\rho''$$

$$\times \langle G(x,\rho;x',\rho')G^*(x,\rho'';x',\rho')u(x,\rho'')\rangle. \quad (2.31)$$

For the model of the delta-correlated field $\varepsilon(x,\rho)$ given by eqs. (2.28) and (2.28′) we obtain the closed equation for the mean field:

$$\left(2ik\frac{\partial}{\partial x}+\Delta_\rho\right)\langle u(x,\rho)\rangle + i\frac{k^3}{4}A(0)\langle u(x,\rho)\rangle = 0. \quad (2.32)$$

It is possible to take into account the finiteness of the longitudinal correlation radius of the field $\varepsilon(x,\rho)$, if we suppose that in scales of the order of the longitudinal correlation radius $l_\parallel$, the influence of fluctuations of the parameters of the medium on the dynamics of the wavefield is not important. In this case it is possible to replace G in eq. (2.31) by g from eq. (2.23). This operation results in the closed equation (integrodifferential equation), which corresponds with the diffusion approximation (see, e.g. Klyatskin [1991d]),

$$\left(2ik\frac{\partial}{\partial x}+\Delta_\rho\right)\langle u(x,\rho)\rangle = \int d\rho''\, D(x,\rho;\rho'')\langle u(x,\rho'')\rangle, \quad (2.33)$$

where

$$D(x,\rho;\rho'') = -i\frac{k^3}{2}\int_0^x dx'\int d\rho'$$

$$\times B_\varepsilon(x,\rho;x',\rho')g(x-x';\rho-\rho')g^*(x-x';\rho'-\rho'').$$

The diffusion approximation is very closed with respect to Chernov's local method (Chernov [1975]), and the last equation has a larger field of applicability than the approximation of the delta-correlated random field and describes wave propagation in a medium with extended random irregularities (Saichev and Slavinskij [1985], Virovlyanskij, Saichev and Slavinskij [1985]).

For spatially homogeneous fluctuations of ε for plane incident wave and for sufficiently large distances ($x \gg l_\parallel$), we can solve eq. (2.33) in the following way:

$$\langle u(x, \boldsymbol{\rho})\rangle = \exp\left\{\frac{k^2}{4} x \int_0^\infty \mathrm{d}\xi \int \mathrm{d}\boldsymbol{\rho}' \, B_\varepsilon(\xi, \boldsymbol{\rho}')g(\xi, \boldsymbol{\rho}')\right\}. \tag{2.34}$$

The condition of applicability of the diffusion approximation for homogeneous fluctuations $\varepsilon(x, \boldsymbol{\rho})$ is clearly the condition

$$\frac{k^2}{4} l_\parallel \left| \int_0^\infty \mathrm{d}\xi \int \mathrm{d}\boldsymbol{\rho} \, B_\varepsilon(\xi, \boldsymbol{\rho})g(\xi, \boldsymbol{\rho}) \right| \ll 1.$$

Note that neither the delta-correlated random field $\varepsilon(x, \boldsymbol{\rho})$ nor the diffusion approximation is applicable when $\varepsilon(x, \boldsymbol{\rho}) = \varepsilon(\boldsymbol{\rho})$ or $\varepsilon(z)$, which occurs, for example, in the layered medium with respect to z. In this case the random field $\varepsilon(z)$ formally has the infinite correlation radius along the x axis, as will be discussed next.

§ 3. Plane Waves in Layered Random Media

3.1. FORMULATION OF BOUNDARY-VALUE WAVE PROBLEMS

3.1.1. Wave incidence on the medium layer

Let the layer of an inhomogeneous medium occupy a part of a space $L_0 < x < L$. A plane wave of unit amplitude $u_0(x) = \exp[-ik(x - L)]$ is incident on it from the region $x > L$. The wavefield within the layer is described by the Helmholtz equation

$$\frac{\mathrm{d}^2}{\mathrm{d}x^2} u(x) + k^2[1 + \varepsilon(x)]u(x) = 0. \tag{3.1}$$

We assume $\varepsilon(x) = 0$ outside the layer $\varepsilon(x) = \varepsilon_1(x) + i\gamma$ inside the medium, where $\varepsilon_1(x) = \varepsilon_1^*(x)$ and $\gamma \ll 1$ describes the absorption of the wave in the medium. In the region $x > L$ the wavefield has the structure $u(x) = \mathrm{e}^{-ik(x-L)} + R_L \mathrm{e}^{ik(x-L)}$, where R_L is a complex reflection factor. In the region $x < L_0$ the wavefield has the form $u(x) = T_L \mathrm{e}^{-ik(x-L_0)}$, where T_L is a transmitter factor. Boundary conditions for the problem (3.1) are continuity conditions for $u(x)$ and $du(x)/dx$ on layer boundaries, that is, with the following conditions:

$$\frac{i}{k}\frac{d}{dx}u(x) + u(x)\big|_{x=L} = 2, \qquad \frac{i}{k}\frac{d}{dx}u(x) - u(x)\big|_{x=L_0} = 0. \tag{3.2}$$

If medium parameters $\varepsilon(x)$ are given statistically, the solution of the statistical problem necessitates finding statistical characteristics of the wavefield $u(x)$ and values of $R_L = u(L) - 1$, $T_L = u(L_0)$.

The wavefield in the medium can be given in the form

$$u(x) = a(x)e^{-ikx} + b(x)e^{ikx}, \qquad \frac{d}{dx}u(x) = -ik[a(x)e^{-ikx} - b(x)e^{ikx}],$$

where $a(x)$ and $b(x)$ are complex amplitudes of *waves coming from opposite directions*. Their relation with the wavefield follows from the representation

$$a(x) = \frac{1}{2}\left[u(x) + \frac{i}{k}\frac{d}{dx}u(x)\right]e^{ikx}, \qquad a(L) = e^{ikL};$$

$$b(x) = \frac{1}{2}\left[u(x) - \frac{i}{k}\frac{d}{dx}u(x)\right]e^{-ikx}, \qquad b(L_0) = 0. \tag{3.3}$$

Note that intensities of opposite waves $W_1(x) = |a(x)|^2$, $W_2(x) = |b(x)|^2$ are primary objects of the linear phenomenological theory of radiative transfer (see, e.g. Sobolev [1956], Chandrasekhar [1960]), where they are described by equations

$$\frac{d}{dx}W_1(x) = 2\alpha W_1(x) + \mu(W_1 - W_2),$$

$$\frac{d}{dx}W_2(x) = -2\alpha W_1(x) + \mu(W_1 - W_2).$$

Phenomenological constants α and μ play roles of damping and diffusion factors, respectively. A logical statistical theory can be used to reveal the statistical meaning of these parameters, to state limits to the applicability of the linear-transfer theory, and to describe the radiative transfer in the case in which the linear theory is incorrect.

The boundary problem (3.1) and (3.2) describes not only scattered waves, caused by the medium inhomogeneity $\varepsilon(x)$, but also the regular reflection on the boundary $x = L$, because of the discontinuity of the quantity $k^2(L) = k^2[1 + \varepsilon(L)] \neq k^2$ on this boundary. If no such discontinuity exists, that is, the wavenumber in the free part of the space is $k(L) - $ const., boundary conditions (adjusted boundary condition) for the problem (3.1) will be

$$\frac{i}{k(L)} \frac{d}{dx} u(x) + u(x)|_{x=L} = 2, \qquad \frac{i}{k} \frac{d}{dx} u(x) - u(x)|_{x=L_0} = 0. \tag{3.2'}$$

The quantity $\varepsilon(x)$ in eq. (3.1) describes uniformities of the wave propagation speed (of dielectric permittivity or refractive index). If both $\varepsilon(x)$ and the density of the medium are inhomogeneous (e.g. for acoustic waves), the wave equation appears to be

$$\frac{d^2}{dx^2} u(x) - \frac{\rho'(x)}{\rho(x)} \frac{d}{dx} u(x) + k^2 [1 + \varepsilon(x)] u(x) = 0 \quad \left(\rho'(x) = \frac{d}{dx} \rho(x) \right). \tag{3.4}$$

Its boundary conditions are continuity of $u(x)$ and $[1/\rho(x)] du(x)/dx$.

Equation (3.4) also describes the propagation of diversely polarized electromagnetic waves. If the electric field of the incident wave is normal to the incident plane $\rho(x) = 1$, and if it is parallel to the plane of incidence $\rho(x) = 1 + \varepsilon(x)$ (the magnetic permittivity of the medium is assumed to be equal to unity).

Among other possible extensions of the problem of wave incidence on the layer medium, we note the nonlinear problem, in which the function $\varepsilon(x, wI(x))$ also depends on the intensity of the wavefield $I(x) = |u(x)|^2$ inside the medium (w is the intensity of the incident wave). Its formulation and detailed solution for the deterministic case are described by Babkin and Klyatskin [1980a, 1982a], Klyatskin, Kozlov and Yaroschuk [1982], and Klyatskin and Yaroschuk [1985] (see also Klyatskin [1986]), and for the statistical case by Spigler [1986], Jordan, Papanicolaou and Spigler [1986], Yaroschuk [1986a, b, 1988a, b], and Knapp, Papanicolaou and White [1989].

In this case the stationary nonlinear problem of self-action of the wave is described by the nonlinear boundary-value problem

$$\frac{d^2}{dx^2} u(x) + k^2 [1 + \varepsilon(x, wI(x))] u(x) = 0,$$

$$\frac{i}{k} \frac{d}{dx} u(x) + u(x)|_{x=L} = 2, \qquad \frac{i}{k} \frac{d}{dx} u(x) - u(x)|_{x=L_0} = 0. \tag{3.1'}$$

Note that the problem described by a nonlinear Schrödinger equation is analogous to the boundary problem (3.1') (see, e.g. McLaughlin,

Papanicolaou, Sulem and Sulem [1986], LeMesurier, Papanicolaou, Sulem and Sulem [1987, 1988a, b]).

3.1.2. Source of plane waves inside the medium

Let the source of plane waves be located inside the medium layer at the point $L_0 < x_0 < L$. The wavefield within the layer is then described by the equation

$$\frac{d^2}{dx^2} G(x; x_0) + k^2 [1 + \varepsilon(x)] G(x; x_0) = 2ik\delta(x - x_0). \tag{3.5}$$

Outside the layer the field has the form of transmitter waves

$$G(x; x_0) = \begin{cases} T_1 e^{ik(x-L)}, & x > L, \\ T_2 e^{-ik(x-L_0)}, & x < L_0, \end{cases}$$

and continuity conditions for $G(x; x_0)$ and $\partial G(x; x_0)/\partial x$ on layer boundaries are

$$\frac{i}{k}\frac{d}{dx} G(x; x_0) + G(x; x_0)|_{x=L} = 0, \quad \frac{i}{k}\frac{d}{dx} G(x; x_0) - G(x; x_0)|_{x=L_0} = 0. \tag{3.6}$$

Within the medium layer the field $G(x; x_0)$ is continuous at each point, but its derivative at the source location has a finite discontinuity

$$\frac{d}{dx} G(x; x_0)|_{x=x_0+0} - \frac{d}{dx} G(x; x_0)|_{x=x_0-0} = 2ik. \tag{3.7}$$

Thus, the field of the point source is described by the boundary-value problem (3.5) and (3.6).

Note that the problem (3.1) and (3.2) of the wave incidence on the medium layer corresponds to the location of the source on the boundary $x_0 = L$ in the problem (3.5) and (3.6), that is, $u(x) = G(x; L)$.

The solution of the boundary-value problem (3.5) and (3.6) has the structure

$$G(x; x_0) = G(x_0; x_0) \begin{cases} \exp\left[ik \int_x^{x_0} d\xi \psi_1(\xi) \right], & x_0 \geqslant x, \\ \exp\left[ik \int_{x_0}^{x} d\xi \psi_2(\xi) \right], & x_0 \leqslant x, \end{cases} \tag{3.8}$$

where $G(x_0; x_0) = 2[\psi_1(x_0) + \psi_2(x_0)]^{-1}$, and the function $\psi_i(x)$ satisfies the

Riccati equation

$$\frac{d}{dx}\psi_{1,2}(x) = \pm ik[\psi_{1,2}^2(x) - 1 - \varepsilon(x)], \quad \psi_1(L_0) = 1, \quad \psi_2(L) = 1. \quad (3.9)$$

The scheme of the problem solution is given in fig. 1a. Two equations (3.9) are solved initially, and then the field inside the layer is reconstructed by using the quadratures (3.8). This method is well known, and is called the run, or Gaussian differencing, method; but it is not suitable for analysis of statistical problems.

Let us introduce functions $R_i(x)$ instead of $\psi_i(x)$ with the help of equalities

$$\psi_i(x) = \frac{1 - R_i(x)}{1 + R_i(x)}, \quad i = 1, 2.$$

The wavefield in the region $x < x_0$ then can be written in the form

$$G(x; x_0) = \frac{[1 + R_1(x_0)][1 + R_2(x_0)]}{1 - R_1(x_0)R_2(x_0)} \exp\left\{ik \int_x^{x_0} d\xi \frac{1 - R_1(\xi)}{1 + R_1(\xi)}\right\}, \quad (3.10)$$

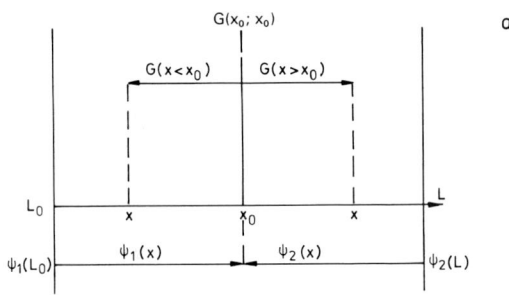

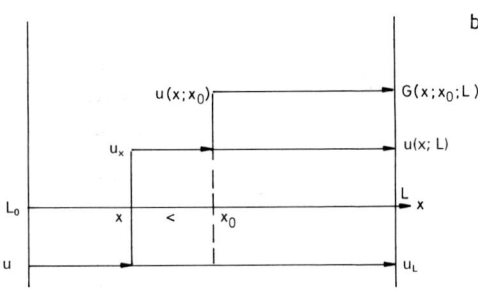

Fig. 1. The scheme of the solution of the boundary problem (3.5): (a) run method; (b) imbedding method.

where $R_1(x)$, being the function of x, is described by the Ricatti equation

$$\frac{d}{dx} R_1(x) = 2ikR_1(x) + i\frac{k}{2}\varepsilon(x)[1 + R_1(x)]^2, \quad R_1(L_0) = 0. \tag{3.11}$$

When $x_0 = L$, expression (3.10) becomes

$$u(x; L) = [1 + R_1(L)] \exp\left\{ik \int_x^L d\xi \frac{1 - R_1(\xi)}{1 + R_1(\xi)}\right\}, \tag{3.12}$$

and, consequently, the quantity $R_1(L) = R_L$ is the reflection factor of the plane wave incident from the region $x > L$ on the medium layer. It is similar for the quantity $R_2(x_0)$, which is the reflection factor of the wave incident from the homogeneous space $x < x_0$ (i.e. $\varepsilon = 0$) on the medium layer (x_0, L).

Using equality (3.12), formula (3.10) can be rewritten as

$$G(x; x_0) = \frac{1 + R_2(x_0)}{1 - R_1(x_0)R_2(x_0)} u(x; x_0), \quad x \leqslant x_0; \tag{3.13}$$

that is, *the field of the point source in the region $x < x_0$ is proportional to the field of the plane wave incident from the homogeneous space $x > x_0$ on the medium layer (L_0, x_0)*. The influence of the part of the layer (x_0, L) is described only by the quantity $R_2(x_0)$.

Problems with absolutely reflecting boundaries, where values G or $\partial G/\partial x$ become zero, have special meaning in physical applications. Thus in the latter case, for the source, placed on this reflecting boundary, we have $R_2(x_0) = 1$, and, consequently,

$$G_{\text{ref}}(x; x_0) = \frac{2}{1 - R_1(x_0)} u(x; x_0). \tag{3.14}$$

Note that the following equality follows from equation (3.5), as $x < x_0$,

$$k\gamma I(x; x_0) = \frac{d}{dx} S(x; x_0), \tag{3.15}$$

where $I(x; x_0) = |G(x; x_0)|^2$ is the wavefield intensity and the quantity $S(x; x_0)$ is the energy-flow density, which is defined by the relation

$$S(x; x_0) = \frac{1}{2ik}\left[G(x; x_0)\frac{d}{dx} G^*(x; x_0) - G^*(x; x_0)\frac{d}{dx} G(x; x_0)\right].$$

Using equality (3.10), it is possible to obtain the expression for $S(x; x_0)$

$$S(x; x_0) = S(x_0; x_0) \exp\left\{-k\gamma \int_x^{x_0} d\xi \frac{|1 + R_1(\xi)|^2}{1 - |R_1(\xi)|^2}\right\}, \tag{3.16a}$$

$$S(x_0; x_0) = \frac{(1 - |R_1(x_0)|^2)|1 + R_2(x_0)|^2}{|1 - R_1(x_0)R_2(x_0)|^2}. \quad (3.16b)$$

Later we shall address statistical problems of the wave incidence on the half-space of a randomly inhomogeneous medium, and of a source situated in infinite space where the attenuation is sufficiently small ($\gamma \Rightarrow 0$). It can be seen from expression (3.15) that these limit transitions are generally not permutable. In fact, if $\gamma = 0$, the conservation of the energy-flow density $S(x; x_0)$ in the whole space $x < x_0$ follows from eq. (3.15). In the presence of small finite attenuation, however, the integration of eq. (3.15) over the half-space $x < x_0$ yields the restriction for the value of energy that is contained in this half-space,

$$k\gamma \int_{-\infty}^{x_0} dx \, I(x; x_0) = S(x_0; x_0). \quad (3.17)$$

3.2. IMBEDDING METHOD IN ONE-DIMENSIONAL WAVE PROBLEMS

As mentioned earlier, the use of the imbedding method permits the reformulation of boundary-value problems into problems with initial conditions. We shall examine such a transition as applied to boundary-value problems that were formulated in § 3.1.

3.2.1. A wave incident on the layer

The solution for the boundary-value problems (3.1) and (3.2) depends on the parameter L which is the position of the right-hand boundary of the layer. The wave is incident on this boundary; that is, $u(x) = u(x; L)$. By differentiating eq. (3.1) and boundary conditions (3.2) with respect to parameter L, it is easy to see that $u(x; L)$ satisfies the equation, being the function of the parameter L:

$$\frac{\partial}{\partial L} u(x; L) = iku(x; L) + \frac{ik}{2} \varepsilon(L)(1 + R_L)u(x; L), \quad u(x; x) = 1 + R_x. \quad (3.18)$$

The initial condition for eq. (3.18) and a coefficient in the equation are determined by the quantity R_L, which is the reflection factor, and is described in turn by the Riccati equation

$$\frac{d}{dL} R_L = 2ikR_L + i\frac{k}{2} \varepsilon(L)(1 + R_L)^2, \quad R_{L_0} = 0. \quad (3.19)$$

Thus, the problem of the reflection factor in dependence on the layer thickness is solved first, and then the field inside the medium is reconstructed with the help of eq. (3.18).

Note that the field $\partial u(x; L)/\partial x$ is linked with the field $u(x; L)$ by the simple equality

$$\frac{\partial}{\partial x} u(x; L) = -\mathrm{i}k \frac{1 - R_x}{1 + R_x} u(x; L).$$

Note also that if the free space $x \leqslant L_0$ is characterized by the wave number k_1, the boundary condition on the boundary $x = L_0$ is

$$\frac{\mathrm{i}}{k_1} \frac{\mathrm{d}}{\mathrm{d}x} u(x) - u(x)|_{x = L_0} = 0.$$

In this case the imbedding equations do not change. Only the initial condition for the reflection coefficient changes and becomes

$$R_{L_0} = \frac{k - k_1}{k + k_1}.$$

Amplitudes of opposite waves $a(x)$ and $b(x)$, defined by equalities (3.2), were introduced in § 3.1.1. These quantities as functions of the parameter L are also described by eq. (3.18), and differ only in the initial conditions at $L = x$. Opposite wave intensities $W_1(x; L) = |a(x; L)|^2$, $W_2(x; L) = |b(x; L)|^2$ are described by equations that result from eq. (3.18):

$$\frac{\partial}{\partial L} W_j(x; L) = -\frac{k\gamma}{2}[2 + R_L + R_L^*]W_j(x; L) + \mathrm{i}\frac{k}{2}\varepsilon_1(L)(R_L - R_L^*)W_j(x; L),$$

$$W_1(x; x) = 1, \quad W_2(x; x) = |R_x|^2. \tag{3.18'}$$

Note that wavefield intensity $I(x; L) = |u(x; L)|^2$ is also described by eq. (3.18') with the initial condition $I(x; L)|_{L=x} = I(x; x) = |1 + R_x|^2$. If dissipation is absent ($\gamma = 0$), the equation for the modulus square of the reflection coefficient $W_L = |R_L|^2$ follows from eq. (3.19):

$$\frac{\mathrm{d}}{\mathrm{d}L} W_L = -\mathrm{i}\frac{k}{2}\varepsilon_1(L)(R_L - R_L^*)(1 - W_L), \quad W_{L_0} = 0. \tag{3.18''}$$

Hence, eq. (3.18') for the intensity can be integrated, and we obtain

$$I(x; L) = |1 + R_x|^2 \frac{1 - W_L}{1 - W_x}. \tag{3.18'''}$$

Thus, the intensity of the wavefield inside the layer in this case is described by the solution of the only equation for the R_L. If the layer thickness changes, the following equality results from eq. (3.18'''):

$$I(x; L') = \frac{1 - W_{L'}}{1 - W_L} I(x; L) \quad (L' \leq L).$$

Hence, a definite *property of the invariance of the intensity distribution of the wavefield exists inside the medium layer with respect to changes in layer thickness.*

If the initial condition for eq. (3.18'') is $R_{L_0} = \pm 1$, for example, giving the reflection boundary ($k_1 \Rightarrow 0$ or $k_1 \Rightarrow \infty$), then $W_L \equiv 1$ and the reflection coefficient has a structure $R_L = e^{\phi_L}$. For the phase ϕ_L and wavefield intensity, we now have the system of imbedding equations:

$$\frac{d}{dL} \phi_L = 2k + k\varepsilon(L)(1 + \cos \phi_L), \quad \phi_{L_0} = \phi_0;$$

$$\frac{\partial}{\partial L} I(x; L) = -k\varepsilon(L) I(x; L) \sin \phi_L, \quad I(x; L)|_{L=x} = 2(1 + \cos \phi_x).$$

Note that for an inclined incident wave under the angle θ (fig. 2), the system of imbedding equations for the wavefield and reflection coefficient is (2.19), where $q = k \sin \theta$, that is,

$$\frac{\partial}{\partial L} u(x; L) = ik(\cos \theta) u(x; L) + \frac{ik}{2 \cos \theta} \varepsilon(L)(1 + R_L) u(x; L),$$

$$u(x; x) = 1 + R_x;$$

$$\frac{d}{dL} R_L = 2ik(\cos \theta) R_L + \frac{ik}{2 \cos \theta} \varepsilon(L)(1 + R_L)^2.$$

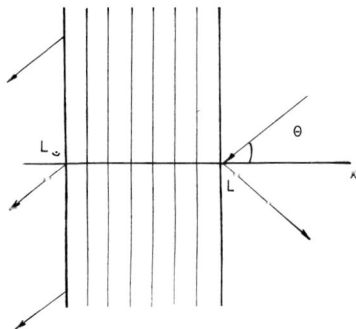

Fig. 2. Inclined incidence of a wave.

Hence, for reflecting boundary L_0 we have an equation for the phase of the reflection coefficient ϕ_L:

$$\frac{d}{dL}\phi_L = 2k\cos\theta + \frac{k}{\cos\theta}\varepsilon(L)(1+\cos\phi_L), \quad \phi_{L_0} = \phi_0. \qquad (3.19')$$

Imbedding equations (3.18) and (3.19) relate to the problem where the presence of the regular reflection from the boundary $x = L$ results from discontinuity of the function $k^2(L)$ on this boundary. If the quantity $k^2(L)$ is continuous on this boundary, the wave problem can be described by the boundary-wave problem (3.2'). In this case the imbedding equations are

$$\frac{\partial}{\partial L}u(x;L) = ik(L)u(x;L) + \frac{1}{2}\frac{k'(L)}{k(L)}(1-R_L)u(x;L), \quad u(x;x) = 1+R_x;$$

$$\frac{d}{dL}R_L = 2ik(L) + \frac{1}{2}\frac{k'(L)}{k(L)}(1-R_L^2), \quad R_{L_0} = \frac{k(L_0)-k}{k(L_0)+k}. \qquad (3.20)$$

If dissipation is absent, the equation for the wave intensity $I(x;L)$ can be integrated analogously with the previous case, giving the equality

$$I(x;L) = \frac{k(L)}{k(x)}|1+R_x|^2\frac{1-W_L}{1-W_x}.$$

Hence, the property of invariance with respect to change of layer thickness also occurs in this case:

$$I(x;L') = \frac{k(L')}{k(L)}\frac{1-W_{L'}}{1-W_L}I(x;L) \quad (L' \leq L).$$

If the density $\rho(x)$ of the medium is also variable inside the layer, that is, the problem is described by eq. (3.4), the imbedding equations will have the form (the density outside the medium layer being taken to be equal to unity) (Klyatskin and Lyubavin [1983a]):

$$\frac{\partial}{\partial L}u(x;L) = ik\rho(L)u(x;L) + \frac{ik}{2\rho(L)}[1+\varepsilon(L)-\rho^2(L)](1+R_L)u(x;L),$$

$$u(x;x) = 1+R_x; \qquad (3.21)$$

$$\frac{d}{dL}R_L = 2ik\rho(L)R_L + \frac{ik}{2\rho(L)}[1+\varepsilon(L)-\rho^2(L)](1+R_L)^2, \quad R_{L_0} = 0.$$

The special feature of the problem (3.21) is that the *problem does not contain the derivative of the density*.

Let us examine the boundary-value problem in the layer (L_0, L)

$$\frac{d}{dx}p(x) = i\omega\rho(x)v(x), \quad \frac{d}{dx}v(x) = \frac{i\omega}{\rho(x)c^2(x)}p(x); \quad p(L) = 1, \quad v(L_0) = 0,$$
(3.22)

which describes the excitation of acoustic waves by external forces applied at the boundary $x = L$ (e.g. by atmospheric pressure). By introducing the displacement of a liquid particle $v(x) = -i\omega\zeta(x)$, the boundary-value problem (3.22) can be reformulated into the initial-value problem

$$\frac{\partial}{\partial L}p(x; L) = -\omega^2 \rho(L)\zeta_L p(x; L), \quad p(x; x) = 1;$$

$$\frac{d}{dL}\zeta_L = -\omega^2 \rho(L)\left[\zeta_L^2 + \frac{1}{\omega^2 \rho^2(L)c^2(L)}\right], \quad \zeta_{L_0} = 0.$$
(3.23)

Detailed discussion of this problem is found in the works of Klyatskin and Lyubavin [1983a, 1984], Gulin and Klyatskin [1986a, b, 1989, 1993], and Gulin [1985, 1987a, b], which describe the distribution in depth of the low-frequency acoustic noise in the ocean for different models of the surface noise, medium stratification, and impedance properties of the ocean bottom.

Note that the problem solution (3.23) has a resonance structure, meaning that poles of the function ζ_L (or zeros of the function $f_L = 1/\zeta_L$) describe eigenvalues and eigenfunctions of the homogeneous boundary-value problem (3.22). This feature can be used directly to find spectral characteristics of a boundary-value problem. In particular, dynamic equations can be obtained for them (the problem with initial conditions), which are suitable for studying statistical characteristics (Scott, Shampine and Wing [1969], Saichev [1980], Goland and Klyatskin [1988, 1989], Goland [1987, 1988]).

The analysis of the eigenvalues is based on the analysis of zeros of the Riccati equation solution. This equation can be written in the general form

$$\frac{d}{dL}f_L = a_L(\lambda) + b_L(\lambda)f_L + c_L(\lambda)f_L^2,$$
(3.24)

where λ is a spectral parameter. The eigenvalues are defined as the solution of the equation

$$f_L(\lambda_L) = 0,$$
(3.24')

where we have introduced the dependence of the spectral parameter on L.

Since these eigenvalues are functions of the parameter L, they satisfy the equation

$$a_L(\lambda_L) + A_L(\lambda_L)\frac{d}{dL}\lambda_L = 0, \qquad (3.25)$$

where $A_L(\lambda) = \partial f_L(\lambda)/\partial \lambda$, and whose initial condition at $L \Rightarrow 0$ (here $L_0 = 0$) is determined by the dynamics of the concrete eigenvalue.

By considering the wavefield $u(x)$ as a function of parameters L and W for the nonlinear boundary-value problem (3.1'), we have imbedding equations in the form

$$\left[\frac{\partial}{\partial L} - wb(L, w)\frac{\partial}{\partial w}\right]u_L(w)$$
$$= 2ik[u_L(w) - 1] + i\frac{k}{2}\varepsilon(L, J_L)u_L^2(w), \quad u_{L_0}(w) = 1;$$
$$\left[\frac{\partial}{\partial L} - wb(L, w)\frac{\partial}{\partial w}\right]u(x; L, w) \qquad (3.26)$$
$$= a(L, w)u(x; L, w), \quad u(x; x, w) = u_x(w),$$

where

$$a(L, w) = ik + i\frac{k}{2}\varepsilon(L, J_L)u_L(w), \quad b(L, w) = a(L, w) + a^*(L, w),$$

$$J_L = w|u_L(w)|^2, \quad u_L(w) = u(L; L, w).$$

The equation for the wave intensity, $J(x; L, w) = w|u(x; L, w)|^2$, follows from eq. (3.26)

$$\left[\frac{\partial}{\partial L} - wb(L, w)\frac{\partial}{\partial w}\right]J(x; L, w) = 0, \quad J(x; x, w) = J_L(w).$$

The reflection coefficient $R_L(w)$ is determined by the equality $R_L(w) = u_L(w) - 1$. The principal feature of the system (3.26) is that it represents an initial-value problem, although it contains partial differential equations. These equations are first-order partial differential equations, and can be solved by the method of characteristics.

If we introduce the characteristic curves $w_L = w(L, w_0)$ according to the equality

$$\frac{d}{dL}w_L = -b(L, w_L)w_L, \quad w_{L_0} = w_0,$$

the field on the layer boundary $u_L(w)$ can be described by the equation

$$\frac{\mathrm{d}}{\mathrm{d}L} u_L = 2\mathrm{i}k(u_L - 1) + \mathrm{i}\frac{k}{2}\varepsilon(L, J_L)u_L^2, \quad u_{L_0} = 1,$$

and the equation for the wave intensity inside the medium is transformed to the equality

$$\frac{\mathrm{d}}{\mathrm{d}L} J(x; L, w_L) = 0, \quad J(x; x, w_x) = J_x(w_x),$$

and, consequently, the intensity of the wavefield inside the medium remains constant on the characteristic, that is,

$$J(x; L, w_L) = J_x(w_x).$$

Hence, the intensity of the wave inside the medium is completely defined by the value of the field on the boundary. If we know the dynamics of the characteristics w_L as a function of L and the distribution of the wave intensity inside the medium for some fixed layer thickness $J(x; L, w_L)$, then for any other layer thickness $L_1 \leqslant L$ the intensity behavior inside the medium does not change, but will correspond to the incident wave intensity w_{L_1}; that is,

$$J(x; L_1, w_{L_1}) = J(x; L, w_L).$$

Thus, *a definite property of the invariance of the intensity distribution of the wavefield exists inside the medium layer with respect to changes in layer thickness and the intensity of the wave incident on the layer.*

When $x = L_0$, $J(L_0; L, w_L) = w_0$. Taking into account that the field on the layer boundary $x = L_0$ defines the complex coefficient of the wave passage through the medium layer $T_L = u(L_0; L, w)$, we obtain for the square of the modulus of the transmitter coefficient

$$|T_L|^2 = J(L_0; L, w_L)/w_L = w_0/w_L,$$

which reveals the physical sense of the characteristic curves $w_L = w(L, w_0)$.

As noted earlier, problems of electromagnetic wave propagation are similar to acoustic problems with variable density. Thus, the preceding equations are also valid for electromagnetic waves. The study of short and ultra-short, radio-wave propagation in the troposphere wave-guide over the ocean surface based on this approach was studied by Bugrov, Klyatskin and Shevtsov [1984, 1985], Koshel' [1986, 1987, 1990a–d, 1992a, b], Goland and Koshel' [1990], Koshel' and Shishkarev [1993a, b].

3.2.2. Source location inside the layer of an inhomogeneous medium

If we let the source of plane waves be placed inside the layer of an inhomogeneous medium at the point x_0, then the wavefield within the layer is described by the boundary-value problem (3.5) and (3.6). Assuming its solution to be a function of the parameter L, we obtain imbedding equations in the form

$$\frac{\partial}{\partial L} G(x; x_0; L) = i\frac{k}{2} \varepsilon(L) u(x; L) u(x_0; L),$$

$$G(x; x_0; L)|_{L=\max(x, x_0)} = \begin{cases} u(x; x_0) & x \leqslant x_0, \\ u(x_0; x), & x \geqslant x_0. \end{cases} \qquad (3.27)$$

Here, $u(x; L)$ is the solution of the problem (3.18), that is, the problem of the wave incidence on the medium layer. Thus, fig. 1b illustrates the scheme of the solution of the problem (3.5) and (3.6), based on the imbedding method. By comparing this scheme with the usual methods (fig. 1a), we see that the direction of the problem solution is changed in the imbedding method, which allows us to carry out the statistical analysis of corresponding boundary-value stochastic problems of waves in randomly inhomogeneous layered media.

We have examined the source of plane waves inside the layer of medium $L_0 < x < L$. Outside this layer the medium is homogeneous with $\varepsilon(x) = 0$. If, outside the layer, the medium is homogeneous with the wave parameter k_2 if $x > L$, and the medium is homogeneous with the wave parameter k_1 if $x < L_0$, then the wavefield inside the layer should be described by the boundary problem

$$\frac{d^2}{dx^2} \tilde{G}(x; x_0) + k^2[1 + \varepsilon(x)] \tilde{G}(x; x_0) = 2ik\delta(x - x_0);$$

$$\frac{i}{k_2} \frac{d}{dx} \tilde{G}(x; x_0) + \tilde{G}(x; x_0)|_{x=L} = 0, \quad \frac{i}{k_1} \frac{d}{dx} \tilde{G}(x; x_0) - \tilde{G}(x; x_0)|_{x=L_0} = 0. \qquad (3.28)$$

The use of the Green formulas allows us to express the solution of the boundary problem (3.28) in the form (Babkin, Klyatskin and Lyubavin [1982a])

$$\tilde{G}(x; x_0) = G_1(x; x_0) + G_2(x; x_0), \qquad (3.29)$$

where $G_1(x; x_0)$ is the solution of the boundary problem

$$\frac{d^2}{dx^2} G_1(x; x_0) + k^2[1 + \varepsilon(x)] G_1(x; x_0) = 2ik\delta(x - x_0), \qquad (3.30a)$$

$$\frac{\mathrm{i}}{k}\frac{\mathrm{d}}{\mathrm{d}x}G_1(x;x_0) + G_1(x;x_0)|_{x=L} = 0, \qquad (3.30\mathrm{b})$$

$$\frac{\mathrm{i}}{k_1}\frac{\mathrm{d}}{\mathrm{d}x}G_1(x;x_0) - G_1(x;x_0)|_{x=L_0} = 0, \qquad (3.30\mathrm{c})$$

and, as a function of the parameter L, is described by the system of imbedding equations

$$\begin{aligned}
&\frac{\partial}{\partial L}G_1(x;x_0;L) = \mathrm{i}\frac{k}{2}\varepsilon(L)u(x;L)u(x_0;L), \\
&G_1(x;x_0;L)|_{L=\max(x,x_0)} = \begin{cases} u(x;x_0), & x \leqslant x_0, \\ u(x_0;x), & x \geqslant x_0, \end{cases} \\
&\frac{\partial}{\partial L}u(x;L) = \mathrm{i}ku(x;L) + \frac{\mathrm{i}k}{2}\varepsilon(L)u_L u(x;L), \quad u(x;x) = u_x; \\
&\frac{\mathrm{d}}{\mathrm{d}L}u_L = 2\mathrm{i}k(u_L - 1) + \mathrm{i}\frac{k}{2}\varepsilon(L)u_L^2, \quad u_{L_0} = \frac{2k}{k+k_1}.
\end{aligned} \qquad (3.31)$$

The function $G_2(x;x_0)$ describes the boundary condition at $x = L$, and has a structure

$$G_2(x;x_0) = \frac{1}{G-u_L}u(x;L)u(x_0;L), \quad G = \frac{2k}{k-k_2}, \qquad (3.32)$$

where u_L, $u(x;L)$ are solutions of the problem (3.31). If the source of plane waves is located on the boundary $x_0 = L$, we have

$$\tilde{G}(x;L) = \frac{G}{G-u_L}u(x;L),$$

and this quantity does not determine the wavefield from the incident wave $\mathrm{e}^{-\mathrm{i}k_2(x-L)}$ inside the medium. The wavefield of this incident wave is expressed by the quantity

$$\tilde{u}(x;L) = \frac{k_2}{k}\tilde{G}(x;L) = \frac{k_2}{k}\frac{G}{G-u_L}u(x;L). \qquad (3.33)$$

In particular, by setting $x = L$ in eq. (3.33), we obtain the relation between the reflection coefficients $\tilde{R}_L = \tilde{u}(L;L) - 1$ and $R_L = u_L - 1$ in the form

$$\tilde{R}_L = \frac{(k_2-k)+(k_2+k)R_L}{(k_2+k)+(k_2-k)R_L}. \qquad (3.33')$$

If the boundary $x = L$ is a reflection boundary with a boundary condition

$$\frac{d}{dx}\tilde{G}(x; x_0)|_{x=L} = 0 \quad (k_2 \Rightarrow 0),$$

the constant $G = 2$ in eq. (3.32), and the Green function has a structure

$$\tilde{G}(x; x_0) = G_1(x; x_0) + \frac{1}{1 - R_L} u(x; L) u(x_0; L), \quad \tilde{G}(x; L) = \frac{2}{1 - R_L} u(x; L),$$

which coincides with the formula (3.14).

If we have a reflection boundary $x = L$ with a boundary condition

$$\tilde{G}(L; x_0) = 0 \quad (k_2 \Rightarrow \infty),$$

the constant $G = 0$ and

$$\tilde{G}(x; x_0) = G_1(x; x_0) - \frac{1}{1 + R_L} u(x; L) u(x_0; L), \quad \tilde{G}(x; L) = 0.$$

For the case $k_2 = k$, the constant $G = \infty$, and $\tilde{G}(x; x_0) \equiv G_1(x; x_0)$.

3.3. STATISTICAL DESCRIPTION

In the simplest problem formulation the random function $\varepsilon_1(x)$ can be regarded as a Gaussian random function, which is *delta-correlated in the space* (approximation of the white noise), with the parameters

$$\langle \varepsilon_1(x) \rangle = 0, \quad \langle \varepsilon_1(x) \varepsilon_1(x') \rangle = 2\sigma_\varepsilon^2 l_0 \delta(x - x'), \tag{3.34}$$

where $\sigma_\varepsilon^2 = \langle \varepsilon_1^2(x) \rangle \ll 1$ is the variance of ε_1 and l_0 is the correlation radius. This approximation means that the passage to the asymptotic case $l_0 \Rightarrow 0$ in the exact problem solution with finite radius of correlation l_0 gives the result, which coincides with the solution of the statistical problem (3.34) (Klyatskin [1975, 1980a, 1985, 1991d]).

Because of the smallness of σ_ε^2, all statistical effects can be separated into local and accumulative types, according to the effect of multiple wave re-reflection in the medium. We are interested in the latter effects.

From the statement of boundary-value wave problems, we saw that two types of wavefield characteristics occur which are of direct physical interest. The first type is connected with field magnitudes on layer boundaries (factors R_L and T_L), with field at the point of source location $G(x_0; x_0)$, with energy flow density at the point of source location $S(x_0; x_0)$, and so on. The second type is connected with the determination of statistical wavefield characteristics inside the layer of randomly inhomogeneous medium, and is the object of the statistical theory of radiative transfer (Kohler and Papanicolaou [1973, 1974], Babkin and Klyatskin [1982b], Klyatskin [1980a, 1985, 1986, 1991a–c], Doucot and Rammal [1987], Rammal and Doucot [1987]).

3.3.1. Coefficients of reflection and transmission

Let the layer of randomly inhomogeneous medium occupy a part of space $L_0 < x < L$. Then the reflection factor of a plane wave incident on the medium layer is R_L. This is described by the Riccati equation (3.19). Let us introduce the quantity $W_L = |R_L|^2$, which satisfies the equation following from eq. (3.19),

$$\frac{d}{dL} W_L = -2k\gamma W_L - i\frac{k}{2}\varepsilon_1(L)(R_L - R_L^*)(1 - W_L), \quad W_{L_0} = 0. \tag{3.35}$$

Fast oscillating functions that do not contribute to accumulative effects have been omitted in the dissipative term. By introducing the function $\Phi_L(W) = \delta(W_L - W)$, which satisfies the Liouville equation

$$\frac{\partial}{\partial L}\Phi_L(W) = 2k\gamma \frac{\partial}{\partial W} W\Phi_L(W) + i\frac{k}{2}\varepsilon_1(L)\frac{\partial}{\partial W}(R_L - R_L^*)(1 - W)\Phi_L(W),$$

and averaging the latter over the ensemble of realizations of random functions $\varepsilon_1(L)$, we obtain the Fokker–Planck equation for the probability density of the modulus square of the reflection factor $P_L(W) = \langle \Phi_L(W) \rangle$

$$\frac{\partial}{\partial L} P_L(W) = 2k\gamma \frac{\partial}{\partial W} W P_L(W) - 2D \frac{\partial}{\partial W} W(1 - W) P_L(W)$$

$$+ D \frac{\partial}{\partial W}(1 - W)^2 W \frac{\partial}{\partial W} P_L(W), \quad P_{L_0}(W) = \delta(W), \tag{3.36}$$

where $D = k^2 \sigma_\varepsilon^2 l_0 / 2$ is the diffusion coefficient.

In some cases it is convenient to express the quantity W_L in the form

$$W_L = \frac{u_L - 1}{u_L + 1}, \quad u_L = \frac{1 + W_L}{1 - W_L}, \quad u_L \geq 1.$$

Then, for the probability density of the random quantity $u_L - P_L(u) = \langle \delta(u_L - u) \rangle$, we have

$$\frac{\partial}{\partial L} P_L(u) = D \frac{\partial}{\partial u}(u^2 - 1)\frac{\partial}{\partial u} P_L(u) + k\gamma \frac{\partial}{\partial u}(u^2 - 1)P_L(u), \tag{3.36'a}$$

$$P_{L_0}(u) = \delta(u-1). \tag{3.36'b}$$

During the process of derivation of eqs. (3.36) and (3.36′) the additional averaging over fast oscillations can be used.

3.3.1.1. *Nondissipative medium.* If wave attenuation in the medium is absent (i.e. $\gamma = 0$), eq. (3.36′) in the dimensionless form ($DL \Rightarrow L$) is

$$\frac{\partial}{\partial L} P_L(u) = \frac{\partial}{\partial u}(u^2-1)\frac{\partial}{\partial u} P_L(u), \quad P_{L_0}(u) = \delta(u-1). \tag{3.37}$$

Its solutions can be easily found using the integral Meller–Fock transformation (see, e.g. Didkin and Prudnikov [1974])

$$F(\mu) = \int_1^\infty dx\, f(x) P_{-1/2+i\mu}(x) \quad (\mu \geq 0),$$

$$f(x) = \int_0^\infty d\mu\, \mu \tanh(\mu\pi)\, F(\mu) P_{-1/2+i\mu}(x) \quad (x \geq 1),$$

where $P_{-1/2+i\mu}(x)$ is the Legendre function of the first kind (cone function). As a result, we obtain

$$P_L(W) = \int_0^\infty d\mu\, \mu \tanh(\mu\pi)\, e^{-(\mu^2+1/4)(L-L_0)} P_{-1/2+i\mu}(x). \tag{3.38}$$

By using the representation (3.38), one can calculate statistical characteristics of reflection and transmission coefficients $|T_L|^2 = 1 - |R_L|^2 = 2/(1+u_L)$ (when $\gamma = 0$), and, in particular (Klyatskin [1980a, 1985, 1986]), the expressions

$$\langle |T_L|^{2n} \rangle = 2^n \pi \int_0^\infty d\mu\, \frac{\mu \sinh(\mu\pi)}{\cosh^2(\mu\pi)} K_n(\mu) e^{-(\mu^2+1/4)(L-L_0)}, \tag{3.39}$$

where

$$K_{n+1}(\mu) = \frac{1}{2n}[\mu^2 + (n-1/2)^2] K_n(\mu), \quad K_1(\mu) = 1.$$

The asymptotic formula follows from eq. (3.39) as $(L-L_0) \gg 1$,

$$\langle |T_L|^{2n} \rangle \approx \frac{[(2n-3)!!]^2 \pi^{5/2}}{2^{2n-1}(n-1)!} (L-L_0)^{-3/2} e^{-(L-L_0)/4}. \tag{3.40}$$

Therefore, the asymptotic dependence on the layer thickness is universal for any momentum of the quantity $|T_L|$; only the numerical factor changes. The tendency to zero of the quantity $|T_L|$ with the increase of layer thickness means that $|R_L| \Rightarrow 1$ with unit probability; that is, *the half-space of randomly inhomogeneous nondissipative medium completely reflects the wave incident on it.*

The same situation applies to the sliding incidence of a wave on the randomly layered half-space. In this case the reflection coefficient has the structure $R_L = \exp(i\phi_L)$, where the phase ϕ_L can be described by the imbedding equation (3.19')

$$\frac{d}{dL}\phi_L = 2k\cos\theta + \frac{k}{\cos\theta}\varepsilon(L)(1+\cos\phi_L).$$

Let us introduce a new function $z_L = \tan(\phi_L/2)$ instead of ϕ_L, which has singular points. The dynamic equation for z_L is

$$\frac{d}{dL}z_L = p(1+z_L^2) + \frac{k^2}{p}\varepsilon(L), \quad p = k\cos\theta.$$

As before, the function $\varepsilon(L)$ is supposed to be the Gaussian random delta-correlated process. The corresponding Fokker–Planck equation for the probability distribution $P_L(z) = \langle \delta(z - z_L) \rangle$ defined on the whole line $(-\infty, +\infty)$ then has the form

$$\frac{\partial}{\partial L}P_L(z) = -p\frac{\partial}{\partial z}(1+z^2)P_L(z) + \frac{2k^2}{p^2}D\frac{\partial^2}{\partial z^2}P_L(z), \quad D = k^2\sigma_\varepsilon^2 l_0/2.$$

The steady-state probability density $P(z)$ under the limit transition $L_0 \Rightarrow -\infty$ (the half-space) is described by the equation

$$-\kappa\frac{d}{dz}(1+z^2)P(z) + \frac{d^2}{dz^2}P(z) = 0,$$

where the parameter κ is

$$\kappa = p^3/2k^2D = \frac{\alpha}{2}\cos^3\theta, \quad \alpha = k/D.$$

Further analysis of this equation largely depends on the formulation of a boundary condition with respect to z, which specifies the type of problems under study. Thus, if we consider the function z_L to be discontinuous and defined for all values of L, although as mentioned, its conversion into $-\infty$ in some points is immediately accompanied by its origin in the same points with the value equal to $+\infty$, then its boundary condition is the condition

$$J(z)|_{z=-\infty} = J(z)|_{z=+\infty},$$

where

$$J(z) = -\kappa(1+z^2)P(z) + \frac{d}{dz}P(z)$$

is the probability flux density.

Hence, its solution is (Guzev, Klyatskin and Popov [1992])

$$P(z) = J(\kappa) \int_0^\infty d\xi \, e^{-\kappa\xi[1+\xi^2/3+z(z+\xi)]}, \qquad (3.41)$$

where the steady state of the probability flux density is defined from the equality

$$J^{-1}(\kappa) = \left(\frac{\pi}{k}\right)^{1/2} \int_0^\infty d\xi \, \xi^{-1/2} e^{-\kappa(\xi+\xi^3/12)}.$$

The corresponding steady-state probability distribution of the phase ϕ_L defined over the interval $(-\pi, \pi)$ is described by the formula

$$P(\phi) = \frac{1+z^2}{2} P(z)|_{z=\tan(\phi/2)}.$$

On condition that $\kappa \gg 1$, we have an asymptotic solution $P(z) = 1/[\pi(1+z^2)]$ corresponding to the uniform probability distribution of $\phi : P(\phi) = 1/(2\pi)$.

In the limit case $\kappa \ll 1$ responsible for the inclined wave incidence on the half-space ($\theta \Rightarrow \pi/2$), we have

$$P(z) = \kappa^{1/3} \left(\frac{3}{4}\right)^{1/6} \frac{\pi^{-1/2}}{\Gamma(1/6)} \Gamma\left(\frac{1}{3}, \frac{\kappa z^3}{3}\right),$$

where $\Gamma(\mu, y)$ is the incomplete gamma function. The asymptotic expression for $P(z)$ follows from eq. (3.41) at $\kappa|z|^3/3 \gg 1$ and $|z| \Rightarrow \infty$, namely

$$P(z) = \kappa^{1/3} \left(\frac{3}{4}\right)^{1/6} \frac{\pi^{-1/2}}{\Gamma(1/6)} \left(\frac{3}{\kappa z^3}\right)^{2/3}.$$

The results of numerical calculations of the probability density are presented graphically in fig. 3a for $\kappa = 0.1; 1.0; 10$.

The probability distribution (3.41) allows us to calculate various quantities connected with fluctuations of the reflection coefficient phase; for example, on the boundary $x = L$ the mean value of the wavefield intensity is

$$\langle I(L; L)\rangle = 2\langle 1 + \cos \phi_L \rangle = \begin{cases} 2 & \kappa \gg 1, \\ 2(3)^{1/6}\Gamma(2/3)\kappa^{1/3} & \kappa \ll 1. \end{cases}$$

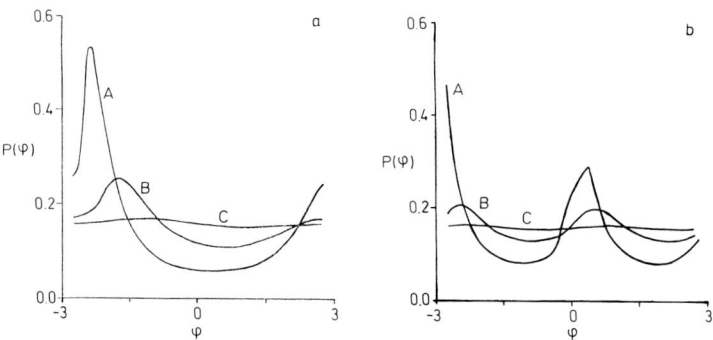

Fig. 3. Steady-state probability density $P(\phi)$: (a) slab with increase $\varepsilon(x)$ on the boundary; (b) slab without increase $\varepsilon(x)$ on the boundary. Curves correspond to A: $\kappa = 0.1$; B: $\kappa = 1$; C: $\kappa = 10$.

Numerical simulation results of the quantity $\langle I(L; L) \rangle = 2 + \langle R_L + R_L^* \rangle$ as a function of θ at $\alpha = 25$ are illustrated by Goland, Klyatskin and Yaroschuk [1991] and with asymptotic formulas (3.41). The transition from values at $\kappa \gg 1$ to values at $\kappa \ll 1$ is accomplished at the value $\kappa \approx 1$, corresponding with an angle of about 65 degrees. For the sliding incidence, when $\theta \to \pi/2$, the quantity $R_L \to -1$ and the wavefield $u(L) = 2 + R_L$ tend to zero at the boundary $x = L$.

This result indicates that the medium manifests itself as a mirror. This effect is linked with discontinuity of the function $\varepsilon(x)$ at the boundary $x = L$. This increase, being small, contributes little to the statistics of the phase on the incidence angles θ, when they are small. For the sliding incidence, on the other hand, this increase appears to be like an infinite barrier and contributes significantly to the statistics. Hence, the probability distribution of the reflection coefficient phase contains information about how the wave scatters on an increase of $\varepsilon(L)$ and on random inhomogeneities inside the medium without separating this effect.

When we consider the medium with an adjusted right hand boundary, that is, $k_L^2 = p^2$, the effect made by the discontinuity at $\varepsilon(L)$ can be excluded. Corresponding boundary conditions for the wavefield $u(x)$ have the form

$$\frac{i}{k_L}\frac{d}{dx}u(x) + u(x)|_{x=L} = 2, \quad \frac{i}{p}\frac{d}{dx}u(x) - u(x)|_{x=L_0} = 0.$$

The boundary problem is then reduced to the problem with initial values with respect to L, eq. (3.20), that is,

$$\frac{\partial}{\partial L} u(x; L) = ik_L u(x; L) + \frac{k'_L}{2k_L}(1 - R_L)u(x; L), \quad u(x; x) = 1 + R_x;$$

$$\frac{d}{dL} R_L = 2ik_L R_L + \frac{k'_L}{2k_L}(1 - R_L^2), \quad R_{L_0} = 0.$$

where $k'_L/k_L = k^2 \xi(L)/2p^2 [1 + k^2 \varepsilon(L)/p^2]$ and $\xi(L) = \partial \varepsilon(L)/\partial L$. Because of the smallness of the fluctuations, it is possible to replace k'_L/k_L and k_L by $k^2 \xi(L)/2p^2$ and p, respectively; then

$$\frac{\partial}{\partial L} u(x; L) = ipu(x; L) + \frac{k^2 \xi(L)}{2p^2}(1 - R_L)u(x; L), \quad u(x; x) = 1 + R_x;$$

$$\frac{d}{dL} R_L = 2ip R_L + \frac{k^2 \xi(L)}{2p^2}(1 - R_L^2), \quad R_{L_0} = 0.$$

A Gaussian random process $\xi(x)$ with parameters is

$$\langle \xi(x) \rangle = 0, \quad B_{\xi\xi}(x - x') = -\frac{\partial^2}{\partial x^2} B_{\varepsilon\varepsilon}(x - x').$$

It is easy to see that the nonlinear term changes in the equation of the reflection coefficient. In the case of a random medium occupying the half-space $x < L$ with the adjusted right-hand boundary, the quantity $R_L^2 \to 1$ and $R_L \to \pm 1$ for the sliding incidence. Hence, these points contribute most to the statistical characteristics linked with the phase.

Substituting $R_L = \exp[i\phi_L]$, we have the equation for ϕ_L

$$\frac{d}{dL} \phi_L = 2p - \frac{k^2 \xi(L)}{2p^2} \sin \phi_L.$$

Let us now introduce the new variable $z_L = \tan(\phi_L/2)$ in place of the phase as we did previously. We have the dynamic equation for z_L

$$\frac{d}{dL} z_L = p(1 + z_L^2) - \frac{k^2 \xi(L)}{2p^2} z_L.$$

The corresponding probability density $P_L(z) = \langle \delta(z_L - z) \rangle$, defined over the whole line $z_L \in (-\infty, \infty)$ now satisfies the Fokker–Planck equation (Guzev and Klyatskin [1993])

$$\frac{\partial}{\partial L} P_L(z) = -p \frac{\partial}{\partial z}(1 + z^2) P_L(z) + \frac{2k^2}{p^2} D \frac{\partial}{\partial z} z \frac{\partial}{\partial z} z P_L(z),$$

where the quantity $D = k^2 \sigma_\varepsilon^2 l_0/2$ defines the diffusion coefficient at the normal incidence of the wave on the medium slab as was done previously.

In the case of the half-space ($L_0 \to -\infty$), the steady-state (independent of L) probability density $P(z)$ satisfies the equation

$$\kappa \frac{d}{dz}(1+z^2)P(z) + \frac{d}{dz} z \frac{d}{dz} zP(z) = 0,$$

where $\kappa = p^3/2k^2 D = (\alpha/2)\cos^3\theta$, $\alpha = k/D$, as before. The solution for $P(z)$ has the form

$$P(z) = -\frac{J(\kappa)}{z}\int_{z_0}^{z}\frac{dz_1}{z_1}\exp\left[\kappa\left(z - \frac{1}{z} - z_1 + \frac{1}{z_1}\right)\right].$$

The constant $J(\kappa)$ is defined from the condition $\int_{-\infty}^{\infty} dz\, P(z) = 1$, and the value of z_0 should be chosen such that the quadrature is finite for z throughout $(-\infty, \infty)$; then

$$P(z) = \theta(z)P_+(z) + \theta(-z)P_-(z),$$

$$P_+(z) = \frac{J(\kappa)}{z}\int_0^\infty \frac{ds}{1+s}\exp\left[-\kappa s\left(z + \frac{1}{z(1+s)}\right)\right], \quad z > 0,$$

$$P_-(z) = -\frac{J(\kappa)}{z}\int_0^1 \frac{ds}{1-s}\exp\left[\kappa s\left(z + \frac{1}{z(1-s)}\right)\right], \quad z < 0.$$

(3.41′)

The probability density $P(z)$ is a continuous function, and $P_+(z = +0) = P_-(z = -0) = J(\kappa)/\kappa$, where

$$[J(\kappa)]^{-1} = \pi^2[J_0^2(2\kappa) + N_0^2(2\kappa)] = \begin{cases} \pi, & \kappa \gg 1, \\ \pi^2\left[1 + \frac{4}{\pi^2}(\ln\kappa + C)^2\right], & \kappa \ll 1. \end{cases}$$

Here, $J_0(x)$ is the Bessel function, $N_0(x)$ is the Neumann function, and C is the Euler constant.

On condition that $\kappa \gg 1$, we have the asymptotic solution $P(z) = 1/[\pi(1+z^2)]$ corresponding to the uniform probability distribution of the phase $P(\phi) = 1/(2\pi)$. For $\kappa \ll 1$ we cannot write an asymptotic expression of $P(\phi)$ uniform over ϕ. The numerical results are shown graphically in fig. 3b for $\kappa = 0.1;\ 1.0;\ 10$.

In the limit case $\kappa \ll 1$, let us consider wave characteristics linked with the phase fluctuations.

The mean value of the wavefield intensity on the boundary $x = L$ is

$$\langle I(L; L) \rangle = 2 + \langle R_L \rangle + \langle R_L^* \rangle = 2\langle 1 + \cos \phi_L \rangle = 4 \int_{-\infty}^{\infty} \frac{dz}{1+z^2} P(z).$$

As a result, for $\kappa \ll 1$ we have $\langle I(L; L) \rangle = 2$, which means that statistical weights of phase values for which $R_L = +1$ and $R_L = -1$ are equal, although the probability distribution $P(\phi)$ differs considerably from the uniform one.

Note that in the case of normal wave incidence ($\theta = 0$), the parameter $\kappa = k/D$ and describes the role of the wavenumber in problems of wave propagation in random medium (Yaroschuk [1984], Popov and Yaroschuk [1988, 1990]).

3.3.1.2. Dissipative medium. In the presence of absorption eqs. (3.36) and (3.36′) cannot be solved for the layer. For the half-space ($L_0 \Rightarrow -\infty$) however, a steady state exists independent of L probability distributions for W_L and u_L (Abramovich and Dyatlov [1975], Kohler and Papanicolaou [1976]):

$$P_\infty(W) = \frac{2\beta}{(1-W)^2} \exp\left(-\frac{2\beta W}{1-W}\right), \quad P_\infty(u) = \beta e^{-\beta(u-1)}, \tag{3.42}$$

where $\beta = k\gamma/D$. By using eq. (3.42), all moments of the quantity $W_L = |R_L|^2$ can be calculated. In particular, we have

$$\langle W_L \rangle = \begin{cases} 1 - 2\beta \ln(1/\beta), & \beta \ll 1, \\ 1/(2\beta), & \beta \gg 1, \end{cases}$$

and the recurrent equality for moments of higher order

$$n\langle W_L^{n+1} \rangle - 2(\beta + n)\langle W_L^n \rangle + n\langle W_L^{n-1} \rangle = 0, \quad n = 1, 2, \ldots$$

Note that for the problem under study mean values of the energy-flux density and wavefield intensity on the layer boundary are defined by expressions such as ($\beta \ll 1$)

$$\langle S(L; L) \rangle = 1 - \langle W_L \rangle = 2\beta \ln(1/\beta), \quad \langle I(L; L) \rangle = 1 + \langle W_L \rangle = 2. \tag{3.43}$$

3.3.2. Source inside the layer of a medium

A plane wave source located inside the layer of a medium, the wavefield, and the energy-flux density at the point of the source location are described by formulas (3.10) and (3.16). For the model of delta-correlated fluctuations $\varepsilon_1(x)$, the quantities $R_1(x_0)$ and $R_2(x_0)$ are statistically independent because of being described by dynamic equations not overlapping in spaces. Then, for the infinite space ($L_0 \Rightarrow -\infty$, $L \Rightarrow \infty$) by using the distribution (3.41), we have (Klyatskin [1980a, 1985, 1986])

$$\langle I(x_0; x_0) \rangle = 1 + 1/\beta, \qquad \langle S(x_0; x_0) \rangle = 1. \tag{3.44}$$

Thus, *the mean energy-flux density at the point of a source location does not depend on fluctuations of medium parameters,* and coincides with the density of a source placed in free space.

Similarly, we use the formula (3.14) for a source placed on the reflective boundary $x_0 = L$, obtaining

$$\langle I_{\text{ref}}(L; L) \rangle = 4[1 + 2/\beta], \qquad \langle S_{\text{ref}}(L; L) \rangle = 4; \tag{3.45}$$

that is, the mean energy-flux density on the reflective boundary is also independent of fluctuations of medium parameters, and coincides with a value obtained in the problem for free space.

Note that the mean intensity of wavefield $\langle I(x; x_0; L) \rangle = \langle |G(x; x_0; L)|^2 \rangle$ satisfies the imbedding equation, which follows from eq. (3.24) after additional averaging over fast oscillations ($x \leqslant x_0$),

$$\frac{\partial}{\partial L} \langle I(x; x_0; L) \rangle = D \langle I(x; L) I(x_0; L) \rangle, \qquad \langle I(x; x_0; x_0) \rangle = \langle I(x; x_0) \rangle,$$

where $I(x; L)$ is the intensity of wavefield in the problem of a wave incident on the layer of medium, and hence (Klyatskin [1980a, 1985, 1986]),

$$\langle I(x; x_0; L) \rangle = \langle I(x; x_0) \rangle + D \int_{x_0}^{L} d\xi \, \langle I(x; \xi) I(x_0; \xi) \rangle.$$

By setting $x = x_0$ in this quantity, we have for the half-space as $\beta \ll 1$

$$D \int_{-\infty}^{L} d\xi \, \langle I^2(\xi; L) \rangle = 1/\beta.$$

3.3.3. Energy localization in the statistical problem

Expressions obtained earlier and connected with field magnitudes in fixed points of the space (on the layer boundary and at the point of source location) permit, due to equality (3.17), however, reaching a common conclusion on the behavior of mean intensity inside the randomly inhomogeneous medium. The equality for the mean energy contained in the half-space $(-\infty, x)$ follows from eq. (3.17),

$$\langle E \rangle = D \int_{-\infty}^{x_0} dx \, \langle I(x; x_0) \rangle = \frac{1}{\beta} \langle S(x_0; x_0) \rangle. \tag{3.46}$$

Thus, for the wave incident on the half-space $x \leq L$, we obtain, due to eq. (3.43) as $\beta \ll 1$,

$$\langle E \rangle = 2\ln(1/\beta), \quad \langle I(L;L) \rangle = 2.$$

Therefore, most of the energy is concentrated in the space region

$$l_\beta \simeq \ln(1/\beta);$$

that is, the localization exists that is linked with the attenuation. Note that if there are no fluctuations of medium parameters, the energy localization also occurs along the absorption length $l_{\text{abs}} \simeq 1/\beta \gg l_\beta$ as $\beta \ll 1$. If $\beta \to 0$, the quantity $l_\beta \to \infty$, and in the limit case of the absorption's absence the localization is absent.

For a source in the infinite space, we have

$$\langle E \rangle = 1/\beta, \quad \langle I(x_0;x_0) \rangle = 1 + 1/\beta,$$

and consequently, unlike the previous case, the energy localization occurs in the space region $D|x - x_0| \simeq 1$ if $\beta \to 0$.

Similarly for a source on the reflective boundary, we have

$$\langle E \rangle = 4/\beta, \quad \langle I_{\text{ref}}(L;L) \rangle = 4(1 + 2/\beta),$$

and therefore the energy localization occurs in the space region $D(L - x) \simeq 1/2$ for small β.

This difference in the behavior of mean intensity for different wave problems is especially caused by statistical averaging. For a separate realization of the random function $\varepsilon_1(x)$ wavefield realizations have, referring to eqs. (3.13) and (3.14), a space structure similar to the field structure of an incident wave, and differ only in the constant factor, which is different, however, for different realizations. Thus, the apparent difference in the mean wavefield intensity results from the correlation of these constants with the main space structure. Two typical realizations of wavefield intensity $I(x; L) = |u(x;L)|^2$ for a sufficiently thick layer ($D(L - L_0) = 5$) are shown in fig. 4 (Yaroschuk [1986a]) (parameters are $\beta = 0.08$ and $k/D = 25$) for a single realization of random function $\varepsilon_1(x)$ (white circles). Black circles correspond with the replacement of the function $\varepsilon_1(x)$ by $-\varepsilon_1(x)$ in the middle of the layer over the section of the wavelength order. The continuous curve corresponds with the absence of fluctuations $\varepsilon_1(x)$; that is, $\varepsilon_1 = 0$. This figure gives only qualitative information, since realization dots $I(x; L)$ are too far apart on the figure (distance of 10 wavelengths order). In fact, the plot of $I(x; L)$ is much more indented and has many more increases. These increases become larger with the decrease of parameter β. The total energy E enclosed in the

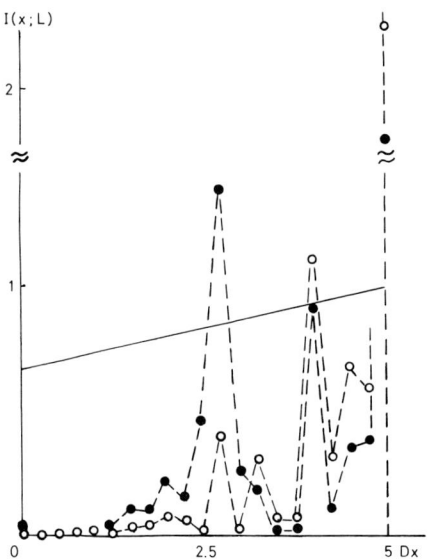

Fig. 4. The dependence of the distribution of the wavefield intensity $I(x; L)$ for the separate realization $\varepsilon(x)$ when $\beta = 0.08$.

half-space $(-\infty, L)$, however, depends weakly on the parameter β. In fact, by using the formula (3.17) for total energy, one can calculate the distribution density for quantity E; that is, the area under the curves in fig. 4

$$P(E) = 2E^{-2} e^{-2(1-\beta E)/E} \theta(1 - \beta E),$$

which allows the limiting transition, $\beta \Rightarrow 0$,

$$P(E) = 2E^{-2} e^{-2/E}.$$

The characteristic feature of this distribution is its attenuation according to the power law when E is large, which brings the divergence at infinity when moments of the quantity E are calculated.

By analogy, we can obtain the following result for a source placed on the reflective boundary

$$P_{\text{ref}}(E) = \sqrt{\frac{2}{\pi}} E^{-3/2} e^{-2(1-\beta E/4)^2/E}.$$

3.3.4. Diffusive approximation

When deducing eqs. (3.36) and (3.36'), we used the model of Gaussian delta-correlated fluctuations $\varepsilon_1(x)$ of medium parameters and additional averaging over fast oscillations. In general cases of a one-scale random process $\varepsilon_1(x)$ with an arbitrary correlation function $B_\varepsilon(x - x') = \langle \varepsilon_1(x)\varepsilon_1(x') \rangle$ ($\sigma_\varepsilon^2 \ll 1$), the given theory also remains valid. The diffusion factor is the only value that changes. It can be described by the formula

$$D(k) = \frac{k^2}{4} \int_{-\infty}^{\infty} d\xi\, B_\varepsilon(\xi) \cos 2k\xi = \frac{k^2}{4} \Phi_\varepsilon(2k),$$

where

$$\Phi_\varepsilon(q) = \int_{-\infty}^{\infty} d\xi\, B_\varepsilon(\xi) e^{iq\xi}$$

is the spectral function of the random process $\varepsilon_1(x)$. This is the diffusive approximation, which corresponds with the assumption that fluctuations of $\varepsilon_1(x)$ do not influence wavefield dynamics over scales of the order of l_0, and the wavefield, as a function of the parameter L, is the Markov random process. The condition of applicability of this approximation is

$$Dl_0 \ll 1, \qquad \alpha = k/D \gg 1.$$

An origin of a spectrum on the twice-spaced harmonic is linked physically with the known Bragg condition for the diffraction on the space structures. In this case the process $\varepsilon_1(x)$ can be regarded as the white noise with effective parameters $\langle \varepsilon_1(x) \rangle = 0$ and $\langle \varepsilon_1(x)\varepsilon_1(x') \rangle = \delta(x - x')\Phi_\varepsilon(2k)$.

The structure of the diffusion coefficient depends on the parameter kl_0. If $kl_0 \ll 1$, the delta-correlated approximation (white noise) results, in which case the diffusion coefficient does not depend on the statistical model, but is determined by the expression

$$D(k)l_0 = \frac{k^2 l_0}{4} \Phi_\varepsilon(0), \qquad \Phi_\varepsilon(0) = 2 \int_0^{\infty} d\xi\, B_\varepsilon(\xi/l_0).$$

In another limit case $kl_0 \gg 1$, the diffusion coefficient largely depends on the model of $\varepsilon_1(x)$; that is, it depends on the function $B_\varepsilon(\xi)$. For the model of the Gaussian Markov random process $\varepsilon_1(x)$ with the correlation function

(a) $B_\varepsilon(\xi) = \sigma_\varepsilon^2 e^{-|\xi|/l_0}$,

we have

$$D(k)l_0 = \frac{\sigma_\varepsilon^2}{2} \frac{(kl_0)^2}{1 + 4(kl_0)^2},$$

and the quantity $Dl_0 = \sigma_\varepsilon^2/8$ does not depend on the wavelength if $kl_0 \gg 1$. If the correlation function is

(b) $B_\varepsilon(\xi) = \sigma_\varepsilon^2 e^{-\xi^2/l_0^2}$,

we have

$$D(k)l_0 = \frac{\pi^{1/2}}{4} \sigma_\varepsilon^2 (kl_0)^2 e^{-(kl_0)^2},$$

and the quantity Dl_0 decreases rapidly when kl_0 increases. Note that Dl_0 has the maximum at $kl_0 = 1$.

Thus, the diffusive approximation can be applied for sufficiently small fluctuations: $\sigma_\varepsilon^2 \ll 1$.

This analysis, however, can be applied only in the case of one-scale fluctuations of $\varepsilon(x)$. Thus, if the correlation function of the random quantity $\varepsilon(x)$ has the form

$$B_\varepsilon(x) = \sigma_\varepsilon^2 e^{-\lambda|x|} \cos 2kx,$$

the spectral function $\Phi_\varepsilon(q)$ is defined by the expression

$$\Phi_\varepsilon(q) = 2\lambda \sigma_\varepsilon^2 \frac{\lambda^2 + 4k^2 + q^2}{(\lambda^2 + q^2 - 4k^2)^2 + 16\lambda^2 k^2}.$$

In fact, it has the delta shape when $\lambda \ll 2k$ in the vicinity of $q \approx 2k$, where

$$\Phi_\varepsilon(2k) = \sigma_\varepsilon^2/\lambda.$$

This circumstance shows that this problem resembles the problem of wave propagation in the medium with periodical inhomogeneities; the role of fluctuations has to be considerably less, since it is given by the diffusion approximation. In this case, over scales of the order of the correlation radius $l \approx 1/\lambda$, it is now impossible to neglect the effect related to the influence of the medium on the wavefield. It is necessary to consider the parametric influence (Guzev and Klyatskin [1991c]).

We should note that examination of stationary problems of wave propagation in periodical media on the basis of the imbedding method was carried out by Klyatskin and Koshel' [1983, 1984]. The propagation of temporal pulses in periodical media was discussed by Burridge, Papanicolaou and White [1988] and Gulin and Temchenko [1990, 1992].

3.4. STATISTICAL DESCRIPTION OF WAVEFIELDS

3.4.1. Wave incidence on the layer of the medium

We should first note that for the medium layer occupying a part of space $L_0 < x < L$ (its thickness is $D(L - L_0) \gg 1$ but is finite; the absorption is

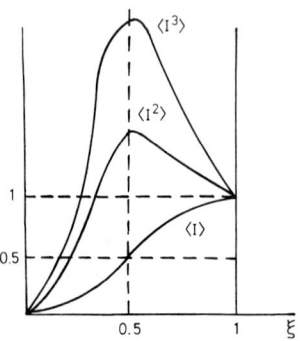

Fig. 5. A schematic presentation of the behavior of the intensity moments inside a medium.

absent), the phenomenon of stochastic wave parametric resonance occurs (Klyatskin [1979a, b, 1980a, 1985, 1986]). It consists of the exponential growth of quantities $\langle I^n(x; L)\rangle$ from boundaries deep in the layer as $n > 2$. The schematic behavior of intensity moments normalized by their value when $x = L$, as functions of the parameter $\xi = (L - x)/(L - L_0)$, is shown in fig. 5. In the limit of the half-space ($L_0 \Rightarrow -\infty$), the region of the exponential moment's growth, starting from the second moment, occupies the whole half-space from the boundary $x = L$ and $\langle I(x; L)\rangle = 2$.

Let us examine the case of the half-space in greater detail. In § 3.1.1 we introduced amplitudes and intensities of opposite waves in the medium layer, which satisfy imbedding equations (3.18'). They can be rewritten in the form (oscillating terms being omitted, since they give no cumulative contribution to statistics)

$$\frac{\partial}{\partial L} W_j(x; L) = -k\gamma W_j(x; L) + i\frac{k}{2} \varepsilon_1(L)(R_L - R_L^*)W_j(x; L),$$

$$W_1(x; x) = 1, \quad W_2(x; x) = |R_x|^2. \tag{3.47}$$

Let us introduce the function

$$Q_{\lambda,\mu}(x; L; W) = \langle W_1^{\lambda-\mu}(x; L)W_2^{\mu}(x; L)\delta(|R_L|^2 - W)\rangle,$$

which describes correlations of opposite wave intensities with the modulus of the reflection factor. For the half-space ($L_0 \Rightarrow -\infty$) this function satisfies the equation (Babkin and Klyatskin [1980b, c, 1982b],Klyatskin [1980a, 1985, 1986])

$$\frac{\partial}{\partial \xi} Q_{\lambda,\mu}(\xi, W) = -\beta \left(\lambda - 2\frac{\partial}{\partial W} W \right) Q_{\lambda,\mu}(\xi, W)$$

$$- \left[\lambda + \frac{\partial}{\partial W}(1-W) \right] Q_{\lambda,\mu}(\xi, W) \qquad (3.48)$$

$$+ \left[\lambda + \frac{\partial}{\partial W}(1-W) \right]^2 W Q_{\lambda,\mu}(\xi, W),$$

$$Q_{\lambda,\mu}(0, W) = W^\lambda P_\infty(W),$$

where $\xi = D(L - x)$ and function $P_\infty(W)$ is described by the formula (3.42).

3.4.1.1. *Nondissipative medium.* When $\beta = 0$, the solution of eq. (3.48) has the form $Q_{\lambda,\mu}(\xi, W) = \delta(W - 1)e^{\lambda(\lambda-1)\xi}$, and consequently,

$$\langle W_1^{\lambda-\mu}(x; L) W_2^\mu(x; L) \rangle = e^{\lambda(\lambda-1)\xi}. \qquad (3.49)$$

Because of the arbitrary nature of parameters λ and μ, the quantity (3.49) testifies to $W_1(x; L) = W_2(x; L)$ with probability equal to unity. Quantities $\chi(x; L) = \ln W(x; L)$ are distributed according to the Gaussian law with parameters

$$\langle \chi(x; L) \rangle = -\xi, \qquad \sigma_\chi^2 = 2\xi. \qquad (3.49')$$

Consequently, quantities $W_j(x; L)$ are distributed by the lognormal law, and for integer values of parameters λ and μ the exponential growth into the medium is

$$\langle W(x; L) \rangle = 1, \qquad \langle W^n(x; L) \rangle = e^{n(n-1)\xi}, \quad n = 2, \ldots$$

From the physical viewpoint the lognormal law means that within the medium in some points in each realization there are rare but strong discontinuities of the wavefield intensity (see fig. 4). Figure 4 also shows that increases of wavefield intensity take place against the background of exponential fall, which is described by the function $\chi(x; L)$,

$$I(x; L) = e^{-(L-x)/l_{\text{loc}}},$$

where

$$l_{\text{loc}} = -\frac{\partial}{\partial L} \langle \chi(x; L) \rangle.$$

Usually this is associated with the property of localization of disordered systems of physics, that is, dynamic localization (see, e.g. Anderson [1958],

Lifshits, Gredescul and Pastur [1988], Freilikher and Gredescul [1992], Sheng, White, Zhang and Papanicolaou [1986a, 1990], Sheng [1990], Klyatskin and Saichev [1992]), and the quantity l_{loc} is the localization length. In this case

$$l_{loc} = D^{-1}.$$

However, from the standpoint of energetic characteristics analysis, it is clear that all statistics of quantities $W_j(x; L)$ are caused by increases. With reference to results in § 3.3.3 no statistical energetic localization occurs in this case.

Section 3.2.2 indicates that the intensity of the wavefield $I(x; L)$ has a structure

$$I(x; L) = 2W(x; L)(1 + \cos \phi_L),$$

where the intensity of the opposite waves $W_1(x; L) = W_2(x; L) = W(x; L)$ is

$$W(x; L) = e^{-[q(L) - q(x)]}, \quad (3.50)$$

and $q(L)$ is described by a stochastic equation

$$\frac{d}{dL} q(L) = k\varepsilon_1(L) \sin \phi_L. \quad (3.50')$$

The initial condition for eq. (3.50') is in an arbitrary point B, located far enough from the boundary $L: q(L)|_{L=B} = 0$, so that the quantity ϕ_B has a uniform distribution.

In this case random functions $q(L)$ and ϕ_L are statistically independent. The probability density of the quantity $q(L) - P_L(q) = \langle \delta(q(L) - q) \rangle$ is described by the Fokker–Planck equation averaged over fast oscillations

$$\frac{\partial}{\partial L} P_L(q) = -D \frac{\partial}{\partial q} P_L(q) + D \frac{\partial^2}{\partial q^2} P_L(q), \quad P_B(q) = \delta(q).$$

Therefore, the random quantity $q(L)$ is a Gaussian Markov random process with parameters

$$\langle q(L) \rangle = D(L - B), \quad \langle \tilde{q}(L) \tilde{q}(L') \rangle = 2D \min \{(L - B, L' - B)\},$$

where $\tilde{q}(L) = q(L) - \langle q(L) \rangle$.

Note that the random process $q(L)$ is statistically equivalent to the random process

$$\frac{d}{dL} q(L) = D + \tilde{\varepsilon}(L), \quad q(B) = 0. \quad (3.50'')$$

where a random quantity $\tilde{\varepsilon}(L)$ is a random Gaussian function with parameters $\langle \tilde{\varepsilon}(L) \rangle = 0$ and $\langle \tilde{\varepsilon}(L) \tilde{\varepsilon}(L') \rangle = 2D\delta(L - L')$.

The presentation (3.50) is suitable for a detailed study of the structure of independent realizations of the wavefield intensity (see Appendix B) (Klyatskin and Saichev [1992]). Thus, for the quantity

$$S_n(L) = D \int_{-\infty}^{L} d\xi \, W^n(\xi; L),$$

describing the area under the curve $W^n(x; L)$ in the whole half-space, we obtain the stochastic equation

$$\frac{d}{dL} S_n(L) = D(1 - nS_n) - n\tilde{\varepsilon}(L)S_n(L).$$

Therefore its probability density $P_L^{(n)}(S) = \langle \delta(S_n(L) - S) \rangle$ is described by the equation

$$\frac{\partial}{\partial L} P_L^{(n)}(S) + D \frac{\partial}{\partial S}(1 - nS)P_L^{(n)}(S) = n^2 D \frac{\partial}{\partial S} S \frac{\partial}{\partial S} S P_L^{(n)}(S).$$

Its steady-state solution independent of L has the form

$$P^{(n)}(S) = \frac{1}{n^{2/n} \Gamma(1/n)} S^{-(1+1/n)} e^{-1/(n^2 S)}, \qquad (3.51)$$

where $\Gamma(x)$ is the gamma function.

By comparing the expression $P^{(1)}S = S^{-1} e^{-1/S}$ with the expression corresponding to the total energy E, which is enclosed in the half-space, we can see that they coincide when $E = 2S$. In other words, the presence of the fast-oscillating term connected with $\cos \phi_L$ in the expression for the wave intensity does not contribute to the total energy.

From eq. (3.50) it also follows that the part of the energy enclosed in the region $(-\infty, x)$, that is, the area under the curve $W(x; L)$,

$$S(x; L) = D \int_{-\infty}^{x} d\xi \, W(\xi; L),$$

is defined by the expression

$$S(x; L) = W(x; L)S(x; x) - W(x; L)S_1(x),$$

where quantities $W(x; L)$ and $S_1(x)$ are statistically independent, because they are functionals of the random process $\tilde{\varepsilon}(L)$ in nonoverlapping spaces. Therefore, $W(x; L)$ is defined by the function $\tilde{\varepsilon}(L)$ in the region (x, L), and the quantity $S_1(x)$ in the region $(-\infty, x)$. Thus, most of the energy is concentrated in the region near the boundary L.

As has just been shown, the wavefield intensity $W(x; L)$ for each realization represents an exponentially decreasing function (for the present model $\exp(\langle \ln W(x; L)\rangle) = e^{-D(L-x)}$) with very large increases. Despite this, the probability of the inequality $W(\xi) < \exp(-\xi)$, being valid for any fixed value $\xi = D(L - x)$, is defined by the equality

$$P(W(\xi) < e^{-\xi}) = \int_{-\infty}^{-\xi} d\chi\, P(\chi) = 1/2.$$

The function $\exp(-\xi)$ is called a typical realization of the random function $W(x; L)$ in the physics of disordered systems. The term *typical realization*, as applied to the function $\exp(-\xi)$, is justified because it is the isoprobable curve of the random function $W(x; L)$, which corresponds to the value of probability $p = 1/2$ (see Appendixes A and B). In other words, for any interval on the axis $\xi = D(L - x)$, the function $W(x; L)$ passes an average of one half of the interval over the typical realization and one half of the interval below it (Klyatskin and Saichev [1992]).

It is also possible to make a majorant evaluation of the wavefield intensity $W(x; L)$ valid for all values of ξ (see Appendix B). Therefore, the inequality

$$W(x; L) < 4e^{-\xi/2}$$

is valid with probability $p = 1/2$ (Klyatskin and Saichev [1992]).

Note that the correlation function of the wavefield intensity as $x > x'$

$$\langle W(x; L)W(x'; L)\rangle = \exp[2D(L - x)]$$

does not depend on the point x'; that is, from a formal standpoint it has infinite correlation radius.

Let us discuss the quantity $l_{\text{loc}}(k) = D^{-1}(k)$ on the basis of diffusional approximation.

The quantity $l_{\text{loc}}(k)$, as mentioned earlier, characterizes an exponential fall of the wavefield intensity in realizations. If $kl_0 \ll 1$ (low frequencies), the localization length $l_{\text{loc}}(k) \sim k^{-2}$ is independent in the statistical model. If $kl_0 \gg 1$ (high frequencies), the behavior of $l_{\text{loc}}(k)$ with respect to k depends on the statistical model. The model (a) in § 3.2.2 results in $l_{\text{loc}} \sim \text{const.}$; in the case of model (b), the function $l_{\text{loc}}(k)$ increases and reaches the minimum value at $kl_0 = 1$. Sheng, White, Zhang and Papanicolaou [1986b] drew attention to the possibility of this effect. Note that moments of intensity of the wavefield have the exponential growth deep in the medium.

The situation is completely different for the inclined incidence of a plane wave on the slab of the medium. The equation for the function $\langle \chi(x; L)\rangle =$

$\langle \ln I(x; L) \rangle$ can now be written

$$\frac{\partial}{\partial L} \langle \chi(x; L) \rangle = -\frac{2k^2}{p^2} D \langle \cos \phi_L (1 + \cos \phi_L) \rangle.$$

Hence, the localization length has the form

$$Dl_{\text{loc}}(\theta) = \frac{\cos^2 \theta}{2 \langle \cos \phi_L (1 + \cos \phi_L) \rangle}.$$

The mean value $\langle \cos \phi_L (1 + \cos \phi_L) \rangle$ is defined with the help of the distribution (3.41), obtaining

$$Dl_{\text{loc}}(\theta) = \begin{cases} \cos^2 \theta, & \kappa \gg 1 \quad (\theta \Rightarrow 0), \\ \dfrac{\Gamma(1/6)}{2\pi^{1/2} 3^{1/3} \alpha^{2/3}}, & \kappa \ll 1 \quad (\theta \Rightarrow \pi/2); \end{cases}$$

that is, the localization length does not depend on an incidence angle for the sliding incidence of a wave on the randomly layered half-space (Guzev, Klyatskin and Popov [1992]). Note that this effect is entirely due to an increase of $\varepsilon(x)$ at the boundary $x = L$. If we have an adjusted boundary, the localization length is given by (Guzev and Klyatskin [1993])

$$Dl_{\text{loc}}(\theta) = \frac{\cos^2 \theta}{2 \langle \sin^2 \phi_L \rangle}.$$

Numerical calculations show that there is a significant difference between values of $Dl_{\text{loc}}(\theta)$ and values of $Dl_{\text{loc}}(\theta) = \cos^2 \theta$, corresponding to the uniform probability density of the phase at $\theta = 80°$, $\alpha = 25$ (the sliding incidence). In this region the asymptotic behavior of $\langle \sin^2 \phi_L \rangle$ has the form

$$\langle \sin^2 \phi_L \rangle = \frac{4}{\pi^2} \left[\ln\left(\frac{1}{\kappa}\right) - C \right] \left[1 + \frac{4}{\pi^2} \left[\ln\left(\frac{1}{\kappa}\right) - C \right]^2 \right]^{-1}.$$

Finally, we consider statistical wavefield characteristics $\langle I^n(x, L) \rangle = \langle |u(x; L)|^{2n} \rangle$.

Let us introduce the function $\Phi_L(x, z) = I^n(x; L) \delta(z_L - z)$, for which we have the stochastic equation if the right-hand boundary is not adjusted

$$\frac{\partial}{\partial L} \Phi_L(x, z) = -p \frac{\partial}{\partial z}(1 + z^2) \Phi_L(x, z) - \frac{k^2}{p} \varepsilon(L) \left[\frac{2nz}{1 + z^2} + \frac{\partial}{\partial z} \right] \Phi_L(x, z)$$

(3.52)

and the initial condition is

$$\Phi_x(x, z) = 4^n \delta(z_x - z)/(1 + z^2)^n. \tag{3.52'}$$

Then, by averaging eqs. (3.52) and (3.52') over the realization ensemble of the random process $\varepsilon(L)$ and supposing that the medium occupies the half-space ($L_0 \Rightarrow -\infty$), we obtain an evolutionary equation for $\Phi_\xi(z) = \langle \Phi_L(x, z) \rangle$

$$\frac{\partial}{\partial \xi} \Phi_\xi(z) = -\kappa \frac{\partial}{\partial z}(1 + z^2)\Phi_\xi(z) + \frac{1}{(1+z^2)^n} \frac{\partial^2}{\partial z^2}(1+z^2)^n \Phi_\xi(z),$$

$$\Phi_0(z) = 4^n P(z)/(1+z^2)^n, \tag{3.53}$$

where $\xi = 2k^2 D(L-x)/p^2 = 2D(L-x)/\cos^2\theta$, and $P(z)$ is described by eq. (3.41). For the sliding incidence the first term on the right-hand size of eq. (3.53) is not essential in contrast to another term. The solution then has the form

$$\Phi_\xi(z) = \frac{4^n}{(1+z^2)^n} e^{\xi \partial^2/\partial z^2} P(z),$$

and the quantity of interest is

$$\langle I^n(\xi) \rangle = \int_{-\infty}^{\infty} dz\, \Phi_\xi(z) = \frac{4^n}{2(\pi\xi)^{1/2}} \int_{-\infty}^{\infty} \frac{dz}{(1+z^2)^n} \int_{-\infty}^{\infty} dt\, e^{-(z-t)^2/4\xi} P(t). \tag{3.54}$$

From eq. (3.54) the following asymptotical behavior is obtained:

$$\langle I^n(\xi) \rangle = \begin{cases} \langle I^n(0) \rangle, & \xi \ll 1, \\ 2^n \pi^{1/2} \dfrac{(2n-3)!!}{(n-1)!} \xi^{-1/2}, & \xi \gg 1. \end{cases}$$

Hence, for the sliding incidence of the plane wave on a layered medium, the quantities $\langle I^n(\xi) \rangle$ decrease according to the universal $\xi^{-1/2}$ law at a large distance inside the medium (Guzev, Klyatskin and Popov [1992]).

For the adjusted boundary the stochastic equation for the function $\Phi_L(z) = I^n(x; L)\delta(z_L - z)$ is

$$\frac{\partial}{\partial L} \Phi_L(z) = -p \frac{\partial}{\partial z}(1+z^2)\Phi_L(z) + \frac{k^2}{2p^2} \xi(L) \left[\frac{2nz^2}{1+z^2} + \frac{\partial}{\partial z} z \right] \Phi_L(z),$$

$$\Phi_L(z)|_{L=x} = 4^n \delta(z_x - z)/(1+z^2)^n.$$

Averaging the equation over realizations of the random process $\varepsilon(L)$ yields for the half-space ($L_0 \to -\infty$)

$$\frac{\partial}{\partial \xi}\Phi_\xi(z) = -\kappa \frac{\partial}{\partial z}(1+z^2)\Phi_\xi(z) - \frac{n}{(1+z^2)^n}\frac{\partial}{\partial z}z(1+z^2)^n\Phi_\xi(z)$$

$$+ \frac{1}{(1+z^2)^n}\frac{\partial}{\partial z}z\frac{\partial}{\partial z}z(1+z^2)^n\Phi_\xi(z), \quad (3.53')$$

$$\Phi_0(z) = 4^n P(z)/(1+z^2)^n,$$

where $\xi = 2D(L-x)/\cos^2\theta$, $P(z)$ is the steady-state probability density of the phase (3.41'). For the sliding incidence we neglect the first term of the right-hand side of eq. (3.53') and then the function $Q_\xi(z) = z(1+z^2)^n \Phi_\xi(z)$ satisfies the equation

$$\frac{\partial}{\partial \xi}Q_\xi(z) = \left[z\frac{\partial}{\partial z}z\frac{\partial}{\partial z} - nz\frac{\partial}{\partial z}\right]Q_\xi(z), \quad Q_0(z) = 4^n z P(z). \quad (3.53'')$$

The function Q_ξ has the structure $Q_\xi(z) = \theta(z)Q_\xi^+(z) + \theta(-z)Q_\xi^-(z)$, whereas $Q_\xi^\pm(z)$ is a solution of eq. (3.53'') with the corresponding function $P_\pm(z)$. By changing the variable $t = \ln|z|$, we find this solution and the integral representation for the moments of the wavefield intensity

$$\langle I^n(\xi) \rangle = \frac{4^n}{2\sqrt{\pi\xi}} \int_0^\infty \frac{dz}{z(1+z^2)^n} \int_0^\infty dz_1 \, [P_+(z_1) + P_-(-z_1)]$$

$$\times \exp\left[-\frac{1}{4\xi}\left(n\xi + \ln\left|\frac{z_1}{z}\right|\right)^2\right]. \quad (3.54')$$

Let us discuss the asymptotic behavior of eq. (3.54') at $\xi \ll 1$ and $\xi \gg 1$. By interchanging the orders of the integrals and changing the variable, we transform eq. (3.54') to

$$\langle I^n(\xi) \rangle = I_+ + I_-,$$

$$I_\pm = \frac{4^n}{2\sqrt{\pi\xi}} \int_0^\infty dp \, P_\pm(\pm p) \int_{-\infty}^\infty \frac{dt \, e^{2tn}}{(p^2 + e^{2t})^n} \exp\left[-\frac{1}{4\xi}(n\xi + t)^2\right].$$

The asymptotic expression of $\langle I^n(\xi)\rangle = \langle I^n(0)\rangle$ at $\xi \ll 1$ follows from the preceding, since $n\xi \to 0$, $\exp[-t^2/4\xi]/2(\pi\xi)^{1/2} \to \delta(t)$. For $\xi \gg 1$ we have (Guzev and Klyatskin [1993])

$$I_\pm \simeq 4^n \varepsilon^{1/2} \int_0^\infty \frac{dt \, P_\pm(\pm \varepsilon^{1/2} t)}{(1+t^2)^n},$$

where $\varepsilon = \exp[-2n\xi]$. By changing $P_\pm(\pm\varepsilon^{1/2}t)$ to $P_\pm(\pm 0) = J(\kappa)/\kappa$, the following final expression is obtained for the asymptotic behavior

$$\langle I^n(\xi)\rangle \simeq \frac{2^{n+1}(2n-3)!!}{(n-1)!} \frac{e^{-n\xi}}{\pi[1+4\pi^{-2}(\ln\kappa+C)^2]\kappa}. \tag{3.54''}$$

As we have seen, if the boundary $x = L$ is not adjusted, the moments of intensity decrease according to the $\xi^{-1/2}$ law. This primary difference in behavior of statistical characteristics is explained by the discontinuity of $\varepsilon(L)$. Its contribution to the statistics has provided the power decrease of characteristics. Thus, the asymptotic behavior of eq. (3.54'') is a "pure" statistical one. We excluded the influence of the boundary $x = L$ linked with the increase of $\varepsilon(x)$, and have taken into account the scattering of the wave on random inhomogeneities. Note that at $\xi \gg 1$ the dominant structure of the random function $I(\xi)$ is $I(\xi) = Ae^{-\xi}$, where A is the random magnitude; its moments are given by the expression (3.54'').

Note that in the other limit case ($\kappa \gg 1$) the quantity $I(x; L)$ has a log-normal probability distribution and $\langle I(x; L)\rangle = 2$, and the higher-intensity moments exponentially increase inside the slab, as was mentioned earlier.

3.4.1.2. Dissipative medium. In the presence of finite attenuation (however small), the exponential growth of intensity moments has to stop and be transformed into abrupt attenuation. Figure 6 shows an example of the quantities $\langle W_1(x)\rangle$ and $\langle W_1^3(x)\rangle$ obtained by numerical integration of eq. (3.48) for different values of the parameter β (Babkin, Klyatskin, Kozlov and Yaroschuk [1981], Babkin and Klyatskin [1982b], Klyatskin and Yaroschuk [1983a], Klyatskin [1985, 1986]).

In discussing the limit transition $\beta \Rightarrow 0$, let us introduce the function $Q_{\lambda,\mu}(x; L; u) = \langle W_1^{\lambda-\mu}(x; L)W_2^\mu(x; L)\delta(u_L - u)\rangle$, which for the half-space is described by the equation (Guzev and Klyatskin [1991a])

$$\frac{\partial}{\partial \xi} Q_{\lambda,\mu}(\xi; W) = \left[-\lambda\beta + \beta\frac{\partial}{\partial u}(u^2-1) + \lambda(\lambda+1) - \frac{2\lambda^2}{u+1} \right.$$

$$\left. + 2\lambda(u-1)\frac{\partial}{\partial u} + \frac{\partial}{\partial u}(u^2-1)\frac{\partial}{\partial u} \right] Q_{\lambda,\mu}(\xi; u), \tag{3.55}$$

with the initial condition

$$Q_{\lambda,\mu}(0; u) = \left(\frac{u-1}{u+1}\right)^\mu P_\infty(u),$$

where $\xi = D(L-x)$, and $P_\infty(u)$ is the steady-state probability distribution

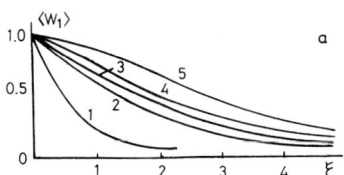

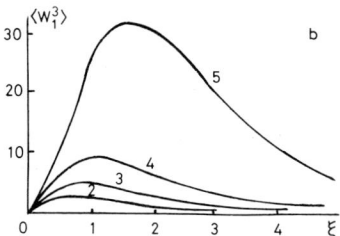

Fig. 6. The average intensity distribution for a transmitted wave inside the layer. The curves 1 to 5 correspond respectively to $\beta = 1.0, 0.1, 0.06, 0.04$ and 0.02: (a) is the first moment; (b) is the third moment.

(3.42). The quantities of interest are

$$\langle W_1^{\lambda-\mu}(x;L) W_2^{\mu}(x;L) \rangle = \int_1^\infty du \, P_\infty(u).$$

We change the variable $u \Rightarrow \beta(u-1)$ in eq. (3.55), and go to the limit $\beta \Rightarrow 0$. This results in a more simple equation:

$$\frac{\partial}{\partial \xi} Q_{\lambda,\mu}(\xi;u) = \left[\lambda(\lambda+1) - \frac{2\lambda^2 \beta}{u} + 2\lambda u \frac{\partial}{\partial u} + \frac{\partial}{\partial u} u^2 \frac{\partial}{\partial u} \right] Q_{\lambda,\mu}(\xi;u),$$
$$Q_{\lambda,\mu}(0;u) = e^{-u}. \qquad (3.56)$$

The solution of eq. (3.56) must have a singular feature with respect to the variable u (and with respect to the parameter β), with arbitrary small but finite absorption, and can be expressed by the Kontorovich–Lebedev transformation, which is defined by the integrals

$$\hat{\psi}(\tau) = \int_0^\infty dx \, K_{i\tau}(x) \psi(x), \qquad \psi(x) = \frac{2}{\pi^2 x} \int_0^\infty d\tau \, \tau \sinh \pi\tau \, K_{i\tau}(x) \hat{\psi}(\tau).$$

As a result, the solution of the problem (3.56) for integers $\lambda = n$, $\mu = m$ can be written in the form

$$\langle W_1^{n-m}(\xi)W_2^m(\xi)\rangle = \frac{4}{\pi(\varepsilon n)^{2n-1}}\int_0^\infty d\tau\,\tau\sinh(\pi\tau/2)\,e^{-\xi(1+\tau^2)/4}g_n(\tau)\hat{\psi}_0(\tau),$$

$$\hat{\psi}_0(\tau) = \int_0^\infty \frac{dy}{y^{2(\lambda+1)}}\,\frac{e^{-1/y^2}}{(1+2\beta y^2)^\mu}K_{i\tau}(\varepsilon\lambda y),\quad \varepsilon = (8\beta)^{1/2},$$

$$g_n(\tau) = [(2n-3)^2 + \tau^2]g_{n-1}(\tau),\quad g_1(\tau) = 1.$$

It is seen from here that as $\beta \ll 1$, the intensities of the reciprocal waves are equal with unit probability, and at small ξ this solution coincides with that of the stochastic wave parametric resonance.

Note that for sufficiently large values of ξ, namely,

$$\xi \gg \xi_{sc} = 4(n-1/2)\ln(n/\beta),$$

quantities $\langle W^n(\xi)\rangle$ go into universal localization dependence (Guzev and Klyatskin [1991a]),

$$\langle W^n(\xi)\rangle \simeq A_n\beta^{-(n-1/2)}\ln(1/\beta)\xi^{-3/2}e^{-\xi/4},$$

which is conditioned by the operator $\partial/\partial u\,(u^2 - 1)\,\partial/\partial u$ in corresponding equations (see, e.g. eq. (3.36')), and coincides with the asymptotic behavior (3.40) of the averaged moments of the transmission coefficient through the slab of thickness ξ at $\beta = 0$ (the distinction is in the numerical coefficient).

Thus the behavior of wavefield moments differs in three regions: in the first region, corresponding to the stochastic wave parametric resonance, the role of absorption is not essential; in the second region, the influence of absorption is most important because it stops an exponential increase of the moments; in the third region, the localization of energy not dependent on the absorption takes place. Boundaries between the regions are defined by the absorption and tend to infinity at $\beta \to 0$.

Note that, in general, for an arbitrary value of the parameter β, there is, for example, (Klyatskin [1980a, 1985])

$$\langle \chi_1(x;L)\rangle = -(1+\beta)\xi,\quad \sigma_{\chi_1}^2(x;L) = 2\langle|R|^2\rangle\xi.$$

For parameter values $\beta \gg 1$, quantities $\langle W_j\rangle$ and corresponding equations are transformed into the equations of the linear theory of the radiative transfer (see § 3.1.1) with parameters $2\alpha = k\gamma$, $\mu = D$. In this case no correlation occurs between opposite wave intensities and the reflection factor.

3.4.2. Source in infinite space

Let us now consider a source in the infinite space ($L_0 \Rightarrow -\infty$, $L \Rightarrow \infty$). A relation for the mean intensity in the region $x < x_0$ follows from representations (3.15) and (3.16)

$$\beta \langle I(x; x_0) \rangle = \frac{1}{D} \frac{\partial}{\partial x} \langle \psi(x; x_0) \rangle, \qquad (3.57)$$

where $\psi(x; x_0)$, being a function of parameter x_0, satisfies the stochastic equation

$$\frac{1}{D} \frac{\partial}{\partial x_0} \psi(x; x_0) = -\beta \frac{|1 + R_1(x_0)|^2}{1 - |R_1(x_0)|^2} \psi(x; x_0), \quad \psi(x; x) = 1. \qquad (3.58)$$

By introducing the function $\Phi(x; x_0; u) = \langle \psi(x; x_0) \delta(u_{x_0} - u) \rangle$, where $u_L = (1 + W_L)/(1 - W_L)$, we obtain the equation

$$\frac{\partial}{\partial \xi} \Phi_\xi(u) = -\beta u \Phi_\xi(u) + \beta \frac{\partial}{\partial u} (u^2 - 1) \Phi_\xi(u) + \frac{\partial}{\partial u} (u^2 - 1) \frac{\partial}{\partial u} \Phi_\xi(u),$$

$$\Phi_0(u) = P_\infty(u) = \beta e^{-\beta(u-1)}, \quad \xi = D|x - x_0|. \qquad (3.59)$$

The mean intensity can now be expressed by equalities

$$\beta \langle I(x; x_0) \rangle = -\frac{\partial}{\partial \xi} \int_1^\infty du\, \Phi_\xi(u) = \beta \int_1^\infty du\, u \Phi_\xi(u). \qquad (3.60)$$

It is possible to pass to the limit $\beta \Rightarrow 0$ in eq. (3.59). As a result, we obtain the simpler equation,

$$\frac{\partial}{\partial \xi} \tilde{\Phi}_\xi(u) = -u \tilde{\Phi}_\xi(u) + \frac{\partial}{\partial u} u^2 \tilde{\Phi}_\xi(u) + \frac{\partial}{\partial u} u^2 \frac{\partial}{\partial u} \tilde{\Phi}_\xi(u), \quad \tilde{\Phi}_0(u) = e^{-u}. \qquad (3.59')$$

Hence, the localization distribution of the mean intensity is found as a solution for eq. (3.59'), with the help of the equality

$$\Phi_{\text{loc}}(\xi) = \int_0^\infty du\, u \tilde{\Phi}_\xi(u), \qquad (3.61)$$

where

$$\Phi_{\text{loc}}(\xi) = \lim_{\beta \Rightarrow 0} \beta \langle I(x; x_0) \rangle = \lim_{\beta \Rightarrow 0} \frac{\langle I(x; x_0) \rangle}{\langle I(x_0; x_0) \rangle}.$$

Consequently, the mean intensity of the point source has a structure for

$\beta \ll 1$

$$\langle I(x; x_0)\rangle = \beta^{-1} \Phi_{\text{loc}}(\xi). \qquad (3.62)$$

The solution of eq. (3.59') can be obtained easily by using the Kontorovich–Lebedev integral transformation, and as result, we obtain the expression for the localization curve $\Phi_{\text{loc}}(\xi)$ (Klyatskin [1991a–c])

$$\Phi_{\text{loc}}(\xi) = 2\pi \int_0^\infty d\tau\, \tau(\tau^2 + 1/4) \frac{\sinh \pi\tau}{\cosh^2 \pi\tau} e^{-(\tau^2 + 1/4)\xi}. \qquad (3.63)$$

Note that $\Phi_{\text{loc}}(\xi) = -\partial/\partial\xi \langle |T_\xi|^2 \rangle$, where $|T_\xi|^2$ is the modulus square of the transmitter factor of the layer of thickness ξ, when the plane wave is incident on it.

When values of ξ are small, the localization curve decreases rather quickly as $e^{-2\xi}$, and if ξ are large ($\xi \gg \pi^2$), it decreases much more slowly according to the universal law

$$\Phi_{\text{loc}}(\xi) \approx \tfrac{1}{8}\pi^{5/2} \xi^{-3/2} e^{-\xi/4}, \qquad (3.64)$$

while the total value of the integral $\int_0^\infty d\xi\, \Phi_{\text{loc}}(\xi) = 1$. The plot of $\Phi_{\text{loc}}(\xi)$ is given in fig. 7; asymptotic curves are plotted to make a comparison possible.

The localization curve corresponds to the double limit

$$\Phi_{\text{loc}}(\xi) = \lim_{\beta \Rightarrow 0} \lim_{\substack{L_0 \Rightarrow -\infty \\ L \Rightarrow \infty}} \frac{\langle I(x; x_0)\rangle}{\langle I(x_0; x_0)\rangle}.$$

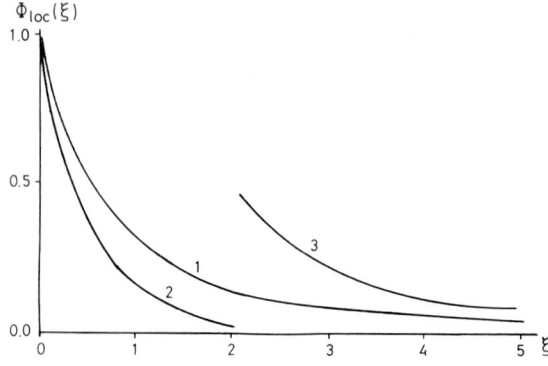

Fig. 7. The plot of the localization curve 1. Curve 2 is the function $e^{-2\xi}$, curve 3 is calculated by the formula (3.64).

However, it is easy to see that these limits are not permutable, and

$$\lim_{\substack{L_0 \Rightarrow -\infty \\ L \Rightarrow \infty}} \lim_{\beta \to 0} \frac{\langle I(x; x_0) \rangle}{\langle I(x_0; x_0) \rangle} = \frac{2}{3} \quad (D|x - x_0| \gg 1).$$

This is analogous to the case of the plane-wave incidence on the medium layer, for which the limits are identical. From a physical standpoint the first passage to the limit appears to be more valid.

The same situation occurs in the case of a source location on the reflective boundary, although in this case

$$\lim_{\beta \to 0} \frac{\langle I_{\text{ref}}(x; L) \rangle}{\langle I_{\text{ref}}(L; L) \rangle} = \frac{1}{2} \Phi_{\text{loc}}(\xi) \quad [\xi = D(L - x)]. \tag{3.65}$$

This result is true in the region $\xi > \frac{1}{3}$, as it corresponds to the neglect of correlations $|\langle R_x R_L^* \rangle| = e^{-3\xi}$, unlike the case of a source in infinite space.

3.4.3. Numerical modelling

The preceding theory is based on two assumptions: the diffusive approximation and the possibility of averaging over fast oscillations. Numerical modelling allows verification of the validity of these assumptions, and also the ability to obtain results in more complicated cases without theoretical solutions. This numerical modelling, in principle, can be carried out by multiple solution of boundary problems for different realizations of the function $\varepsilon_1(x)$, with additional averaging over the ensemble of realizations. However, this approach appears to be of little advantage. A better approach is based on the use of the ergodicity property of boundary problems with respect to the shift parameter of the completed problem along one realization of the function $\varepsilon_1(x)$, specified along the semi-axis (L_0, ∞) (fig. 8). In this case statistical characteristics are calculated by the formula

$$\langle F(L_0; x, x_0; L) \rangle = \lim_{\delta \to \infty} F_\delta(L_0; x, x_0; L),$$

where

$$F_\delta(L_0; x, x_0; L) = \frac{1}{\delta} \int_0^\delta d\Delta \, F(L_0 + \Delta; x + \Delta, x_0 + \Delta; L + \Delta).$$

In problems connected with the half-space ($L_0 \Rightarrow -\infty$), statistical characteristics do not depend on L_0. Research problems have the ergodicity property with respect to imbedding parameter L, the location of the right-hand layer boundary, because the shift parameter is identified with L in this case.

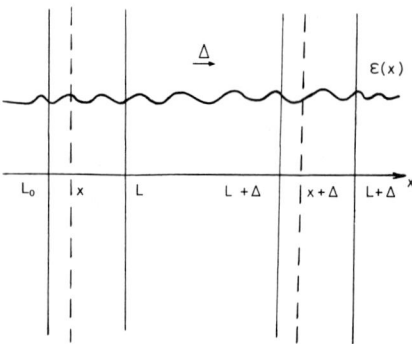

Fig. 8. The scheme of the averaging over the shear parameter Δ, based on the property of the problem ergodicity.

This permits the solution of imbedding equations with respect to parameter L for one realization $\varepsilon_1(x)$, and also obtaining all statistical quantities of interest. Many such calculations for linear wave problems are given by Klyatskin and Yaroschuk [1983b, c], Yaroschuk [1984, 1986a, b] and Klyatskin [1985, 1986]. Here, as an example, fig. 9 shows the results of numerical modelling of the quantity $2\langle I_{\text{ref}}(x; L)\rangle/\langle I_{\text{ref}}(L; L)\rangle$ as $\beta = 0.08$ and $k/D = 25$ (Klyatskin [1991b]), where $I_{\text{ref}}(x; L)$ is the wavefield intensity of a point source located on the reflecting boundary $x_0 = L$ $(\partial/\partial x\, G(x; x_0)|_{x=L} = 0)$. Oscillations are present in fig. 9 in the region $\xi = D(L-x) < 0.3$, with the period $\pi k/D = 0.13$, and with good agreement with the localization function $\Phi_{\text{loc}}(\xi)$ (3.63) for larger ξ.

As a second example let us consider the nonlinear boundary-value problem

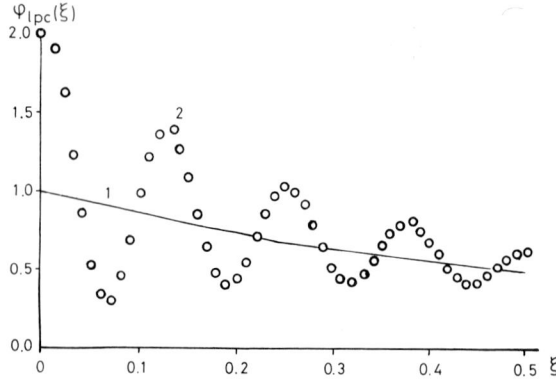

Fig. 9. Results of the numerical modelling of the quantity distribution $2\langle I_{\text{ref}}(x; L)\rangle/\langle I_{\text{ref}}(L; L)\rangle$ in half-space (circles 2). 1 is the localization curve.

(3.1′) and (3.26) (Yaroschuk [1986a, 1988a, b], Goland, Klyatskin and Yaroschuk [1991]).

Let us examine, as an example, the model $\varepsilon_1(J) = J + \tilde{\varepsilon}_1(x)$, where $\tilde{\varepsilon}_1(x)$ is a Gaussian delta-correlated random process. The distinctive feature of the deterministic problem is the unique and smooth solution of the wave problem at any damping value β. However, taking into account the medium fluctuations leads to the appearance of the multiplicities in the solution of the problem, and their origin primarily depends on the parameter β. Figure 10 presents the results of the modelling of the modulus square of the reflection coefficient $\langle |R_\infty|^2 \rangle$ and the intensity of the wavefield on the boundary normalized by $w - \langle J_\infty(w) \rangle / w$ for the half-space of the inhomogeneous medium ($L_0 \Rightarrow -\infty$), as the functions of the intensity of the incident wave. Parameter β is chosen to be equal to unity, which corresponds to the "moderate" influence of the statistics in the linear problem and the multiplicities do not appear. As shown in fig. 10 in the discussed range $w < 2$, the medium weakly reflects the incidence wave ($\langle |R_\infty|^2 \rangle$ is sufficiently small). Similarly, the reflective properties of the medium are defined mainly by the fluctuation of inhomogeneities, which is why $\langle |R_\infty(w)|^2 \rangle$ is close to the diffusive approximation. In contrast, for wavefield intensity on the boundary the nonlinear effect becomes apparent even for small w, and the quantity $\langle J \rangle / w$ tends to the values of the deterministic problem. A more complex effect of statistics and nonlinearity can be observed in fig. 11, which shows results of the modelling $\langle J(\xi, w) \rangle / w$ (circles, $\xi = D(L - x)$) and corresponding functions for the deterministic nonlinear problem and linear stochastic prob-

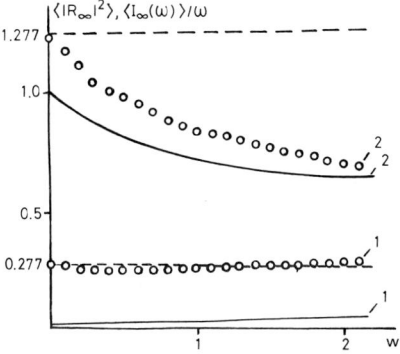

Fig. 10. The results of the modelling of $\langle |R|^2 \rangle$ and $\langle J(w) \rangle / w$ as a function of w. Circles, 1 $\langle |R|^2 \rangle$, 2 $\langle J(w) \rangle / w$; curves, 1 $|R|^2$, 2 $J(w)/w$ correspond to the problem with absence of fluctuation ε.

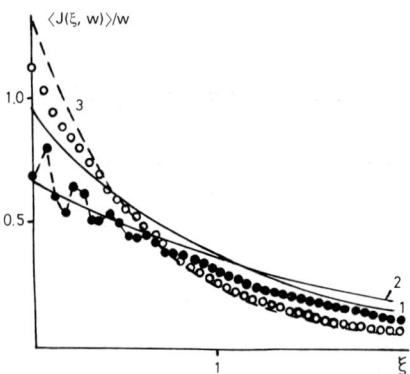

Fig. 11. The results of modelling $\langle J(\xi, w)\rangle/w$ (circles). Solid lines are the solution of the deterministic problem (1 corresponds to $w = 0.2$; 2 corresponds to $w = 2$); dashed line corresponds to the linear stochastic problem; $\bigcirc$, corresponds to $w = 0.2$; $\bullet$, $w = 2$.

lem, when the intensity values of the incident wave are small ($w = 0.2$) and large ($w = 2$). When $w = 0.2$, the wavefield inside the medium is primarily defined by inhomogeneity fluctuations, and the function $\langle J(\xi, w)\rangle/w$ is sufficiently close to the solution of the linear random problem. The wave incidence with an intensity $w = 2$ leads to oscillations of the function $\langle J(\xi, w)\rangle/w$ about the solution of the deterministic problem (caused by interference of direct and reflected waves) near the boundary and to the further fall of the plot between these limiting solutions. This indicates that in some medium layers (until the intensity of the wave in the medium is sufficiently high, $\langle J(\xi, w)\rangle \gg 1$) nonlinear effects play the main role, whereas the role of statistics is thought to be "suppressed", leading to interference in that layer. If $\langle J\rangle$ becomes small (the wave attenuation is sufficient), the behavior of the wavefield inside the layer can be solely defined by medium fluctuations.

This example shows that the character of intensity moments behavior is generally defined by nonlinear effects, and only the wave intensity in a layer is sufficiently high. The influence of statistics becomes important when the wave penetrates sufficiently deeply into the layer of the inhomogeneous medium.

For a layer of finite thickness the problem is no longer ergodic with respect to parameter L. In this case, however, its solution can be expressed by two independent solutions of the problem of the half-space (Klyatskin and Koshel' [1986]) and, therefore, we can consider the preceding case.

3.5. EIGENVALUE STATISTICS

The problem of statistical characteristics of the wavefield in a randomly inhomogeneous medium has been examined in detail. We have investigated problems of the wave incidence on the layer (half-space) and of a source inside the layer. In addition to these problems, in the physics of disordered systems (see, e.g. Lifshits, Gredeskul and Pastur [1988], Anzygina, Pastur and Slusarev [1981]) one pays much attention to the study of statistics of eigenvalues for the Helmholtz equation (energy levels for the Schrödinger equation in limited randomly inhomogeneous systems). The propagation of waves of a different nature in waveguides concerns problems such as these. In the general case of multidimensional systems, analysis of the statistics of eigenvalues involves great difficulties, but it is easier in the one-dimensional case (plane-layered media).

Section 3.2.1 describes the system of dynamic equations and dynamics of eigenvalues (in connection with a change in the thickness of the layer), which enables a study of their statistical characteristics.

3.5.1. Common relationships

The analysis of eigenvalues is based on the analysis of zeros of the Riccati equation solution, which can be written in the general form (3.24) and (3.25)

$$\frac{\mathrm{d}}{\mathrm{d}L} f_L = a_L(\lambda) + b_L(\lambda) f_L(\lambda) + c_L(\lambda) f_L^2(\lambda). \tag{3.66}$$

Eigenvalues, being functions of the parameter L, then satisfy the equation

$$f_L(\lambda_L) = 0, \quad a_L(\lambda_L) + A_L(\lambda_L) \frac{\mathrm{d}}{\mathrm{d}L} \lambda_L = 0, \quad A_L(\lambda) = \frac{\partial}{\partial \lambda} f_L(\lambda), \tag{3.67}$$

with given behavior as $L \Rightarrow L_0$. Hence, the probability density of the quantity $\lambda_L^{(n)}$,

$$\psi_L(\lambda) = \delta(\lambda_L^{(n)} - \lambda),$$

is described by the Liouville equation

$$\frac{\partial}{\partial L} \psi_L(\lambda) = \frac{\partial}{\partial L} \frac{a_L(\lambda)}{A_L(\lambda)} \psi_L(\lambda), \tag{3.68}$$

whose initial condition at $L \Rightarrow L_0$ is determined by the dynamics of the concrete eigenvalue.

The identification of the eigenvalue for this one-dimensional problem can be carried out on the basis of so-called phase formalism (see, e.g. Lifshits, Gredeskul and Pastur [1988]).

Taking into account that the solution of the Riccati equation (3.66) changes in limits from $-\infty$ to $+\infty$, it is possible to change variables

$$f_L = f(\theta_L),$$

where $f \sim \tan \theta_L$ or $f \sim \cot \theta_L$ depending on the initial condition for eq. (3.66). Then values

$$\theta_n = n\pi \quad \text{or} \quad \theta_n = (n + 1/2)\pi$$

will correspond to the eigenvalue $\lambda_L^{(n)}$.

Let us introduce the probability density for the quantity $\theta_L(\lambda)$

$$\psi_L(\lambda, \theta) = \delta(\theta_L(\lambda) - \theta).$$

Taking into account that the function $\psi_L(\lambda, \theta)$ is linked to $\psi_L(\lambda)$ by the equality

$$\psi_L(\lambda, \theta_n) \frac{\partial}{\partial L} \theta_L(\lambda) = \psi_L(\lambda),$$

it is possible to rewrite eq. (3.68) in the form of the equality

$$\frac{\partial}{\partial L} \psi_L(\lambda) = \frac{\partial}{\partial \lambda} \frac{a_L(\lambda)}{\partial f/\partial \theta|_{\theta_n}} \psi_L(\lambda, \theta_n). \tag{3.69}$$

Thus the probability density of eigenvalues is expressed by the probability density of the solution of the Riccati equation (3.66) with the use of the quadrature.

3.5.2. Statistical averaging

If medium parameters have fluctuations, all preceding expressions and relationships should be averaged over the ensemble of realizations of fluctuating parameters.

Let us examine the dynamic problem of eigenvalues as a concrete example,

$$\frac{d^2}{dx^2} u(x) + \lambda u(x) = \varepsilon(x) u(x), \quad u(0) = 0, \quad u(L) = 0. \tag{3.70}$$

In this case the eigenvalues are defined by zeros of the function f_L described by the Riccati equation

$$\frac{d}{dL} f_L = 1 + [\lambda - \varepsilon(L)] f_L^2, \quad f_0 = 0. \tag{3.71}$$

Taking into account that, if $\varepsilon = 0$, the solution of eq. (3.71) has the structure $f_L = \lambda^{-1/2} \tan(\lambda^{1/2} L)$, we can change the variables

$$f_L = \lambda^{-1/2} \tan \theta_L.$$

We then obtain for function θ_L the equation

$$\frac{d}{dL} \theta_L = \lambda^{1/2} - \lambda^{-1/2} \varepsilon(L) \sin^2 \theta_L, \qquad \theta_0 = 0, \tag{3.72}$$

and the value

$$\theta_n = n\pi \quad (n = 1, 2, \ldots)$$

corresponds to the eigenvalue $\lambda^{(n)}$.

The probability density of solution (3.72) $P_L(\lambda, \theta) = \langle \psi_L(\lambda, \theta) \rangle$ for the Gaussian delta-correlated random process satisfies the Fokker–Planck equation

$$\frac{\partial}{\partial L} P_L(\lambda, \theta) = -\lambda^{1/2} \frac{\partial}{\partial \theta} P_L(\lambda, \theta) + \sigma^2 \lambda^{-1} \frac{\partial}{\partial \theta} \sin^2 \theta \frac{\partial}{\partial \theta} \sin^2 \theta \, P_L(\lambda, \theta). \tag{3.73}$$

Hence, the probability density of the nth eigenvalue λ_n is $P_L^{(n)}(\lambda) = \langle \psi_L(\lambda) \rangle$, and, accordingly, eq. (3.69) will be defined by the solution of eq. (3.73) using the formula

$$\frac{\partial}{\partial L} P_L^{(n)}(\lambda) = \frac{\partial}{\partial \lambda} \lambda^{1/2} P_L(\lambda, \theta_n),$$

whose integration gives the expression

$$P_L^{(n)}(\lambda) = \frac{\partial}{\partial \lambda} \lambda^{1/2} \int_0^L d\xi \, P_\xi(\lambda, \theta_n). \tag{3.74}$$

Note also that it is possible to obtain another expression, equivalent to eq. (3.74), for $P_L^{(n)}(\lambda)$:

$$P_L^{(n)}(\lambda) = -\frac{\partial}{\partial \lambda} \int_{-\infty}^{\theta_n} d\theta \, P_L(\lambda, \theta). \tag{3.74'}$$

Thus, to find the function $P_L^{(n)}(\lambda)$, it is necessary to know the solution of eq. (3.73). This equation cannot be solved in the general case. For sufficiently large values of λ, namely for $\lambda \gg \sigma^{4/3}$, one can use the method of averaging over fast oscillations, which is specified by problem solution when fluctuations are absent ($\theta_n = \lambda^{1/2} L$). As a result, for slow changes of the function θ_L

stimulated by the presence of small fluctuations, we obtain the simpler equation

$$\frac{\partial}{\partial L} P_L(\lambda, \theta) = -\lambda^{1/2} \frac{\partial}{\partial \theta} P_L(\lambda, \theta) + \frac{3}{8} \sigma^2 \lambda^{-1} \frac{\partial^2}{\partial \theta^2} P_L(\lambda, \theta), \qquad (3.73')$$

whose solution with the initial condition $P_0(\lambda, \theta) = \delta(\theta)$ has the form of Gaussian distribution of probabilities

$$P_L(\lambda, \theta) = \left(\frac{2\lambda}{3\pi\sigma^2 L}\right)^{1/2} \exp\left\{-\frac{2\lambda}{3\sigma^2 L}(\theta - \lambda^{1/2} L)^2\right\};$$

that is, the quantity θ_L is the Gaussian random function with respect to parameter L with characteristics

$$\langle \theta_L \rangle = \lambda^{1/2} L, \qquad \sigma_\theta^2 = \frac{3}{4\lambda} \sigma^2 L.$$

In this case for equality (3.74') we obtain the density of probabilities for the nth eigenvalue in the form

$$P_L^{(n)}(\lambda) = \left(\frac{L}{6\pi\sigma^2}\right)^{1/2} \left[2 - \left(\frac{\lambda_{0n}}{\lambda}\right)^{1/2}\right] \exp\left\{-\frac{2\lambda L}{3\sigma^2}(\lambda^{1/2} - \lambda_{0n}^{1/2})^2\right\}, \qquad (3.75)$$

where λ_{0n} denotes the eigenvalue of the problem (3.70) in the case of the absence of fluctuations, $\lambda_{0n} = \theta_n^2 L^{-2}$ (Goland and Klyatskin [1988, 1989]).

Note that eq. (3.75) cannot be the formal expression for probability density, because it becomes negative when $\lambda < \lambda_{0n}/4$. This results from using the method of averaging over fast oscillations.

For a sufficiently small σ^2, the function $P_L^{(n)}(\lambda)$ will be concentrated in the vicinity of $\lambda \approx \lambda_{0n}$, where it can be presented in the form

$$P_L^{(n)}(\lambda) = \left(\frac{L}{6\pi\sigma^2}\right)^{1/2} \exp\left\{-\frac{L}{3\sigma^2}(\lambda - \lambda_{0n})^2\right\}. \qquad (3.76)$$

Thus statistical characteristics of eigenvalues are characterized by the dimensionless coefficient of diffusion for the nth mode

$$D_n = \frac{3\sigma^2 L}{8\lambda_{0n}},$$

and the condition of applicability for all expressions is the condition

$$D_n \ll 1.$$

In the limits of this approximation the eigenvalues of the problem are distributed according to Gaussian law, with parameters

$$\langle \lambda_n \rangle = \lambda_{0n}, \qquad \sigma_{\lambda_n}^2 = 3\sigma^2/L.$$

Note that these results also follow from the first approximation of the perturbation theory.

Our discussion of the problem of eigenvalues with concrete boundary conditions (3.70) shows that the results are also valid for other boundary conditions. The expressions for λ_{0n} are the only ones that change. Thus, for the boundary-value problem, for example,

$$\frac{d^2}{dx^2} u(x) + \lambda u(x) = \varepsilon(x) u(x), \qquad \frac{d}{dx} u(0) = 0, \qquad u(L) = 0 \qquad (3.77)$$

all results are valid, and the quantity λ_{0n} is

$$\lambda_{0n} = (n + 1/2)^2 \pi^2 / L^2.$$

3.5.3. Numerical modelling

The preceding results were shown to be valid if the dimensionless coefficient of diffusion D_n was small. This occurs if the parameter σ^2 is sufficiently small or for a large number n. The numerical modelling permits verification of the validity of the preceding results, so long as $D_n > 1$. This modelling was implemented by Goland and Klyatskin [1988] and Goland [1988].

The boundary problem (3.77) was examined. Numerical modelling shows that statistical characteristics of the eigenvalues are well described by the distribution of probabilities (3.76), even for the diffusion coefficient $D_0 \approx 5$. This occurs except when there is a mean value of zero mode, for which

$$\langle \lambda_0 \rangle - \lambda_{00} \approx -D_0.$$

However, this result corresponds to the perturbation theory of the second order, or is responsible for consideration of the next terms in the expansion (3.75) with respect to $(\lambda - \lambda_{0n})$. Coefficients of correlation between different λ_n are close to the values given by the perturbation theory (2/3) even for the case $D_0 \approx 5$.

The results of modelling for statistical characteristics of the eigenfunctions are shown in fig. 12, in which the mean values are plotted of the normalized eigenfunctions $\langle \Phi_n(x) \rangle$ of the first two modes, and their variances $\sigma_\phi = (\langle \phi_n^2 \rangle - \langle \phi_n \rangle^2)^{1/2}$ are calculated for the case $D_0 = 4.67$. Variances of eigenfunctions σ_ϕ have the same order as $\langle \phi_n(x) \rangle$. Figure 13 shows the profiles of the coefficients of eigenfunction correlations for the first two

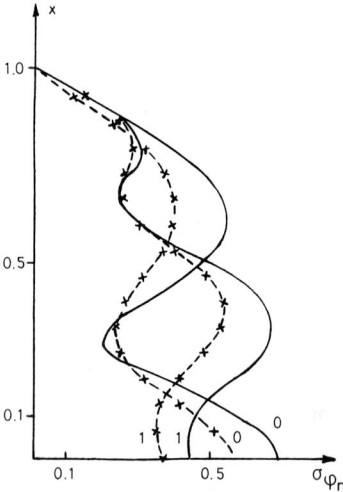

Fig. 12. The variance of eigenfunction $n = 0$; 1; $D_0 = 4.67$. ×, Results of the numerical modelling; solid lines, calculations by the perturbation theory.

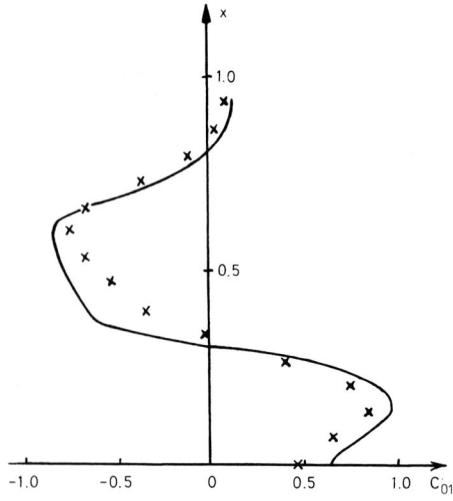

Fig. 13. The profiles of correlation coefficients of zero and first modes. Solid lines, $D_0 = 10^{-2}$; ×, $D_0 = 4.67$.

modes,

$$c_{01} = f_{01}/(f_{00}f_{11})^{1/2}, \qquad f_{mn} = \langle \phi_m \phi_n \rangle - \langle \phi_m \rangle \langle \phi_n \rangle.$$

These are calculated for two values of D_0. From these plots we can see that modes are highly correlated; in some points x the coefficients of correlation are close to unity. The correlation weakens a little with the growth of D_0.

Thus, the results of numerical modelling testify that the results of asymptotic analysis implemented earlier have greater application than the conditions $D_n \ll 1$.

The problems of plane-wave propagation in layered media have been examined. Attempts to construct a theory of wave propagation in quasilayered media based on the imbedding methods have been made by Bugrov [1988, 1989] and Koshel' [1990d].

§ 4. Multidimensional Wave Problems in Layered Random Media

4.1. NONSTATIONARY PROBLEMS

4.1.1. Formulation of boundary-wave problems

Let us now examine the nonstationary problem of incidence of the plane wave $f[t + (x - L)/c_0]$ from the region $x > L$ on the medium layer occupying the part of the space $L_0 < x < L$. The wavefield inside the layer is described by the wave equation

$$\left[\frac{\partial^2}{\partial x^2} - \frac{1}{c^2(x)}\frac{\partial}{\partial t}\left(\frac{\partial}{\partial t} + \tilde{\gamma}\right)\right]u(x, t) = 0, \quad (4.1)$$

with boundary conditions

$$\left(\frac{\partial}{\partial x} + \frac{1}{c_0}\frac{\partial}{\partial t}\right)u(x, t)|_{x=L} = \frac{2}{c_0}\frac{\partial}{\partial t}f(t), \quad \left(\frac{\partial}{\partial x} - \frac{1}{c_0}\frac{\partial}{\partial t}\right)u(x, t)|_{x=L_0} = 0. \quad (4.1')$$

Similarly, for a plane-wave source located within the medium layer at the point x_0, we have the boundary-value problem

$$\left[\frac{\partial^2}{\partial x^2} - \frac{1}{c^2(x)}\frac{\partial}{\partial t}\left(\frac{\partial}{\partial t} + \tilde{\gamma}\right)\right]u(x, x_0; t) = -\frac{2}{c_0}\delta(x - x_0)\frac{\partial}{\partial t}f(t),$$

$$\left(\frac{\partial}{\partial x} + \frac{1}{c_0}\frac{\partial}{\partial t}\right)u(x, x_0; t)|_{x=L} = 0, \quad \left(\frac{\partial}{\partial x} - \frac{1}{c_0}\frac{\partial}{\partial t}\right)u(x, x_0; t)|_{x=L_0} = 0. \quad (4.2)$$

Note that the boundary-value problem (4.1) and (4.1') corresponds to the problem (4.2) if $x_0 = L$. The solution of problem (4.2) can be written in the form of the Fourier integral (the parameter $\tilde{\gamma}$ is supposed to be small)

$$u(x, x_0; t) = \frac{1}{2\pi} \int_{-\infty}^{\infty} d\omega\, G_\omega(x, x_0)\hat{f}(\omega)e^{-i\omega t},$$

$$G_\omega(x, x_0) = \int_{-\infty}^{\infty} dt\, G(x, x_0; t)e^{i\omega t}, \tag{4.2'}$$

where

$$\hat{f}(\omega) = \int_{-\infty}^{\infty} dt\, f(t)e^{i\omega t}.$$

The function $G_\omega(x; x_0)$ is the solution of the stationary problem in the field of plane-wave point source

$$\frac{d^2}{dx^2} G_\omega(x; x_0) + k^2[1 + \varepsilon(x)]G_\omega(x; x_0) = 2ik\delta(x - x_0), \tag{4.3}$$

where $c^{-2}(x) = c_0^{-2}[1 + \varepsilon(x)]$, $k = \omega/c_0$, $\varepsilon(x) = \varepsilon_1(x) + i\tilde{\gamma}/\omega$. This problem was examined earlier. The parameter $\tilde{\gamma}$, which characterizes the absorption of the wave in the medium, is related to the parameter γ, previously introduced by the relation $\gamma = \tilde{\gamma}/2c_0$.

The boundary problem (4.1) and (4.1') describes not only scattered waves caused by the medium inhomogeneity $c(x)$, but also the regular reflection on the boundary $x = L$ because of the discontinuity of the quantity $c(L) \neq c_0$ on this boundary. If no such discontinuity exists (the adjusted boundary), that is, the wave-speed propagation in the free part of the space is constant and equal to $c(L)$, the boundary conditions for the problem (4.1) will be

$$\left(\frac{\partial}{\partial x} + \frac{1}{c(L)}\frac{\partial}{\partial t}\right)u(x,t)|_{x=L} = \frac{2}{c(L)}\frac{\partial}{\partial t}f(t),$$

$$\left(\frac{\partial}{\partial x} - \frac{1}{c_0}\frac{\partial}{\partial t}\right)u(x,t)|_{x=L_0} = 0.$$

Note also the formulation of the nonlinear boundary-wave problem about self-action, which is described by the nonlinear wave equation

$$\frac{\partial^2}{\partial x^2}u(x,t) = \frac{\partial^2}{\partial t^2}\left[\frac{1}{c^2(x, u(x,t))}u(x,t)\right], \tag{4.1''}$$

with boundary condition (4.1'). This problem describes the time-pulse $f(t + (x - L)/c_0)$ incidence on a slab of nonlinear medium.

4.1.2. Imbedding equations

Let us introduce the nonstationary Green function $G(x; L; t)$. The wave $f(t + (x - L)/c_0)$ incident on the layer creates a distribution of sources $\hat{f}(t_0)$ on the layer boundary $x = L$, such that

$$f(t) = \frac{1}{2c_0} \int_{-\infty}^{\infty} dt_0\, \theta(t - t_0) \hat{f}(t_0), \quad \hat{f}(t_0) = 2c_0 \frac{\partial}{\partial t_0} f(t_0).$$

The wavefield inside the medium can then be given in the form

$$u(x; t) = \int_{-\infty}^{\infty} dt_1\, G(x; L; t - t_1) \frac{\partial}{\partial t_1} f(t_1),$$

where the function $G(x; L; t - t_0)$ satisfies the wave equation (4.1) with the boundary condition at the boundary $x = L$

$$\left(\frac{\partial}{\partial x} + \frac{1}{c_0} \frac{\partial}{\partial t} \right) G(x; L; t - t_0)|_{x=L} = \frac{2}{c_0} \delta(t - t_0).$$

The boundary problem to determine $G(x; L; t)$ (without absorption for simplicity) can be reformulated in the form of the problem with initial values over parameter L (we assume $t_0 = 0$) (Babkin, Klyatskin and Lyubavin [1982b], Klyatskin [1986])

$$\left(\frac{\partial}{\partial L} + \frac{1}{c_0} \frac{\partial}{\partial t} \right) G(x; L; t) = -\frac{1}{2c_0} \varepsilon(L) \int_{-\infty}^{\infty} dt_1$$

$$\times \frac{\partial}{\partial t} G(x; L; t - t_1) \frac{\partial}{\partial t_1} H(L; t_1),$$

$$G(x; L; t)|_{L=x} = H(x; t).$$

The function $H(L; t) = G(L; L; t)$, being the field on the layer boundary, satisfies the problem with initial conditions

$$\left(\frac{\partial}{\partial L} + \frac{2}{c_0} \frac{\partial}{\partial t} \right) H(L; t) = \frac{2}{c_0} \delta(t) - \frac{1}{2c_0} \varepsilon(L) \int_{-\infty}^{\infty} dt_1$$

$$\times \frac{\partial}{\partial t} H(L; t - t_1) \frac{\partial}{\partial t_1} H(L; t_1), \quad (4.4)$$

$$H(L; t)|_{L=L_0} = \theta(t).$$

The function $G(x; L; t)$ describes the wavefield inside the medium when the wave of the form $\theta(t + (x - L)/c_0)$ is incident on it. The function $H(L; t)$ has the structure $H(L; t) = \theta(t)H_L(t)$; by substituting it into eq. (4.4) and separating the singular ($\sim \delta(t)$) and regular ($\sim \theta(t)$) terms in eq. (4.4), we obtain (Bugrov and Klyatskin [1989])

$$\left(\frac{\partial}{\partial L} + \frac{2}{c(L)}\frac{\partial}{\partial t}\right) H_L(t) = -\frac{1}{2c_0}\varepsilon(L) \int_0^t dt_1 \frac{\partial}{\partial t} H_L(t - t_1) \frac{\partial}{\partial t_1} H_L(t_1),$$

$$H_{L_0}(t) = 1, \quad H_L(+0) = \frac{2c(L)}{c_0 + c(L)}. \tag{4.5}$$

We shall not dwell on the analysis of the problem (4.5) and the field inside the layer, but simply note that eq. (4.5) enables solution of the inverse problem. Thus, if we know the function $H_L(t)$ as a function of time t (i.e. the field reflected backwards), the profile of the wave-propagation speed $c(x)$ in the medium can be obtained. This inverse problem permits an analytical solution when the function $H_L(t)$ is exponential or linearly dependent (Bugrov and Klyatskin [1989], Kreider [1989]). Note also, the field amplitude both inside the layer and on layer boundaries experiences discontinuities when waves reflected from layer boundaries arrive. However, the magnitudes of these discontinuities are defined only by local values of the function $c(x)$ at this point. When $t \Rightarrow \infty$, the solution for eq. (4.5) goes to the stationary value $H_L(t) = 1$.

Imbedding equations (4.4) and (4.5) describe the problem where the quantity $c(x)$ experiences an increase on boundaries $x = L_0, L$. If no increase occurs, the equations having $c'(L)/c(L)$ into integral term can be written in a way similar to that used for deduction of eq. (3.20) (Corones, Davison and Krueger [1983], Kriestensson and Krueger [1986a, b, 1987, 1989], Krueger and Ochs [1989]). Instead of eq. (4.5), we have

$$\left(\frac{\partial}{\partial L} + \frac{2}{c(L)}\frac{\partial}{\partial t}\right) H_L(t) = -\frac{c'(L)}{2c(L)} H_L(t)$$

$$+ \frac{c'(L)}{2c(L)} \int_0^t dt_1 \, H_L(t - t_1) \frac{\partial}{\partial t_1} H_L(t_1), \tag{4.5'}$$

$$H_L(+0) = 1, \quad H_{L_0}(t) = \frac{2c(L_0)}{c_0 + c(L_0)}.$$

Examination of a nonlinear medium is carried out in the same ideological framework of the imbedding method as used for a linear medium. The essential "technical" difference is that the boundary problem's (4.1) and

(4.1′) solution $u(x, t) = u(x, L; t)$ is a nonlinear functional of source distribution $\hat{f}(t)$ on the slab boundary. The imbedding equations now contain variational derivatives (Klyatskin [1986], Guzev [1991])

$$\left(\frac{\partial}{\partial L} + \frac{1}{c_0}\frac{\partial}{\partial t}\right) u(x, L; t) = -\int_{-\infty}^{\infty} dt_1 \left(\frac{\delta}{\delta \hat{f}(t_1)} u(x, L; t)\right)$$

$$\times \frac{\partial^2}{\partial t_1^2} F(L, H(L; t_1)),$$

$$\left(\frac{\partial}{\partial L} + \frac{2}{c_0}\frac{\partial}{\partial t}\right) H(L; t) = \frac{\hat{f}(t)}{c_0^2} - \int_{-\infty}^{\infty} dt_1 \left(\frac{\delta}{\delta \hat{f}(t_1)} H(L; t)\right) \quad (4.6)$$

$$\times \frac{\partial^2}{\partial t_1^2} F(L, H(L; t_1)),$$

$$F(L, H(L; t)) = \left[\frac{c_0^2}{c^2(L; H(L; t))} - 1\right] H(L; t),$$

$$H(L; t) = u(L, L; t).$$

Let $f(t + (x - L)/c_0) = (1/2c_0)\theta(t + (x - L)/c_0)$ be the wave incident on the slab of the nonlinear medium. The wave's front achieves the boundary $x = L$ at moment $t = 0$. The field in the slab is defined by a solution of the system (4.6) with $\hat{f}(t) = \delta(t)$. This problem was analyzed by Guzev [1991].

Note that in this case the wavefield $H_L(0)$ on the boundary $x = L$ at the moment $t = 0$ satisfies the transcendent equation

$$c_0^2 F(L, H_L(0))H_L(0) + 2c_0 H_L(0) = 1,$$

provided the unique solution is determined by the demand of continuous limit transition to the linear problem solution. In the simplest case of self-action, the function $F(x, u(x, t)) = z(x)u^{n+1}(x, t)$ $(n > 0)$, the wave amplitude on the wave's front depends on $z(x)$ locally, which is analogous to the linear problem.

4.1.3. Statistical description

Let us now examine the statistical characteristics of the solution of a nonstationary problem describing the time-pulse field generated inside the randomly inhomogeneous medium. This problem is described by eq. (4.2), and its solution can be given in the form of the Fourier integral (4.2′). We are interested in limit values of the mean magnitude of the wavefield intensity

$$I(x; x_0; t) = u^2(x; x_0; t),$$

as $t \Rightarrow \infty$ and $\tilde{\gamma} \Rightarrow 0$. The mean intensity is determined by the quantity

$$\langle I(x; x_0; t)\rangle = \frac{1}{(2\pi)^2} \int_{-\infty}^{\infty} \int d\omega \, d\psi$$
$$\times \langle I_{\omega,\psi}(x; x_0)\rangle \hat{f}(\omega + \psi/2)\hat{f}^*(\omega - \psi/2)e^{-i\psi t}.$$

When $t \Rightarrow \infty$, the value of this integral is defined by the integrand's behavior at small values of ψ, that is, by the quantity

$$\langle I(x; x_0; t)\rangle|_{t \Rightarrow \infty} = \frac{1}{(2\pi)^2} \int_{-\infty}^{\infty} d\omega \, |\hat{f}(\omega)|^2 \int_{-\infty}^{\infty} d\psi \, \langle I_{\omega,\psi}(x; x_0)\rangle e^{-i\psi t}. \tag{4.7}$$

The two-frequency analogy of the plane-wave intensity is introduced in formula (4.7)

$$I_{\omega,\psi}(x; x_0) = G_{\omega+\psi/2}(x; x_0) G^*_{\omega-\psi/2}(x; x_0).$$

Proceeding from eq. (4.3), when ψ and $\tilde{\gamma}$ are small, we obtain for it the equality ($x \leqslant x_0$)

$$\frac{d}{dx} S_{\omega,\psi}(x; x_0) = c_0^{-1}(\tilde{\gamma} - i\psi) I_{\omega,\psi}(x; x_0). \tag{4.8}$$

Here, $S_{\omega,\psi}(x; x_0)$ is the two-frequency analogy of the energy-flux density

$$S_{\omega,\psi}(x; x_0) = \frac{c_0}{2i\psi} \left[G_{\omega+\psi/2}(x; x_0) \frac{d}{dx} G^*_{\omega-\psi/2}(x; x_0) \right.$$
$$\left. - G^*_{\omega-\psi/2}(x; x_0) \frac{d}{dx} G_{\omega+\psi/2}(x; x_0) \right].$$

By integrating equality (4.8) over a half-space $-\infty < x < x_0$, we obtain

$$c_0^{-1}(\tilde{\gamma} - i\psi) \int_{-\infty}^{x_0} dx \, I_{\omega,\psi}(x; x_0) = S_{\omega,\psi}(x_0; x_0).$$

Hence, by integrating eq. (4.7) over the half-space, for the value of mean energy enclosed in this half-space we have

$$E(t) = \int_{-\infty}^{x_0} dx \, \langle I(x; x_0; t)\rangle$$
$$= \frac{c_0}{(2\pi)^2} \int_{-\infty}^{\infty} d\omega \, |\hat{f}(\omega)|^2 \int_{-\infty}^{\infty} \frac{d\psi}{\tilde{\gamma} - i\psi} \langle S_{\omega,\psi}(x_0; x_0)\rangle e^{-i\psi t}. \tag{4.9}$$

Let us now study the statistical description of quantities $S_{\omega,\psi}(x;x_0)$ and $I_{\omega,\psi}(x;x_0)$. According to the corresponding expressions for a stationary problem, they will be defined by the quantity $W_{\omega,\psi} = R_{\omega+\psi/2} R^*_{\omega-\psi/2}$, which is the two-frequency analogy of the modulus square of the reflection coefficient $W = |R|^2$. When $\psi = 0$, expressions for $S_{\omega,\psi}$ and $I_{\omega,\psi}$ become expressions corresponding to one-frequency characteristics of the stationary problem. Thus, for the calculation of mean values of $S_{\omega,\psi}$ and $I_{\omega,\psi}$, it is necessary to know the statistics of the quantity $W_{\omega,\psi}$.

The function $R_\omega(x)$, being a function of x, satisfies the stochastic Riccati equation, which is written in this case in the form

$$\frac{d}{dx} R_\omega(x) = \frac{2i}{c_0}\left(\omega + i\frac{\tilde{\gamma}}{2}\right) R_\omega + \frac{\omega}{2c_0} \varepsilon_1(x)(1 + R_\omega)^2, \quad R_\omega(-\infty) = 0.$$

Hence, the function $W_{\omega,\psi}$ is described by the equation

$$\frac{d}{dx} W_{\omega,\psi}(x) = -\frac{2}{c_0}(\tilde{\gamma} - i\psi) W_{\omega,\psi}$$

$$- i\frac{\omega}{2c_0} \varepsilon_1(x)(R_{\omega+\psi/2} - R^*_{\omega-\psi/2})(1 - W_{\omega,\psi}).$$

We obtain recurrent equality for statistical moments $W^{(n)}_{\omega,\psi} = \langle [W_{\omega,\psi}]^n \rangle$ for the model $\varepsilon_1(x)$ of "white noise" ($\langle \varepsilon_1(x) \rangle = 0$, $\langle \varepsilon_1(x)\varepsilon_1(x') \rangle = 2\sigma_\varepsilon^2 l_0 \delta(x-x')$) by the standard way

$$\frac{d}{dx} W^{(n)}_{\omega,\psi}(x) = -\frac{2n}{c_0}(\tilde{\gamma} - i\psi) W^{(n)}_{\omega,\psi}(x)$$

$$+ D(\omega)\{W^{(n+1)}_{\omega,\psi}(x) - 2W^{(n)}_{\omega,\psi}(x) + W^{(n-1)}_{\omega,\psi}(x)\}.$$

Here, as before, $D(\omega) = \sigma_\varepsilon^2 l_0 \omega^2 / 2c_0^2$. Hence, the solution independent of x and corresponding to the half-space satisfies the recurrent equality

$$\frac{2}{c_0}(\tilde{\gamma} - i\psi) W^{(n)}_{\omega,\psi}(x) = D(\omega)n\{W^{(n+1)}_{\omega,\psi}(x) - 2W^{(n)}_{\omega,\psi}(x) + W^{(n-1)}_{\omega,\psi}(x)\}. \quad (4.10)$$

When $\psi = 0$, equality (4.10) turns into equality (3.42′). The probability density (3.42) corresponds to it. The equality (4.10) can be considered as an analytical prolongation into the complex region of equality (3.42′) with respect to parameter $\tilde{\gamma}$. Hence, all statistical characteristics obtained in the stationary problem, which were prolonged analytically in the complex plane with respect to parameter $\tilde{\gamma}$, will define the corresponding two-frequency statistical characteristics (Shevtsov [1982]). Thus, if there is no damping

($\tilde{\gamma} = 0$), it is necessary to replace the parameter $\tilde{\gamma}$ by $0 - i\psi$ in the corresponding statistical characteristics of the plane-wave problem in order to obtain expressions for the two-frequency statistical characteristics; that is,

$$\beta(\omega) = \frac{1}{c_0 D(\omega)}(0 - i\psi).$$

As a result, we obtain as $t \Rightarrow \infty$ and $\tilde{\gamma} = 0$; that is, when values ψ are small,

$$\langle S_{\omega,\psi}(x_0; x_0) \rangle = 1, \qquad \langle I_{\omega,\psi}(x_0; x_0) \rangle = i\frac{D(\omega)c_0}{\psi + i0}.$$

Hence, formulas (4.7) and (4.9) become expressions corresponding to the asymptotic $t \Rightarrow \infty$ (after being integrated with respect to ψ):

$$\langle I(x_0; x_0; \infty) \rangle = \frac{c_0}{2\pi} \int_{-\infty}^{\infty} d\omega\, D(\omega) |\hat{f}(\omega)|^2,$$

$$E(\infty) = \frac{c_0}{2\pi} \int_{-\infty}^{\infty} d\omega\, |\hat{f}(\omega)|^2.$$

(4.11)

Thus, if integrals (4.11) exist, the mean field intensity in the source point and the total wave energy in the half-space appear to be finite. This fact proves the presence of space localization for the mean intensity. Obviously, the localization length will be given by equality

$$l_{\text{loc}} = \int_{-\infty}^{\infty} d\omega\, |\hat{f}(\omega)|^2 \bigg/ \int_{-\infty}^{\infty} d\omega\, D(\omega) |\hat{f}(\omega)|^2.$$

This property of localization is stipulated by the finite nature of the total energy in the half-space of the randomly inhomogeneous medium, which is stipulated, in turn, by the independence of the mean energy flux on the medium parameter fluctuations for the plane-wave stationary problem. We obtain the form of the localization curve by using equality (3.62)

$$\langle I(x; x_0; \infty) \rangle = \frac{c_0}{2\pi} \int_{-\infty}^{\infty} d\omega\, D(\omega) |\hat{f}(\omega)|^2 \Phi_{\text{loc}}(\xi) \quad (\xi = D(\omega)|x - x_0|),$$

(4.12)

where $\Phi_{\text{loc}}(\xi)$ is the localization curve (3.63) for the stationary problem. Its dependence on ω enters here only as the diffusive coefficient $D(\omega)$.

If the pulse $f(t)$ is characterized by the only parameter (the width of the pulse) we can obtain from eq. (4.12) the asymptotic dependence for the white-noise model $\varepsilon_1(x)$ for large values of $|x - x_0|$

$$\langle I(x; x_0; \infty) \rangle \sim |x - x_0|^{-3/2}.$$

If we have the pulse of a high-frequency package (the frequency is ω_0), the asymptotic dependence will be

$$\langle I(x; x_0; \infty) \rangle \sim \Phi_{\text{loc}}(\xi) \quad (\xi = D(\omega_0)|x - x_0|).$$

By analogy we obtain expressions for the case of a pulse source located on the reflective boundary:

$$\langle I_{\text{ref}}(x; L; \infty) \rangle = \frac{2c_0}{\pi} \int_{-\infty}^{\infty} d\omega \, D(\omega) |\hat{f}(\omega)|^2 \Phi_{\text{loc}}(\xi)$$

$(\xi = D(\omega)(L-x))$,

$$\langle I_{\text{ref}}(L; L; \infty) \rangle = \frac{4c_0}{\pi} \int_{-\infty}^{\infty} d\omega \, D(\omega) |\hat{f}(\omega)|^2,$$

$$E_{\text{ref}}(\infty) = \frac{2c_0}{\pi} \int_{-\infty}^{\infty} d\omega \, |\hat{f}(\omega)|^2.$$

Hence, the wavefield is also localized on the half-size smaller distance.

In the case of the time-pulse incidence on the half-space of randomly inhomogeneous medium $x < L$, we have $(I(L; t) = u^2(L; t))$

$$\langle I(L, t) \rangle|_{t \Rightarrow \infty} = \frac{1}{(2\pi)^2} \int_{-\infty}^{\infty} d\omega \, |\hat{f}(\omega)|^2 \int_{-\infty}^{\infty} d\psi \, \{1 + W_{\omega,\psi}^{(1)}\} e^{-i\psi t},$$

$$E(t) = \int_{-\infty}^{L} dx \, \langle I(x; L; t) \rangle|_{t \Rightarrow \infty} \qquad (4.13)$$

$$= \frac{c_0}{(2\pi)^2} \int_{-\infty}^{\infty} d\omega \, |\hat{f}(\omega)|^2 \int_{-\infty}^{\infty} \frac{d\psi}{0 - i\psi} \{1 - W_{\omega,\psi}^{(1)}\} e^{-i\psi t},$$

where $W_{\omega,\psi}^{(1)} = -\beta(\omega) \int_0^\infty du \, [u/(u+2)] e^{-i\beta(\omega)u}$ is the analytical prolongation of the corresponding expression for $\langle |R_L|^2 \rangle$ with respect to the parameter β. By accomplishing the integration with respect to ω and u in eq. (4.13), we obtain the following asymptotic expressions when t is sufficiently large (Burridge, Papanicolaou, Sheng and White [1989]):

$$\langle I(L, t) \rangle = \frac{c_0}{\pi} \int_{-\infty}^{\infty} d\omega \, |\hat{f}(\omega)|^2 \frac{D(\omega)}{[2 + D(\omega)c_0 t]^2},$$

$$E(t) = \frac{c_0}{\pi} \int_{-\infty}^{\infty} d\omega \, |\hat{f}(\omega)|^2 \frac{1}{2 + D(\omega)c_0 t}. \qquad (4.14)$$

Formulas (4.14) give the asymptotic dependence on time for the mean intensity behavior of the pulse reflected from the layer, and the mean energy contained inside the layer of the randomly inhomogeneous medium. From here on we obtain expressions for pulses with and without the high-frequency package

$$\langle I(L;t)\rangle \sim t^{-2} \quad \text{or} \quad \langle I(L;t)\rangle \sim t^{-3/2}.$$

It follows from eq. (4.13) that for $t \Rightarrow \infty$, all waves are entirely radiated out of the randomly inhomogeneous medium.

4.1.4. Layer of finite thickness

In the case of a layer of finite thickness (Guzev and Klyatskin [1991a]), the primary difficulty for statistical analysis of the wavefield $u(x;t)$ involves the influence of a boundary condition at $x = L_0$. Our task is to find the wavefield $H_L(t)$, and to calculate its statistics for $t \Rightarrow \infty$. It is clear that initial values of $H_L(t)$ at $t = 0$ do not influence the statistics of $t \Rightarrow \infty$. Then, by carrying out the Laplace transformation of eq. (4.5) and neglecting the terms including the initial values, we have an equation whose solution has the form

$$H_L(t) = \frac{1}{2\pi i} \int_{-i\infty+\sigma}^{i\infty+\sigma} dp\, e^{p(t-2\tau_L)} \left\{ p\left[1 + \frac{p}{2c_0} \int_{L_0}^{L} d\xi\, \varepsilon(\xi) e^{-2p\tau_\xi} \right] \right\}^{-1},$$

$$\tau_L = \int_{L_0}^{L} \frac{d\eta}{c(\eta)}. \tag{4.15}$$

The quantity of interest is the averaged wavefield intensity $\langle I(L;t)\rangle$ under the condition $t \Rightarrow \infty$. The integral representation for $\langle I(L;t)\rangle$ follows from eq. (4.2') at $x = x_0 = L$

$$\langle I(L;t)\rangle = \frac{1}{\pi} \int_{-\infty}^{\infty} d\omega\, |\hat{f}(\omega)|^2 \int_{-\infty}^{\infty} d\tau\, R_L(t;\tau) e^{-2i\omega\tau},$$

$$R_L(t;\tau) = \left\langle \frac{\partial}{\partial t} H_L(t-\tau) \frac{\partial}{\partial t} H_L(t+\tau) \right\rangle. \tag{4.16}$$

The calculations yield the final expression for $R_L(t;\tau)$ for $t \Rightarrow \infty$

$$R_L(t;\tau) = \frac{1}{16\pi t(\alpha c_0 t)^{1/2}} \sum_{\kappa = \pm 1} \theta(\kappa\tau) \left(1 - \frac{\kappa\tau}{(\alpha c_0 t)^{1/2}} \right)$$

$$\times \exp[-2\kappa\tau(\alpha c_0 t)^{-1/2}],$$

$$\alpha = \sigma_\varepsilon^2 l_0 / 2c_0^2.$$

At $t \gg 2(L-L_0)/c_0$, the asymptotic behavior of the averaged intensity for a slab of finite thickness coincides with the corresponding asymptotic of eq. (4.14) for a half-space. Thus the wavefield discontinuities on the boundaries of the slab are not essential for statistics formation; it is specified by scattering on random inhomogeneities.

We have examined the statistical description of the wave pulse in a randomly inhomogeneous medium. It is also possible to study in the same way the problem of the wave-space package in a randomly inhomogeneous medium (Asch, Papanicolaou, Postel, Sheng and White [1990], Burridge, Papanicolaou and White [1988], Papanicolaou, Postel, Sheng and White [1990], Asch, Kohler, Papanicolaou, Postel and White [1991], Kohler, Papanicolaou, Postel and White [1991], Kohler, Papanicolaou and White [1991]). The property of localization obtained above can also obviously be conserved for this case. It can be treated as the existence of the statistical waveguide in directions normal to the x axis (Gredescul and Freilikher [1988, 1990], Freilikher and Gredescul [1990, 1992]).

4.2. POINT SOURCE INSIDE A LAYERED RANDOM MEDIUM

4.2.1. *Factorization of the wave equation*

Let us now examine the wavefield of a source inside a multidimensional randomly inhomogeneous layered medium following the work of Guzev and Klyatskin [1991b], for which the Green function satisfies the equation

$$\left\{ \frac{\partial^2}{\partial z^2} + \Delta_\perp + k^2[1+\varepsilon(z)] \right\} G(x, \boldsymbol{\rho}; z_0) = \delta(z-z_0)\delta(\boldsymbol{\rho}), \qquad (4.17)$$

where $\boldsymbol{\rho} = (x, y)$, $\Delta_\perp = \partial^2/\partial x^2 + \partial^2/\partial y^2$.

We have an integral representation of the Green function in one-, two- and three-dimensional spaces, respectively:

$$G^{(1)}(z; z_0) = \frac{1}{2ik} \int_0^\infty dt \, e^{ikt/2} \psi(t; z; z_0),$$

$$G^{(2)}(x, z; z_0) = \frac{1}{2ik} \left(\frac{k}{2\pi i}\right)^{1/2} \int_0^\infty dt \, t^{-1/2} e^{ik(x^2+t^2)/2} \psi(t; z; z_0), \qquad (4.18)$$

$$G^{(3)}(\boldsymbol{\rho}, z; z_0) = -\frac{1}{4\pi} \int_0^\infty dt \, t^{-1} e^{ik(\rho^2+t^2)/2t} \psi(t; z; z_0),$$

where $\psi(t, z; z_0)$ is defined from the equation

$$\frac{\partial}{\partial t}\psi(t, z; z_0) = \frac{i}{2k}\left[\frac{\partial^2}{\partial z^2} + k^2\varepsilon(z)\right]\psi(t, z; z_0), \quad \psi(0, z; z_0) = \delta(z - z_0).$$
(4.19)

These formulas express the factorization property of the Helmholtz equation in a layered medium (see Appendixes C and D).

The evolutional problem (4.19) will be supplemented with boundary conditions along z. We consider the following boundary problems:

a. A source is located in the unbounded space with a radiation condition for $|z| \Rightarrow \infty$.

b. A source is located on the reflective boundary for which $\partial\psi/\partial z|_{z=z_0-0} = 0$, and there is a radiation condition for $z \Rightarrow \infty$.

c. A source is located on the boundary of a homogeneous half-space with a radiative condition for $|z| \Rightarrow \infty$.

When $x, |\boldsymbol{\rho}| \Rightarrow \infty$, we have the asymptotic

$$G^{(2)}(x, z; z_0) \simeq \frac{1}{2ik} e^{ik|x|}\psi(|x|, z; z_0),$$

$$G^{(3)}(\boldsymbol{\rho}, z; z_0) \simeq \frac{1}{4\pi}\left(\frac{2\pi i}{k\rho}\right)^{1/2} e^{ik\rho}\rho^{-1/2}\psi(\rho, z; z_0),$$
(4.20)

if the function $\psi(t, z; z_0)$ has no exponential behavior with respect to t. Formulas (4.20) correspond to a small-angle scattering (the parabolic approximation). To analyze the contribution of scattering at wide angles to the wavefield statistical behavior, the exact representation (4.18) of the Green function should be used.

We can write $\psi(t, z; z_0)$ by means of the Fourier transformation

$$\psi_\omega(z; z_0) = \int_{-\infty}^{\infty} dt\, e^{-i\omega t}\psi(t, z; z_0),$$

and we see that $\psi_\omega(z; z_0)$ satisfies the equation

$$\left[\frac{d^2}{dz^2} - 2k\omega + k^2\varepsilon(z)\right]\psi_\omega(z; z_0) = 2ik\delta(z - z_0). \tag{4.21}$$

The solution $\psi_{\omega<0}(z; z_0)$ corresponds to propagating waves when $\omega < 0$, and to evanescent decreasing waves when $\omega > 0$.

We are interested in the asymptotic behavior of Green functions for

$x, |\boldsymbol{\rho}| \Rightarrow \infty$. We then have

$$G^{(1)}(z; z_0) = \frac{1}{2ik} \psi_{-k/2-i0}(z; z_0),$$

$$G^{(2)}(x, z; z_0) = \frac{1}{4i\pi k} \int_{-\infty}^{\infty} \frac{d\omega}{(1+2\omega/k)^{1/2}} \psi_{\omega-i0}(z; z_0)$$
$$\times \exp[ikx(1+2\omega/k)^{1/2}], \qquad (4.22)$$

$$G^{(3)}(\boldsymbol{\rho}, z; z_0) = -\frac{1}{8\pi^2} \left(\frac{2\pi i}{k\rho}\right)^{1/2} \int_{-\infty}^{\infty} \frac{d\omega}{(1+2\omega/k)^{1/4}} \psi_{\omega-i0}(z; z_0)$$
$$\times \exp[ikx(1+2\omega/k)^{1/2}].$$

Formulas (4.22) are reduced to (4.20) under the condition $2\omega/k \ll 1$ in the integrals.

4.2.2. Parabolic equation

Let us first examine the statistical behavior of the Green functions under a parabolic approximation. The representation of the function $\psi(t, z; z_0)$ can then be written in the form

$$\psi(t, z; z_0) = \frac{1}{2\pi} \int_0^\infty d\omega \left(\frac{k}{2\omega}\right)^{1/2} e^{-i\omega t} \tilde{\psi}_\omega(z; z_0)$$
$$+ \frac{1}{2\pi} \int_0^\infty d\omega \, e^{i\omega t} \psi_\omega(z; z_0), \qquad (4.23)$$

where $\tilde{\psi}_\omega(z; z_0)$ is described by the equation

$$\left[\frac{d^2}{dz^2} + 2k\omega + k^2\varepsilon(z)\right] \tilde{\psi}_\omega(z; z_0) = 2ik(2k\omega)^{1/2}\delta(z-z_0). \qquad (4.24)$$

For evanescent waves the influence of fluctuations $\varepsilon(z)$ is not essential, and $\tilde{\psi}_{\omega<0}(z, z_0) = -i(k/2\omega)^{1/2} \exp\{-(2k\omega)^{1/2}|z-z_0|\}$ corresponds to a free space field. The unknown function $\tilde{\psi}_\omega(z; z_0)$ is a solution of eq. (4.24), the statistical analysis of which was done in § 3.

The most important feature of a one-dimensional statistical problem is that any small but finite absorption $\gamma \ll 1$ has to be taken into account. We introduced it by means of the function $\varepsilon(z) = \varepsilon_1(z) + i\gamma$, where $\varepsilon_1(z)$ is a real random function. The quantity of interest is the average intensity $\langle I(t, z; z_0)\rangle = \langle \psi(t, z; z_0)\psi^*(t, z; z_0)\rangle$ at sufficiently large values of t. The representation for this quantity is

$$\langle I(t, z; z_0)\rangle = I_{\text{fluc}}(t, z; z_0) + I_1(t, z; z_0) + I_2(t, z; z_0), \qquad (4.25)$$

where functions $I_{\text{fluc}}(t, z; z_0)$, $I_2(t, z; z_0)$ are contributions of the first and second terms in the right-hand side of eq. (4.23) to the average intensity, and $I_1(t, z; z_0)$ takes into account crossed terms $\langle \tilde{\psi}_{\omega_1}(z; z_0)\psi^*_{\omega_2}(z; z_0)\rangle$.

Let us consider first

$$I_{\text{fluc}}(t, z; z_0) = \frac{k}{2(2\pi)^2} \int_0^\infty d\Omega \int_{-2\Omega}^{2\Omega} \frac{d\omega}{(\Omega^2 - \omega^2/4)^{1/2}} e^{-i\omega t} \langle I_{\Omega,\omega}(z; z_0)\rangle,$$

where $\langle I_{\Omega,\omega}(z; z_0)\rangle = \langle \tilde{\psi}_{\Omega+\omega/2}(z; z_0)\tilde{\psi}^*_{\Omega-\omega/2}(z; z_0)\rangle$ is the two-frequency correlator of the corresponding boundary-problem solution. When $t \Rightarrow \infty$, the values of the expression under the integral have to be taken for $\omega \Rightarrow 0$.

The statistical characteristics of the solution of eq. (4.24) are defined by the statistics of the reflection coefficient $R_{z_0}(\omega)$ of the plane wave incident on the half-space $z_0 > z$ from a homogeneous space. The function $R_{z_0}(\omega)$ now satisfies the Riccati equation

$$\frac{d}{dz_0} R_{z_0}(\omega) = [2i(2k\omega)^{1/2} - k(k/2\omega)^{1/2}\gamma] R_{z_0}(\omega)$$

$$+ i\frac{k}{2}\left(\frac{k}{2\omega}\right)^{1/2} \varepsilon_1(z_0)[1 + R_{z_0}(\omega)]^2.$$

This yields the following equation for the two-frequency function $W_{z_0}(\Omega, \omega) = R_{z_0}(\Omega + \omega/2)R^*_{z_0}(\Omega - \omega/2)$ at $\omega \Rightarrow 0$:

$$\frac{d}{dz_0} W_{z_0}(\Omega, \omega) = -2\left(\frac{k}{2\Omega}\right)^{1/2}(k\gamma - i\omega)W_{z_0}(\Omega, \omega)$$

$$- i\frac{k}{2}\left(\frac{k}{2\Omega}\right)^{1/2} \varepsilon_1(z_0)[R_{z_0}(\Omega) - R^*_{z_0}(\Omega)][1 - W_{z_0}(\Omega, \omega)].$$

(We have neglected terms proportional to $\omega\gamma$, $\omega\varepsilon_1$.) For simplicity we assume $\varepsilon_1(z)$ is the Gaussian white noise, as before.

The one-frequency statistic characteristics are then expressed by the non-dimensional parameters

$$\beta = \frac{k\gamma}{D}\left(\frac{k}{2\Omega}\right)^{1/2}, \quad D = D_0 \frac{k}{2\Omega}, \quad D_0 = k^2\sigma_e^2 l_0/2,$$

and the functional dependence on β is defined by the boundary conditions.

In this case there is a probability distribution $P(u)$ of $u = (1+W)/(1-W)$ independent of z_0 (half-space) (3.42):

$$P(u) = \beta e^{-\beta(u-1)}. \tag{4.26}$$

The process of determination of the two-frequency function $W_{z_0}(\Omega, \omega)$ at small ω is reduced to the analytical prolongation of the corresponding one-frequency solution with respect to β:

$$\beta \Rightarrow \left(\frac{k}{2\Omega}\right)^{1/2} \frac{k\gamma - i\omega}{D} = -i\left(\frac{2\Omega}{k}\right)^{1/2} \frac{\omega + ik\gamma}{D_0}. \tag{4.27}$$

Let us investigate some problems for different boundary conditions.

a. Source location in the unbounded space.

One assumes that the average intensity at $z = z_0$. The one-frequency characteristic is $\langle I_{\Omega,0}(z_0; z_0)\rangle = \langle \tilde{\psi}_\Omega(z_0; z_0)\tilde{\psi}_\Omega^*(z_0; z_0)\rangle = 1 + 1/\beta$. By accomplishing the analytical prolongation with respect to β, one has

$$I_{\text{fluc}}(t, z_0; z_0) = \frac{kD_0}{2i(2\pi)^2} \int_0^\infty d\Omega \int_{-2\Omega}^{2\Omega} \frac{d\omega}{(\Omega^2 - \omega^2/4)^{1/2}}$$

$$\times e^{-i\omega t} \left(\frac{k}{2\Omega}\right)^{1/2} \frac{1}{\omega + i\gamma}. \tag{4.28}$$

Expression (4.28) for $I_{\text{fluc}}(t, z_0; z_0)$ exists for a point source only in the case of a finite absorption γ, and if $\gamma \Rightarrow 0$

$$I_{\text{fluc}}(t, z_0; z_0) \sim D_0 k/\gamma^{1/2}.$$

Hence, the presence of a small but finite absorption is important for a description of the wavefield characteristics (the result is identical to that of a one-dimensional problem).

b. Source location on the reflective boundary.

For this case we see that all formulated conclusions for a source in the unbounded space are valid for a source on the reflective boundary.

c. Source location on the half-space boundary.

If a source is placed on the boundary $z = z_0$ of the medium, the one-frequency characteristic has the form

$$\langle I_{\Omega,0}(z_0; z_0)\rangle = 1 + \langle W_{\Omega,0}\rangle = 1 + \langle |R_{z_0}(\Omega)|^2\rangle,$$

where the mean value is calculated with the help of the statistical probability distribution (4.26). Then the averge intensity

$$\langle I(t, z_0; z_0)\rangle = I_{\text{free}}(t, z_0; z_0) + \tilde{I}_{\text{fluc}}(t, z_0; z_0),$$

$$\tilde{I}_{\text{fluc}}(t, z_0; z_0) = \frac{k}{2(2\pi)^2} \int_0^\infty d\Omega \int_{-2\Omega}^{2\Omega} \frac{d\omega}{(\Omega^2 - \omega^2/4)^{1/2}}$$

$$\times e^{-i\omega t} \beta(\Omega, \omega) \int_0^\infty \frac{u\, du}{u+2} e^{-\beta(\Omega,\omega)},$$

and the parameter $\beta(\Omega, \omega)$ is given by the formula (4.27). Integration with respect to ω and u yields the expression ($\gamma \Rightarrow 0$)

$$\tilde{I}_{\text{fluc}}(t, z_0; z_0) = \frac{k^{3/2} D_0}{2(2\pi)^{1/2}} \int_0^\infty d\Omega\, \Omega^{-3/2} \left[2 + D_0 \left(\frac{k}{2\Omega}\right)^{1/2} t\right]^{-2}.$$

In the limit $t \Rightarrow \infty$, we have

$$\tilde{I}_{\text{fluc}}(t, z_0; z_0) \simeq I_{\text{free}}(t, z_0; z_0) = k/2\pi t,$$

and the average intensity of the wavefield is redoubled. This is analogous to the result for a one-dimensional problem.

4.2.3. General case

For an exact description of the problem of a point source, we consider the function $G^{(2)}(x, z; z_0)$ (4.22). We divide the integration region into three parts $(-\infty, -k/2)$, $(-k/2, 0)$ and $(0, +\infty)$. The contribution of the first region to the Green function is approximately $\psi_{-k/2}(z_0; z_0)/kx$, because of a decreasing exponent for $x \Rightarrow +\infty$. It gives rise to the term

$$\langle I_2^{(2)}(x)\rangle \sim D_0/(kx)^2 k\gamma$$

in the average intensity. In the third region the field $\psi_\omega(z_0; z_0)$ corresponds to a wave propagating in free space, and generates the contribution $\langle I_3^{(2)}(x)\rangle \sim 1/kx$ to the average intensity. To evaluate the contribution of a region $(-k/2, 0)$, we use a method analogous to that for the parabolic equation. Then the corresponding term in the average intensity is

$$\langle I_1^{(2)}(x)\rangle \sim D_0/k\gamma^{1/2} \qquad (4.29)$$

for sufficiently large values of x, but $k\gamma x \ll 1$. Note that crossed terms occur, which are produced by fields from different regions. However, there are no powers of γ in denominators of corresponding asymptotic expressions.

A combination of these terms yields the condition $\gamma^{3/4} \ll k\gamma x \ll 1$, defining the leading order of $\langle I_2^{(2)}(x)\rangle$ in the average intensity.

For the three-dimensional case the results are analogous to those for the two-dimensional case:

$$\langle I_1^{(3)}(\rho)\rangle \sim D_0 k/\gamma(k\rho)^4, \qquad \langle I_2^{(3)}(\rho)\rangle \sim D_0/\rho\gamma^{1/2}, \qquad \langle I_3^{(3)}(\rho)\rangle \sim 1/(k\rho)^2,$$

when $\rho \Rightarrow \infty$, $k\gamma\rho \ll 1$. Under the condition $k\gamma\rho \gg \gamma^{5/6}$, the function $\langle I_2^{(3)}(\rho)\rangle$ contributes most to the average intensity of a point source at $z = z_0$; it corresponds to a cylindrical wave, and is specified by the absorption γ.

Analogous results are true for a point source on the reflective boundary. If a source is located on the boundary of a randomly inhomogeneous half-space, the average intensity is

$$\langle I(x, z_0; z_0)\rangle = I_{\text{free}}(x, z_0; z_0)\left(1 + 2D_0 x \int_0^1 \frac{\mathrm{d}s}{[D_0 x + 2s(1 - s^2)^{1/2}]^2}\right).$$

We are interested in the limit asymptotic cases: $D_0 x \ll 1$, but $kx \gg 1$ and $D_0 x \gg 1$. In the first case

$$\langle I(x, z_0; z_0)\rangle = 2I_{\text{free}}(x, z_0; z_0).$$

This result is analogous to that obtained for the parabolic equation, and shows that scattering at wide angles has no influence on the statistics. When $D_0 x \gg 1$, the average intensity

$$\langle I(x, z_0; z_0)\rangle = I_{\text{free}}(x, z_0; z_0)(1 + 2/D_0 x),$$

and scattering at wide angles is essential in the formation of statistics. It appears to be an additional attenuating factor for the intensity in empty space. The same result occurs for the three-dimensional case.

We saw earlier that the principal feature of the one-dimensional problem of plane waves in layered random media is that a small but finite absorption has to be taken into account (a parameter γ). The statistics of the wavefield are caused by an interference of multiple reflected waves in the medium giving rise to dependence of the average intensity $\langle I\rangle$ on γ, for example, $\langle I\rangle \sim 1/\gamma$ for a point source in unbounded space. In multidimensional layered problems, the outcome for spatial diffraction is like the absorption effect, which permits the calculation of statistical characteristics by means of analytic prolongation with respect to γ. I had hoped that the diffraction effects would remove the dependence of the wavefield statistics on γ, and the finite limit at $\gamma \Rightarrow 0$ would exist in multidimensional problems. However, this did not occur. The diffraction effects diminish the power of γ (e.g. $\langle I\rangle \sim 1/\gamma^{1/2}$ for a point source in unbounded space), and the absorption is important for multidimensional problems of waves in layered random media.

This raises doubt concerning the results of work by Freilikher and Gredescul [1990] and Gredescul and Freilikher [1988, 1990]) based on an analogy to quantum-mechanics problems that does not consider the effect of absorption.

Acknowledgements

I am grateful to A. G. Bugrov, V. I. Goland, O. E. Gulin, M. A. Guzev, K. V. Koshel', A. I. Saichev, B. M. Shevtsov, E. V. Yaroschuk and I. O. Yaroschuk (Russia) for collaboration in the field of wave propagation in layered random media, to L. A. Pastur (Ukraine), V. D. Freilikher (Israel) and W. Kohler and G. Papanicolaou (USA) for helpful discussions of the problems of radiative localization in random media, and to B. White (USA) for helpful discussion of the problem of caustics formation in random media.

I would like to thank especially professor E. Wolf and his colleagues during the process of publishing this volume of Progress in Optics for the titanic work done to edit the submitted manuscript.

Appendix
A. Statistical Description of Dynamical Systems

A.1. THE FOKKER–PLANCK EQUATION AND ITS BOUNDARY CONDITIONS

Let the vector function $\xi(t) = \{\xi_1(t), \ldots, \xi_n(t)\}$ satisfy the system of dynamical equations

$$\frac{d}{dt}\xi(t) = v(\xi, t) + f(\xi, t), \qquad \xi(0) = \xi_0, \tag{A.1}$$

where $v(\xi, t)$ denote deterministic functions, and $f(x, t)$ denote a Gaussian random field with correlation tensor

$$\langle f(x, t) \rangle = 0, \qquad B_{ij}(x, t; x', t') = \langle f_i(x, t) f_j(x', t') \rangle.$$

Here the parentheses indicate averaging over the realization ensemble of the random field f. The approximation of the delta-correlated random field f leads to the introduction of the effective correlation tensor

$$B_{ij}^{\text{eff}}(x, t; x', t') = 2\delta(t - t') F_{ij}(x, x'; t) \tag{A.2}$$

instead of the tensor B_{ij}, and the quantity F_{ij} is determined by

$$F_{ij}(x, x'; t) = \frac{1}{2} \int_{-\infty}^{\infty} dt' \, B_{ij}(x, t; x', t'). \tag{A.3}$$

In this case the probability density of the solution $\xi(t)$ of the system (A.1)

$$P_t(x) = \langle \delta(x - \xi(t)) \rangle$$

is described by the Fokker–Planck equation (see, e.g. Klyatskin [1975, 1980a, 1985])

$$\frac{\partial}{\partial t} P_t(x) + \frac{\partial}{\partial x_k} \{[v_k(x, t) + A_k(x, t)] P_t(x)\} - \frac{\partial^2}{\partial x_k \partial x_l} [F_{kl}(x, x; t) P_t(x)] = 0,$$

$$P_0(x) = \delta(x - \xi_0), \quad A_k(x, t) = \frac{\partial}{\partial x'_l} F_{kl}(x, x')|_{x'=x}. \tag{A.4}$$

In this approximation the solution of the system equations (A.1) is a Markov random process, and its transition probability density

$$p(x, t \mid x_0, t_0) = \langle \delta(x - \xi(t)) \mid \xi(t_0) = x_0) \rangle$$

is also described by eq. (A.4), with initial condition

$$p(x, t \mid x_0, t_0)|_{t \Rightarrow t_0} = \delta(x - x_0).$$

The Fokker–Planck equation (A.4) is a partial differential equation, for which further analysis greatly depends on the formulation of boundary conditions with respect to x, which specify the type of problems under study.

The solution of the Fokker–Planck equation (A.4) specifies not only a complete single-point (in time), statistical description of the dynamical system (A.1), but also gives some information about time behavior for individual specific solution realizations for the system (A.1) (Klyatskin and Saichev [1992]). In fact, we can discuss an arbitrary random process $y(t)$ with an integral distribution function

$$F(y; t) = \int_{-\infty}^{y} d\tilde{y} \, P_t(\tilde{y}) = \langle \theta(y - y(t)) \rangle, \tag{A.5}$$

where $\theta(z)$ is a unity function equal to zero for $z < 0$, and to unity for $z > 0$. Let us name the deterministic function $Z(t; p)$ as the isoprobable curve of the process $y(t)$. Its value for every given moment of time t is defined from the equation

$$F(Z(t; p); t) = p. \tag{A.6}$$

By integrating this equality over an arbitrary interval, we obtain

$$\int_{t_1}^{t_2} dt\, F(Z(t; p); t) = p(t_2 - t_1). \tag{A.7}$$

On the other hand, it follows from the definition (A.5) of the integral distribution function that the integral in the left-hand side of this equality is

$$\int_{t_1}^{t_2} dt\, F(Z(t; p); t) = \langle T(t_1, t_2) \rangle, \tag{A.8}$$

where $T(t_1, t_2) = \sum_1^N \Delta t_k$ is the total time that the realization of the process $y(t)$ inside the interval (t_1, t_2) is under the isoprobable curve. By comparing equalities (A.7) and (A.8), we find that

$$\langle T(t_1, t_2) \rangle = p(t_2 - t_1)$$

is the mean time that the process $y(t)$ inside the interval (t_1, t_2) is under the isoprobable curve $Z(t; p)$. It is proportional to the duration of this interval $t_2 - t_1$. The proportionality factor p is equal to the fraction of the time when the inequality $y(t) < Z(t; p)$ is true. Thus, if p is close enough to unity, the plots of the realization of the process $y(t)$ are almost always below the isoprobable curve inside an interval (t_1, t_2). If, for example, $p = \frac{1}{2}$, the realization of the process $y(t)$ twines around the isoprobable curve, being on average half of the time above and half of the time below this curve. Thus it is natural to describe the isoprobable curve $Z(t; \frac{1}{2})$ as a typical realization of the process $y(t)$, although the plot $Z(t; \frac{1}{2})$ can differ significantly from the plot of any realization of the process $y(t)$.

Interpretation of an isoprobable curve as a typical realization, which characterizes the dynamical behavior of realizations of the corresponding random processes, is confirmed by a limit property of isoprobable curves. As we know, a quantitative measure of a random process $y(t)$ is its variance $\sigma^2(t) = \langle y^2(t) \rangle - \langle y(t) \rangle^2$. When $\sigma \Rightarrow 0$, the process $y(t)$ tends to some deterministic function $y_0(t) = \langle y(t) \rangle$. It follows from the definition of isoprobable curves that in the limit $\sigma \Rightarrow 0$ and at any $p < 1$

$$\lim_{\sigma \Rightarrow 0} Z(t; p) = y_0(t);$$

that is, the isoprobable curves tend to the deterministic function.

If fluctuations of the dynamic system parameters are small, the perturbation theory can be formulated with respect to these parameters. The scheme of this construction is as follows.

Let us write the Fokker–Planck equation in the following form:

$$\frac{\partial}{\partial t} P_t(x) + A(x, t; \varepsilon) P_t(x) + B_i(x, t; \varepsilon) \frac{\partial}{\partial x_i} P_t(x)$$

$$- \varepsilon^2 D_{ij}(x, t; \varepsilon) \frac{\partial^2}{\partial x_i \partial x_j} P_t(x) = 0, \qquad P_0(x) = p(x), \qquad \text{(A.9)}$$

where the parameter ε^2 characterizes the intensity of fluctuations of the dynamical system parameters. We express the solution of eq. (A.9) in the form

$$P_t(x) = C(\varepsilon) \exp\left\{-\frac{1}{\varepsilon^2} \phi(t, x; \varepsilon)\right\}. \qquad \text{(A.10)}$$

Then we obtain the equation for the function $\phi(x, t; \varepsilon)$

$$\frac{\partial}{\partial t} \phi(x, t; \varepsilon) - \varepsilon^2 A(x, t; \varepsilon) + B_i(x, t; \varepsilon) \frac{\partial}{\partial x_i} \phi(x, t; \varepsilon) - \varepsilon^2 D_{ij}(x, t; \varepsilon) \frac{\partial^2}{\partial x_i \partial x_j}$$

$$\times \phi(x, t; \varepsilon) + D_{ij}(x, t; \varepsilon) \frac{\partial}{\partial x_i} \phi(x, t; \varepsilon) \frac{\partial}{\partial x_j} \phi(x, t; \varepsilon) = 0. \qquad \text{(A.9')}$$

Its solution can be expanded into a series of ε^2

$$\phi(x, t; \varepsilon) = \phi_0(x, t) + \varepsilon^2 \phi_1(x, t) + \ldots.$$

By substituting this expansion into eq. (A.9'), expanding coefficients in eq. (A.9') into a series of ε^2, and equating the corresponding factor at the powers to zero, we obtain the successive perturbation theory. Thus for the function $\phi_0(x, t)$ we have the equation

$$\frac{\partial}{\partial t} \phi_0(x, t) + B_i(x, t) \frac{\partial}{\partial x_i} \phi_0(x, t) + D_{ij}(x, t) \frac{\partial}{\partial x_i} \phi_0(x, t) \frac{\partial}{\partial x_j} \phi_0(x, t) = 0,$$

which is a first-order partial differential equation, and can be solved, for example, by the characteristic method. The first term of the series describes the primary singular feature of the Fokker–Planck equation solution with respect to the parameter ε. The next term of the expansion, $\phi_1(x, t)$, describes the factor at the exponent, and the constant $C(\varepsilon)$ in eq. (A.10) is determined by the behavior of the solution (A.9) as $t \Rightarrow 0$ and by the initial condition for this equation.

The constructed perturbation theory clearly describes the dynamics of the nonlinear system in the finite time interval (no matter how long), and does not allow the limit transition at $t \Rightarrow \infty$. For the analysis of this limit case eq. (A.9) is usually transformed to the form containing the self-adjoint operator with respect to space variables, which has a discrete spectrum.

Equation (A.4) is usually called a forward Fokker–Planck equation. It is also easy to obtain a backward Fokker–Planck equation, which describes the evolution of the transition probability density with respect to initial condition t_0, x_0 (see, e.g. Gardiner [1985]):

$$\frac{\partial}{\partial t_0} p(x, t | x_0, t_0) = -[v_i(x_0, t_0) + F_i(x_0, x_0; t_0)] \frac{\partial}{\partial x_{0i}} p(x, t | x_0, t_0)$$

$$- F_{ij}(x_0, x_0; t_0) \frac{\partial^2}{\partial x_{0i} \partial x_{0j}} p(x, t | x_0, t_0). \quad (A.4')$$

The forward and backward Fokker–Planck equations are equivalent. The former is more suitable when studying the time evolution of statistical characteristics of the (A.1) problem solution. The latter is more suitable for the study of statistical characteristics associated with the time of the presence of a random process $\xi(t)$ in any space area, with the time of reaching the area's boundary, and so on. In fact, the probability of the presence of the random process $\xi(t)$ in the space area V is described by the integral

$$G(t; x_0, t_0) = \int_V dx\, p(x, t | x_0, t_0), \quad (A.11)$$

which according to eq. (A.4) will be described by the closed equation

$$\frac{\partial}{\partial t_0} G(t; x_0, t_0) = -[v_i(x_0, t_0) + F_i(x_0, x_0; t_0)] \frac{\partial}{\partial x_{0i}} G(t; x_0, t_0)$$

$$- F_{ij}(x_0, x_0; t_0) \frac{\partial^2}{\partial x_{0i} \partial x_{0j}} G(t; x_0, t_0), \quad (A.12)$$

$$G(t; x_0, t_0) = \begin{cases} 1 & (x_0 \in V) \\ 0 & (x_0 \in V). \end{cases}$$

It is necessary to formulate the additional boundary condition for eq. (A.12), which will be dependent on the character of the space area V and its boundaries.

As an example, let us consider the stochastic equation derived from work by Kulkarny and White [1982]

$$\frac{d}{dt} x(t) = -\lambda x^2(t) + f(t), \quad x(0) = x_0, \quad \lambda > 0, \quad (A.13)$$

where $f(t)$ is the Gaussian delta-correlated process. When no fluctuations occur, the solution of eq. (A.13) has the form

$$x(t) = \frac{1}{\lambda} \frac{1}{t + 1/\lambda x_0}.$$

When $x_0 > 0$, the problem solution tends to zero monotonically. In the case of $x_0 < 0$, the solution becomes $-\infty$ during the finite time $t_0 = -1/\lambda x_0$; that is, it has an explosive character. In this case the influence of a random force $f(t)$ on the system's dynamics is not important. The influence of a random force is important only in the case of positive x_0. The solution decreases monotonically with time. Having reached sufficiently small values of $x(t)$, it will be thrown into the negative area by the action of the force $f(t)$. Thus, in the statistical case the solution of problem (A.13) will have an explosive character at any values of x_0, and it will become $-\infty$ during the finite time t_0. Let us estimate the mean value of this time (more detailed analysis occurs in problems regarding the origin of caustics).

The solution of the stochastic problem is described by forward and backward Fokker–Planck equations $(t - t_0 \Rightarrow t)$

$$\frac{\partial}{\partial t} p(x, t | x_0) = \lambda \frac{\partial}{\partial x} x^2 p(x, t | x_0) + \sigma^2 \frac{\partial^2}{\partial x^2} p(x, t | x_0),$$

$$\frac{\partial}{\partial t} p(x, t | x_0) = -\lambda x_0^2 \frac{\partial}{\partial x_0} p(x, t | x_0) + \sigma^2 \frac{\partial^2}{\partial x_0^2} p(x, t | x_0), \quad (A.14)$$

$$p(x, 0 | x_0) = \delta(x - x_0).$$

In this case the mean time $\langle T(x_0) \rangle$ of the system transition from the state x_0 into the state $-\infty$ is described by the equation (see, e.g. Gardiner [1985])

$$-1 = \sigma^2 \frac{\partial^2}{\partial x_0^2} \langle T(x_0) \rangle - \lambda x_0^2 \frac{\partial}{\partial x_0} \langle T(x_0) \rangle, \quad (A.15)$$

with boundary conditions $\langle T(x_0) \rangle \Rightarrow 0$ as $x_0 \Rightarrow -\infty$ and $\langle T(x_0) \rangle$ is limited as $x_0 \Rightarrow \infty$.

By renormalizing the variables in eq. (A.15) $x_0 = (2\sigma^2/\lambda)^{1/3} y$, $\langle T(x_0) \rangle = (2\lambda^2 \sigma^2)^{-1/3} \tau(y)$, the equation can be rewritten in the form

$$-1 = \frac{1}{2} \frac{d^2}{dy^2} \tau(y) - y^2 \frac{d}{dy} \tau(y).$$

This equation can be easily integrated, and we obtain

$$\tau(y) = 2 \int_{-\infty}^{y} d\xi \, e^{2\xi^3/3} \int_{\xi}^{\infty} d\eta \, e^{-2\eta^3/3}. \tag{A.16}$$

Asymptotic relations follow from the quadrature (A.16):

$$\tau(y) = \tau(\infty) - \frac{1}{y} + \frac{1}{4y^4} + O(y^{-7}), \quad y \Rightarrow \infty,$$

$$\tau(y) = -\frac{1}{y} + \frac{1}{4y^4} + O(y^{-7}), \quad y \Rightarrow -\infty,$$

where $\tau(\infty) = 2^{2/3} 3^{-5/6} \pi^{1/2} \Gamma(1/2) \approx 6.27$. Note also that the quantity $\tau(0)$ is defined by the value of $\tau(0) = \frac{2}{3}\tau(\infty) \approx 4.18$, and consequently the mean time of the transition from state $x_0 = \infty$ to state $x_0 = 0$ will be equal to $\tau = \tau(\infty)/3 \approx 2.09$.

The example under discussion is similar to the problem of caustics formation in random media (Kulkarny and White [1982], White [1983], Zwillinger and White [1985], Klyatskin [1993]). Let us examine this case more carefully, since it is typical for a comprehensive analysis of the corresponding boundary-value Fokker–Planck equation.

A.2. CAUSTICS IN RANDOM MEDIA

A.2.1. General remarks

In the parabolic equation approximation the wavefield is described by eq. (2.22)

$$2ik \frac{\partial}{\partial x} u(x, \boldsymbol{\rho}) + \Delta_\perp u(x, \boldsymbol{\rho}) + k^2 \varepsilon(x, \boldsymbol{\rho}) u(x, \boldsymbol{\rho}) = 0$$

$$\left(\boldsymbol{\rho} = \{y, z\}, \quad \Delta_\perp = \frac{\partial^2}{\partial y^2} + \frac{\partial^2}{\partial z^2} \right), \tag{A.17}$$

where $\varepsilon(x, \boldsymbol{\rho})$ is the fluctuating part of the dielectric permittivity or of the refractive index.

Let us introduce the wavefield in the form

$$u(x, \boldsymbol{\rho}) = A(x, \boldsymbol{\rho}) e^{iS(x, \boldsymbol{\rho})},$$

where A is the wave amplitude and S denotes the fluctuation of the wave phase with respect to the wave phase of the incident wave (kx).

APPENDIX A. STATISTICAL DESCRIPTION OF DYNAMICAL SYSTEMS

In the geometric optics approximation, in which $k \Rightarrow \infty$, the analysis of amplitude and phase fluctuations is greatly simplified. In this case the equation for phase assumes the form ($\nabla_\perp = \partial/\partial \boldsymbol{\rho}$)

$$2k \frac{\partial}{\partial x} S(x, \boldsymbol{\rho}) + (\nabla_\perp S)^2 = k^2 \varepsilon(x, \boldsymbol{\rho}), \tag{A.18}$$

and the wavefield intensity $I(x, \boldsymbol{\rho}) = A^2(x, \boldsymbol{\rho})$ is described by the equation

$$k \frac{\partial}{\partial x} I(x, \boldsymbol{\rho}) + \nabla_\perp (I \nabla_\perp S) = 0. \tag{A.19}$$

The quantity ∇S is now described by the closed equation that follows from eq. (A.18)

$$\frac{\partial}{\partial x} \nabla_\perp S + \frac{1}{k} (\nabla_\perp S \nabla_\perp) \nabla_\perp S = \frac{k}{2} \nabla_\perp \varepsilon(x, \boldsymbol{\rho}). \tag{A.20}$$

This is a partial differential equation for the function $\boldsymbol{p}(x, \boldsymbol{\rho}) = (1/k) \nabla_\perp S(x, \boldsymbol{\rho})$. If we introduce a characteristic curve (rays) $\boldsymbol{R}(x)$ according to the equation

$$\frac{\mathrm{d}}{\mathrm{d}x} \boldsymbol{R}(x) = \boldsymbol{p}(x), \tag{A.21}$$

eq. (A.20) along these characteristics will be written in the form

$$\frac{\mathrm{d}}{\mathrm{d}x} \boldsymbol{p}(x) = \frac{1}{2} \frac{\partial}{\partial \boldsymbol{R}} \varepsilon(x, \boldsymbol{R}). \tag{A.21'}$$

The system of eqs. (A.21) and (A.21') is closed. It must be solved with the initial condition at $x = 0$ determining the parameterization of characteristic curves.

The equation for the intensity (A.19) can be rewritten in the form of the equation along the characteristics

$$\frac{\mathrm{d}}{\mathrm{d}x} I(x) = -\frac{1}{k} I(x) \Delta_R S(x, \boldsymbol{R}), \tag{A.22}$$

which contains a new unknown function $\Delta_R S(x, \boldsymbol{R})$. To find the equation for it, let us introduce a new function

$$u_{ij}(x, \boldsymbol{\rho}) = \frac{1}{k} \frac{\partial^2}{\partial \rho_i \partial \rho_j} S(x, \boldsymbol{\rho}),$$

which describes a curvature of the phase front $S(x, \boldsymbol{\rho}) = \text{const}$. By differentiating eq. (A.20) with respect to $\boldsymbol{\rho}$, we obtain the equation that, along the characteristics, takes the form of the equation linked to the system (A.21) and (A.21′) by means of the argument of the function $\varepsilon(x, \mathbf{R})$,

$$\frac{d}{dx} u_{ij}(x) + u_{ik}(x) u_{kj}(x) = \frac{1}{2} \frac{\partial^2}{\partial R_i \partial R_j} \varepsilon(x, \mathbf{R}). \tag{A.23}$$

In addition, eq. (A.22) takes the form

$$\frac{d}{dx} I(x) = -I(x) u_{ii}(x). \tag{A.23′}$$

Equations (A.21) to (A.23′) become much simpler for a two-dimensional case ($R = y$), and have the form

$$\frac{d}{dx} y(x) = p(x), \qquad \frac{d}{dx} p(x) = \frac{1}{2} \frac{\partial}{\partial y} \varepsilon(x, y),$$

$$\frac{d}{dx} I(x) = -I(x) u(x), \qquad \frac{d}{dx} u(x) = -u^2(x) + \frac{1}{2} \frac{\partial^2}{\partial y^2} \varepsilon(x, y), \tag{A.24}$$

where the quantity $u(x, y) = (1/k) \partial^2/\partial y^2 \, S(x, y)$ characterizes the curvature of the phase curve $S(x, y) = \text{const.}$ in the (x, y) plane.

Note that if there are no medium inhomogeneities ($\varepsilon = 0$), rays appear to be straight lines, and the integration of eqs. (A.24) for quantities $u(x)$ and $I(x)$ gives

$$u(x) = \frac{u_0}{1 + u_0 x}, \qquad I(x) = \frac{I_0}{u_0} u(x). \tag{A.25}$$

If the initial condition is $u_0 < 0$, the quantities $u(x_0) = -\infty$, $I(x_0) = \infty$ in the point $x_0 = -1/u_0$; this means that the solution has an explosive character. The presence of inhomogeneities $\varepsilon(x, y)$ indicates the existence of such singular points for any sign of u_0, which means that in the statistical problem singular points appear at finite distances. They are stipulated by a random focusing of the wavefield based on the fact that the curvature of phase front and the intensity of the wavefield become infinite. This corresponds to random focusing of the wavefield in the randomly inhomogeneous medium, that is, to the appearance of caustics.

A.2.2. Statistical description

In a two-dimensional case a curvature of the phase curve in the (x, y) plane is described by eq. (A.24) for the function $u(x, y)$ along the ray

$$\frac{d}{dx} u(x) + u^2(x) = f(x), \qquad u(0) = u_0, \tag{A.26}$$

where $f(x) = \frac{1}{2}\partial^2/\partial y^2\, \varepsilon(x, y(x))$, and the transversal displacement of the ray $y(x)$ is described by the system of eqs. (A.21) and (A.21′). For the homogeneous isotropic delta-correlated fluctuation $\varepsilon(x, y)$ ($\langle \varepsilon(x, y)\varepsilon(x', y') \rangle = \delta(x - x')A(y - y')$), the probability density of the stochastic equation (A.26) is described by the Fokker–Planck equation, which can be written in the form

$$\frac{\partial}{\partial x} P_x(u) = \frac{\partial}{\partial u} u^2 P_x(u) + \frac{D}{2} \frac{\partial^2}{\partial u^2} P_x(u), \qquad P_0(u) = \delta(u - u_0), \qquad (\text{A.27})$$

where the diffusion coefficient D is defined for this case by the expression

$$D = \frac{1}{4} \frac{\partial^4}{\partial y^4} A(0) = \pi \int_0^\infty d\kappa\, \kappa^4\, \Psi_\varepsilon(0, \kappa).$$

Here, $\Psi_\varepsilon(\boldsymbol{\kappa})$ is the two-dimensional spectrum of the random field $\varepsilon(x, y)$.

Equation (A.27) has yet been considered. It was shown that the random process $u(x)$ becomes infinite at the finite distance $x(u_0)$, which is defined by the initial condition u_0. The mean value $\langle x(u_0) \rangle$ is determined, for this case, by the equality

$$\langle x(u_0) \rangle = \frac{2}{D} \int_{-\infty}^{u_0} d\xi\, e^{2\xi^3/3D} \int_\xi^\infty d\eta\, e^{-2\eta^3/3D}, \qquad (\text{A.28})$$

and consequently,

$$D^{1/3}\langle x(\infty) \rangle = 6.27, \qquad D^{1/3}\langle x(0) \rangle = \tfrac{2}{3} D^{1/3}\langle x(\infty) \rangle = 4.18.$$

The quantity $\langle x(0) \rangle$ describes the mean distance of the appearance of the focus for the initial plane wave, and the quantity $\langle x(\infty) \rangle$ describes the mean distance between two subsequent focuses. Taking into account the variance of the amplitude level in the first approximation of the method of smooth perturbation $\sigma_0^2(x) \sim Dx^3$ (see, e.g. Tatarskii [1971], Klyatskin [1980a, 1985], Rytov, Kravtsov and Tatarskii [1987–1989]), we see that the random focusing takes place in the region of strong-intensity fluctuation when $\sigma_0^2 \geqslant 1$.

Further analysis of eq. (A.27) greatly depends on the formulation of a boundary condition with respect to u, which specifies the type of problems under study. Thus, if we consider the function $u(x)$ to be discontinuous and defined for all values of x, while its conversion into $-\infty$ for $x \Rightarrow x_0 - 0$ is immediately accompanied by its origin in the point $x_0 + 0$ with the value equal to $+\infty$, then the boundary condition for eq. (A.27) is the condition

$$J(x, u)|_{u \Rightarrow \infty} - J(x, u)|_{u \to -\infty}, \qquad (\text{A.29})$$

where

$$J(x, u) = u^2 P_x(u) + \frac{D}{2} \frac{\partial}{\partial u} P_x(u) \qquad (A.30)$$

is the probability flux density. This type of problem was examined in § 3.

Another type of boundary condition arises if we suppose that the curve $u(x)$ is broken, having approached the point x_0, where it becomes $-\infty$. This corresponds to the condition that the probability flux must be equal to zero when $u \Rightarrow \infty$, that is, to conditions

$$J(x, u) \Rightarrow 0 \quad \text{as } u \Rightarrow \infty; \qquad P_x(u) \Rightarrow 0 \quad \text{as } u \Rightarrow -\infty. \qquad (A.31)$$

In this case the quantity $C(x) = \int_{-\infty}^{\infty} du\, P_x(u) \neq 1$ determines the probability of the random function $u(x)$ being finite along the whole axis $(-\infty, \infty)$, which is the probability of having no focus at the distance $x - C(x) = P(x_0 > x)$. Hence, the probability of the appearance of the focus at the distance x is determined by the equality

$$P(x > x_0) = 1 - \int_{-\infty}^{\infty} du\, P_x(u),$$

and its probability density is

$$p(x) = \frac{\partial}{\partial x} P(x > x_0) = -\frac{\partial}{\partial x} \int_{-\infty}^{\infty} du\, P_x(u) = \lim_{u \Rightarrow -\infty} J(x, u). \qquad (A.32)$$

To derive the asymptotic dependence of the probability density $p(x)$ on the parameter $D \Rightarrow 0$, we shall use the standard procedure for analysis of the parabolic equation (A.27) with a small parameter for the highest derivative.

Let us write the solution of eq. (A.27) in the form

$$P_x(u) = C(D) \exp\left\{-\frac{1}{D} A(x, u) - B(x, u)\right\}. \qquad (A.33)$$

By substituting eq. (A.33) in eq. (A.27) and marking out terms of the order D^{-1} and D^0, we obtain partial differential equations for functions $A(x, u)$ and $B(x, u)$. The constant $C(D)$ is defined from the condition that when $x \Rightarrow 0$, the probability distribution for the plane incident wave must have the form

$$P_x(u) = \frac{1}{(2\pi Dx)^{1/2}} \exp\left\{-\frac{u^2}{2Dx}\right\}; \qquad (A.33')$$

that is,

$$C(D) \sim D^{-1/2}.$$

By substituting eq. (A.33) in eq. (A.32), we obtain the expression for the probability density of the focus appearance in the random medium according to eq. (A.30):

$$p(x) = \lim_{u \Rightarrow -\infty} P_x(u) \left[u^2 - \frac{1}{2} \frac{\partial}{\partial u} A(x, u) \right]. \tag{A.34}$$

Note that the expression of $P_x(u)$ in the form of eq. (A.34) immediately permits showing the structural dependence of $p(x)$ on x, based on consideration of the dimension (Klyatskin [1993]). In fact, it follows from eq. (A.27) that quantities u, D and $P_x(u)$ have the following dimensions:

$$[u] = x^{-1}, \quad [D] = x^{-3}, \quad [P_x(u)] = x.$$

Consequently, according to eqs. (A.33) and (A.34) the function $p(x)$ has a structure

$$p(x) = C_1 D^{-1/2} x^{-5/2} \exp(-C_2/Dx^3),$$

and the problem is to calculate the positive constants C_1 and C_2. These constants were calculated by Kulkarny and White [1982], and

$$p(x) = 3\alpha^2 (2\pi D)^{-1/2} x^{-5/2} \exp\{-\alpha^4/6Dx^3\}, \tag{A.35}$$

where $\alpha = \int_0^\infty d\xi \, (1 + \xi^4)^{-1/2} = K(\tfrac{1}{2}) = 1.85$ (K is a complete elliptic integral).

The condition of the applicability of eq. (A.35) is the condition $Dx^3 \ll 1$. However, as was shown by Kulkarny and White [1982], by the numerical modelling, the expression (A.35) accurately describes the probability distribution of the appearance of a random focus when $Dx^3 \approx 1$.

Note that for a three-dimensional problem we obtain the law for probability density of the appearance of a caustic on the basis of dimension consideration (Klyatskin [1993]),

$$p(x) = \alpha D^{-1} x^{-4} e^{-\beta/Dx^3},$$

where α and β are numerical constants. This law with $\alpha = 1.74$ and $\beta = 0.66$ was obtained by White [1983].

General concepts of statistical descriptions for dynamical systems, which were illustrated in detail by the asymptotic problem of the origin of a caustic in media with small random inhomogeneities, are listed at the end of this appendix. Simpler stochastic equations enable a more complete analysis.

We shall consider more closely the simplest stochastic equations, which specify Wiener and lognormal random processes, since they are especially important in physics.

B. Dynamical Properties of the Wiener and Lognormal Random Processes

B.1. GENERAL REMARKS

The Wiener random process is determined as a solution of the stochastic equation

$$\frac{d}{dt} w(t) = z(t), \qquad w(0) = 0, \tag{B.1}$$

where $z(t)$ is the Gaussian process, delta correlated on time with parameters

$$\langle z(t) \rangle = 0, \qquad \langle z(t)z(t') \rangle = 2\sigma^2 \delta(t - t').$$

The solution of eq. (B.1),

$$w(t) = \int_0^t d\tau \, z(\tau),$$

is a continuous Gaussian nonstationary random process, described by parameters

$$\langle w(t) \rangle = 0, \qquad \langle w(t)w(t') \rangle = 2\sigma^2 \min(t, t'). \tag{B.2}$$

Thus its characteristic functional has a structure

$$\Phi_t[v(\tau)] = \left\langle \exp\left\{ i \int_0^t d\tau \, w(\tau)v(\tau) \right\} \right\rangle$$

$$= \exp\left\{ -\sigma^2 \int_0^t \int_0^t d\tau_1 \, d\tau_2 \min\{\tau_1, \tau_2\} v(\tau_1) v(\tau_2) \right\}. \tag{B.3}$$

Let us study the increment of the process $w(t)$ on the interval (t_1, t_2)

$$\Delta w(t_1, t_2) = w(t_2) - w(t_1) = \int_{t_1}^{t_2} d\tau \, z(\tau).$$

Similar to the process $w(t)$ itself, the increments have Guassian statistics. Their statistical properties are defined by two first moments:

$$\langle \Delta w(t_1, t_2) \rangle = 0, \qquad \langle [\Delta w(t_1, t_2)]^2 \rangle = 2\sigma^2 |t_2 - t_1|.$$

The Wiener process $w(t)$ is the Gaussian continuous random process with independent increments. This means that if intervals (t_1, t_2) and (t_3, t_4) do not overlap, the increments of the process $w(t)$ on these intervals are statistically independent.

If we observe the characteristic functional of the process $\Delta w(t_0, t_0 + t) = \int_{t_0}^{t_0+t} d\tau\, z(\tau)$, it is easy to see that it coincides with the characteristic functional of the process $w(t)$. Thus the realizations of processes $w(t)$ and $\Delta w(t_0, t_0 + t)$ are *statistically equivalent* at any t. So if we know the process realizations only, we cannot say to what process they belong. The processes $w(t)$ and $w(-t)$ are also statistically equivalent; that is, the Wiener process is *reversible in time* in the sense of the above definition.

One more unusual property, the fractal property, applies to the Wiener process. According to this property the realizations of the Wiener process $w(at)$ compressed in time (with $a > 1$) are statistically equivalent to realizations of $a^{1/2}w(t)$ stretched along the vertical axis. The fractal property of the Wiener process can also be treated as the statistical equivalence of the realizations of $w(t)$ and $w(at)/a^{1/2}$, because of the coincidence of their characteristic functionals according to eq. (B.3).

If we discuss a more general process that depends on the parameter α

$$w(t; \alpha) = -\alpha t + w(t),$$

it satisfies the stochastic equation

$$\frac{d}{dt} w(t; \alpha) = -\alpha + z(t), \qquad w(0; \alpha) = 0. \tag{B.1'}$$

The process $w(t; \alpha)$ is the Markov process, and its probability density

$$P_t(w) = \langle \delta(w(t; \alpha) - w) \rangle$$

satisfies the Fokker–Planck equation

$$\frac{\partial}{\partial t} P_t(w) = \alpha \frac{\partial}{\partial w} P_t(w) + \sigma^2 \frac{\partial^2}{\partial w^2} P_t(w), \qquad P_0(w) = \delta(w). \tag{B.4}$$

Its solution has the form of the Gaussian distribution

$$P_t(w) = \frac{1}{2(\pi\sigma^2 t)^{1/2}} \exp\left[-\frac{(w + \alpha t)^2}{4\sigma^2 t} \right]. \tag{B.5}$$

Furthermore, we shall assume the time t to be a dimensionless quantity (i.e. $\sigma^2 = 1$).

The corresponding integral distribution function equal to the probability of $w(t; \alpha) < w$ is

$$F(w; t, \alpha) = P(w(t; \alpha) < w) = \int_{-\infty}^{w} dw\, P_t(w) = \Phi\left(\frac{w}{(2t)^{1/2}} + \alpha \left(\frac{w}{2}\right)^{1/2} \right), \tag{B.6}$$

where the function $\Phi(z)$ is a probability integral

$$\Phi(z) = \frac{1}{(2\pi)^{1/2}} \int_{-\infty}^{z} dy \exp\left(-\frac{y^2}{2}\right). \tag{B.6'}$$

In addition to the initial condition, let us complete eq. (B.5) with a boundary condition

$$P_t(w = h) = 0 \quad (t > 0), \tag{B.7}$$

which contradicts the realizations of the process $w(t; \alpha)$ at the moment when they reach the boundary h. The solution of the boundary problem (B.4) and (B.7) will be denoted as $P_t(w; h)$. When $w < h$, it describes the probability distribution of values of those realizations of the process $w(t; \alpha)$ that have *survived* by the time t; that is, during the entire time they have never reached the boundary h. Correspondingly, the probability density $P_t(w; h)$ is normalized by the probability of $t > t^*$, where t^* is the time when the process $w(t; \alpha)$ reaches the boundary h for the first time, rather than by unity,

$$\int_{-\infty}^{h} dw\, P_t(w; h) = P(t < t^*). \tag{B.8}$$

Let us introduce an integral distribution function and probability density of the random time moment when the boundary is attained for the first time:

$$\begin{aligned} F(t; \alpha, h) &= 1 - P(t < t^*) = 1 - \int_{-\infty}^{h} dw\, P_t(w; h), \\ P(t; \alpha, h) &= \frac{\partial}{\partial t} F = -\frac{\partial}{\partial w} P_t(w; h)|_{w=h}. \end{aligned} \tag{B.9}$$

On average if $\alpha > 0$, the process $w(t; \alpha)$ moves away from the boundary h with the growth of t; when $t \Rightarrow \infty$, the probability $P(t < t^*)$ (B.8) tends to the probability that the process $w(t; \alpha)$ never reaches the boundary h. In other words,

$$\lim_{t \Rightarrow \infty} \int_{-\infty}^{h} dw\, P_t(w; h) = P(w_m < h) \tag{B.10}$$

is equal to the *probability of the absolute maximum* of the process $w(t; \alpha)$,

$$w_m(\alpha) = \max_{t \in (0, \infty)} w(t; \alpha) \tag{B.11}$$

to be less than h. Thus, it follows from eqs. (B.10) and (B.8) that the integral distribution function of the absolute maximum w_m values is equal,

$$F(h, \alpha) = P(w_m < h) = \lim_{t \Rightarrow \infty} \int_{-\infty}^{h} dw \, P_t(w; h). \tag{B.12}$$

For example, having solved the boundary problem (B.4) and (B.7) by the reflection method, we obtain

$$P_t(w; h) = \frac{1}{2(\pi t)^{1/2}} \left\{ \exp\left[-\frac{(w + \alpha t)^2}{4t} \right] - \exp\left[-h\alpha - \frac{(w - 2h + \alpha t)^2}{4t} \right] \right\}. \tag{B.13}$$

By substituting this expression into eq. (B.9), we find the probability density of the time t^* which is the first attainment of the boundary h by the process $w(t; \alpha)$

$$P(t; \alpha, h) = \frac{1}{2t(\pi t)^{1/2}} \exp\left[-\frac{(h + \alpha t)^2}{4t} \right].$$

By integrating eq. (B.13) with respect to w and allowing $t \Rightarrow \infty$, we finally obtain, according to eq. (B.12), an integral distribution function of the absolute maximum w_m values (Klyatskin and Saichev [1992]),

$$F(h; \alpha) = 1 - e^{-\alpha h}. \tag{B.14}$$

Let us now look at the description of the statistical properties of the lognormal process constructed on the basis of the Wiener process

$$y(\tau, t; \alpha) = e^{w(\tau; \alpha) - w(t; \alpha)} = e^{\Delta w(\tau, t) - \alpha(\tau - t)}. \tag{B.15}$$

It can also be given in the form

$$y(\tau, t; \alpha) = y(\tau; \alpha)/y(t; \alpha), \tag{B.16}$$

where

$$y(t; \alpha) = e^{w(t; \alpha)}. \tag{B.16'}$$

The process $w(t; \alpha)$, which is in the index of the exponent power, has independent increments. In physics this property is called an additive property of the process $w(t; \alpha)$. Correspondingly, the process $y(t; \alpha)$ has a multiplicative property, according to which, in particular, the process $y(t, \alpha)$ can be represented in the form of a product of the statistically independent processes

$$y(\tau; \alpha) = y(t; \alpha) y(\tau, t; \alpha). \tag{B.17}$$

As the function of the argument $\kappa = \tau - t$, the realizations of the process $y(\tau, t; \alpha)$ are statistically equivalent to the realizations of the process $y(\kappa; \alpha)$.

Let us discuss in more detail the lognormal process $y(t; \alpha)$ (B.16'). It satisfies the stochastic equation

$$\frac{d}{dt} y(t; \alpha) + \alpha y(t; \alpha) = f(t) y(t; \alpha), \qquad y(0; \alpha) = 1.$$

From this equation it follows that the probability density of this process

$$P_t(y; \alpha) = \langle \delta(y - y(t; \alpha)) \rangle$$

satisfies the Fokker–Planck equation

$$\frac{\partial}{\partial t} P_t(y; \alpha) = \alpha \frac{\partial}{\partial y} y P_t(y; \alpha) + \frac{\partial}{\partial y} y \frac{\partial}{\partial y} P_t(y; \alpha), \qquad P_0(y; \alpha) = \delta(y - 1).$$
(B.18)

Its solution is the lognormal probability density. It is possible to find the solution if we note that the probability of the inequality $y(t; \alpha) < y$ is exactly equal to the probability of the inequality $w(t; \alpha) < \ln y$. Taking into account eq. (B.6), we obtain the integral distribution function of the process $y(t; \alpha)$:

$$F(y; t, \alpha) = \Phi\left[\frac{\ln y}{(2t)^{1/2}} + \alpha \left(\frac{\tau}{2}\right)^{1/2}\right] = \Phi\left[\frac{1}{(2t)^{1/2}} \ln^2(y e^{\alpha t})\right].$$
(B.19)

Differentiating it with respect to t and taking into account the definition (B.6'), we find the solution of eq. (B.18):

$$P_t(y; \alpha) = \frac{1}{2(\pi t)^{1/2} y} \exp\left[-\frac{1}{4t} \ln^2(y e^{\alpha t})\right].$$
(B.20)

Using this probability density or, more easily, directly from eq. (B.18) it is possible to find process moments

$$\langle y^n(t; \alpha) \rangle = e^{n(n-\alpha)t}.$$

In particular, for the process

$$y(t) = y(t; 1) = e^{w(t) - t},$$
(B.21)

moments are equal to

$$\langle y^n(t) \rangle = e^{n(n-1)t}.$$

The mean value of the process $y(t)$ is $\langle y(t) \rangle = 1$, and is the same for any t whereas all other moments of $y(t)$ grow exponentially with the growth of t.

The exponential increase of highest moments of the lognormal process $y(t)$ can be explained by means of the slow decrease of tails of the probability density (B.20) as $y \gg 1$. With respect to the realization, this means that the rarer but also the higher increases will occur in the realization of the process $y(t)$ while t is growing. These increases must provide the exponential growth of $y(t)$ moments. At the same time, as seen from eq. (B.20), the main probability mass of the probability density of the process $y(t)$,

$$P_t(y) = P_t(y; \alpha = 1),$$

is concentrated in the area of small values of y. In fact, according to eq. (B.19) the probability of the inequality $y(t) < 1$ is

$$P(y(t) < 1) = F(1; t, 1) = \Phi(\sqrt{t/2}),$$

and when $t \gg 1$, it tends exponentially to unity

$$P(y(t) < 1) = 1 - \frac{1}{(\pi t)^{1/2}} e^{-t/4}.$$

Thus the plots of realizations of the process are below the level of its mean value $\langle y(t) \rangle = 1$ most of the time, although statistical moments of the process $y(t)$ are specified in the main by its large discontinuities.

B.2. MAJORANT CURVES

A contradiction that was discovered between the character of the behavior of statistical moments of the process $y(t)$ and its realization motivates us to study more closely the dynamics of realizations of the process $y(t)$ and the more general process $y(t; \alpha)$. Let us introduce the notion of majorant curves. We shall name a curve $M(t, p; \alpha)$ for which the inequality

$$y(t; \alpha) < M(t, p; \alpha)$$

is satisfied with the probability p for any t as a majorant curve. In other words, $100p$ percent of all realizations of the process $y(t; \alpha)$ is located below the majorant curve $M(t, p, \alpha)$. The statistic of the absolute maximum (B.11) of the process $w(t; \alpha)$ studied earlier permits us to indicate a sufficiently large class of majorant curves.

Let p be the probability of the absolute maximum $w_m(\beta)$ of the auxiliary process $w(t; \beta)$ to satisfy the inequality

$$w_m(\beta) < h = \ln A,$$

with an arbitrary value of the parameter β lying within the limits $0 < \beta < \alpha$. Then it is obvious that the whole realization of the process $y(t; \alpha)$ with the same probability p will be below the majorant curve

$$M(t, p, \alpha, \beta) = Ae^{(\beta - \alpha)t}, \tag{B.22}$$

because

$$y(t; \alpha) = e^{w(t;\alpha)} = e^{w(t;\beta) + (\beta - \alpha)t} < e^{h + (\beta - \alpha)t} = Ae^{(\beta - \alpha)t}$$

is valid with the probability p.

It is seen from eq. (B.14) that the probability of the process $y(t; \alpha)$ never exceeding the majorant curve (B.22) depends on its parameters as follows:

$$p = 1 - A^{-\beta}, \quad A = (1 - p)^{-1/\beta}. \tag{B.23}$$

Having applied the previous approach to the process $y(t)$, we can indicate that its realizations are bounded from above by the majorant curve

$$M(t, p, \beta) = (1 - p)^{-1/\beta} e^{-(1 - \beta)t}, \tag{B.24}$$

with the probability p.

Note that, in spite of the constancy of the statistical mean of $\langle y(t) \rangle = 1$ and the exponential growth of highest moments of the process $y(t)$, it is always possible to find the majorant curve (B.24) decreasing exponentially (as $\beta < 1$) such that realizations of the process $y(t)$ will be below it with any previously set probability $p < 1$. In particular, one-half of realizations of $y(t)$ lie below the exponentially decreasing majorant curve

$$M(t, \tfrac{1}{2}, \tfrac{1}{2}) = 4e^{-t/2}. \tag{B.24'}$$

From the existence of exponentially decreasing majorant curves we can conclude a corollary that is useful for statistical and dynamical understanding of the behavior of the process $y(t; \alpha)$ realization.

First, whereas the behavior of the highest moments of these processes is specified by the presence of large discontinuities in their realizations, these discontinuities themselves are not observed in all realizations of the process. This means, for example, that the mean value of $\langle y(t) \rangle = 1$ and the exponential growth of the highest moments $y(t)$ are pure statistical effects, caused by averaging over the whole ensemble of realizations.

Second, the area under the exponentially decreasing majorant curve is finite. Thus the very large discontinuities causing the exponential increase of the highest moments do not influence the area under the realization plot, which is also in practice finite for all realizations of processes $y(t; \alpha)$.

B.3. STATISTICS OF A RANDOM AREA

Related to the preceding findings, it is interesting to study the statistics of a random area under the realization plot of the process $y(t; \alpha)$

$$S(\alpha) = \int_0^\infty dt\, y(t; \alpha).$$

Let us discuss an auxiliary random process $S(t; \alpha)$, satisfying the stochastic equation

$$\frac{d}{dt} S(t; \alpha) = 1 - \alpha S(t; \alpha) + f(t) S(t; \alpha), \qquad S(0; \alpha) = 0. \tag{B.25}$$

Its solution is

$$S(t; \alpha) = \int_0^t d\tau\, y(\tau, t; \alpha). \tag{B.26}$$

The process under the integral is specified by the equality (B.15). It follows from the reversibility in time of the Wiener process, that the process $S(t; \alpha)$ (B.26) possesses a single-point probability density, which coincides with the probability density of a random function

$$S(t; \alpha) = \int_0^t dt\, y(t; \alpha)$$

equal to the area under the realization plot $y(t; \alpha)$ in the interval $(0, t)$. Therefore, in particular, if we find the probability density

$$P_t(S; \alpha) = \langle \delta(S - S(t; \alpha)) \rangle \tag{B.27}$$

in the limit $t \Rightarrow \infty$, it will coincide with the probability density of the area under the plot of the whole realization of the process $y(t; \alpha)$,

$$P(S; \alpha) = \lim_{t \Rightarrow \infty} P_t(S; \alpha). \tag{B.28}$$

When $\alpha = 1$, the above probability density coincides with the probability density of the area under the realization plot of the process $y(t)$ (B.21):

$$P(S) = P(S; \alpha = 1). \tag{B.29}$$

If we know $P(S; \alpha)$, it is easy to find the probability density $P_n(S)$ of random integrals

$$S_n = \int_0^\infty dt\, y^n(t).$$

In fact, it follows from the fractal property of the Wiener process that the process $y^n(t)$ is statistically equivalent to the process $y(n^n t; 1/n)$ compressed in time. It means, in particular, that

$$P_n(S) = n^2 P(n^2 S; 1/n).$$

The probability density $P(S; \alpha)$ (B.27) determined above satisfies the Fokker–Planck equation, following from eq. (B.25)

$$\frac{\partial}{\partial t} P_t(S; \alpha) + \frac{\partial}{\partial S} P_t(S; \alpha) = \alpha \frac{\partial}{\partial S} S P_t(S; \alpha) + \frac{\partial}{\partial S} S \frac{\partial}{\partial S} S P_t(S; \alpha),$$

$$P_0(S; \alpha) = \delta(S).$$

When $t \Rightarrow \infty$, the solution of this equation tends to the steady-state probability density $P(S; \alpha)$

$$P(S; \alpha) = \Gamma^{-1}(\alpha) S^{-\alpha-1} e^{-1/S}. \tag{B.30}$$

Then, by setting $\alpha = 1$, we obtain the probability density (B.29) of a random area under the realization plot of the process $y(t)$

$$P(S) = S^{-2} e^{-1/S}.$$

The corresponding integral distribution function is equal to

$$F(S) = e^{-1/S}.$$

The dependence of the probability distribution of the random process

$$\int_{-\infty}^{\tau-\Delta} dt\, y(\tau, t; \alpha) \quad (\Delta > 0) \tag{B.31}$$

on the time Δ gives additional information on the dynamics of the realization behavior of the processes $y(\tau, t; \alpha)$ and $y(t; \alpha)$ in time. It is easy to show that these processes are statistically equivalent to the process

$$\int_{\Delta}^{\infty} dt\, y(t; \alpha). \tag{B.31'}$$

It follows from the multiplicative property of the process $y(t; \alpha)$ (B.17) that the single-point probability distribution of processes (B.31) and (B.31') coincides with the probability density of the process

$$S(\Delta; \alpha) = y(\Delta; \alpha) S(\alpha),$$

where $y(t; \alpha)$ and $S(\alpha)$ are statistically independent. Their probability densities are specified by expressions (B.20) and (B.30), respectively. In particular it follows from here that the integral distribution function of a random process

$$S(t) = y(t)S,$$

which coincides with the distribution function of the area under the realization plot of the process $y(t)$ inside an infinite interval (t, ∞), is equal to

$$F(S; t) = \frac{1}{(\pi t)^{1/2}} \int_0^\infty \frac{dy}{y} \exp\left[-\frac{y}{S} - \frac{1}{4t} \ln^2(ye^t)\right].$$

While t increases, the probability of the inequality $S(t) < S$ tends to the unity monotonically at any S. This proves again the tendency to zero for every independent realization of the process $y(t)$ while t grows.

B.4. ISOPROBABLE CURVE

Appendix A introduced the notions of isoprobable curves and typical realization. Having applied these notions to the lognormal random process $y(t; \alpha)$ just discussed, we obtain from the formula (B.19), as $\alpha = 1$, that the isoprobable curves of the random function $y(t)$ at any $p < 1$ are determined by the equality

$$Z(t; p) = \exp(-t + r(2t)^{1/2}), \tag{B.32}$$

where r is the solution of the equation $\Phi(r) = p$, and $\Phi(r)$ is defined by the formula (B.6'). It can be seen from eq. (B.32) that at any given value of $p < 1$, however close to unity, the isoprobable curve tends exponentially to zero if t is large enough. When $p = \frac{1}{2}$, the isoprobable curve

$$Z(t; \tfrac{1}{2}) = e^{-t}$$

is a typical realization of the random process $y(t)$ according to the meaning described above.

In conclusion, we should note that in practice lognormal processes similar to those discussed here arise in all fields of physics where it is necessary to describe characteristics of positively defined physical quantities, processes and fields. This includes the description of intensity fluctuations of optical and radio waves in turbulent media and the analysis of the behavior of the amplitudes of radio physical systems subject to fluctuations in the parameter. For these physical problems a lognormal process emerges as the simplest adequate model correctly accounting for principal properties of the phenomena under investigation, such as positive definiteness, conservation laws and parametric instability. To illustrate this, we shall consider some simple examples that are of particular interest for problems of wave localization in random media.

B.5. SOME EXAMPLES

B.5.1. Stochastic parametric resonance

We shall discuss the stochastic parametric excitation in an oscillatory system with small linear friction due to parameter fluctuations. The dynamical system under question is described by the equation

$$\frac{d^2}{dt^2} x(t) + 2\gamma \frac{d}{dt} x(t) + \omega_0^2 [1 + z(t)] x(t) = 0,$$

with initial values $x(0) = x_0$ and $dx/dt|_{t=0} = y_0$. Here, $z(t)$ denotes a random process. This equation appears in many branches of physics, and many publications have been devoted to it (see, e.g. Klyatskin [1975, 1980a, 1985]). From the physical standpoint, it is obvious that this dynamical system must admit parametric excitation because the random process $z(t)$ contains the harmonic components of all frequencies, including the values $2\omega_0/n$, $n = 1, 2, 3, ...$, which exactly correspond to parametric resonance in the system with periodical function $z(t)$ ($\gamma = 0$).

Rewriting this equation in the form of a system of equations

$$\frac{d}{dt} x(t) = y(t), \qquad \frac{d}{dt} y(t) = -2\gamma y(t) - \omega_0^2 [1 + z(t)] x(t), \qquad (B.33)$$

let us introduce new functions $A(t)$ and $\phi(t)$, that is, the oscillation amplitude and the phase, instead of $x(t)$ and $y(t)$:

$$x(t) = A(t) \sin(\omega_0 t + \phi(t)), \qquad y(t) = \omega_0 A(t) \cos(\omega_0 t + \phi(t)). \qquad (B.34)$$

Substituting eq. (B.34) into eq. (B.33) leads to a system of equations for the functions $A(t)$ and $\phi(t)$:

$$\frac{d}{dt} A(t) = -2\gamma A(t) \cos^2 \psi(t) - \tfrac{1}{2} \omega_0 z(t) A(t) \sin 2\psi(t),$$

$$\frac{d}{dt} \phi(t) = \gamma \sin 2\psi(t) + \omega_0 z(t) \sin^2 \psi(t), \qquad (B.34')$$

where $\psi(t) = \omega_0 t + \phi(t)$. If we make the substitution $A(t) = e^{u(t)}$, we obtain

$$\frac{d}{dt} u(t) = -2\gamma \cos^2 \psi(t) - \tfrac{1}{2} \omega_0 z(t) \sin 2\psi(t), \qquad (B.35a)$$

$$\frac{d}{dt}\phi(t) = \gamma \sin 2\psi(t) + \omega_0 z(t) \sin^2 \psi(t). \tag{B.35b}$$

We suppose that the process $z(t)$ is Gaussian and delta correlated in time with parameters

$$\langle z(t) \rangle = 0, \qquad \langle z(t)z(t') \rangle = 2\sigma^2 \tau_0 \delta(z - z') \rangle,$$

where τ_0 is the correlation radius. Then, for the joint probability density for the solution of the system (B.35),

$$P_t(u, \phi) = \langle \delta(u(t) - u)\delta(\phi(t) - \phi) \rangle,$$

the Fokker–Planck equation can be written as

$$\frac{\partial}{\partial t} P_t(u, \phi) = 2\gamma \frac{\partial}{\partial u}(\cos^2 \psi P_t) - \gamma \frac{\partial}{\partial \phi}(\sin 2\psi P_t)$$

$$+ D \frac{\partial}{\partial u}(\cos 2\psi \sin^2 \psi P_t) - 2D \frac{\partial}{\partial \phi}(\sin^3 \psi \cos \psi P_t)$$

$$+ \frac{1}{4} D \frac{\partial^2}{\partial u^2}(\sin^2 2\psi P_t) + D \frac{\partial^2}{\partial \phi^2}(\sin^4 \psi P_t)$$

$$- D \frac{\partial}{\partial u} \frac{\partial}{\partial \phi}(\sin 2\psi \sin^2 \psi P_t), \tag{B.36}$$

where $D = \sigma^2 \omega_0^2 \tau_0$ is the diffusion coefficient.

If the parameter $\omega_0/D \gg 1$ is large (i.e. $\sigma^2 \omega_0 \tau_0 \ll 1$), the relatively slow changing of the statistical characteristics of the system (B.33) solution with respect to t is accompanied by ordinary oscillations with frequency ω_0. The latter can be eliminated if the appropriate statistical characteristics are averaged over the time period $T = 2\pi/\omega_0$. The averaging (B.36) over the oscillation period T then yields

$$\frac{\partial}{\partial t} P_t(u, \phi) = \gamma \frac{\partial}{\partial u} P_t - \frac{1}{4} \frac{\partial}{\partial u} P_t + \frac{1}{8} D \frac{\partial^2}{\partial u^2} P_t + \frac{3}{8} D \frac{\partial^2}{\partial \phi^2} P_t. \tag{B.37}$$

Here we assume that the variations of the statistical parameters over a time in an interval of order T are small, and hence a sufficient condition for carrying out this operation is that the trigonometric functions in the right-hand side of eq. (B.36) are averaged.

It follows from eq. (B.37) that the statistical characteristics of the oscillation amplitude and the oscillation phase averaged over the period of oscillation are statistically independent. The probability densities that describe them are Gaussian with parameters

$$\langle u(t)\rangle = u_0 - \gamma t + \tfrac{1}{4}Dt, \qquad \sigma_u^2(t) = \tfrac{1}{4}Dt,$$

$$\langle \phi(t)\rangle = 0, \qquad \sigma_\phi^2(t) = \tfrac{3}{4}Dt.$$

Thus the probability density of the oscillation amplitude $A(t)$ is lognormal, and for its momentum functions we have the following equality:

$$\langle A^n(t)\rangle = \langle \exp[nu(t)]\rangle = \exp[n\langle u(t)\rangle + \tfrac{1}{2}n^2\sigma_u^2(t)]$$
$$= A_0^n \exp[-n\gamma t + \tfrac{1}{8}n(2+n)Dt]. \tag{B.38}$$

If the condition

$$8\gamma < (2+n)D$$

is true, the stochastic system (B.33) is subjected to stochastic parametric excitation (starting from the momentum function of order n).

Note that the typical realization function for the random process $A(t)$ is

$$A(t) \approx A_0 e^{-(\gamma - D/4)t}.$$

Thus, if friction in the dynamical system (B.33) is small, but

$$D < 4\gamma < (1 + n/2)D,$$

which is valid for sufficiently large values of n, the typical realization function has an exponential fall. At the same time, however, the momentum functions starting from the order n must have the behavior of exponential growth.

B.5.2. Wave-beam propagation in a random parabolic waveguide

Wave-beam propagation in the direction of the x axis will be described on the basis of a parabolic-equation approximation (quasioptics) (2.22)

$$2ik\frac{\partial}{\partial x}u(x,\pmb{\rho}) + \Delta_\perp u(x,\pmb{\rho}) + k^2\varepsilon(x,\pmb{\rho})u(x,\pmb{\rho}) = 0, \qquad u(0,\pmb{\rho}) = u_0(\pmb{\rho}),$$

$$\left(\pmb{\rho} = \{y,z\}, \quad \Delta_\perp = \frac{\partial^2}{\partial y^2} + \frac{\partial^2}{\partial z^2}\right), \tag{B.39}$$

where $\varepsilon(x,\pmb{\rho})$ is the fluctuating part of the dielectric permittivity or of the refractive index, and $\pmb{\rho}$ designates the coordinates of the plane perpendicular to the x axis.

An important case that describes wave propagation in a waveguide corresponds to $\varepsilon(x, \rho) = -\alpha\rho^2$. Such a function ε occurs in the study of acoustic waves in a natural waveguide (the ocean) or of radio waves. In the absence of random inhomogeneities the solution of eq. (B.39) can be presented in the form

$$u_0(x, \rho) = f(x, \rho)v(x, \rho),$$

where

$$f(x, \rho) = \frac{1}{\cos \alpha x} \exp\{-\tfrac{1}{2}i\alpha k\rho^2 \tan \alpha x\},$$

$$v(x, \rho) = \int d\kappa \, \hat{u}_0(\kappa) \exp\left\{-i\frac{\kappa^2}{2k\alpha} \tan \alpha x + \frac{i\kappa\rho}{\cos \alpha x}\right\}$$

with

$$u_0(\rho) = \int d\kappa \, \hat{u}_0(\kappa) e^{i\kappa\rho}.$$

The function $f(x, \rho)$ describes the propagation of a plane wave. At the point $x_n = (2n + 1)\pi/2\alpha$, the function $f(x, \rho)$ becomes infinite. This corresponds to the focusing of the plane wave. In general, the wavefield $u_0(x, \rho)$ remains finite. As an example, we consider a wave beam for which

$$u_0(\rho) = u_0 e^{-\rho^2/2a^2}. \tag{B.40}$$

Then the wavefield $u_0(x, \rho)$ equals

$$u_0(x, \rho) = u_0 \left[\cos \alpha x \left(1 + \frac{i}{k\alpha a^2} \tan \alpha x\right)\right]^{-1} \exp\left\{-\frac{1}{2}\rho^2 \left[i\alpha k \tan \alpha x + \left[a^2 \cos^2 \alpha x \left(1 + \frac{i}{k\alpha a^2} \tan \alpha x\right)\right]^{-1}\right]\right\}, \tag{B.41}$$

and its intensity is given by

$$I_0(x, \rho) = u_0(x, \rho)u_0^*(x, \rho) = \frac{|u_0|^2}{g_\alpha^2(x)} \exp\left\{-\frac{\rho^2}{a^2 g_\alpha^2(x)}\right\},$$

where

$$g_\alpha^2(x) = \cos^2 \alpha x + \frac{1}{k^2\alpha^2 a^4} \sin^2 \alpha x.$$

If the wave beam (B.40) matches the inhomogeneous medium, that is, if the parameter a satisfies

$$k\alpha a^2 = 1,$$

then the wavefield $u_0(x, \rho)$ specified by eq. (B.41) assumes the form

$$u_0(x, \rho) = u_0 \exp\left\{-\frac{1}{2a^2}\rho^2 - i\alpha x\right\},$$

and, therefore, its amplitude does not vary during the wave propagation. This wavefield corresponds to an eigenmode of the wave equation.

We suppose now that $\varepsilon(x, \rho)$ is given by

$$\varepsilon(x, \rho) = -\alpha^2 \rho^2 + z(x)\rho^2,$$

where α is a nonrandom parameter and $z(x)$ is a Gaussian delta-correlated random process, with parameters

$$\langle z(x) \rangle = 0, \qquad \langle z(x)z(x') \rangle = 2\sigma^2 l\delta(x - x').$$

The solution of eq. (B.39) with initial condition (B.40) can be written as

$$u(x, \rho) = u_0 \exp\left\{-\frac{\rho^2}{2a^2} A(x) + B(x)\right\}.$$

Here, the functions $A(x)$ and $B(x)$ satisfy

$$\begin{aligned}\frac{d}{dx}A(x) &= -\frac{i}{ka^2}[A^2(x) - \alpha^2 k^2 a^4] - ika^2 z(x), \\ \frac{d}{dx}B(x) &= -\frac{i}{ka^2}A^2(x),\end{aligned} \qquad (B.42)$$

with $A(0) = 1$ and $B(0) = 0$. Therefore,

$$B(x) = -\frac{i}{ka^2}\int_0^x d\xi \, A(\xi),$$

and the wavefield intensity equals

$$I(x, \rho) = |u_0|^2 \exp\left\{-\frac{\rho^2}{2a^2}[A(x) + A^*(x)] - \frac{i}{ka^2}\int_0^x d\xi \, [A(\xi) - A^*(\xi)]\right\}. \qquad (B.43)$$

It is possible to eliminate the imaginary part of the function A in expression (B.43) by applying the first equation of eqs. (B.42):

$$-\frac{i}{ka^2}[A(x) - A^*(x)] = \frac{d}{dx}\ln[A(x) + A^*(x)].$$

This leads to the following expression for the field intensity ($|u_0|^2 = 1$):

$$I(x, \rho) = I(x, 0) \exp\left[-\frac{\rho^2}{a^2} I(x, 0)\right],$$

where

$$I(x, 0) = \tfrac{1}{2}[A(x) + A^*(x)]$$

is the variation of the wave intensity along the x axis of the unperturbed waveguide.

Thus the statistical characteristics of the wave intensity are determined by the statistical characteristics of the quantity A, that is, the solution of eq. (B.42). The system (B.42) is similar to that describing the wave-reflection coefficient in a one-dimensional problem with layered inhomogeneities, which was discussed in § 3 (Papanicolaou, McLaughlin and Burridge [1973], Klyatskin [1980a, b, 1985]).

We write the function $A(x)$ as

$$A(x) = k\alpha a^2 \frac{1 + \psi(x) \exp\{-2i\alpha x\}}{1 - \psi(x) \exp\{-2i\alpha x\}}.$$

The function $\psi(x)$ then satisfies the equation

$$\frac{d}{dx}\psi(x) = -\frac{i}{2\alpha k}(e^{i\alpha x} - e^{-i\alpha x})^2 z(x), \qquad \psi(0) = \frac{1 - k\alpha a^2}{1 + k\alpha a^2}.$$

We now assume that

$$\psi(x) = \sqrt{\frac{w(x) - 1}{w(x) + 1}} e^{i\{\phi(x) - 2\alpha x\}}, \qquad w \geq 1.$$

The functions w and ϕ then satisfy the system of equations

$$\frac{d}{dx}w(x) = -\frac{1}{\alpha k} z(x) \sqrt{w^2 - 1} \sin\{\phi(x) - 2\alpha x\},$$

$$w(0) = \frac{1}{2k\alpha a^2}[1 + k^2\alpha^2 a^4],$$

$$\frac{d}{dx}\phi(x) = \frac{1}{\alpha k} z(x)\left[1 - \frac{w}{\sqrt{w^2 - 1}} \cos\{\phi(x) - 2\alpha x\}\right], \qquad \phi(0) = 0.$$

Correspondingly, the expression for $I(x, 0)$ equals

$$I(x, 0) = \frac{\alpha k a^2}{w(x) + \sqrt{w^2(x) - 1} \cos\{\phi(x) - 2\alpha x\}}. \tag{B.44}$$

Furthermore, we assume that the intensity of the fluctuations $z(x)$ is sufficiently small. Therefore, the statistical characteristics of $w(x)$ and $\phi(x)$ vary slowly over a scale of the order $1/\alpha$, and in order to determine the statistical characteristics of the wave intensity (B.44), one has to average over rapidly oscillating functions. In this case the probability distribution for $w - P_x(w) = \langle \delta(w(x) - w) \rangle$ is determined by the equation

$$\frac{\partial}{\partial x} P_x(w) = D \frac{\partial}{\partial w}(w^2 - 1) \frac{\partial}{\partial w} P_x(w), \quad D = \sigma^2 l/2\alpha^2 k^2.$$

We should also assume that the quantity $\phi(x)$ is statistically independent on $w(x)$, and that it is uniformly distributed over the interval $[0, 2\pi]$ ($Dx \gg 1$).

Under the foregoing assumptions we calculate the wave intensity moments $\langle I^n(x) \rangle$. The averaging is carried out in two steps. In the first step we average over the angle (viz. ϕ). We obtain the expression

$$\left\langle \left(\frac{I}{\alpha k a^2}\right)^n \right\rangle_\phi = P_{n-1}(w). \tag{B.45}$$

Here, $P_n(w)$ denotes the Legendre polynomial of order n.

In the second step we average eq. (B.45) with respect to w. As a result, we obtain the expression

$$\left\langle \left(\frac{I}{\alpha k a^2}\right)^n \right\rangle = P_{n-1}(w_0) e^{Dn(n-1)x}.$$

If the parameters of the beam are matched to the parameters of the waveguide, and so $\alpha k a^2 = 1$, we obtain

$$\langle I^n(x, 0) \rangle = e^{Dn(n-1)x}. \tag{B.46}$$

This means that the quantity $I(x, 0)$ is distributed accordingly to a lognormal law. In particular, it follows from eq. (B.46) that

$$\langle I(x, 0) \rangle = 1 \quad \text{and} \quad \langle I^2(x, 0) \rangle = e^{2Dx}.$$

Thus, the wave intensity moments along the axis of the waveguide have an exponential growth. As we saw earlier, however, the typical realization of wavefield intensity is

$$I(x, 0) \approx e^{-Dx} \quad \text{as } Dx \gg 1.$$

This means that radiation goes out from the waveguide for each concrete realization; that is, we have the property of dynamic localization in the x direction.

C. Fundamental Solutions of Wave Problems

This appendix will discuss some properties of the fundamental solutions (Green functions) of wave equations.

First, let us consider the Green function for the one-dimensional Helmholtz equation

$$\frac{d^2}{dx^2} g(x - x_0) + k^2 g(x - x_0) = \delta(x - x_0). \tag{C.1}$$

The solution of eq. (C.1) satisfying radiation conditions when $x \to \pm\infty$ has the form

$$g(x - x_0) = \frac{1}{2ik} e^{ik|x - x_0|}. \tag{C.2}$$

The appearance of modulus $|x - x_0|$ in the right-hand part of eq. (C.2) is because eq. (C.1) is the equation of second order on x. If we fix mutual positions of the observation points and source, however, the Green function satisfies the following equality (we suppose that $x_0 > x$ for clarity)

$$\frac{\partial}{\partial x_0} g(x - x_0) = ikg(x - x_0), \tag{C.3}$$

which can be considered as the first-order equation, provided that it is completed by the initial condition

$$g(x - x_0)|_{x_0 = x} = g(0) = \frac{1}{2ik}. \tag{C.3'}$$

Thus, having fixed the source position with respect to the point of observation, we can reduce the order of the equation for the Green function. This result is common for wave problems (factorization property), and it corresponds to the fact that the wave radiated in the direction $x < x_0$ (or $x > x_0$) in free space will propagate without any direction changes.

In the general case the Green function satisfies the following operator equation of the first order ($x_0 > x$):

$$\frac{\partial}{\partial x_0} g(x - x_0; \eta - \eta_0) = \hat{L}^+(\eta) g(x - x_0; \eta - \eta_0) = \hat{L}^-(\eta_0) g(x - x_0; \eta - \eta_0),$$

$$g(x - x_0; \eta - \eta_0)|_{x_0 = x} = g(0; \eta - \eta_0) = g(\eta - \eta_0), \tag{C.4}$$

where operator $\hat{L}$ affects time- and space-dependent variables designated by η.

Hence, the Green function structure is analogous to eq. (C.2):

$$g(x - x_0; \eta - \eta_0) = \exp\{|x - x_0|\hat{L}^+(\eta)\}g(\eta - \eta_0). \tag{C.5}$$

By differentiating eq. (C.5) twice with respect to x, we can obtain the wave equation

$$\left\{\frac{\partial^2}{\partial x^2} - \hat{L}^2(\eta)\right\} g(x - x_0; \eta - \eta_0) = 2\delta(x - x_0)\hat{L}^+(\eta)g(\eta - \eta_0). \tag{C.6}$$

Therefore, the operator $\hat{L}$ for eq. (C.1) is simply given by the number given by $\hat{L}^2 = -k^2$ and $g(\eta) = 1/2ik$.

Generally, the operator $\hat{L}$ can be considered as the integral ones. In fact, the effect of the operator $\hat{L}$ on the arbitrary function $f(\eta)$ can be presented in the form

$$\hat{L}^+(\eta)f(\eta) = \int d\xi\, \hat{L}^+(\eta)\delta(\eta - \xi)f(\xi) = \int d\xi\, L(\eta - \xi)f(\xi), \tag{C.7}$$

where the kernel at the integral operator is determined by the equality

$$L(\eta - \xi) = \hat{L}^+(\eta)\delta(\eta - \xi). \tag{C.7'}$$

By analogy,

$$\hat{L}^-(\eta_0)f(\eta_0) = \int d\xi\, f(\xi)L(\xi - \eta_0).$$

One can also introduce the inverse operator $\hat{L}^{-1}(\eta)$ with the corresponding kernel $\tilde{L}(\eta - \xi)$.

Usually a $\delta(x - x_0)\delta(\eta - \eta_0)$ term is in the right-hand side of a wave equation, where it corresponds to a spacetime point source, and the operator $\hat{L}^2(\eta)$ is simply the differential form. Hence,

$$2\hat{L}^+(\eta)g(\eta - \eta_0) = \delta(\eta - \eta_0). \tag{C.8}$$

By acting on eq. (C.8) with the operator $\hat{L}^+(\eta)$, according to eq. (C.7') we obtain the following kernel of the integral operator

$$L(\eta - \eta_0) = 2\hat{L}^2(\eta)g(\eta - \eta_0), \tag{C.9}$$

which can be rewritten as

$$L(\eta - \eta_0) = 2 \lim_{x \to 0} \frac{\partial^2}{\partial x^2} g(x; \eta - \eta_0). \tag{C.9'}$$

Taking into account that the Green function for wave equations is of

$g(x^2; \eta - \eta_0)$ type, the latter equality can be reduced to

$$L(\eta - \eta_0) = 4 \lim_{x \to 0} \frac{\partial}{\partial x^2} g(x^2; \eta - \eta_0). \tag{C.9''}$$

By acting on eq. (C.8) with the inverse operator $\hat{L}^{-1}(\eta)$, we obtain the kernel of the inverse integral operator

$$\tilde{L}(\eta - \eta_0) = \hat{L}^{-1}(\eta)\delta(\eta - \eta_0) = 2g(\eta - \eta_0). \tag{C.10}$$

Thus, the kernels of integral operators $\hat{L}(\eta)$ and $\hat{L}^{-1}(\eta)$ are simply defined by the fundamental solutions of wave equations themselves.

To consider some concrete wave problems, let us write a Helmholtz equation in the form

$$\left\{ \frac{\partial^2}{\partial x^2} + \Delta_\rho + k^2 \right\} g(x - x_0; \boldsymbol{\rho} - \boldsymbol{\rho}_0) = \delta(x - x_0)\delta(\boldsymbol{\rho} - \boldsymbol{\rho}_0), \tag{C.11}$$

where ρ designates the coordinates of the plane perpendicular to the x axis.

The solution of eq. (C.11) with the condition of radiation at infinity has the form

$$g(x - x_0; \boldsymbol{\rho} - \boldsymbol{\rho}_0) = -\frac{1}{4\pi|\boldsymbol{r} - \boldsymbol{r}_0|} e^{ik|\boldsymbol{r} - \boldsymbol{r}_0|} \quad (\boldsymbol{r} = \{x, \boldsymbol{\rho}\}). \tag{C.12}$$

Function $g(\boldsymbol{r})$ can be presented by the integral

$$g(x, \boldsymbol{\rho}) = \frac{1}{8i\pi^2} \int \frac{d\boldsymbol{q}}{(k^2 - q^2)^{1/2}} \exp[i(k^2 - q^2)^{1/2}|x| + i\boldsymbol{q}\boldsymbol{\rho}]. \tag{C.12'}$$

Therefore, operators $\hat{L}^\pm$ in the given case have the form

$$\hat{L}^+(\boldsymbol{\rho}) = i(k^2 + \Delta_\rho)^{1/2}, \qquad \hat{L}^-(\boldsymbol{\rho}_0) = i(k^2 + \Delta_{\rho_0})^{1/2},$$

and corresponding kernels of the integral operators are determined according to eqs. (C.9) to (C.9'') as

$$L(\boldsymbol{\rho}) = -2(k^2 + \Delta_\rho)g(\boldsymbol{\rho}) = \frac{1}{2\pi\rho^2}\left(\frac{1}{\rho} - ik\right)e^{ik\rho}, \qquad \tilde{L}(\boldsymbol{\rho}) = -\frac{1}{2\pi\rho}e^{ik\rho}. \tag{C.13}$$

For a two-dimensional case we have, correspondingly,

$$g(\boldsymbol{r} - \boldsymbol{r}_0) = -\frac{i}{4} H_0^{(1)}(k|\boldsymbol{r} - \boldsymbol{r}_0|) \quad (\boldsymbol{r} = \{x, y\}),$$

where $H_0^{(1)}$ is the Hankel function of the first order, and thus the kernels of corresponding operators $\hat{L}(y) = i(k^2 + \partial^2/\partial y^2)^{1/2}$ and $\hat{L}^{-1}(y)$ are determined as

$$L(y) = \frac{ik}{2|y|} H_1^{(1)}(k|y|), \qquad \tilde{L}(y) = -\frac{i}{2} H_0^{(1)}(k|y|). \tag{C.14}$$

For a one-dimensional case, as was pointed out earlier, the operators $\hat{L}$ and $\hat{L}^{-1}$ are simply numbers.

We have considered some properties of the fundamental solutions (Green functions) of wave equations describing the field of a point source in unbounded free space, but this analysis will not change essentially for problems describing the field of a point source inside a limited layer of free or layered space.

D. Factorization of the Wave Equation in a Layered Medium

Wave equations are factorized if waves propagating in the (x, y) plane are not back-scattered in layered media with $\varepsilon(x, y, z) = \varepsilon(z)$.

The wavefield of a point source in two-dimensional space is described by the Green function $G^{(2)}(x, z; z_0)$, satisfying the equation

$$\left[\frac{\partial^2}{\partial x^2} + \frac{\partial^2}{\partial z^2} + k^2(z) \right] G^{(2)}(x, z; z_0) = \delta(x)\delta(z - z_0).$$

The solution $G^{(2)}(x, z; z_0)$ has the form

$$G^{(2)}(x, z; z_0) = e^{i|x|\hat{L}(z)} G^{(2)}(0, z; z_0), \tag{D.1}$$

where an operator $\hat{L}^2(z) = \partial^2/\partial z^2 + k^2(z)$ and a function $G^{(2)}(0, z; z_0)$ describe the wavefield on the axis $x = 0$. The increase discontinuity of $\partial G^{(2)}(x, z; z_0)/\partial x$ at the point $x = 0$ gives the equation

$$2i\hat{L}(z)G^{(2)}(0, z; z_0) = \delta(z - z_0). \tag{D.2}$$

From here it follows that

$$G^{(2)}(0, z; z_0) = \frac{1}{2i} \hat{L}^{-1}(z)\delta(z - z_0). \tag{D.3}$$

By applying the operator $\hat{L}^2(z)$, we have

$$\hat{L}^2(z)G^{(2)}(0, z; z_0) = \frac{1}{2i} \hat{L}(z)\delta(z - z_0). \tag{D.4}$$

The operators $\hat{L}(z)$ and $\hat{L}^{-1}(z)$ can be regarded as integral operators; then equalities (D.3) and (D.4) define their kernels. Taking this into account, we see that eq. (D.2) is the nonlinear integral equation for $G^{(2)}(0, z; z_0)$,

$$\int_{-\infty}^{\infty} d\xi \, G^{(2)}(0, z; \xi) G^{(2)}(0, \xi; z_0) = -\tfrac{1}{4} G^{(1)}(z; z_0), \tag{D.5}$$

where $G^{(1)}(z; z_0) = \hat{L}^{-2}(z)\delta(z - z_0)$ is the Green function of a one-dimensional layered problem.

By differentiating eq. (D.1) at $x > 0$, we obtain the evolutional equation

$$\frac{\partial}{\partial x} G^{(2)}(x, z; z_0) = i\hat{L}(z) G^{(2)}(x, z; z_0).$$

It can be written in the form of an integrodifferential equation if eq. (D.4) is taken into account:

$$\frac{\partial}{\partial x} G^{(2)}(x, z; z_0) = -2\hat{L}^2(z) \int_{-\infty}^{\infty} d\xi \, G^{(2)}(0, z; \xi) G^{(2)}(x, \xi; z_0). \tag{D.6}$$

The integral representation of $G^{(2)}(x, z; z_0)$ by means of a solution of the parabolic equation was obtained by Polyanskii [1972, 1985] and De Santo [1977]:

$$G^{(2)}(x, z; z_0) = x \frac{e^{-i\pi/4}}{2} \sqrt{\frac{k}{\pi}} \int_0^{\infty} d\xi \, \xi^{-3/2} e^{ikx^2/4\xi} P(\xi, z; z_0),$$

$$ik \frac{\partial}{\partial x} P(x, z; z_0) + \hat{L}^2(z) P(x, z; z_0) = 0, \qquad P(0, z; z_0) = G^{(2)}(0, z; z_0).$$

(D.7)

It is observed that eq. (D.7) contains an unknown wavefield on the axis $x = 0$ as an initial condition, which is not convenient for the solution of wave problems.

In a three-dimensional case the Green function $G^{(3)}(x, y, z; z_0)$ is determined from the equation

$$\left[\frac{\partial^2}{\partial x^2} + \frac{\partial^2}{\partial y^2} + \frac{\partial^2}{\partial z^2} + k^2(z) \right] G^{(3)}(x, y, z; z_0) = \delta(x)\delta(y)\delta(z - z_0).$$

We can write the solution in the form

$$G^{(3)}(x, y, z; z_0) = e^{i|x|\hat{L}(y,z)} G^{(3)}(0, y, z; z_0),$$

where $\hat{L}^2(y, z) = \partial^2/\partial y^2 + \hat{L}^2(z)$ and a function $G^{(3)}(0, y, z; z_0)$ describe the wavefield on the (y, z) plane. The discontinuity of $\partial G^{(3)}(x, y, z; z_0)/\partial x$ at the point $x = 0$ yields

$$G^{(3)}(0, y, z; z_0) = \frac{1}{2i}\hat{L}^{-1}(y, z)\delta(y)\delta(z - z_0).$$

This equality has the operator form by means of the Hankel function of the first-order $H_0^{(1)}(z)$:

$$G^{(3)}(0, y, z; z_0) = -\frac{i}{4}H_0^{(1)}[y\hat{L}(z)]\delta(z - z_0).$$

By using an integral representation of

$$H_0^{(1)}(\beta\mu) = \frac{1}{i\pi}\int_0^\infty \frac{dx}{x}\exp\left[i\frac{\mu}{2}\left(x + \frac{\beta^2}{x}\right)\right],$$

we obtain the connection of $G^{(3)}(0, y, z; z_0)$ with the solution of the parabolic equation

$$\frac{\partial}{\partial t}u(t, z; z_0) = \frac{i}{2k}\hat{L}^2(z)u(t, z; z_0), \qquad u(0, z; z_0) = \delta(z - z_0)$$

on the foundation of the formula

$$G^{(3)}(0, y, z; z_0) = -\frac{1}{4\pi}\int_0^\infty \frac{dt}{t}\exp\left[\frac{ik}{2t}y^2\right]u(t, z; z_0)$$

$$= -\frac{1}{4\pi}\int_0^\infty \frac{dt}{t}\exp\left[\frac{ik}{2t}(y^2 + t^2)\right]\psi(t, z; z_0).$$

The function $\psi(t, z; z_0)$ is the solution of the equation

$$\frac{\partial}{\partial t}\psi(t, z; z_0) = \frac{i}{2k}\left[\frac{\partial^2}{\partial z^2} + k^2(z) - k^2\right]\psi(t, z; z_0),$$

$$\psi(0, z; z_0) = \delta(z - z_0).$$
(D.8)

Since the x axis is arbitrary at $y > 0$, then $G^{(3)}(0, y, z; z_0) = G^{(3)}(\rho, z; z_0)$, $\rho^2 = x^2 + y^2$ defines the Green function in the whole space:

$$G^{(3)}(\rho, z; z_0) = -\frac{1}{4\pi}\int_0^\infty \frac{dt}{t}\exp\left[\frac{ik}{2t}(\rho^2 + t^2)\right]\psi(t, z; z_0). \qquad (D.9)$$

By integrating eq. (D.9) with respect to y and x, we obtain the functions $G^{(2)}(x, z; z_0)$, $G^{(1)}(z; z_0)$ coinciding with eq. (4.18).

The representation (D.9) of the field of a point source is more convenient than eq. (D.7), because it is defined by the solution of the parabolic equation (D.8) with the fixed initial condition. Equation (D.8) coincides with a parabolic equation for a small-angle approximation in a two-dimensional problem for a layered medium, and only differs in boundary conditions along the transverse coordinate.

References

Abramovich, B.S., and A.I. Dyatlov, 1975, On the theory of wave propagation in a one-dimensional randomly inhomogeneous absorptive medium, Izv. Vyssh. Uchebn. Zaved. Radiofiz. **18**(8), 1222–1224.
Ambartsumian, V.A., 1943, Diffuse reflection of light by a foggy medium, Comptes Rendus (Doklady) de l'USSR **38**(8), 229–232.
Ambartsumian, V.A., 1944, On the problem of diffuse reflection of light, J. Phys. USSR **8**(1), 65.
Ambartsumian, V.A., 1989, On the principle of invariance and its applications, in: Principle of Invariance and its Applications, eds M.A. Mnatsakarian and H.V. Pickichian (Armenian SSR Acad. Sci., Yerevan) pp. 9–18.
Anderson, P.W., 1958, Absence of diffusion in certain random lattices, Phys. Rev. **109**, 1492–1505.
Anzygina, T.N., L.A. Pastur and V.A. Slusarev, 1981, Localization of states and the kinetic properties of one-dimensional disordered systems, Fiz. Niz. Temp. **7**(1), 5–44.
Apresyan, L.A., and Yu.A. Kravtsov, 1983, Theory of Radiative Transfer. Statistical and Wave Aspects (Nauka, Moscow). In Russian.
Asch, M., W. Kohler, G. Papanicolaou, M. Postel and B. White, 1991, Frequency content of randomly scattered signals, SIAM Rev. **33**(4), 519–625.
Asch, M., G. Papanicolaou, M. Postel, P. Sheng and B. White, 1990, Frequency content of randomly scattered signals, Part I, Wave Motion **12**(4), 429–450.
Babkin, G.I., and V.I. Klyatskin, 1980a, Theory of wave propagation in nonlinear inhomogeneous media, J. Teor. & Exper. Fiz. **79**(3), 817–827.
Babkin, G.I., and V.I. Klyatskin, 1980b, Wave intensity fluctuations inside a one-dimensional randomly inhomogeneous medium. III. The effect of absorption and transfer equation, Izv. Vyssh. Uchebn. Zaved. Radiofiz. **23**(4), 432–441.
Babkin, G.I., and V.I. Klyatskin, 1980c, Wave intensity fluctuations inside a one-dimensional randomly inhomogeneous medium. IV. Invariant imbedding and probability distribution for intensity, Izv. Vyssh. Uchebn. Zaved. Radiofiz. **23**(10), 1185–1194.
Babkin, G.I., and V.I. Klyatskin, 1982a, Invariant imbedding method for wave problems, Wave Motion **4**(2), 195–207.
Babkin, G.I., and V.I. Klyatskin, 1982b, Statistical theory of radiative transfer in layered media, Wave Motion **4**(3), 327–339.
Babkin, G.I., V.I. Klyatskin, V.F. Kozlov and E.V. Yaroschuk, 1981, Wave intensity fluctuations inside a one-dimensional randomly inhomogeneous medium. V. Numerical solution of radiative transfer equations, Izv. Vyssh. Uchebn. Zaved. Radiofiz. **24**(8), 952–959.
Babkin, G.I., V.I. Klyatskin and L.Ya. Lyubavin, 1980, Invariant imbedding theory and waves in randomly inhomogeneous media, Dokl. Akad. Nauk **250**(5), 1112–1115.
Babkin, G.I., V.I. Klyatskin and L.Ya. Lyubavin, 1982a, Theory of sound propagation in the ocean, Akust. Zh. **28**(3), 310–315.

Babkin, G.I., V.I. Klyatskin and L.Ya. Lyubavin, 1982b, Boundary-value problems for wave equation, Akust. Zh. **28**(1), 1–7.
Barabanenkov, Yu.N., and D.I. Kryukov, 1992, Functional Fokker–Planck formalism for wave propagation in random media under delocalized and weak localized regimes, Waves in Random Media **2**(1), 1–6.
Barabanenkov, Yu.N., Yu.A. Kravtsov, V.D. Ozrin and A.I. Saichev, 1991, Enhanced backscattering in optics, in: Progress in Optics, Vol. 29, ed. E. Wolf (North-Holland, Amsterdam) pp. 67–197.
Bellman, R., and G.M. Wing, 1975, An Introduction to Invariant Imbedding (Wiley, New York).
Bugrov, A.G., 1988, On theory of sound propagation in quasi-layered ocean, Dokl. Akad. Nauk SSSR **301**(6), 1472–1474.
Bugrov, A.G., 1989, Quasi-layered medium approximation and imbedding equations in three-dimensional wave propagation problem, Izv. Vyssh. Uchebn. Zaved. Radiofiz. **32**(4), 468477.
Bugrov, A.G., and V.I. Klyatskin, 1989, Imbedding method and inverse problems for a layered medium, Izv. Vyssh. Uchebn. Zaved. Radiofiz. **32**(3), 321–330.
Bugrov, A.G., V.I. Klyatskin and B.M. Shevtzov, 1984, On the theory of radio wave propagation over the sea, Dokl. Akad. Nauk SSSR **275**(6), 1372–1376.
Bugrov, A.G., V.I. Klyatskin and B.M. Shevtzov, 1985, On the theory of short radio wave propagation in spherically-layered media, Radiotekh. & Elektr. **30**(4), 684–690.
Burridge, R., G. Papanicolaou, P. Sheng and B. White, 1989, Probing a random medium with a pulse, SIAM J. Appl. Math. **49**, 582–607.
Burridge, R., G. Papanicolaou and B. White, 1988, One-dimensional wave propagation in a highly discontinuous medium, Wave Motion **10**(1), 19–44.
Casti, J., and R. Calaba, 1973, Imbedding Methods in Applied Mathematics (Addison-Wesley, Reading, MA).
Chandrasekhar, S., 1960, Radiative Transfer (Dover, New York).
Chernov, L.A., 1975, Waves in Random Media (Nauka, Moscow). In Russian.
Corones, J.P., 1975, Bremmer series that correct parabolic approximations, J. Math. Anal. Appl. **50**, 361–370.
Corones, J.P., M.E. Davison and R.J. Krueger, 1983, Direct and inverse scattering in the time domain via invariant imbedding equations, J. Acoust. Soc. Am. **74**(5), 1535–1541.
De Santo, J.A., 1977, Relation between the solution of Helmholtz and parabolic equations for sound propagation, J. Acoust. Soc. Am. **62**(2), 295–297.
Ditkin, V.A., and A.P. Prudnikov, 1974, Integral Operations and Integral Calculations (Nauka, Moscow). In Russian.
Doucot, B., and R. Rammal, 1987, Invariant imbedding approach to localization. II. Nonlinear random media, J. Phys. **48**(2), 527–546.
Fortus, V.M., and B.M. Shevtzov, 1986, About the statistical approach for the light backscattering description, Opt. & Spektrosk. **60**(3), 578–582.
Freilikher, V.D., and S.A. Gredeskul, 1990, Randomly layered media: fluctuating waveguide, J. Opt. Soc. Am. A **7**(5), 868–874.
Freilikher, V.D., and S.A. Gredeskul, 1992, Localization of waves in media with one-dimensional disorder, in: Progress in Optics, Vol. 30, ed. E. Wolf (North-Holland, Amsterdam) pp. 137–203.
Furutsu, K., 1993, Random Media and Boundaries. Unified Theory, Two-Scale Method, and Applications (Springer, New York).
Gardiner, C.W., 1985, Handbook of Stochastic Methods for Physics, Chemistry and the Natural Sciences (Springer, Berlin).
Goland, V.I., 1987, The imbedding method in the problem of determining dispersion curves for internal waves in a stratified ocean, Morsk. Gidrofiz. Zh. **1**, 45–51.
Goland, V.I., 1988, Statistical characteristics of the normal modes of acoustical field in randomly inhomogeneous ocean, Akust. Zh. **34**(6), 1020 1022.

Goland, V.I., and V.I. Klyatskin, 1988, On the statistics of eigen-value and eigen-functions in one-dimensional boundary-value wave problem. Akust. Zh. **34**(5), 828–833.
Goland, V.I., and V.I. Klyatskin, 1989, On asymptotical methods of stochastic analysis of Shturm–Liouville problem, Akust. Zh. **35**(5), 942–944.
Goland, V.I., and K.V. Koshel', 1990, Method of the spectral parameter evolution in the problem with horizontal propagation of ultrashort waves, Radiotekh. & Elektr. **35**(9), 1805–1809.
Goland, V.I., V.I. Klyatskin and I.O. Yaroschuk, 1991, Some aspects of wave propagation theory in layer random media, Lectures Appl. Math. **27**, 477–486.
Gredeskul, S.A., and V.D. Freilikher, 1988, Waveguiding properties of randomly stratified media, Izv. Vyssh. Uchebn. Zaved. Radiofiz. **31**(10), 1210–1217.
Gredeskul, S.A., and V.D. Freilikher, 1990, Localization and wave propagation in randomly layered media, Sov. Phys. Usp. **33**(1), 134–136.
Gulin, O.E., 1984, On vector characteristics in statistical nonhomogeneous waveguides, Akust. Zh. **30**(4), 460–466.
Gulin, O.E., 1985, On the theory of acoustical noise in deep layered ocean, Akust. Zh. **31**(4), 524–527.
Gulin, O.E., 1987a, Numerical modelling of low-frequency acoustical noise in layered ocean, Akust. Zh. **33**(1), 113–116.
Gulin, O.E., 1987b, Spectrum of low-frequency acoustical noise in plane-layered ocean with impedance bottom, Akust. Zh. **33**(4), 618–623.
Gulin, O.E., and V.I. Klyatskin, 1986a, On the resonant structure of the spectral components of the acoustic field in the ocean when acted upon by atmospheric pressure, Izv. Akad. Nauk SSSR, Fiz. Atmosph. & Okeana **22**(3), 282–291.
Gulin, O.E., and V.I. Klyatskin, 1986b, On the theory of acoustical noise in randomly inhomogeneous ocean, Dokl. Akad. Nauk SSSR **288**(1), 226–228.
Gulin, O.E., and V.I. Klyatskin, 1989, Atmospheric excitation of low-frequency acoustic noise in a stratified ocean under various stratification models, in: Acoustics of the Ocean, ed. L.M. Brekhovskikh (Nauka, Moscow) pp. 133–140. In Russian.
Gulin, O.E., and V.I. Klyatskin, 1993, Generation of low-frequency acoustic noise in a stratified ocean, in: Book Natural Physical Sources of Underwater Acoustical Noise, ed. B. Kerman (Kluwer, Amsterdam).
Gulin, O.E., and V.V. Temchenko, 1990, On some analytical solutions of a one-dimensional problem of a pulse scattering on an inhomogeneous layer, Akust. Zh. **36**(4), 644–648.
Gulin, O.E., and V.V. Temchenko, 1992, Pulse scattering on periodic inhomogeneous media. Results of numerical simulation, Akust. Zh. **38**(3), 450–455.
Guzev, M.A., 1991, To the theory of wave fronts propagation in nonlinear layered medium, J. Prikl. Mat. & Tekn. Fiz. **4**(188), 63–67.
Guzev, M.A., and V.I. Klyatskin, 1991a, Plane waves in a layered weakly dissipative randomly inhomogeneous medium, Waves in Random Media **1**(1), 7–19.
Guzev, M.A., and V.I. Klyatskin, 1991b, Approximation of the parabolic equation and the wavefield of a point source in a layered random medium, Waves in Random Media **1**(4), 275–286.
Guzev, M.A., and V.I. Klyatskin, 1991c, To the theory of wave propagation in random media with two-scale inhomogeneities, Izv. Vyssh. Uchebn. Zaved. Radiofiz. **34**(3), 274–279.
Guzev, M.A., and V.I. Klyatskin, 1993, Influence of boundary conditions on statistical characteristics of wavefield in layered randomly inhomogeneous medium, Waves in Random Media **3**, 307–315.
Guzev, M.A., V.I. Klyatskin and G.V. Popov, 1992, Phase fluctuations and localization length in layered randomly inhomogeneous media, Waves in Random Media **2**(2), 117–123.
Hudson, J.A., 1980, A parabolic approximation for elastic waves, Wave Motion **2**(2), 207–217.
Jordan, K.E., G. Papanicolaou and R. Spigler, 1986, On the numerical solution of a nonlinear

stochastic Helmholtz equation with a multigrid preprocessor, Appl. Math. & Comp. **19**, 145–157.

Kagiwada, H.H., and R. Kalaba, 1974, Integral Equations via Imbedding Methods (Addison-Wesley, Reading, MA).

Klyatskin, V.I., 1975, Statistical Description of Dynamical Systems with Fluctuating Parameters (Nauka, Moscow). In Russian.

Klyatskin, V.I., 1979a, Stochastic wave parametric resonance (wave intensity fluctuations) inside a one-dimensional randomly inhomogeneous medium, Izv. Vyssh. Uchebn. Zaved. Radiofiz. **22**(2), 180–191.

Klyatskin, V.I., 1979b, Wave intensity fluctuations inside a one-dimensional randomly inhomogeneous layer of the medium. II. Izv. Vyssh. Uchebn. Zaved. Radiofiz. **22**(5), 591–597.

Klyatskin, V.I., 1980a, Stochastic Equations and Waves in Randomly Inhomogeneous Medium (Nauka, Moscow). In Russian.

Klyatskin, V.I., 1980b, On statistical theory of wave propagation in parabolic waveguide, Akust. Zh. **26**(2), 207–213.

Klyatskin, V.I., 1985, Ondes et Equations Stochastiques dans les Milieus Aleatoirement non Homogenes (Editions de Physique, Besançon). In French.

Klyatskin, V.I., 1986, The Imbedding Method in Wave Propagation Theory (Nauka, Moscow). In Russian.

Klyatskin, V.I., 1991a, Plane waves in layered random media. The role of boundary conditions, in: Applied and Industrial Mathematics, ed. R. Spigler (Kluwer, Amsterdam) 291–299.

Klyatskin, V.I., 1991b, The statistical theory of radiative transfer in layered random media, Izv. Akad. Nauk SSSR, Fiz. Atmosph. & Okeana **27**(1), 4546.

Klyatskin, V.I., 1991c, A statistical theory of radiative transfer in layered random media, in: Mathematical and Numerical Aspects of Wave Propagation Phenomena, eds G. Cohen, L. Halpern and L. Joly (SIAM, Philadelphia, PA) 595–608.

Klyatskin, V.I., 1991d, Approximations by delta-correlated random processes and diffusive approximation in stochastic problem, Lectures Appl. Math. **27**, 447–476.

Klyatskin, V.I., 1993, Caustics in random media, Waves in random media **3**(2), 93–100.

Klyatskin, V.I., and K.V. Koshel', 1983, Numerical simulation of wave propagation in periodic media, Zh. Eksp. & Teor. Fiz. **84**(3), 2092–2098.

Klyatskin, V.I., and K.V. Koshel', 1984, Numerical simulation of Bragg resonator in nonhomogeneous media, Izv. Vyssh. Uchebn. Zaved. Radiofiz. **27**(2), 263–265.

Klyatskin, V.I., and K.V. Koshel', 1986, The field of a point source in a layered media, Dokl. Akad. Nauk SSSR **288**(6), 1478–1481.

Klyatskin, V.I., and L.Ya. Lyubavin, 1983a, Theory of sound propagation in a layered ocean with varying density, Akust. Zh. **29**(1), 64–68.

Klyatskin, V.I., and L.Ya. Lyubavin, 1983b, On the boundary-value problems of the theory of internal wave propagation in a stratified ocean, Dokl. Akad. Nauk SSSR **271**(6), 1496–1498.

Klyatskin, V.I., and L.Ya. Lyubavin, 1984, On the theory of the excitation and propagation of acoustic-gravity waves in a stratified ocean, Izv. Akad. Nauk SSSR, Fiz. Atmosph. & Okeana **20**(5), 422 430.

Klyatskin, V.I., and A.I. Saichev, 1992, Statistical and dynamical localization of plane waves in randomly layered media, Sov. Phys. Usp. **35**(3), 231–247.

Klyatskin, V.I., and I.O. Yaroschuk, 1983a, Wave intensity fluctuations inside a one-dimensional randomly inhomogeneous medium. VI. An account of the boundary influence, Izv. Vyssh. Uchebn. Zaved. Radiofiz. **26**(9), 1092–1099.

Klyatskin, V.I., and I.O. Yaroschuk, 1983b, Wave intensity fluctuations inside a one-dimensional randomly inhomogeneous medium. VII. Numerical modeling wave propagation in random media, Izv. Vyssh. Uchebn. Zaved. Radiofiz. **26**(10), 1241–1250.

Klyatskin, V.I., and I.O. Yaroschuk, 1983c, Wave intensity fluctuations inside a one-dimensional randomly inhomogeneous medium. VIII. Influence of the medium model, Izv. Vyssh. Uchebn. Zaved. Radiofiz. **27**(11), 1395–1402.

Klyatskin, V.I., and E.V. Yaroschuk, 1985, Numerical solution on one-dimensional problem of self-action of wave in nonlinear medium layer, Izv. Vyssh. Uchebn. Zaved. Radiofiz. **28**(3), 320329.

Klyatskin, V.I., V.F. Kozlov and E.V. Yaroschuk, 1982, On the reflection coefficient in a one-dimensional problem of self-influence of waves, J. Teor. & Exper. Fiz. **82**(2), 386–396.

Knapp, R., G. Papanicolaou and B. White, 1989, Nonlinearity and localization in one-dimensional random media, in: Nonlinearity and Disorder, eds A.R. Bishop, D.K. Campbell and S. Pnevmaticos, Springer Series in Solid-State Science (Springer, Berlin).

Kohler, W., and G. Papanicolaou, 1973, Power statistics for wave propagation in one dimension and comparison with radiative transport theory. I. J. Math. Phys. **14**(12), 1733–1745.

Kohler, W., and G. Papanicolaou, 1974, Power statistics for wave propagation in one dimension and comparison with radiative transport theory. II. J. Math. Phys. **15**(12), 2186–2197.

Kohler, W., and G. Papanicolaou, 1976, Power reflection from a lossy one-dimensional random medium, SIAM J. Appl. Math. **30**(2), 263–267.

Kohler, W., G. Papanicolaou, M. Postel and B. White, 1991a, Reflection of waves generated by a point source over a randomly layered medium, Wave Motion **13**(1), 53–87.

Kohler, W., G. Papanicolaou, M. Postel and B. White, 1991b, Reflection of pulsed electromagnetic waves from a randomly stratified half space, J. Opt. Soc. Am. A **8**(7), 1109–1125.

Koshel', K.V., 1986, Numerical solution of short radio wave propagation problem in tropospheric waveguide, Radiotekh. & Elektr. **31**(12), 2313–2318.

Koshel', K.V., 1987, Numerical solution of the tropospheric propagation problem of short radio waves. Point of receiving over the surface, Radiotekh. & Elektr. **32**(6), 13051308.

Koshel', K.V., 1990a, On the effect of the elevated tropospheric layers on long-range short radio wave propagation, Radiotekh. & Elektr. **35**(3), 647–649.

Koshel', K.V., 1990b, Quantity analysis of the problem of short waves tropospheric propagation with the finite impedance of the surface, Radiotekh. & Elektr. **35**(6), 1326–1329.

Koshel', K.V., 1990c, On horizon short radio wave propagation in tropospheric waveguide with layered fluctuations of refraction index, Radiotekh. & Elektr. **35**(12), 2502–2507.

Koshel', K.V., 1990d, Short radio waves propagation about horizon in nonuniform tropospheric waveguide, Izv. Akad. Nauk SSSR, Fiz. Atmosph. & Okeana **26**(10), 1069–1077.

Koshel', K.V., 1992a, The influence of refractive index fluctuations on the beyond-the-horizon SHF propagation above the sea in the evaporation duct, Izv. Akad. Nauk SSSR, Fiz. Atmosph. & Okeana **28**(10–11), 1054–1061.

Koshel', K.V., 1992b, Application of invariant imbedding method to simulate numerically beyond-the-horizon propagation of SHF over the sea, J. Electromag. Waves and Appl. **6**(10), 1433–1453.

Koshel', K.V., and A.A. Shishkarev, 1993a, Influence of layer and anisotropic fluctuations of the refractive index on the beyond-the-horizon SHF propagation in the troposphere over the sea when there is evaporation duct, Waves in Random Media **3**(1), 25–38.

Koshel', K.V., and A.A. Shishkazev, 1993b, Influence of anisotropic fluctuations of the refractive index on the beyond-the-horizon SHF propagation in the troposphere, Izv. Akad. Nauk, Fiz. Atmosph. & Okeana **29**(1), 86–91.

Kravtsov, Yu.A., and A.I. Saichev, 1982, Effects of double passage of waves in randomly inhomogeneous media, Sov. Phys. Usp. **25**(7), 494–508.

Kravtsov, Yu.A., and A.I. Saichev, 1985, Properties of coherent waves reflected in a turbulent medium, J. Opt. Soc. Am. A **2**(12), 2100–2105.

Kreider, K.L., 1989, Time-dependent inverse scattering from gradient-type interfaced using an exact solution, J. Math. Phys. **30**(1), 53–58.

Kristensson, G., and R.J. Krueger, 1986a, Direct and inverse scattering in the time domain for a dissipative wave equation. I. Scattering operators, J. Math. Phys. **27**(6), 1667–1682.

Kristensson, G., and R.J. Krueger, 1986b, Direct and inverse scattering in the time domain for a dissipative wave equation. II. Simultaneous reconstruction of dissipation and phase velocity profiles, J. Math. Phys. **27**(6) 1683–1693.
Kristensson, G., and R.J. Krueger, 1987, Direct and inverse scattering in the time domain for a dissipative wave equation. III. Scattering operators in the presence of a phase velocity mismatch, J. Math. Phys. **28**(2), 360–370.
Kristensson, G., and R.J. Krueger, 1989, Direct and inverse scattering in the time domain for a dissipative wave equation. IV. Use of the phase velocity mismatches to simplify inversions, Inverse Problems **5**(2), 375–388.
Krueger, R.J., and R.L. Ochs Jr, 1989, A Green's function approach to the determination of internal fields, Wave Motion **11**(6), 525–543.
Kulkarny, V.A., and B.S. White, 1982, Focusing of rays in a turbulent inhomogeneous medium, Phys. Fluids **25**, 1770–1784.
LeMesurier, B., G. Papanicolaou, C. Sulem and P.-L. Sulem, 1987, The focusing singularity of the nonlinear Schroedinger equation, in: Directions in Partial Differential Equations, eds M.G. Crandall, P.H. Rainowitz and R.E.L. Turner (Academic Press, New York) pp. 159–201.
LeMesurier, B., G. Papanicolaou, C. Sulem and P.-L. Sulem, 1988a, Focusing and multifocusing solutions of the nonlinear Schroedinger equation, Physica D **31**(5), 78–102.
LeMesurier, B., G. Papanicolaou, C. Sulem and P.-L. Sulem, 1988b, Local structure of the self focusing singularity of the nonlinear Schroedinger equation, Physica D **32**, 210–226.
Lifshits, I.M., S.A. Gredeskul and L.A. Pastur, 1988, Introduction to the Theory of Disordered Solids (Wiley, New York).
Malakhov, A.N., and A.I. Saichev, 1979, Representation of a wave reflected from a randomly inhomogeneous layer in the form of a series satisfying the causality condition, Izv. Vyssh. Uchebn. Zaved. Radiofiz. **22**(11), 1324–1333.
Manning, R.M., 1993, Stochastic Electromagnetic Image Preparation and Adaptive Compensation (McGraw-Hill, New York).
McDaniel, S.T., 1975, Parabolic approximation for underwater sound propagation, J. Acoust. Soc. Am. **58**, 1178–1185.
McLaughlin, D., G. Papanicolaou, P.-L. Sulem and C. Sulem, 1986, Focusing singularity of the nonlinear Schroedinger equation, Phys. Rev. A **34**(2), 1200–1210.
Papanicolaou, G., D. McLaughlin and R. Burridge, 1973, A stochastic Gaussian beam, J. Math. Phys. **14**(1), 84–89.
Papanicolaou, G., M. Postel, P. Sheng and B. White, 1990, Frequency content of randomly scattered signals. Part II: Inversion, Wave Motion **12**(4), 527–549.
Polyanskii, E.A., 1972, On connection between the solutions of Helmholtz equation and Schroedinger type equation, Zh. Vichisl. Mat. & Mat. Fiz. **12**(1), 241–249.
Polyanskii, E.A., 1985, Correction Method for the Solution of Parabolic Equation in Inhomogeneous Waveguide (Nauka, Moscow). In Russian.
Popov, G.V., and I.O. Yaroschuk, 1988, On the oblique plane wave incidence on a stratified stochastic medium, Izv. Vyssh. Uchebn. Zaved. Radiofiz. **31**(10), 1266–1267.
Popov, G.V., and I.O. Yaroschuk, 1990, Point source radiation in random layered media (spectral analysis of wave field), Izv. Vyssh. Uchebn. Zaved. Radiofiz. **33**(11), 1232–1240.
Rammal, R., and B. Doucot, 1987, Invariant imbedding approach to localization. I. General framework and basic equations, J. Phys. **48**(2), 509–526.
Rytov, S.M., Yu.A. Kravtsov and V.I. Tatarskii, 1987–1989, Principles of Statistical Radiophysics, Volumes 1–4 (Springer, Berlin).
Saichev, A.I., 1980, On the statistics of the eigen-values of a one-dimensional randomly inhomogeneous boundary-value problem, Izv. Vyssh. Uchebn. Zaved. Radiofiz. **23**(2), 183–188.
Saichev, A.I., and M.M. Slavinskij, 1985, Equations for moment functions of waves propagation in a medium with extended random irregularities, Izv. Vyssh. Uchebn. Zaved. Radiofiz. **28**(1), 75–83.

Scott, M.R., L.F. Shampine and G.M. Wing, 1969, Invariant imbedding and the calculation of eigen-values for Sturm–Liouville systems, Computing **4**(1), 10–17.
Sheng, P., ed., 1990, Scattering and Localization of Classical Waves in Random Media (World Scientific, Singapore).
Sheng, P., B. White, Z. Zhang and G. Papanicolaou, 1986a, Multiple scattering noise in one dimension: universality through localization length scales, Phys. Rev. Lett. **57**(8), 1000–1003.
Sheng, P., B. White, Z. Zhang and G. Papanicolaou, 1986b, Minimum wave localization length in a one dimensional random medium, Phys. Rev. B **34**(7), 4757–4761.
Sheng, P., B. White, Z. Zhang and G. Papanicolaou, 1990, Wave localization and multiple scattering in randomly layered media, in: Scattering and Localization of Classical Waves in Random Media, ed. P. Sheng (World Scientific, Singapore) pp. 563–619.
Shevtsov, B.M., 1981, To the statistical theory of backscattering in randomly inhomogeneous media, Izv. Vyssh. Uchebn. Zaved. Radiofiz. **24**(11), 1351–1355.
Shevtsov, B.M., 1982, Three-dimensional problem of backscattering in stratified randomly inhomogeneous media, Izv. Vyssh. Uchebn. Zaved. Radiofiz. **25**(9), 1032–1040.
Shevtsov, B.M., 1983, A problem of backscattering in three-dimensional randomly inhomogeneous media, Izv. Vyssh. Uchebn. Zaved. Radiofiz. **26**(4), 434–439.
Shevtsov, B.M., 1985, Statistical characteristics of the backscattering field, Izv. Vyssh. Uchebn. Zaved. Radiofiz. **28**(6), 717–724.
Shevtsov, B.M., 1987, Statistical characteristics for wave packet scattering in a layered randomly inhomogeneous medium above a reflecting surface, Izv. Vyssh. Uchebn. Zaved. Radiofiz. **30**(8), 1007–1012.
Shevtsov, B.M., 1989, Backscattering of a wave in layered regularly and randomly inhomogeneous media, Izv. Vyssh. Uchebn. Zaved. Radiofiz. **32**(9), 1079–1083.
Shevtsov, B.M., 1990, The statistical characteristics of the wave in the layer regular and random medium, Izv. Vyssh. Uchebn. Zaved. Radiofiz. **33**(2), 191–195.
Sobolev, V.V., 1956, Radiative Transfer in Atmospheres of Stars and Planets (Gostekhizdat, Moscow). In Russian.
Spigler, R.J., 1986, Mean power reflection from a one-dimensional nonlinear random medium, J. Math. Phys. **27**(7), 1760–1771.
Tappert, F.D., 1977, The parabolic approximation method in wave propagation, in: Underwater Acoustics, eds J.B. Keller and J.S. Papadakis, Lecture Notes in Physics, Vol. 70 (Springer, New York).
Tatarskii, V.I., 1971, The Effects of the Turbulent Atmosphere on Wave Propagation (National Technical Information Service, Springfield, VA).
Virovlyanskij, A.L., A.I. Saichev and M.M. Slavinskij, 1985, Momentum functions of waves propagation in waveguide channels with extended chaotic irregularities of refractive index, Izv. Vyssh. Uchebn. Zaved. Radiofiz. **28**(9), 1149–1159.
White, B.S., 1983, The stochastic caustic, SIAM J. Appl. Math. **44**, 127–149.
Yaroschuk, I.O., 1984, On numerical simulation of one-dimensional stochastic wave problems, Zh. Vichisl. Mat. & Mat. Fiz. **24**(11), 1748–1751.
Yaroschuk, I.O., 1986a, Numerical Modeling of the Propagation of Plane Waves in Randomly Layered and Nonlinear Media, Dissertation (Pacific Oceanological Institute, Vladivostok).
Yaroschuk, I.O., 1986b, On the role of wave number in problems of the wave propagation in a stochastic medium, Izv. Vyssh. Uchebn. Zaved. Radiofiz. **29**(11), 1392–1394.
Yaroschuk, I.O., 1988a, Numerical solution on one-dimensional problem of self-action of a wave in a stochastic nonlinear medium, Izv. Vyssh. Uchebn. Zaved. Radiofiz. **31**(1), 53–60.
Yaroschuk, I.O., 1988b, On a method for numerical simulation of wave propagation in one-dimensional nonlinear media with stochastic nonhomogeneities, Zh. Vichisl. Mat. & Mat. Fiz. **28**(5), 760–764.
Zwillinger, D., and B.S. White, 1985, Propagation of initially plane waves in the region of random caustics, Wave Motion **7**, 207–227.

II

QUANTUM STATISTICS OF DISSIPATIVE NONLINEAR OSCILLATORS

BY

V. Peřinová and A. Lukš

Laboratory of Quantum Optics,
Natural Science Faculty,
Palacký University,
Olomouc,
Czech Republic

CONTENTS

	PAGE
§ 1. INTRODUCTION	131
§ 2. DISSIPATIVE THIRD-ORDER NONLINEAR OSCILLATOR	132
§ 3. HIGHER-ORDER NONLINEAR OSCILLATORS	184
§ 4. COUPLED DISSIPATIVE THIRD-ORDER NONLINEAR OSCILLATORS	185
§ 5. USE OF A KERR MEDIUM FOR GENERATION AND DETECTION OF NONCLASSICAL STATES	194
ACKNOWLEDGEMENT	198
REFERENCES	198

§ 1. Introduction

Among the nonlinear optical phenomena, those effects which are connected with the dependence of the complex dielectric constant on the light intensity have drawn special attention. They are the so-called "self-action" effects of strong light beams. Unlike other nonlinear optical effects, where waves mix at several very different frequencies, the process of self-action assumes and exhibits quasimonochromaticity. The self-action effect then results in a change in the form of the amplitude (self-focusing, self-steepening), in a change in the phase (self-phase modulation), and in the state of polarization of the wave. The theoretical feasibility is related to an isotropic and third-order nonlinear medium or, more concisely, an optical Kerr medium. In quantum optics, both single-mode models and multi-mode situations are studied.

Many self-action nonlinear phenomena have been studied theoretically and observed experimentally, such as self-focusing, self-trapping, and self-phase modulation (Akhmanov, Khokhlov and Sukhorukov [1972], Svelto [1974], Shen [1976]). Stationary and nonstationary cases of self-focusing have been discerned with respect to the relationship between the time duration of laser pulses and the relaxation time of the medium. The self-focusing of the light beam is described by the wave equation:

$$\Delta E - \frac{1}{c^2}\frac{\partial^2}{\partial t^2}E = \frac{1}{c^2}\frac{\partial^2}{\partial t^2}[\chi^{(1)}E + \chi^{(3)}(E\cdot E)E], \qquad (1.1)$$

where E represents the electric field vector, $\chi^{(1)}$ and $\chi^{(3)}$ are the susceptibilities of order 1 and 3 respectively, c is the velocity of light in vacuum, and Δ is the Laplace operator. Using the relation

$$E = E^{(+)}\exp(-i\omega t) + \text{c.c.}, \qquad (1.2)$$

where c.c. stands for a complex conjugate, performing a slowly varying envelope approximation (considering only one polarization component $E^{(+)} = E^{(+)}(z)\mathbf{e}$, where $\mathbf{e}$ is the unit polarization vector), and using a paraxial approximation, we obtain the nonlinear Schrödinger equation. Even the solution of this idealized description is not quite simple and it leads to the soliton solution (Infeld and Rowlands [1990]). Solitons have been generated in a nonlinear planar waveguide (Zakharov and Shabat [1971], Barthelemy, Maneuf and Froehly [1985], Maneuf, Desailly and Froehly [1988], Hasegawa [1990]); quantum solitons have been studied by many authors (Thacker [1981], Singer, Potasek, Fang and Teich [1992]). A final simplification which neglects self-focusing yields the solution (Yariv [1967]):

$$E^{(+)}(z) = E^{(+)}(0) \exp\left(-i \frac{k}{4n^2} \chi^{(3)} |E^{(+)}(0)|^2 z \right), \tag{1.3}$$

where $n = (1 + \chi^{(1)})^{1/2}$ is the refractive index and k is the wave number. This solution represents a good analogue to that for the quantum conservative third-order nonlinear oscillator.

The use of Kerr media in photonics is well known (Saleh and Teich [1991]). The nonlinear properties of fibers may be used to cancel the fiber dispersion so that pulses behave as if they were travelling through a linear nondispersive medium. The spreadless pulses are known as optical solitons.

Optical bistability has been observed in a number of materials that exhibit the optical Kerr effect (Gibbs [1985]). The application is important in the digital circuits used in communications, signal processing, and computing. Bistable devices are used as switches, logical gates, and memory elements (Saleh and Teich [1991]).

Quantum nondemolition measurement of photon number based on the optical Kerr effect has been proposed by Imoto, Haus and Yamamoto [1985], Kitagawa, Imoto and Yamamoto [1987], and Imoto and Saito [1989]. Kerr media are used for the generation of nonclassical states (Kitagawa and Yamamoto [1986], Yamamoto, Machida, Imoto, Kitagawa and Björk [1987]) and their detection (Agarwal [1989], Hillery [1991]).

§ 2. Dissipative Third-Order Nonlinear Oscillator

In quantum optics a Kerr medium is modelled as the one-mode quantum third-order nonlinear oscillator (Selloni and Schwendimann [1979], Drummond and Walls [1980], Tanaś [1984], Collett and Walls [1985], Milburn [1986]). From quantum mechanics, the one-dimensional anharmonic oscillator is known to be connected with a special potential of force, but the nonlinear optical oscillator assumes the Hamiltonian to be a nonlinear function of the photon number.

The third-order nonlinear oscillator has stimulated an interest from the viewpoint of quantum statistical physics (Haake, Risken, Savage and Walls [1986]), and it has also been developed within the framework of nonlinear quantum optics. The statistical properties of this nonlinear oscillator were first studied by Drummond and Walls [1980]. Neglecting losses, the dynamics of the system can be described exactly (Drummond and Walls [1980], Tanaś [1984], Milburn [1986]). Including losses, the dynamics of the system has until recently been treated exactly for a "quiet" reservoir of zero temperature only (Milburn and Holmes [1986], Peřinová and Lukš [1988], Milburn, Mecozzi and Tombesi [1989]), and a general initial state has been considered by Peřinová and Lukš [1988]. The effect of quantum fluctuations has been evaluated with the help of approximate methods (Peřinová and Lukš [1988]). The assumption of zero reservoir temperature has been removed (Daniel and Milburn [1989], Peřinová and Lukš [1990], Chaturvedi and Srinivasan [1991a]), and the dynamics of the nonlinear oscillator with a noisy reservoir has been determined exactly for an initial coherent state (Daniel and Milburn [1989]) and for an initial general state (Peřinová and Lukš [1990], Chaturvedi and Srinivasan [1991a]).

2.1. QUANTUM DYNAMICS

Modelling dissipation by coupling the third-order nonlinear oscillator to a reservoir of (linear) oscillators, we can formulate the Hamiltonian of the compound optical system (Peřinová and Lukš [1988]):

$$\hat{H} = \hat{H}_F + \hat{H}_N + \hat{H}_R + \hat{H}_I, \tag{2.1}$$

where

$$\hat{H}_F = \hbar\omega(\hat{a}^+ \hat{a} + \tfrac{1}{2}), \tag{2.2}$$

$$\hat{H}_N = \hbar\kappa \hat{a}^{+2} \hat{a}^2, \tag{2.3}$$

$$\hat{H}_R = \hbar \sum_j \psi_j (\hat{c}_j^+ \hat{c}_j + \tfrac{1}{2}), \tag{2.4}$$

and

$$\hat{H}_I = \hbar \sum_j (\eta_j \hat{c}_j \hat{a}^+ + \text{H.c.}); \tag{2.5}$$

where H.c. stands for a Hermitian conjugate. In eqs. (2.2), (2.3), and (2.5), $\hat{a}$ ($\hat{a}^+$) is the photon annihilation (creation) operator describing the radiation field of the frequency ω, κ is a real constant for the intensity dependence, $\hat{c}_j$ ($\hat{c}_j^+$) in eqs. (2.4) and (2.5) are the boson annihilation (creation) operators of the reservoir oscillators with the frequencies ψ_j, and η_j are the coupling constants of the radiation to the reservoir. Some papers (Milburn [1986], Milburn and Holmes [1986], Milburn, Mecozzi and Tombesi [1989]) involve the Kerr nonlinearity in the form:

$$\hat{H}_S = \hbar\kappa(\hat{a}^+\hat{a})^2. \tag{2.6}$$

Comparisons of squeezing properties and quantum coherence on the assumption of either ordering have been performed by Tanaś [1989] and by Peřinová and Lukš [1990], respectively.

We assume that the reservoir is chaotic and has a broad-band spectrum. We further assume that the mean number of reservoir oscillators in the jth mode does not depend on j; $\bar{n}_d = \langle \hat{c}_j^+(0)\hat{c}_j(0)\rangle$. In the standard treatments of the quantum theory of damping (Peřina [1991]), the master equation for the reduced density operator $\hat{\rho}_r$ in the interaction picture ($\hat{a} \to \exp(i\omega t)\hat{a}$) can be derived as:

$$\frac{\partial}{\partial t}\hat{\rho}_r = -i\kappa(\hat{a}^{+2}\hat{a}^2\hat{\rho}_r - \hat{\rho}_r\hat{a}^{+2}\hat{a}^2) + \frac{\gamma}{2}(2\hat{a}\hat{\rho}_r\hat{a}^+ - \hat{a}^+\hat{a}\hat{\rho}_r - \hat{\rho}_r\hat{a}^+\hat{a})$$

$$+ \gamma\bar{n}_d(\hat{a}^+\hat{\rho}_r\hat{a} + \hat{a}\hat{\rho}_r\hat{a}^+ - \hat{a}^+\hat{a}\hat{\rho}_r - \hat{\rho}_r\hat{a}\hat{a}^+), \tag{2.7}$$

where γ is a damping constant determined by the reservoir correlation function, $\bar{n}_d = [\exp(\hbar\omega/KT) - 1]^{-1}$, T is the absolute temperature of the reservoir, and K is the Boltzmann constant. This way of combining the Hamilton term and the non-Hamilton terms in the master equation holds for very low temperatures KT. The generalization of master eq. (2.7) to finite temperatures and nonlinearity has been performed with the restriction to the limit of small damping (Haake, Risken, Savage and Walls [1986]). In that case, the quantities $\Delta\omega$ and $\bar{n}_d$, obtained by the elimination of reservoir variables, are not constant but depend on the intensity, whereas γ remains constant. Then it is necessary to interpret the quantities $\Delta\omega$ and $\bar{n}_d$ as operators in the master equation. In this case, the standard methods cannot be applied when deriving the generalized Fokker–Planck equation.

Considering the classical–quantum correspondence (Lukš and Peřinová [1987]):

$$\tilde{C}\phi_{\mathscr{A}} = \pi^{-1}\hat{\rho}_r, \tag{2.8}$$

related to the quantum correspondence $C^{-1}(\hat{a}^k\hat{a}^{+l}) = \alpha^k\alpha^{*l}$, where the complex amplitude α corresponds to the operator $\exp(i\omega t)\hat{a}$, we obtain the generalized Fokker–Planck equation for the quasidistribution $\phi_{\mathscr{A}}(\alpha, t)$

appropriate to the antinormal ordering of field operators:

$$\frac{\partial}{\partial t}\phi_{\mathscr{A}}(\alpha,t) = \left\{ i\kappa \left[2\alpha|\alpha|^2 \frac{\partial}{\partial \alpha} + \alpha^2 \frac{\partial^2}{\partial \alpha^2} - \text{c.c.} \right] \right.$$

$$+ \gamma + \frac{\gamma}{2}\left(\alpha \frac{\partial}{\partial \alpha} + \alpha^* \frac{\partial}{\partial \alpha^*} \right)$$

$$\left. + \gamma(\bar{n}_d + 1) \frac{\partial^2}{\partial \alpha \partial \alpha^*} \right\} \phi_{\mathscr{A}}(\alpha,t). \tag{2.9}$$

We assume that the initial state is described by a quasidistribution $\phi_{\mathscr{A}}(\alpha,0)$, and we define

$$f_{mn}(0) = \frac{1}{m!n!} \frac{\partial^{m+n}}{\partial \alpha^m \partial \alpha^{*n}} \left[\exp(|\alpha|^2) \phi_{\mathscr{A}}(\alpha,0) \right] \Bigg|_{\substack{\alpha=0\\ \alpha^*=0}}. \tag{2.10}$$

On expressing the quasidistribution $\phi_{\mathscr{A}}(\alpha,t)$ in the form:

$$\phi_{\mathscr{A}}(\alpha,t) = \exp(-|\alpha|^2)\phi'_{\mathscr{A}}(\alpha,t), \tag{2.11}$$

and using the rules

$$\pi^{-1}\tilde{C}^{-1}(\hat{a}^+\hat{\rho}_r) = \exp(-|\alpha|^2)\alpha^*\phi'_{\mathscr{A}},$$

$$\pi^{-1}\tilde{C}^{-1}(\hat{\rho}_r\hat{a}^+) = \exp(-|\alpha|^2)\frac{\partial}{\partial \alpha}\phi'_{\mathscr{A}},$$

$$\pi^{-1}\tilde{C}^{-1}(\hat{a}\hat{\rho}_r) = \exp(-|\alpha|^2)\frac{\partial}{\partial \alpha^*}\phi'_{\mathscr{A}}, \tag{2.12}$$

$$\pi^{-1}\tilde{C}^{-1}(\hat{\rho}_r\hat{a}) = \exp(-|\alpha|^2)\alpha\phi'_{\mathscr{A}},$$

eq. (2.7) becomes the equation for $\phi'_{\mathscr{A}}(\alpha,t)$,

$$\frac{\partial}{\partial t}\phi'_{\mathscr{A}}(\alpha,t) = \left\{ \gamma\bar{n}_d|\alpha|^2 + \left[i\kappa D\left(\alpha, \frac{\partial}{\partial \alpha}\right) - \frac{\gamma}{2} - \gamma\bar{n}_d \right]\left(\alpha \frac{\partial}{\partial \alpha} + \text{c.c.} \right) \right.$$

$$\left. + \gamma(\bar{n}_d + 1)\frac{\partial^2}{\partial \alpha \partial \alpha^*} - i\kappa D\left(\alpha, \frac{\partial}{\partial \alpha}\right) - \gamma\bar{n}_d \right\} \phi'_{\mathscr{A}}(\alpha,t), \tag{2.13}$$

where, from group-theoretical or Lie-algebraic considerations (Chaturvedi and Srinivasan [1991a], Ban [1992]), we introduce the symbol

$$D\left(\alpha, \frac{\partial}{\partial \alpha}\right) = \alpha \frac{\partial}{\partial \alpha} - \alpha^* \frac{\partial}{\partial \alpha^*}. \tag{2.14}$$

We suppose the solution of eq. (2.13) in the form (Peřinová and Lukš [1988]):

$$\phi'_{\mathscr{A}}(\alpha, t) = \sum_{m=0}^{\infty} \sum_{n=0}^{\infty} \alpha^m \alpha^{*n} f_{mn}(t). \tag{2.15}$$

Following Daniel and Milburn [1989] and Peřinová and Lukš [1990], we express $\phi'_{\mathscr{A}}(\alpha, t)$ as:

$$\phi'_{\mathscr{A}}(\alpha, t) = \exp(|\alpha|^2 g_D(t)) H(\alpha, t), \tag{2.16}$$

where

$$D \equiv D\left(\alpha, \frac{\partial}{\partial \alpha}\right). \tag{2.17}$$

The assumption (2.16) provides the evolution equation for the differential operator $g_D(t)$,

$$\frac{d}{dt} g_D(t) = \gamma(\bar{n}_d + 1) g_D^2(t) + \left[-\gamma(2\bar{n}_d + 1) + 2i\kappa D\left(\alpha, \frac{\partial}{\partial \alpha}\right) \right] g_D(t) + \gamma \bar{n}_d, \tag{2.18}$$

with the initial condition $g_D(0) = 0$, and the partial differential equation for the function $H(\alpha, t)$,

$$\frac{\partial}{\partial t} H(\alpha, t) = \left\{ A\left(\alpha, \frac{\partial}{\partial \alpha}\right) + g_D(t) B\left(\alpha, \frac{\partial}{\partial \alpha}\right) + C\left(\frac{\partial}{\partial \alpha}\right) \right\} H(\alpha, t), \tag{2.19}$$

where

$$A\left(\alpha, \frac{\partial}{\partial \alpha}\right) = i\kappa D\left(\alpha, \frac{\partial}{\partial \alpha}\right)\left(\alpha \frac{\partial}{\partial \alpha} + \text{c.c.} - 1\right) - \frac{\gamma}{2}\left(\alpha \frac{\partial}{\partial \alpha} + \text{c.c.}\right)$$

$$- \gamma \bar{n}_d \left(\alpha \frac{\partial}{\partial \alpha} + \text{c.c.} + 1\right), \tag{2.20}$$

$$B\left(\alpha, \frac{\partial}{\partial \alpha}\right) = \gamma(\bar{n}_d + 1)\left(\alpha \frac{\partial}{\partial \alpha} + \text{c.c.} + 1\right), \quad C\left(\frac{\partial}{\partial \alpha}\right) = \gamma(\bar{n}_d + 1)\frac{\partial^2}{\partial \alpha \partial \alpha^*},$$

with the initial condition

$$H(\alpha, 0) = \exp(|\alpha|^2) \phi_{\mathscr{A}}(\alpha, 0). \tag{2.21}$$

The initial-value problem [eq. (2.18)] has for $\gamma > 0$ the solution,

$$g_D(t) = \frac{2\bar{n}_d}{\Omega + \Delta \coth\left(\frac{\gamma}{2} \Delta t\right)}, \tag{2.22}$$

where

$$\Omega \equiv \Omega_D = 1 + 2\bar{n}_d + i\frac{2}{\gamma}\kappa D\left(\alpha, \frac{\partial}{\partial\alpha}\right),$$
$$\Delta \equiv \Delta_D = [\Omega^2 - 4\bar{n}_d(\bar{n}_d + 1)]^{1/2}.$$
(2.23)

For $\gamma = 0$, we have

$$g_D(t) = \exp\left[2i\kappa D\left(\alpha, \frac{\partial}{\partial\alpha}\right)t\right].$$
(2.24)

The initial-value problem [eq. (2.19)] yields the solution

$$H(\alpha, t) = \exp\left\{\left[-2i\kappa D\left(\alpha, \frac{\partial}{\partial\alpha}\right) + \frac{\gamma}{2}\right]t\right\} \exp\left\{\left(\alpha\frac{\partial}{\partial\alpha} + \alpha^*\frac{\partial}{\partial\alpha^*}\right)\ln E_D(t)\right\}$$
$$\times \exp\left\{C\left(\frac{\partial}{\partial\alpha}\right)\frac{1}{\gamma\bar{n}_d}g_D(t)\right\}\phi_{\mathscr{A}}(\alpha, 0),$$
(2.25)

where

$$E_D(t) = \frac{\Delta}{\Omega \sinh\left(\frac{\gamma}{2}\Delta t\right) + \Delta \cosh\left(\frac{\gamma}{2}\Delta t\right)}.$$
(2.26)

With respect to eqs. (2.11) and (2.10), we obtain the functions $f_{mn}(t)$ in the form:

$$f_{mn}(t) = \exp\left\{\left[-2i\kappa(m-n) + \frac{\gamma}{2}\right]t\right\} E_{m-n}^{m+n+1}(t)$$
$$\times \sum_{j=0}^{\min(m,n)} \frac{1}{j!}\left[\frac{g_{m-n}(t)}{E_{m-n}^2(t)}\right]^j \sum_{l=0}^{\infty} \frac{1}{l!}\left[\frac{(\bar{n}_d + 1)}{\bar{n}_d}g_{m-n}(t)\right]^l$$
$$\times \frac{(m-j+l)!(n-j+l)!}{(m-j)!(n-j)!} f_{m-j+l,n-j+l}(0),$$
(2.27)

where

$$g_{m-n}(t) = g_D(t)|_{D \to (m-n)}, \qquad E_{m-n}(t) = E_D(t)|_{D \to (m-n)}.$$
(2.28)

The optical nonlinearity κ appears in $f_{mn}(t)$ multiplied by a factor $(m-n)$ which indicates the off-diagonality.

Let us note that for $\bar{n}_d = 0$ we derive formula (12) in the paper by Peřinová and Lukš [1988] using the limit:

$$\lim_{\bar{n}_d \to 0} \frac{1}{\bar{n}_d} g_{m-n}(t) = \frac{\gamma}{\gamma - 2i\kappa(m-n)} [1 - \exp\{[-\gamma + 2i\kappa(m-n)]t\}]. \tag{2.29}$$

Equations (2.11) and (2.15) yield the quasidistribution $\phi_{\mathscr{A}}(\alpha, t)$:

$$\phi_{\mathscr{A}}(\alpha, t) = \exp(-|\alpha|^2) \sum_{m=0}^{\infty} \sum_{n=0}^{\infty} \alpha^m \alpha^{*n} f_{mn}(t). \tag{2.30}$$

Similarly,

$$\rho_{nm}(t) = \langle n|\hat{\rho}_r(t)|m\rangle = \pi(n!m!)^{1/2} f_{mn}(t). \tag{2.31}$$

The photon-number distribution $p(n, t)$ is

$$p(n, t) = n! f_{nn}(t). \tag{2.32}$$

The exact formula for the quasidistribution $\phi_{\mathscr{A}}(\alpha, t)$ is highly instructive in revealing specific quantum features of the evolution.

Following previous studies of Fokker–Planck equations, an eigenvalue analysis of the evolution operators [eq. (2.27)] has been provided by Kärtner and Schenzle [1993].

2.2. PHOTON STATISTICS

The photon-number distribution of the dissipative third-order nonlinear oscillator obeys the rate equation for the dissipative linear (harmonic) oscillator,

$$\frac{d}{dt} p(n, t) = [-\gamma \bar{n}_d - \gamma(2\bar{n}_d + 1)n] p(n, t) + \gamma \bar{n}_d n p(n-1, t)$$

$$+ \gamma(\bar{n}_d + 1)(n + 1) p(n+1, t), \tag{2.33}$$

which implies that the photon statistics of this nonlinear oscillator coincide with those of the linear oscillator under the same initial condition.

The factorial moments of the distribution $p(n, t)$ are expressed as:

$$\left\langle \frac{n(t)!}{(n(t)-k)!} \right\rangle = \langle W^k(t) \rangle_{\mathscr{N}} = \sum_{n=k}^{\infty} \frac{n!}{(n-k)!} p(n, t), \tag{2.34}$$

where $\langle W^k(t) \rangle_{\mathscr{N}}$ is the kth moment of the integrated intensity.

For the third-order nonlinear and for the harmonic oscillator, eqs. (2.27) and (2.32) in the "diagonal" limit provide (Peřinová and Lukš [1988]):

$$p(n, t) = \frac{1}{\bar{n}(t) + 1} \sum_{m=0}^{\infty} p(m, 0)(\bar{n}(t) + 1)^{-m}$$

$$\times \sum_{j=0}^{\min(m,n)} \frac{(m+n-j)!}{j!(n-j)!(m-j)!} (-1)^j [\bar{n}(t) - \exp(-\gamma t)]^j$$

$$\times \left[\frac{\bar{n}(t)}{\bar{n}(t) + 1}\right]^{n-j} [\bar{n}(t) + 1 - \exp(-\gamma t)]^{m-j}, \quad (2.35)$$

where

$$\bar{n}(t) = \bar{n}_d [1 - \exp(-\gamma t)]. \quad (2.36)$$

Then eq. (2.34) simplifies to the form:

$$\langle W^k(t) \rangle_{\mathcal{N}} = \pi \sum_{m=0}^{\infty} p(m, 0) \sum_{j=0}^{\min(k,m)} \binom{k}{j} \frac{(m+k-j)!}{(m-j)!} (-1)^j$$

$$\times [\bar{n}(t) - \exp(-\gamma t)]^j (\bar{n}(t))^{k-j}. \quad (2.37)$$

It is usual to determine the nonclassical character of the photon statistics by means of the second reduced factorial moment:

$$f \equiv f(t) = \frac{\langle (\Delta n(t))^2 \rangle - \langle n(t) \rangle}{\langle n(t) \rangle^2} = \frac{\langle (\Delta W(t))^2 \rangle_{\mathcal{N}}}{\langle W(t) \rangle_{\mathcal{N}}^2}, \quad (2.38)$$

which is known to be zero for Poissonian light; nonclassical (sub-Poissonian) behavior occurs for $f < 0$.

2.3. QUANTUM COHERENCE

The quasidistribution $\phi_{\mathcal{A}}(\alpha, t)$ illustrates a number of important properties of the third-order nonlinear oscillator model. It is known that the initial state of the lossless nonlinear oscillator revives after a certain time interval (the period). Milburn and Holmes [1986] have explained the role of higher-order derivatives in the equation of motion for the Q-function ($Q = \pi \phi_{\mathcal{A}}$) with respect to the revivals of the initial state. Daniel and Milburn [1989] have pointed out the interference in the phase space as a possible source of this quantum effect. Another explanation of the origin of the periodicity of the initial state has been offered with the aid of an analysis of the quasidistribution $\phi_{\mathcal{A}}(\alpha, t)$ oriented to its relation to the matrix elements of the state in the number-state basis (Peřinová and Lukš [1990]).

When we neglect losses, the functions $f_{mn}(t)$, which determine the quasi-distribution $\phi_{\mathscr{A}}(\alpha, t)$, assume the simple form:

$$f_{mn}(t) = \exp[i\kappa(m-n)(m+n-1)t]f_{mn}(0). \tag{2.39}$$

It is clear that the functions $f_{mn}(t)$ exhibit the shortest common period $\bar{t} = \pi/\kappa$. This period is exemplified by the "fundamental" function:

$$f_{20}(t) = \exp(2i\kappa t)f_{20}(0), \tag{2.40}$$

and the functions (2.39) are the harmonics. All relevant quantum statistics share this period.

Following Milburn [1986], Milburn and Holmes [1986], and Daniel and Milburn [1989], we consider the classical dynamics of the lossless third-order nonlinear oscillator as described by the classical equation for the complex amplitude:

$$\alpha^*(t) = \exp(i2\kappa|\alpha(0)|^2 t)\alpha^*(0). \tag{2.41}$$

A probability distribution of the complex variable $\alpha(t)$ is considered. It is evident that the rotation shear, which occurs in the classical description, may be modified on the assumption of discrete values for the intensity $|\alpha(t)|^2$, and classical revivals result for $|\alpha(t)|^2 = 0, \frac{1}{2}, 1, \frac{3}{2}, 2, \ldots$.

To investigate the dynamics of the quantum conservative third-order nonlinear oscillator, we rewrite eq. (2.30) in the form:

$$\phi_{\mathscr{A}}(|\alpha|\exp(i\varphi), t) = \sum_{l}{}' \frac{2}{\Gamma(l+1)} |\alpha|^{2l} \exp(-|\alpha|^2)\phi(l, \varphi, t), \tag{2.42}$$

where

$$\phi(l, \varphi, t) = \frac{\Gamma(l+1)}{2} \sum_{k=-l}^{l} f_{l+k,l-k}(t)\exp(i2k\varphi), \quad l = 0, \tfrac{1}{2}, 1, \tfrac{3}{2}, 2, \ldots. \tag{2.43}$$

$\sum'$ means that the summation step is $\frac{1}{2}$. The variable l assumes not only the eigenvalues of the number operator but also unphysical half-odd values. Another failure of the function $\phi(l, \varphi, t)$ may be the property

$$\phi(l, \varphi + \pi, t) = (-1)^{2l}\phi(l, \varphi, t), \tag{2.44}$$

which contradicts the expected smoothness of the quasidistribution. As it holds that

$$\sum_{l}{}' \int_0^{2\pi} \phi(l, \varphi, t)\,d\varphi = 1, \tag{2.45}$$

the functions $\phi(l, \varphi, t)$, $l = 0, \frac{1}{2}, 1, \frac{3}{2}, 2, \ldots$ form a quasidistribution for number and phase. By analogy with the generalized Fokker–Planck equation [eq. (2.9)] for $\gamma = 0$ and $\bar{n}_d = 0$, the functions $\phi(l, \varphi, t)$ satisfy the partial differential equations

$$\frac{\partial}{\partial t}\phi(l, \varphi, t) = \frac{\partial}{\partial \varphi}(\kappa(2l - 1))\phi(l, \varphi, t). \tag{2.46}$$

The equations have solutions

$$\phi(l, \varphi, t) = \phi(l, \varphi + \kappa(2l - 1)t, 0), \tag{2.47}$$

which evoke the picture of the quasidistributions $\phi(l, \varphi, t)$ rotating in a clockwise direction with the angular velocity $\kappa(2l - 1)$. The quasidistribution $\phi(l, \varphi, t)$ for $l = \frac{1}{2}$ does not move, and that for $l = 1$ rotates with angular velocity κ, but according to eq. (2.44) it consists of two identical parts. Therefore, the circular frequency is 2κ and the period is π/κ.

The quantum coherence is sensitive to dissipation, with the result that the quasidistribution $\phi(l, \varphi, t)$ may not rotate with velocities which preserve the resonance of motion.

Let us proceed to the more general case of $\bar{n}_d = 0$ and $\gamma \neq 0$ (Peřinová, Lukš and Kárská [1990]). Restricting ourselves to the times $t = k\pi/\kappa$, with k an integer, we have:

$$f_{mn}\left(\frac{k\pi}{\kappa}\right) \equiv f_{mn}\left(\frac{k\pi}{\kappa}, \kappa, \gamma\right)$$

$$= \exp\left[-\frac{\gamma}{2}(m+n)\frac{k\pi}{\kappa}\right] \sum_{l=0}^{\infty} \frac{1}{l!} \frac{(m+l)!(n+l)!}{m!n!}$$

$$\times \left[\frac{\gamma\left[1 - \exp\left(-\gamma k \frac{\pi}{\kappa}\right)\right]}{\gamma - 2i\kappa(m-n)}\right]^l f_{m+l, n+l}(0), \tag{2.48}$$

as a consequence of the even parity of $(m - n)(m + n - 1)$. Numerical calculations based on the solutions $f_{mn}(t)$ [eq. (2.27)] show that for $\gamma \gg 0$ or $\bar{n}_d \gg 0$, the periodic behavior of the system is destroyed. Incorporating the damping, we may ask what quasiperiodic behavior develops from the original periodic recurrences of the initial state. For this purpose we can use the stroboscopic method, which means that we calculate the fundamental relations for the nonlinear oscillator in multiples of the period π/κ and expect an effect of retardation or advancement for the optical system. It can be proved easily that eq. (2.27) decomposes into the three relations

$$f^{(1)}_{mn} = \left[\frac{\gamma}{\gamma - 2i\kappa(m-n)}\right]^{(m+n-1)/2} f_{mn}(0), \tag{2.49}$$

$$f^{(2)}_{mn} = \exp\left[-\frac{\gamma}{2}(m+n)\frac{k\pi}{\kappa}\right] \sum_{l=0}^{\infty} \frac{1}{l!} \frac{(m+l)!(n+l)!}{m!n!}$$

$$\times \left[1 - \exp\left(-\gamma \frac{k\pi}{\kappa}\right)\right]^l f^{(1)}_{m+l,n+l}, \tag{2.50}$$

and

$$f^{(3)}_{mn} = \left[\frac{\gamma - 2i\kappa(m-n)}{\gamma}\right]^{(m+n-1)/2} f^{(2)}_{mn}, \tag{2.51}$$

where $f^{(3)}_{mn} = f_{mn}(k\pi/\kappa, \kappa, \gamma)$.

The following instructive picture emerges for small κ according to eq. (2.49). The initial state undergoes nonlinear oscillations without losses for a time $1/\gamma$. The intermediate state is then attenuated according to eq. (2.50) over a time $k\pi/\kappa$, and this state undergoes time-reversed lossless nonlinear oscillations over a time $1/\gamma$. The relaxation time enters the considerations via the approximate relations:

$$f^{(1)}_{mn} \approx \exp\left[i\kappa(m-n)(m+n-1)\frac{1}{\gamma}\right] f_{mn}(0), \tag{2.52}$$

and

$$f_{mn}\left(\frac{k\pi}{\kappa}, \kappa, \gamma\right) \approx \exp\left[i\kappa(m-n)(m+n-1)\left(-\frac{1}{\gamma}\right)\right] f^{(2)}_{mn}. \tag{2.53}$$

Similarly, it is possible to perform the analysis of the quasidistribution $\phi_{\mathscr{A}}(\alpha, t)$ appropriate to the other model of the third-order nonlinear oscillator with the Hamiltonian $\hat{H}_S$ given in eq. (2.6) (Peřinová and Lukš [1990]). In this case, all relevant quantum statistics repeat after the time interval $2\pi/\kappa$.

2.4. SQUEEZED STATES

The concept of squeezed states has been introduced in connection with a reduction of quantum-mechanical limit in measurement of an observable, at the cost of information about the conjugate observable. More generally, it is possible to consider uncertainty relations for two noncommuting observables, which need not be canonically conjugate (Messiah [1961]). In spite of the prominent role of the uncertainty relations, their mathematical proof rests merely upon the Schwartz inequality. We arrive at quantum applications of the Schwartz inequality as follows. Let us consider a concrete physical system and two observables A and B therein. On the assumption that the system is in the pure state $|\psi\rangle$, $\langle\psi|\psi\rangle = 1$, the operators $\hat{A}, \hat{B}$ assigned to these observables have the expectation values:

$$\langle \hat{A} \rangle = \langle \psi | \hat{A} | \psi \rangle, \qquad \langle \hat{B} \rangle = \langle \psi | \hat{B} | \psi \rangle, \tag{2.54}$$

respectively. A physical interpretation of the following mathematical derivation is based on the mean squares of the operators

$$\Delta \hat{A} = \hat{A} - \langle \hat{A} \rangle, \qquad \Delta \hat{B} = \hat{B} - \langle \hat{B} \rangle. \tag{2.55}$$

Introducing the vectors

$$|\varphi\rangle = \Delta \hat{A} |\psi\rangle, \qquad |\chi\rangle = \Delta \hat{B} |\psi\rangle, \tag{2.56}$$

and considering the Schwartz inequality in the form

$$|\langle \varphi | \chi \rangle|^2 \leq \langle \varphi | \varphi \rangle \langle \chi | \chi \rangle, \tag{2.57}$$

we obtain the inequality

$$|\langle \Delta \hat{A} \Delta \hat{B} \rangle|^2 \leq \langle (\Delta \hat{A})^2 \rangle \langle (\Delta \hat{B})^2 \rangle. \tag{2.58}$$

On the right-hand side there is a product of variances. To interpret the left-hand side more easily, we resolve the constituent product

$$\Delta \hat{A} \Delta \hat{B} = \tfrac{1}{2}(\{\Delta \hat{A}, \Delta \hat{B}\} + i\hat{C}), \tag{2.59}$$

where

$$\{\Delta \hat{A}, \Delta \hat{B}\} = \Delta \hat{A} \Delta \hat{B} + \Delta \hat{B} \Delta \hat{A}, \qquad \hat{C} = -i[\Delta \hat{A}, \Delta \hat{B}] = -i[\hat{A}, \hat{B}]. \tag{2.60}$$

Then eq. (2.58) becomes

$$\langle (\Delta \hat{A})^2 \rangle \langle (\Delta \hat{B})^2 \rangle \geq \operatorname{cov}^2(\hat{A}, \hat{B}) + \tfrac{1}{4} \langle \hat{C} \rangle^2, \tag{2.61}$$

where

$$\operatorname{cov}(\hat{A}, \hat{B}) = \tfrac{1}{2} \langle \{\Delta \hat{A}, \Delta \hat{B}\} \rangle. \tag{2.62}$$

Regardless of the above modifications, the equality in eq. (2.61) is attained on the assumption that either

$$\Delta \hat{B} |\psi\rangle = 0 \tag{2.63}$$

or a suitable complex number $\bar{\kappa}$ can be found such that

$$(\Delta\hat{A} + \bar{\kappa}\Delta\hat{B})|\psi\rangle = 0. \tag{2.64}$$

We rewrite this relation in the form of an eigenvalue problem:

$$(\hat{A} + \bar{\kappa}\hat{B})|\psi\rangle = \lambda|\psi\rangle, \tag{2.65}$$

where

$$\lambda = \langle\hat{A}\rangle + \bar{\kappa}\langle\hat{B}\rangle. \tag{2.66}$$

If the equality sign in eq. (2.61) is achieved, then

$$\bar{\kappa} = \frac{-\text{cov}(\hat{A}, \hat{B}) + \frac{i}{2}\langle\hat{C}\rangle}{\langle(\Delta\hat{B})^2\rangle}. \tag{2.67}$$

For the states $|\psi\rangle$ fulfilling the condition $\langle\hat{C}\rangle = 0$, eqs. (2.61) and (2.67) simplify. Nevertheless, this condition emphasizes unsatisfactorily the quantum nature of the relation for the product of uncertainties. Restricting ourselves to the states $|\psi\rangle$ fulfilling the condition

$$\text{cov}(\hat{A}, \hat{B}) = 0, \tag{2.68}$$

we obtain eq. (2.61) in the form:

$$\langle(\Delta\hat{A})^2\rangle\langle(\Delta\hat{B})^2\rangle \geq \tfrac{1}{4}\langle\hat{C}\rangle^2; \tag{2.69}$$

namely, the Heisenberg uncertainty relation. Equation (2.67) becomes, under the condition of eq. (2.68):

$$\bar{\kappa} = i\bar{\gamma}, \qquad \bar{\gamma} = \text{sgn}(\langle\hat{C}\rangle)\left[\frac{\langle(\Delta\hat{A})^2\rangle}{\langle(\Delta\hat{B})^2\rangle}\right]^{1/2}. \tag{2.70}$$

The uncertainty relation holds generally regardless of condition (2.68). In the case of the equality in eq. (2.69), we speak regularly of the minimum-uncertainty states (Messiah [1961]). For these states the uncertainty product $\langle(\Delta\hat{A})^2\rangle\langle(\Delta\hat{B})^2\rangle$ attains the minimum conditioned by the expectation value of the operator $\hat{C}$. This uncertainty product can be state-dependent, and consequently some authors prefer to speak of intelligent states, if such a dependence does occur, instead of minimum-uncertainty states (Aragone, Chalbaud and Salamo [1976]).

Mathematically well-defined physical quantum states enable us to study a reduction of the quantum-mechanical limit in the measurement. In the following, we recall the basic types of squeezed states. For a single-mode electromagnetic field of frequency ω described by the annihilation (creation) operator $\hat{a}$ ($\hat{a}^+$), we consider the slowly varying operators:

$$\hat{a} \to \exp(i\omega t)\hat{a}, \qquad \hat{a}^+ \to \exp(-i\omega t)\hat{a}^+. \tag{2.71}$$

The operators defined in this interaction picture are:

$$\hat{Q} = \hat{a} + \hat{a}^+, \qquad \hat{P} = -i(\hat{a} - \hat{a}^+), \qquad [\hat{Q}, \hat{P}] = 2i\hat{1}, \tag{2.72}$$

which are proportional to the generalized position operator and the generalized momentum operator, respectively. They obey the appropriate uncertainty relation [eq. (2.69); $\hat{A} = \hat{Q}$, $\hat{B} = \hat{P}$]:

$$\langle (\Delta \hat{Q})^2 \rangle \langle (\Delta \hat{P})^2 \rangle \geq 1, \tag{2.73}$$

where

$$\langle (\Delta_{\hat{P}}^{\hat{Q}})^2 \rangle = -1 + 2[\langle \Delta \hat{a} \Delta \hat{a}^+ \rangle \pm \mathrm{Re} \langle (\Delta \hat{a})^2 \rangle]. \tag{2.74}$$

For $\bar{\kappa} = i$, the solutions of the appropriate problem [eq. (2.64)] are the Glauber coherent states. If the variance of one of the quadrature operators is smaller than the corresponding value for a coherent state; i.e.,

$$\min(\langle (\Delta \hat{Q})^2 \rangle, \langle (\Delta \hat{P})^2 \rangle) < 1, \tag{2.75}$$

the field is in a squeezed state according to the standard definition (Walls [1983]). In this case we speak of the squeezed state in a strong (narrow) sense.

The operators (2.72) are called the quadrature phase-amplitude operators, which is abbreviated as the quadrature amplitudes or, simply, the quadratures.

Squeezing of vacuum fluctuations is usually measured in the homodyne detection process which uses a reference beam of the local oscillator in a coherent state. The measured quadrature variances then depend on the phase of the local oscillator and are not rotationally invariant. A more general definition of the squeezing of vacuum fluctuations as actually measured in the homodyne detection process, when the quadrature correlations $\langle \Delta \hat{Q} \Delta \hat{P} \rangle$ and $\langle \Delta \hat{P} \Delta \hat{Q} \rangle$ are different from zero, assumes the measurement of a set of parametrized operators:

$$\hat{Q}(\tau) = \exp(-i\tau)\hat{a} + \exp(i\tau)\hat{a}^+, \qquad \hat{P}(\tau) = -i\exp(-i\tau)\hat{a} - \exp(i\tau)\hat{a}^+, \tag{2.76}$$

where τ is a real number. The operators $\hat{Q}(\tau)$ and $\hat{P}(\tau)$ are named quadrature operators because $\hat{P}(\tau) = \hat{Q}(\tau + \pi/2)$. From eqs. (2.76) it follows that $[\hat{Q}(\tau), \hat{P}(\tau)] = 2i\hat{1}$. Thus, $\hat{Q}(\tau)$ and $\hat{P}(\tau)$ commute, like canonically conjugate observables, on a c-number and obey the uncertainty relation:

$$\langle(\Delta\hat{Q}(\tau))^2\rangle\langle(\Delta\hat{P}(\tau))^2\rangle \geqslant 1. \tag{2.77}$$

Among the operators (2.76), the quadrature operator $\hat{Q}^{(p)}$ exists whose variance

$$\langle(\Delta\hat{Q}^{(p)})^2\rangle = -1 + 2[\langle\Delta\hat{a}\Delta\hat{a}^+\rangle - |\langle(\Delta\hat{a})^2\rangle|] \tag{2.78}$$

is minimum. When $\langle(\Delta\hat{Q}^{(p)})^2\rangle < 1$, we speak of the principal squeezing or the squeezing in a weak (broad) sense (Lukš, Peřinová and Peřina [1988], Lukš, Peřinová and Hradil [1988]). The use of principal squeezing is advantageous in the case of the nonlinear oscillator because the free-field frequency is modified by the self-interaction here, and depends on the intensity of the field. Thus, whether or not the adoption of the interaction picture [eq. (2.71)] introduces true slowly varying operators is an open question. The principal quadrature variance, which is phase-independent, is not affected by uncertainties regarding the free-field frequency and corresponds to another fixing picture.

It is natural to generalize the definition of principal squeezing to a definition of the $2N$th-order squeezed state in $\hat{Q}(\tau)$, $N = 1, 2, 3, ...$, if a phase τ exists such that the expectation value $\langle(\Delta\hat{Q}(\tau))^{2N}\rangle$ is less than its value in a coherent state of the field (Hong and Mandel [1985]):

$$\langle(\Delta\hat{Q}(\tau))^{2N}\rangle < (2N - 1)!!. \tag{2.79}$$

Although this definition could, of course, be extended to moments of odd order as well, it is not very meaningful for odd powers. The term $\langle(\Delta\hat{Q}(\tau))^{2N+1}\rangle$ vanishes for a coherent state, and for other states the sign of the odd moments depends on the phase τ. Moreover, only for even moments does squeezing imply that the state is purely quantum-mechanical.

Another form of higher-order squeezing can be defined in terms of real and imaginary parts of the square (or higher powers) of the field amplitude (Hillery [1987a, b]). We decompose the square of the annihilation operator $\hat{a}$ in the form:

$$\hat{a}^2 = \hat{Y}_1 + i\hat{Y}_2, \tag{2.80}$$

where

$$\hat{Y}_1 = \tfrac{1}{2}(\hat{a}^2 + \hat{a}^{+2}), \qquad \hat{Y}_2 = \frac{1}{2i}(\hat{a}^2 - \hat{a}^{+2}). \tag{2.81}$$

The component operators $\hat{Y}_1$ and $\hat{Y}_2$ obey the commutation relation:

$$[\hat{Y}_1, \hat{Y}_2] = i(2\hat{n} + \hat{1}), \tag{2.82}$$

where $\hat{n} = \hat{a}^+ \hat{a}$. As a result, their variances

$$\langle(\Delta_{\hat{Y}_2}^{\hat{Y}_1})^2\rangle = \tfrac{1}{2}[1 - 2\langle\hat{a}\hat{a}^+\rangle + \langle\Delta\hat{a}^2\Delta\hat{a}^{+2}\rangle \pm \text{Re}\langle(\Delta\hat{a}^2)^2\rangle], \tag{2.83}$$

satisfy the uncertainty relation [eq. (2.69), $\hat{A} = \hat{Y}_1$, $\hat{B} = \hat{Y}_2$] such that

$$\langle(\Delta\hat{Y}_1)^2\rangle\langle(\Delta\hat{Y}_2)^2\rangle \geq \langle\hat{n} + \tfrac{1}{2}\hat{1}\rangle^2. \tag{2.84}$$

We speak of the amplitude-squared squeezing when it holds that $\min(\langle(\Delta\hat{Y}_1)^2\rangle, \langle(\Delta\hat{Y}_2)^2\rangle) < \langle\hat{n} + \tfrac{1}{2}\hat{1}\rangle$. The uncertainty product [eq. (2.84)] is minimized by the generalized coherent states $|\psi\rangle$, the solutions of the eigenproblem [eq. (2.65)], for which the relation $\langle(\Delta\hat{Y}_1)^2\rangle = \langle(\Delta\hat{Y}_2)^2\rangle = \langle\hat{n} + \tfrac{1}{2}\hat{1}\rangle$ is valid.

Squeezed states of photon number and phase represent another kind of electromagnetic field state with minimum uncertainty (Teich and Saleh [1989], Lukš and Peřinová [1992]). The operator $\hat{A}$ in eq. (2.69) is the photon-number operator and $\hat{B}$ and $\hat{C}$ are the operators corresponding to the sine and cosine of the phase:

$$\hat{A} = \hat{n} = \hat{a}^+\hat{a}, \qquad \hat{B} = \widehat{\sin}\,\varphi = \frac{1}{2i}[(\hat{n} + \hat{1})^{-1/2}\hat{a} - \hat{a}^+(\hat{n} + \hat{1})^{-1/2}],$$

$$\hat{C} = \widehat{\cos}\,\varphi = \tfrac{1}{2}[(\hat{n} + \hat{1})^{-1/2}\hat{a} + \hat{a}^+(\hat{n} + \hat{1})^{-1/2}]. \tag{2.85}$$

Normalizable states of minimum number-phase uncertainty have been constructed by Jackiw [1968] as solutions of the eigenvalue problem

$$(\Delta\hat{n} + i\bar{\gamma}\Delta\,\widehat{\sin}(\delta\varphi))|\psi\rangle = 0, \tag{2.86}$$

where

$$\Delta\hat{n} = \hat{n} - \langle\hat{n}\rangle, \qquad \widehat{\sin}(\delta\varphi) = \frac{1}{2i}[\widehat{\exp}(i\delta\varphi) - \text{H.c.}],$$

$$\widehat{\exp}(i\delta\varphi) = \exp(-i\bar{\varphi})\sum_{n=0}^{\infty}|n\rangle\langle n+1|, \tag{2.87}$$

and $\bar{\varphi}$ is the preferred phase defined for $\bar{\gamma} > 0$ by the relation

$$\bar{\varphi} = \arg(\langle\widehat{\exp}(i\varphi)\rangle). \tag{2.88}$$

In the optical oscillator system, the Jackiw states achieve the equality in the uncertainty relation:

$$\langle(\Delta\hat{n})^2\rangle\langle(\Delta\,\widehat{\sin}(\delta\varphi))^2\rangle \geq \tfrac{1}{4}\langle\widehat{\cos}\,\varphi\rangle^2. \tag{2.89}$$

In addition to eq. (2.89), the following uncertainty relations can be considered (Carruthers and Nieto [1968]):

$$\langle(\Delta\hat{n})^2\rangle\langle(\Delta\widehat{\cos}(\delta\varphi))^2\rangle \geq \tfrac{1}{4}\langle\widehat{\sin}\,\varphi\rangle^2, \tag{2.90}$$

and

$$\langle(\Delta\hat{n})^2\rangle\frac{[\langle(\Delta\widehat{\sin}(\delta\varphi))^2\rangle + \langle(\Delta\widehat{\cos}(\delta\varphi))^2\rangle]}{\langle\widehat{\sin}\,\varphi\rangle^2 + \langle\widehat{\cos}\,\varphi\rangle^2} \geq \frac{1}{4}. \tag{2.91}$$

The problem (2.86) has been solved in full generality, and a numerical analysis of the Jackiw states has been performed (Lukš and Peřinová [1991], Lukš, Peřinová and Křepelka [1992]). The application of the theory of operator ordering to the operators (2.85) has led to a quantum analogue of the measure of the phase dispersion in the classical theory of statistics (Bandilla and Paul [1969]):

$$V = 1 - |\langle\widehat{\exp}(i\varphi)\rangle|^2, \tag{2.92}$$

and an appropriate uncertainty relation

$$[\langle(\Delta\hat{n})^2\rangle + \tfrac{1}{4}][1 - |\langle\widehat{\exp}(i\varphi)\rangle|^2] \geq \tfrac{1}{4} \tag{2.93}$$

has been derived (Newton [1980], Lukš and Peřinová [1991]). The inequality (2.93) has been tested with the aid of the Jackiw states.

In quantum optics it is recognized that a phase stretching leads to the number squeezing when some number-phase intelligence is conserved. The corresponding state is similar to a crescent. The term number-squeezed state denotes any state whose variance of photon number is less than the mean photon number; i.e., the variance of the Poisson distribution with the same mean. Such a state may be called a crescent state in a strong (narrow) sense. The deviation from the Poisson photon distribution is expressed by the photon-number variance normalized to the mean photon number (the Fano factor):

$$d = \frac{\langle(\Delta\hat{n})^2\rangle}{\langle\hat{n}\rangle}. \tag{2.94}$$

The normalized variance d is related to the second reduced factorial moment f [cf. eq.(2.38)]:

$$f = \frac{d-1}{\langle\hat{n}\rangle}. \tag{2.95}$$

For a coherent state we have $d = 1$ and $f = 0$.

Similarly as the study of the quadrature squeezing in the weak sense rests on the set of rotating quadrature operators, the investigation of the crescent states in the weak sense may be based upon the set of displaced number operators:

$$\hat{n}_c = (\hat{a}^+ - c^*\hat{1})(\hat{a} - c\hat{1}) = \hat{a}_c^+ \hat{a}_c, \qquad (2.96)$$

where

$$\hat{a}_c = \hat{a} - c\hat{1}. \qquad (2.97)$$

The value c can be interpreted as the center of curvature of the corresponding contour diagram. The idea of considering a set of interferometrically redefined (displaced) number operators has been proposed by Yamamoto, Imoto and Machida [1986]. For a given state, it is possible to find the optimum value of the output Fano factor,

$$d_c = \frac{\langle (\Delta \hat{n}_c)^2 \rangle}{\langle \hat{n}_c \rangle}, \qquad (2.98)$$

by imposing a constraint on the values of c (Kitagawa and Yamamoto [1986]). The minimum variance has been obtained for the center of curvature (Lukš, Peřinová and Křepelka [1992]):

$$c = \frac{(\langle \Delta \hat{a}^+ \Delta \hat{a} \rangle + \tfrac{1}{2})E - \langle (\Delta \hat{a})^2 \rangle E^*}{(\langle \Delta \hat{a}^+ \Delta \hat{a} \rangle + \tfrac{1}{2})^2 - |\langle (\Delta \hat{a})^2 \rangle|^2}, \qquad (2.99)$$

where

$$E = \tfrac{1}{2}(\langle \Delta \hat{n} \Delta \hat{a} \rangle + \langle \Delta \hat{a} \Delta \hat{n} \rangle). \qquad (2.100)$$

In a complete analogy to the principal squeezing, we obtain the characteristics of the crescent shape by comparing the resulting photon-number variance $\langle (\Delta \hat{n}_c)^2 \rangle$ with the mean photon number $\langle \hat{n}_c \rangle$. For the two-photon coherent states the complex coordinate c is equal to the coherent amplitude, and this comparison results in the super-Poissonian behavior of the squeezed vacuum state. If the sub-Poissonian behavior is obtained, we can characterize the states as crescent in the weak sense.

It can be proved easily that the operators $\hat{n}_c$ and

$$\hat{P}_c(\tau) = -\mathrm{i}[\exp(-\mathrm{i}\tau)\hat{a}_c - \exp(\mathrm{i}\tau)\hat{a}_c^+] \qquad (2.101)$$

are not correlated; i.e., we have

$$\tfrac{1}{2}\langle \Delta \hat{n}_c \Delta \hat{P}_c(\tau) + \Delta \hat{P}_c(\tau) \Delta \hat{n}_c \rangle = 0. \qquad (2.102)$$

This property is shared also by an $\hat{n}_c$–$\hat{P}_c(\tau)$ intelligent state (Hradil [1991]).

This leads to the use of another measure of the qualitative properties of crescent states; namely, the uncertainty product. As the parameter τ is arbitrary, we choose $\tau = \bar{\varphi}_c$, where $\bar{\varphi}_c$ is the preferred phase given by

$$\bar{\varphi}_c = \arg(\langle \widehat{\exp}(i\varphi_c)\rangle). \tag{2.103}$$

We linearize the phase operator as

$$\delta\hat{\varphi}_c = \frac{\Delta\hat{P}_c(\bar{\varphi}_c)}{\langle \hat{Q}_c(\bar{\varphi}_c)\rangle}, \tag{2.104}$$

where

$$\hat{Q}_c(\tau) = \exp(-i\tau)\hat{a}_c + \exp(i\tau)\hat{a}_c^+. \tag{2.105}$$

From eq. (2.104), we obtain the measure of the phase dispersion:

$$\langle (\delta\hat{\varphi}_c)^2 \rangle = \frac{\langle (\Delta\hat{P}_c(\bar{\varphi}_c))^2 \rangle}{\langle \hat{Q}_c(\bar{\varphi}_c)\rangle^2}; \tag{2.106}$$

the corresponding uncertainty product u_c reads:

$$u_c = \langle (\Delta\hat{n}_c)^2 \rangle \langle (\delta\hat{\varphi}_c)^2 \rangle \geqslant \tfrac{1}{4}. \tag{2.107}$$

In spite of the relation being simple, we follow the example of eq. (2.93) and derive the relation

$$\bar{u}_c = (\langle (\Delta\hat{n}_c)^2 \rangle + \tfrac{1}{4})V(c) \geqslant \tfrac{1}{4}, \tag{2.108}$$

where

$$V(c) = \frac{\langle (\delta\hat{\varphi}_c)^2 \rangle}{1 + \langle (\delta\hat{\varphi}_c)^2 \rangle}. \tag{2.109}$$

The minimum value of the uncertainty products u_c and $\bar{u}_c$ equal to $\tfrac{1}{4}$ is achieved by the $\hat{n}_c$–$\hat{P}_c(\bar{\varphi}_c)$ intelligent states. The uncertainty products u_c and $\bar{u}_c$ along with the quantity d_c contribute to an assessment of the crescent shape of a studied state, because both the sub-Poissonian behavior and the $\hat{n}_c$–$\hat{P}_c(\bar{\varphi}_c)$ intelligence characterize the crescent property.

The measure of the phase dispersion [eq. (2.106)] can be expressed explicitly in terms of the operators $\hat{a}$, $\hat{a}^+$, $\hat{a}_c$, $\hat{a}_c^+$ as follows:

$$\langle (\delta\hat{\varphi}_c)^2 \rangle = \frac{(\langle \Delta\hat{a}^+\Delta\hat{a}\rangle + \tfrac{1}{2})|\langle \hat{a}_c\rangle|^2 - \mathrm{Re}(\langle (\Delta\hat{a})^2\rangle\langle \hat{a}_c^+\rangle^2)}{2|\langle \hat{a}_c\rangle|^4}. \tag{2.110}$$

From the considerations above, it follows that the expectation values $\langle \hat{a}^k \hat{a}^{+l}\rangle$ are of importance when determining the squeezing properties of the dissipative third-order nonlinear oscillator. Using the quasidistribution $\phi_{\mathscr{A}}(\alpha, t)$ given in eq. (2.30), we can write the moments of the antinormally ordered field operators in the form (Peřinová and Lukš [1988]):

$$\langle \hat{a}^k \hat{a}^{+l} \rangle = \langle \alpha^k \alpha^{*l} \rangle_{\mathscr{A}} = \int \alpha^k \alpha^{*l} \phi_{\mathscr{A}}(\alpha, t) \, \mathrm{d}^2 \alpha$$

$$= \pi \sum_{n=0}^{\infty} (n+l)! f_{n+l-k,n}(t), \quad k \leq l, \quad (2.111)$$

whereas for $k > l$ we consider the complex-conjugate quantity. To be consistent with the notion of off-diagonality in the elements $f_{mn}(t)$ (introduced in § 2.1), we will discuss the off-diagonality $(l - k)$ of these moments. On substitution of the functions $f_{mn}(t)$ from eq. (2.27) into eq. (2.111), we obtain an explicit formula for the moments $\langle \hat{a}^k \hat{a}^{+l} \rangle$ (Peřinová and Lukš [1990]), which can be simplified in the cases $k = 0$, $l \neq 0$ and $k = 1$, $l = 1$, to give:

$$\langle \hat{a}^{+l} \rangle = \pi \exp\left[\left(-2\mathrm{i}\kappa l + \frac{\gamma}{2}\right)t\right] \left[\frac{E_l(t)}{1 - g_l(t)}\right]^{l+1} \sum_{n=0}^{\infty} (n+l)! f_{n+l,n}(0) G_l^n(t), \quad (2.112)$$

and

$$\langle \hat{a} \hat{a}^+ \rangle = \exp(-\gamma t)(\langle \hat{a}(0) \hat{a}^+(0) \rangle - 1) + \bar{n}(t) + 1, \quad (2.113)$$

where

$$G_l(t) = \frac{(\bar{n}_\mathrm{d} + 1)}{\bar{n}_\mathrm{d}} g_l(t) + \frac{E_l^2(t)}{1 - g_l(t)}. \quad (2.114)$$

2.5. PHASE PROPERTIES

Phase properties of a single-mode optical field can be studied in the framework of the formalism of the Hermitian phase operator (Pegg and Barnett [1988], Barnett and Pegg [1989], Pegg and Barnett [1989]) or in terms of the cosine and sine operators (Susskind and Glogower [1964], Carruthers and Nieto [1968], Lukš and Peřinová [1991], Lukš, Peřinová and Křepelka [1992], Lukš and Peřinová [1993]).

Pegg and Barnett constructed phase operators using Loudon's phase states (Loudon [1973]):

$$|\theta, s\rangle = (s+1)^{-1/2} \sum_{n=0}^{s} \exp(\mathrm{i}n\theta) |n\rangle, \quad (2.115)$$

forming the orthogonal vector system $|\theta_m, s\rangle$,

$$|\theta_m, s\rangle = (s+1)^{-1/2} \sum_{n=0}^{s} \exp(in\theta_m)|n\rangle, \qquad (2.116)$$

where

$$\theta_m \equiv \theta_m^{(s)} = \theta + 2\pi \frac{m}{s+1}, \quad m = 0, 1, \ldots, s, \qquad (2.117)$$

and θ is a given value. Pegg and Barnett [1988] recommend the use of the Hermitian phase operator which can be defined with the aid of the orthogonal vectors (2.116). With these vectors, the following general construction is connected. To each phase function $M(\varphi)$, the following operator is assigned:

$$\hat{M}_{\theta,s} = \sum_{m=0}^{s} M(\theta_m)|\theta_m, s\rangle\langle\theta_m, s|, \qquad (2.118)$$

where θ_m is given in eq. (2.117). The concept of the phase function is related to the given domain $\varphi \in [\theta, \theta + 2\pi)$. When a 2π-periodic continuation of $M(\varphi)$ is assumed, the value of θ is less important because $\hat{M}_{\theta,s} = \hat{M}_{\theta',s}$ for $\theta = \theta'$ modulo $2\pi/(s+1)$. Of course, for $M(\varphi) = \varphi$ the mapping (2.118) yields the phase operator

$$\hat{\phi}_{\theta,s} = \sum_{m=0}^{s} \theta_m |\theta_m\rangle\langle\theta_m|; \qquad (2.119)$$

the value of θ is important, and

$$\hat{M}_{\theta,s} = M(\hat{\phi}_{\theta,s}). \qquad (2.120)$$

Equation (2.119) says that the eigenvalues of the phase operator are $\theta_0, \ldots, \theta_s$ and that the application of the Pegg–Barnett formalism is connected with the fact that these eigenvalues are dense in the domain $[\theta, \theta + 2\pi)$ for s tending to infinity. This formalism yields approximative statistics of quantum phase whose precise values are obtained in the limit $s \to \infty$. If the state of a field is described by the operator $\hat{\rho}$, the quantum average of the phase operator (2.119) is expressed in the form

$$\langle \hat{\phi}_{\theta,s} \rangle = \text{Tr}\{\hat{\rho}\hat{\phi}_{\theta,s}\} = \sum_{m=0}^{s} \theta_m p(\theta_m, s), \qquad (2.121)$$

where the discrete phase distribution reads

$$p(\theta_m, s) = \langle \theta_m, s | \hat{\rho} | \theta_m, s \rangle. \qquad (2.122)$$

The continuous phase distribution

$$P(\varphi) = \lim_{s \to \infty} \frac{s+1}{2\pi} p(\theta_m^{(s)}, s), \tag{2.123}$$

where $\theta_m^{(s)}$ approaches the continuous phase variable φ, $\varphi = \lim_{s \to \infty} \theta_m^{(s)}$. With respect to eq. (2.116), the phase distribution $P(\varphi)$ can be expressed in the form:

$$P(\varphi) \equiv P(\varphi, t) = \frac{1}{2\pi} \sum_{n=0}^{\infty} \sum_{m=0}^{\infty} \rho_{nm}(t) \exp[i(m-n)\varphi]. \tag{2.124}$$

The phase distribution is normalized, so that

$$\int_{\theta}^{\theta+2\pi} P(\varphi) \, d\varphi = 1. \tag{2.125}$$

Using $P(\varphi)$, we calculate the average values of the operators:

$$\hat{\phi}_\theta = \lim_{s \to \infty} \hat{\phi}_{\theta,s}, \quad \widehat{\phi_\theta^2} = \lim_{s \to \infty} \hat{\phi}_{\theta,s}^2, \tag{2.126}$$

$$\langle \hat{\phi}_\theta \rangle = \int_{\theta}^{\theta+2\pi} \varphi P(\varphi) \, d\varphi, \tag{2.127}$$

and

$$\langle \widehat{\phi_\theta^2} \rangle = \int_{\theta}^{\theta+2\pi} \varphi^2 P(\varphi) \, d\varphi. \tag{2.128}$$

The phase dispersion is measured by the phase variance:

$$\langle (\Delta \hat{\phi}_\theta)^2 \rangle = \langle \widehat{\phi_\theta^2} \rangle - \langle \hat{\phi}_\theta \rangle^2. \tag{2.129}$$

The variance of the photon number and that of the phase are subject to the uncertainty relation:

$$\langle (\Delta \hat{n})^2 \rangle \langle (\Delta \hat{\phi}_\theta)^2 \rangle \geq \tfrac{1}{4}. \tag{2.130}$$

The legitimacy of the use of the form of eq. (2.129) for the phase-dispersion measure in the quantum theory has been discussed by Lukš and Peřinová [1991].

Another solution of the quantum phase problem is based on the operators (Susskind and Glogower [1964], Carruthers and Nieto [1968]):

$$\hat{u} = \widehat{\exp}(i\varphi), \quad \hat{u}^+ = \widehat{\exp}(-i\varphi). \tag{2.131}$$

These are defined as

$$\hat{u} = (\hat{n} + \hat{1})^{-1/2} \hat{a}, \quad \hat{u}^+ = \hat{a}^+ (\hat{n} + \hat{1})^{-1/2}, \tag{2.132}$$

with the properties

$$\hat{u}\hat{u}^+ = \hat{1}, \quad \hat{u}^+\hat{u} = \hat{1} - |n=0\rangle\langle n=0|. \tag{2.133}$$

From eqs. (2.133), it is evident that the requirement of the unitarity for the phase operators can be achieved algebraically by using the antinormal ordering of the operators $\hat{u}, \hat{u}^+$ (Lukš and Peřinová [1991], Lukš, Peřinová and Křepelka [1992], Lukš and Peřinová [1993]). This is equivalent, however, to the following classical–quantum correspondence. To every phase function $M(\varphi)$ there is assigned the operator

$$\hat{M} = \frac{1}{2\pi} \int_{\theta}^{\theta+2\pi} M(\varphi)|\varphi\rangle\langle\varphi|\,\mathrm{d}\varphi, \tag{2.134}$$

based on the vectors

$$|\varphi\rangle = \sum_{n=0}^{\infty} \exp(\mathrm{i}n\varphi)|n\rangle, \tag{2.135}$$

which are approximately orthogonal:

$$\langle\varphi|\varphi'\rangle = \pi\delta(\varphi-\varphi') + \{1 - \exp[-\mathrm{i}(\varphi-\varphi')]\}^{-1}. \tag{2.136}$$

The operator $\hat{M}$ does not depend on θ when $M(\varphi)$ has a 2π-periodic continuation. As a special case, the Susskind–Glogower cosine and sine operators are obtained:

$$\widehat{\cos}\,\varphi = \frac{1}{2\pi}\int_{\theta}^{\theta+2\pi} \cos\varphi|\varphi\rangle\langle\varphi|\,\mathrm{d}\varphi,$$

$$\widehat{\sin}\,\varphi = \frac{1}{2\pi}\int_{\theta}^{\theta+2\pi} \sin\varphi|\varphi\rangle\langle\varphi|\,\mathrm{d}\varphi. \tag{2.137}$$

It holds that

$$P(\varphi) = \frac{1}{2\pi}\langle\varphi|\hat{\rho}|\varphi\rangle, \tag{2.138}$$

which means that the role of the normalization constant is played by $(2\pi)^{-1/2}$ as adopted in some part of the literature [see eq. (6.38) in the paper by Carruthers and Nieto [1968]). We follow the rest of the literature, represented, for instance, by Shapiro [1992], his formula (42) is our eq. (2.135). In the computation of the average value of the operator $\hat{M}$, the phase distribution is used as follows:

$$\langle\hat{M}\rangle = \mathrm{Tr}\{\hat{\rho}\hat{M}\} = \int_{\theta}^{\theta+2\pi} P(\varphi)M(\varphi)\,\mathrm{d}\varphi = \langle M(\varphi)\rangle_\mathrm{a}, \tag{2.139}$$

where the subscript a indicates the antinormal ordering of the operators $\hat{u}, \hat{u}^+$. The phase dispersion is not measured by the phase variance, as in the Pegg–Barnett formalism, but rather by the quantity V defined in eq. (2.92) (Lukš and Peřinová [1991]).

The solution of the quantum phase problem in terms of the measured phase operators (Barnett and Pegg [1986]) assumes the operators (2.131) and (2.132) in the form:

$$\hat{u}_M = \hat{a}(\bar{n} + \tfrac{1}{2})^{-1/2}, \qquad \hat{u}_M^+ = \hat{a}^+ (\bar{n} + \tfrac{1}{2})^{-1/2}, \tag{2.140}$$

where $\bar{n}$ is defined as the average value of the operator $\hat{n}$ in the state under consideration; $\bar{n} = \langle \hat{n} \rangle = |\alpha|^2$ for the coherent state $|\alpha\rangle$. The measured cosine and sine operators, $\hat{C}_M$ and $\hat{S}_M$, are then expressed by the relations:

$$\hat{C}_M = \tfrac{1}{2}(\hat{u}_M + \hat{u}_M^+), \qquad \hat{S}_M = \frac{1}{2\mathrm{i}}(\hat{u}_M - \hat{u}_M^+), \tag{2.141}$$

and are used for the determination of measured phase characteristics, such as products and ratios of uncertainties, phase-dispersion measure, and so on.

To determine the phase distribution for the dissipative third-order nonlinear oscillator with an arbitrary initial state, we insert the matrix elements $\rho_{nm}(t)$ according to eqs. (2.31) and (2.27) into eq. (2.124). This nonlinear process offers a wide range of phase behavior. For an initial phase state, the Dirac δ-function is obtained as the phase distribution, which evolves into a multi-peak phase distribution corresponding to a discrete superposition of phase states (Sanders [1992b]).

2.6. SPECIAL INITIAL STATES

The quantum dynamics of statistical properties of the dissipative third-order nonlinear oscillator has been thoroughly investigated for a coherent state, a Gaussian mixed state, a Gaussian pure state, and squeezed and displaced number states as initial states.

In the following discussion, we will review briefly the results for: (i) the coherent state, (ii) the SU(1,1) coherent state, (iii) Gaussian mixed and pure states, (iv) the SU(1,1) two-photon coherent state, (v) the k-photon coherent state, and (vi) the squeezed and displaced number states at the beginning of the dissipative and conservative nonlinear evolutions.

2.6.1. The coherent state

For the initial coherent state $|\xi(0)\rangle$,

$$|\xi(0)\rangle = \exp(-\tfrac{1}{2}|\xi(0)|^2) \sum_{n=0}^{\infty} \frac{\xi^n(0)}{(n!)^{1/2}} |n\rangle, \tag{2.142}$$

where $\xi(0) = |\xi(0)| \exp(i\psi)$ is the complex-field amplitude, it holds that

$$f_{mn}(0) = \frac{1}{\pi} \frac{\xi^{*m}(0)}{m!} \frac{\xi^n(0)}{n!} \exp(-|\xi(0)|^2) \tag{2.143}$$

according to eq. (2.10). The time evolution of this optical system has been studied with the aid of the quasidistribution $\phi_{\mathscr{A}}(\alpha, t)$ [eq. (2.30)] in the conservative case (Miranowicz, Tanaś and Kielich [1990], Tanaś, Miranowicz and Kielich [1991]), in the dissipative case for a quiet reservoir (Milburn and Holmes [1986], Peřinová and Lukš [1988]), and for a noisy reservoir (Peřinová and Lukš [1988], Daniel and Milburn [1989], Peřinová and Lukš [1990]).

The photon statistics, according to eqs. (2.35) and (2.37), are determined by the superposition of coherent and chaotic fields (Peřinová, Křepelka and Peřina [1986], Peřinová and Lukš [1988], Peřina [1991]) with the photon-number distribution:

$$p(n, t) = \frac{1}{n!} \frac{(\bar{n}(t))^n}{(\bar{n}(t) + 1)^{n+1}} \exp\left(-\frac{|\xi(0)|^2 \exp(-\gamma t)}{\bar{n}(t) + 1}\right)$$

$$\times L_n^0\left(-\frac{|\xi(0)|^2 \exp(-\gamma t)}{\bar{n}(t)(\bar{n}(t) + 1)}\right), \tag{2.144}$$

and the corresponding factorial moments

$$\langle W^k \rangle_{\mathscr{N}} = (\bar{n}(t))^k L_k^0\left(-\frac{|\xi(0)|^2 \exp(-\gamma t)}{\bar{n}(t)}\right). \tag{2.145}$$

The Laguerre polynomials are defined as:

$$L_k^l(x) = \Gamma(k+l+1) \sum_{j=0}^{k} \frac{(-x)^j}{j!(k-j)!\Gamma(j+l+1)}, \tag{2.146}$$

where $\bar{n}(t)$ is given in eq. (2.36). Neglecting losses, the photon statistics of the light remains unchanged. But enhanced bunching as well as antibunching of weak fields can be expected after interference with a reference beam (Ritze and Bandilla [1979], cf. Kitagawa and Yamamoto [1986]).

The variances [eqs. (2.74), (2.78), (2.83), (2.94), (2.98), and (2.109)], which enter the definitions of squeezed states, can be expressed in a closed form because eq. (2.111) provides:

$$\langle \hat{a}^k \hat{a}^{+l} \rangle = k! \exp\left\{\left[-2i\kappa(l-k) + \frac{\gamma}{2}\right]t\right\} \exp\{-|\xi(0)|^2[1 - G_{l-k}(t)]\}$$

$$\times \xi^{*l-k}(0) E_{l-k}^{l-k+1}(t)(1 - g_{l-k}(t))^{-(l+1)}$$

$$\times L_k^{l-k}\left(-\frac{|\xi(0)|^2 E_{l-k}^2(t)}{1 - g_{l-k}(t)}\right), \quad k \leq l. \tag{2.147}$$

The standard squeezing has been studied for the conservative system (Tanaś [1984], Milburn [1986], Bužek [1989a], Tanaś [1989], Tanaś, Miranowicz and Kielich [1991]) and for the dissipative system (Peřinová and Lukš [1988]). The principal squeezing has been investigated in the conservative case (Lukš, Peřinová and Hradil [1988], Tanaś, Miranowicz and Kielich [1991]) as well as in the dissipative case (Peřinová and Lukš [1988]).

We will illustrate graphically the explicit equation (2.78). In fig. 1 we can trace the time dependence of the principal quadrature variance $\langle (\Delta \hat{Q}^{(p)})^2 \rangle$ for a prescribed mean photon number. At the times $t = (\pi/2\kappa)k$, where k is an integer, it holds that $\langle (\Delta \hat{Q}^{(p)})^2 \rangle = 0.5157$. This variance does not depend on the phase of the initial field amplitude. The dependence of the principal squeezing on the initial mean numbers of coherent photons has also been discussed. The increase of $|\xi(0)|^2$ leads to the increase of the amount of squeezing. This tendency is obvious from fig. 2.

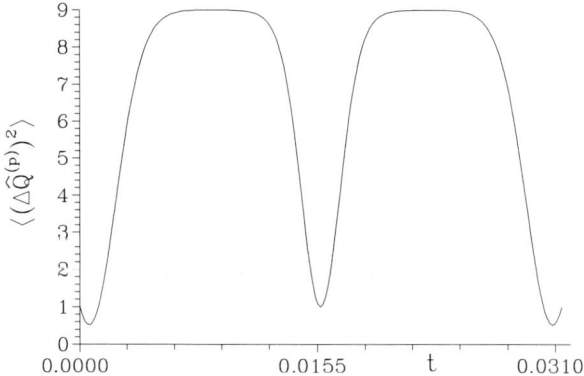

Fig. 1. Time dependence of the principal quadrature variance for $\kappa = 100$, $\bar{n}_d = 0$, $\gamma = 0$, $|\xi(0)| = 2$, and $\psi = 0$.

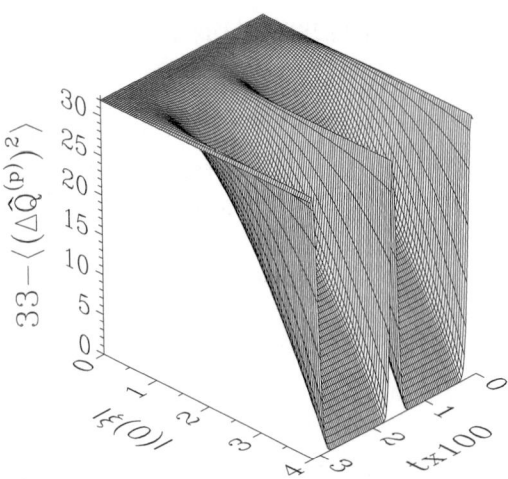

Fig. 2. Dependence on $|\xi(0)| \in [0, 4]$ of the time development of the principal quadrature variance for $\kappa = 100$, $\bar{n}_d = 0$, $\gamma = 0$, and $\psi = 0$. For $|\xi(0)| = 4$, it is valid that $\langle [\Delta \hat{Q}^{(p)}[(\pi/2\kappa)k]]^2 \rangle = 0.4165$.

An interesting connection between the revivals (Rempe, Walther and Klein [1987]) of the coherent part of a field and squeezing has been pointed out (Horák and Peřina [1989], Peřinová and Kárská [1989]); we will now present an analogous property for the nonlinear oscillator under study. The effect of dissipation on the evolution of the real coherent amplitude $|\langle \hat{a} \rangle|$ can be seen in fig. 3. This coherent amplitude collapses and revives during

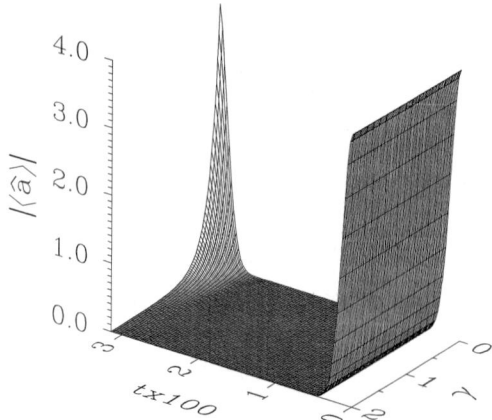

Fig. 3. Effect of dissipation $\bar{n}_d = 5$, $\gamma \in [0, 2]$ on the evolution of $|\langle \hat{a} \rangle|$; $\kappa = 100$, $|\xi(0)| = 4$, and $\psi = 0$.

the interaction with the nonlinear Kerr medium. The revivals occur in a neighborhood of the time points $t = (\pi/\kappa)k$, with k an integer. Their periodicity interval is π/κ; i.e., the same as that for the principal squeezing variance (see figs. 4a and b). Figure 4b shows the influence of quantum fluctuations on $\langle(\Delta\hat{Q}^{(p)})^2\rangle$ under the assumption that γ is constant. The quantum noise attenuates both the revivals and the squeezing, as is obvious from figs. 3 and 4a and b. The limiting value of the principal quadrature variance as t tends to infinity is $(1 + 2\bar{n}_d)$ for $\gamma > 0$, and it does not exist for $\gamma = 0$ (the case of periodicity).

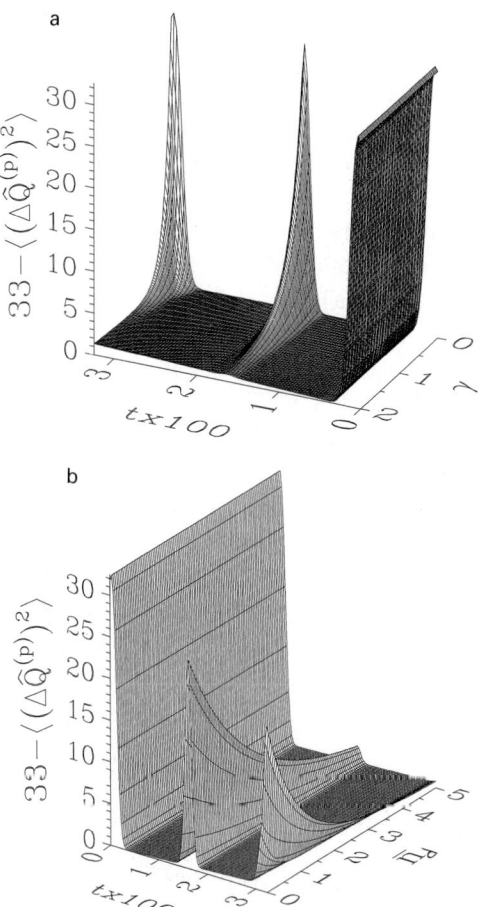

Fig. 4. Attenuation of the principal quadrature variance under the conditions $\kappa = 100$, $|\xi(0)| = 4$, and $\psi = 0$: (a) $\bar{n}_d = 5$ and $\gamma \in [0, 2]$; (b) $\bar{n}_d \in [0, 5]$ and $\gamma = 1$.

Applying the stroboscopic method for $\bar{n}_d = 0$, we can write the complex field amplitude:

$$\langle \hat{a} \rangle(t) = \xi(t) \approx \exp\left[i|\xi(0)|^2 \exp(-\gamma t)\frac{2\kappa}{\gamma}\right]\exp\left(-\frac{\gamma}{2}t\right)$$

$$\times \exp\left(-i|\xi(0)|^2 \frac{2\kappa}{\gamma}\right)\xi(0); \quad t = \frac{\pi}{\kappa}k. \qquad (2.148)$$

This approximate relation means that the average value is attenuated and undergoes lossless nonlinear oscillations during a time $t' = [1 - \exp(-\gamma t)]/\gamma$. Since t' plays the role of the phase of motion, the optical system described by the functions $f_{mn}(t)$ [eq. (2.27)] for $\bar{n}_d = 0$ keeps the same period with the nonlinear oscillator characterized by the same κ and $\gamma = 0$.

In the conservative case it has been shown that the optical system also exhibits higher-order squeezing: 4th and 6th order according to Gerry and Rodriguez [1987a], and 2nd and 6th order (and generally $2N$th order; N odd) by Tanaś [1988]. The degree to which the squeezing occurs increases with order. Amplitude-squared squeezing has been shown to occur under the same conditions as the standard squeezing and the higher-order squeezing (Gerry and Vrscay [1988a], Mir and Razmi [1991]). This squeezing has been interpreted in terms of the Lie algebra of the SU(1,1) group. Physical understanding of the squeezing that sets in at the shorter propagation times and the subsequent desqueezing which occurs as the time of propagation is increased has been provided (Blow, Loudon and Phoenix [1993]). Number squeezing is not present because of the conservation of the photon-number distribution, but a crescent topography of the quasidistribution $\phi_{\mathscr{A}}(\alpha, t)$ suffices for an interferometric generation of number-squeezed states (Kitagawa and Yamamoto [1986]). Number squeezing in both the strong and weak senses has been analyzed (Lukš, Peřinová and Křepelka [1992]).

In the following, we will apply eqs. (2.98), (2.99), (2.107), and (2.108) to illustrate the investigation of crescent states in the weak sense. Let us note that the limit values of d_c, u_c, and $\bar{u}_c$ for t tending to zero are:

$$\lim_{t \to 0} d_c = 2, \quad \lim_{t \to 0} u_c = +\infty, \quad \lim_{t \to 0} \bar{u}_c = \tfrac{1}{4}. \qquad (2.149)$$

The center of curvature and the rotating part of the complex-field amplitude

$$\langle \hat{a} \rangle_r(t) = \xi(0)|\xi(0)|^{-1}\exp[-i|\xi(0)|^2\sin(2\kappa t)], \qquad (2.150)$$

as functions of t are demonstrated for the chosen parameter $\xi(0)$ in fig. 5. The points corresponding to $t_1 = 2.07 \times 10^{-4}$, $t_2 = 3.07 \times 10^{-3}$ (in the

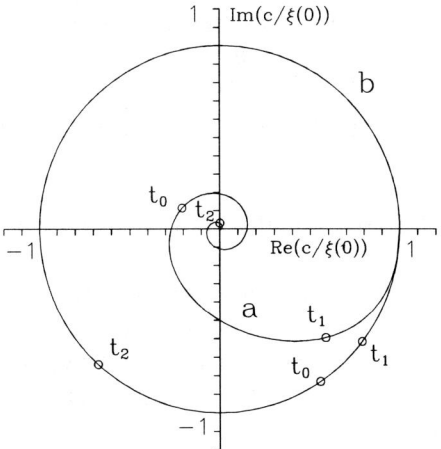

Fig. 5. Plot of the curves for $c(t)|\xi(0)|^{-1}$ (curve a) and $\langle \hat{a} \rangle_r(t)$ (curve b) for $\kappa = 100$, $\bar{n}_d = 0$, $\gamma = 0$, $|\xi(0)| = 4$, and $\psi = 0$.

interval $[t_1, t_2]$ the field is sub-Poissonian) are indicated by the small circles on the appropriate curves. The optimum situation occurs for $t = t_0$, $t_0 = 7.2446 \times 10^{-4}$; in this case, $c(t_0) = -0.8279 + i0.4534$. The value $2\kappa t_0$ corresponds to the preferred phase on the output of the Kerr nonlinear interferometer.

In fig. 6, corresponding to fig. 5 through the choice of the parameter $\xi(0)$, we can follow the distance from the center of curvature to the origin

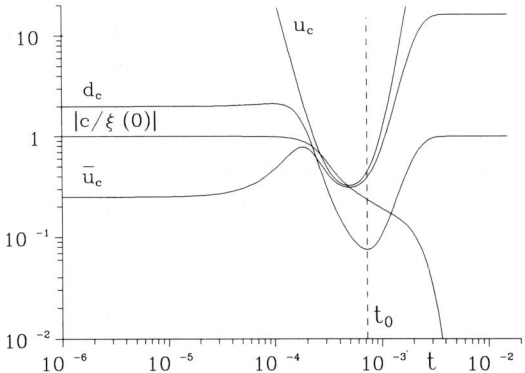

Fig. 6. The center of curvature $|c| |\xi(0)|^{-1}$, the normalized photon-number variance d_c, and the minimum-uncertainty products $u_c, \bar{u}_c$ in the dependence on t for $\kappa = 100$, $\bar{n}_d = 0$, $\gamma = 0$, $|\xi(0)| = 4$, and $\psi = 0$.

$|c||\xi(0)|^{-1}$ and the output Fano factor d_c. The curve d_c indicates the output sub-Poissonian behavior and attains a minimum at t_0, denoted by the vertical dashed line $d_c(t_0) = 7.56 \times 10^{-2}$. The other curve represents the output uncertainty product u_c, which attains its minimum (nearly $\frac{1}{4}$) for a value of t close to t_0. With respect to the length of the period, the effect of the optimum lasts a very short time; it is transient. Another modification of the uncertainty product $\bar{u}_c$ exhibits optimum properties not only for the crescent output state but also for a moderately changed output state. The approximate equality of $\bar{u}_c$ and u_c indicates the strange, typically quantum, stage of the process with an exotic phase distribution for $t_0 < t < (\pi/\kappa) - t_0$. The quasidistribution $\phi_{\mathscr{A}}(\alpha, t_0)$ and its topography are illustrated in figs. 7a and b. Comparing these results with those of the principal squeezing study (fig. 2), we see that the sub-Poissonian statistics can coexist with quadrature squeezing for the chosen range of parameters.

The combination of nonlinearity and displacement, produced by a Mach–Zehnder interferometer with a Kerr medium in one of its arms, generates displaced Kerr states that do not necessarily retain the photon statistics of the initially coherent light. The displaced Kerr states have interesting statistical and phase properties (Wilson-Gordon, Bužek and Knight [1991]). As the coupling constant κ is varied, the output-field statistics vary from sub-Poissonian to super-Poissonian. The range of parameters for which the states are minimum-uncertainty states has been determined.

The phase properties of coherent light interacting with the third-order nonlinear oscillator modelling a lossless Kerr medium have been studied in the framework of the Pegg–Barnett approach (Gerry [1990], Gantsog and Tanaś [1991b, c], Tanaś, Gantsog, Miranowicz and Kielich [1991]), the Susskind–Glogower approach (Gerry [1987a], Gantsog and Tanaś [1991b], Agarwal, Chaturvedi, Tara and Srinivasan [1992]), and the formalism based on the measured phase operators (Lynch [1988], Gantsog and Tanaś [1991b]). The results of the above approaches to the phase problem have been compared (Gantsog and Tanaś [1991b]). The phase properties of coherent light interacting with a lossy Kerr medium have been analyzed in the framework of the Pegg–Barnett and Susskind–Glogower phase formalisms (Gantsog and Tanaś [1991c]). It has been shown that the damping accelerates the randomization of phase at the beginning of the evolution, and removes the quantum periodicity of the process. The damping is useful for illustration of the differences between the two formalisms in the long-time limit. Because the damping diminishes the mean photon number in the

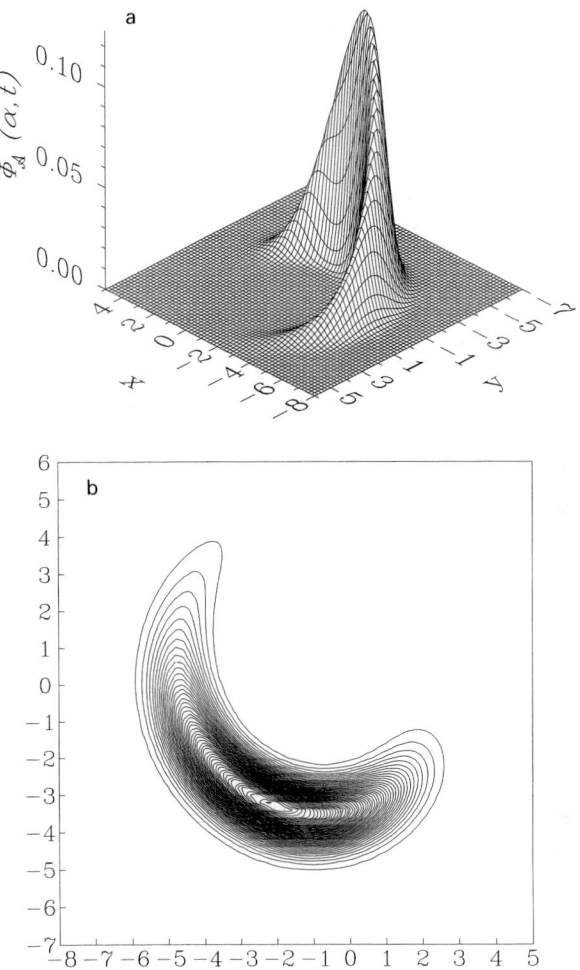

Fig. 7. The quasidistribution $\phi_{\mathscr{A}}(\alpha, t)$ (a) and its topography (b) for $\kappa = 100$, $\bar{n}_d = 0$, $\gamma = 0$, $|\xi(0)| = 4$, $\psi = 0$, and $t_0 = 7.2446 \times 10^{-4}$.

mode and brings the field to a state near to the vacuum, the difference between the results of the two phase approaches is the greatest.

From the Schrödinger equation for a conservative third-order nonlinear oscillator:

$$i\hbar \frac{\partial}{\partial t} |\psi_C(t)\rangle = \hat{H}_C |\psi_C(t)\rangle; \quad C = \text{N, S}, \tag{2.151}$$

describing the evolution of the system in the interaction picture [$\hat{H}_N$, $\hat{H}_S$ are given by eqs. (2.3) and (2.6), respectively], it is possible to obtain, supposing an initial coherent state, a generalized coherent state (Titulaer and Glauber [1966], Stoler [1971], Vourdas and Bishop [1989])

$$|\psi_C(t)\rangle = |\xi(0), \varphi_{Cn}\rangle = \exp(-\tfrac{1}{2}|\xi(0)|^2) \sum_{n=0}^{\infty} \frac{\xi^n(0)}{(n!)^{1/2}} \exp(i\varphi_{Cn})|n\rangle,$$

(2.152)

where

$$\varphi_{Nn} = -\kappa t n(n-1), \qquad \varphi_{Sn} = -\kappa t n^2. \tag{2.153}$$

When the coefficients $\exp(i\varphi_{Cn})$ have the property of k-periodicity,

$$\exp(i\varphi_{C(n+k)}) = \exp(i\varphi_{Cn}) \tag{2.154}$$

for each $n, k > 0$, the generalized coherent state [eq. (2.152)] can be expressed as a discrete superposition of coherent states (Białynicka-Birula [1968], Tombesi and Mecozzi [1987]):

$$|\psi_C(t)\rangle = \sum_{j=1}^{k} a_{Cj}|\exp(i\vartheta_{Cj})\xi(0)\rangle. \tag{2.155}$$

The formulas for the phases ϑ_{Cj} and the coefficients a_{Cj}, $j = 1, ..., k$, are to be found in papers devoted to the propagation of the coherent light in a Kerr medium (Yurke and Stoler [1986], Miranowicz, Tanaś and Kielich [1990], Tanaś [1991]). For a suitable propagation length in the medium (Blow, Loudon and Phoenix [1993]), superpositions of both even and odd numbers of coherent states may arise. General rules for the creation of superpositions of a certain number of coherent states have been derived, and the maximum number of well-separable states for a given initial mean photon number has been estimated. It is amazing that such superpositions of the most classical of field states can, through the action of quantum interference, generate the prototype nonclassical field states, the squeezed vacuum and squeezed coherent states (Bužek and Knight [1991]). Generally, if more states are involved in the superposition, a higher degree of squeezing can be obtained for the same mean value of photons in the mode. The origin of nonclassical effects in a one-dimensional superposition of coherent states has been explained in terms of the Wigner functions and phase-space interference (Schleich, Pernigo and Le Kim [1991], Bužek, Knight and Barranco [1992]).

For the generalized coherent state [eq. (2.152)] we determine the quasi-distribution $\phi_{\mathscr{A}C}(\alpha, t)$:

$$\phi_{\mathscr{A}C}(\alpha, t) = |\langle \alpha | \psi_C(t) \rangle|^2$$

$$= \exp(-|\alpha|^2 - |\xi(0)|^2) \left| \sum_{n=0}^{\infty} \frac{(\alpha^* \xi(0))^n}{n!} \exp(i\varphi_{Cn}) \right|^2. \quad (2.156)$$

This quasidistribution illustrates well the generation of superposition states (Miranowicz, Tanaś and Kielich [1990], Tanaś [1991]). It exhibits regular structures when the component states are entangled. This can be seen in figs. 8a, b, c and d, where the evolution time is taken as a fraction M/N of the period, with M and N being mutually prime integers. Figure 8b shows the growing influence of the interference terms when the number of components becomes larger than $N_{\max}$.

The phase quasidistribution

$$Q_C(\varphi, t) = \int_0^{\infty} \phi_{\mathscr{A}C}(r \exp(i\varphi), t) r \, dr, \quad (2.157)$$

which coincides with the marginal sum $\Sigma_l \phi(l, \varphi, t)$ [eq. (2.43)] and the phase distribution $P(\varphi, t)$ according to eq. (2.124) indicate distinctly the superpositions of coherent states, because both of them exhibit the k-tuple rotational symmetry in the case of superposition of k states (Tanaś, Gantsog, Miranowicz and Kielich [1991], Gantsog and Tanaś [1991b]). The damping of the superpositions is fast, and may be treated adequately through first-order perturbation theory (Braunstein [1992]).

We illustrate graphically in figs. 9a and b the evolution of the phase distribution $P(\varphi, t)$ given in eq. (2.124). The values of parameters correspond to those in fig. 8. This distribution and the quasidistribution $\phi_{\mathscr{A}}(\alpha, t)$ split into the sums of their counterparts for the individual components of the superpositions which may be of value for detecting the superposition states. This is convincingly seen when we plot the contours of the phase distribution versus φ in the polar coordinate system as is obvious from fig. 10.

2.6.2. *The SU(1,1) coherent state*

This state (Perelomov [1986]):

$$|\xi(0)\rangle_P = (1 - |\xi(0)|^2)^{1/2} \sum_{n=0}^{\infty} \xi^n(0) |n\rangle, \quad |\xi(0)| \in [0, 1), \quad (2.158)$$

has been considered as the initial state of the conservative third-order nonlinear oscillator (Gerry [1987c], Bužek [1989b]). Its time evolution,

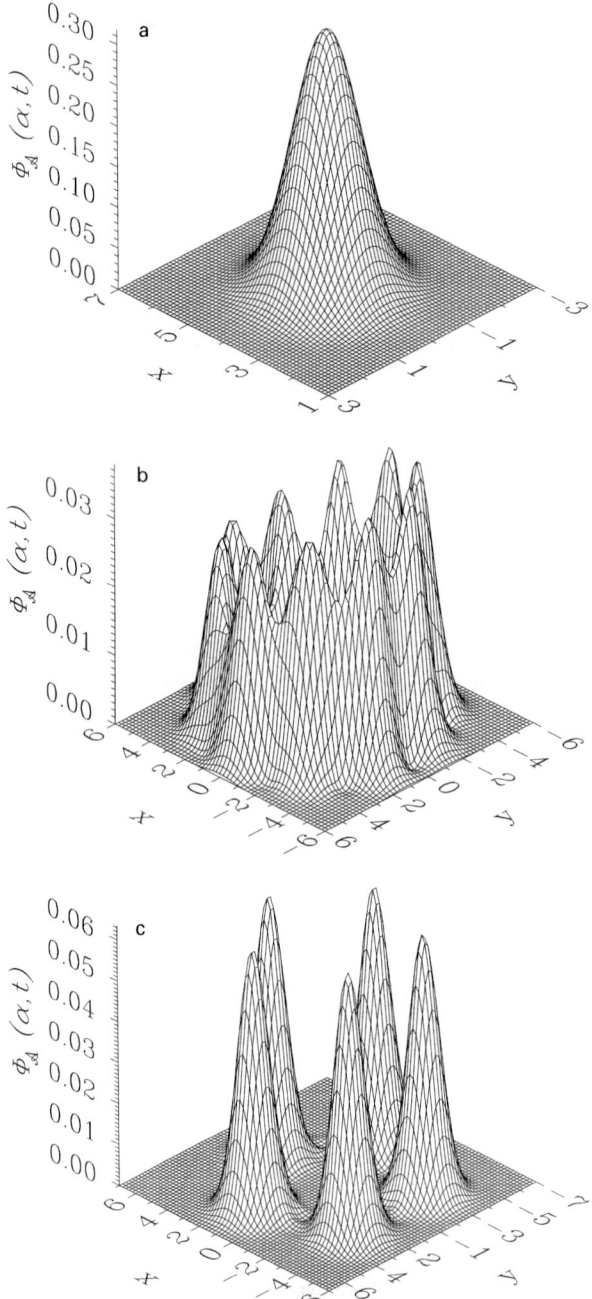

Fig. 8. The quasidistribution $\phi_{\mathscr{A}}(\alpha, t)$ ($\equiv \phi_{\mathscr{A}N}(\alpha, t)$) for $\kappa = 100$, $\bar{n}_d = 0$, $\gamma = 0$, $|\xi(0)| = 4$, and $\psi = 0$: (a) $t = 0$; (b) $t = \pi/10\kappa$; (c) $t = \pi/5\kappa$; (d) $t = \pi/2\kappa$.

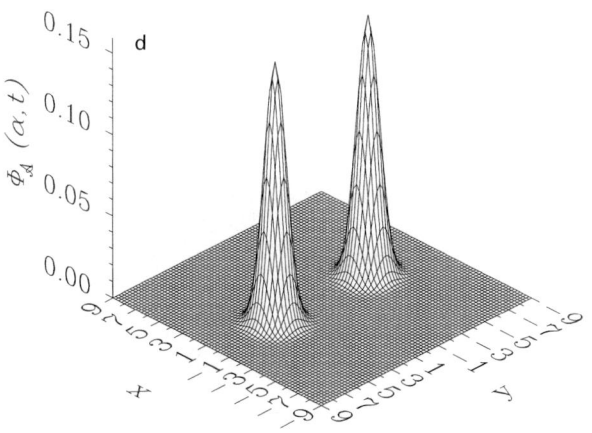

Fig. 8. Continued.

described by the relation

$$|\psi_N(t)\rangle = (1 - |\xi(0)|^2)^{1/2} \sum_{n=0}^{\infty} \xi^n(0) \exp[-i\kappa n(n-1)t]|n\rangle, \qquad (2.159)$$

has provided a basis for an analysis of the standard squeezing. It has been shown that the Kerr medium interacting with squeezed light in the form of an SU(1,1) coherent state generates quantum noise which eventually revokes the squeezing. The greater the initial photon number, the more rapidly the squeezing is revoked.

2.6.3. Gaussian mixed and pure states

A Gaussian mixed state (an alternative name for the generalized superposition of the coherent and chaotic states) has been considered at the beginning of the evolution by Peřinová, Lukš and Kárská [1990]. Its quasidistribution reads (Peřinová, Křepelka and Peřina [1986]):

$$\phi_{\mathscr{A}}(\alpha, 0) = \frac{1}{\pi(K(0))^{1/2}}$$

$$\times \exp\left\{-\frac{B_{\mathscr{A}}(0)}{K(0)}|\alpha - \xi(0)|^2 + \left[\frac{C^*(0)}{2K(0)}(\alpha - \xi(0))^2 + \text{c.c.}\right]\right\},$$

(2.160)

where $K(0) > B_{\mathscr{A}}(0) \geq 1$, and

$$K(0) = B_{\mathscr{A}}^2(0) - |C(0)|^2. \qquad (2.161)$$

A Gaussian pure state [a synonym for the two-photon coherent state $|\beta\rangle_g$ (Yuen [1976])] as a limit of the mixed state has the parameters $B_{\mathscr{A}}(0) =$

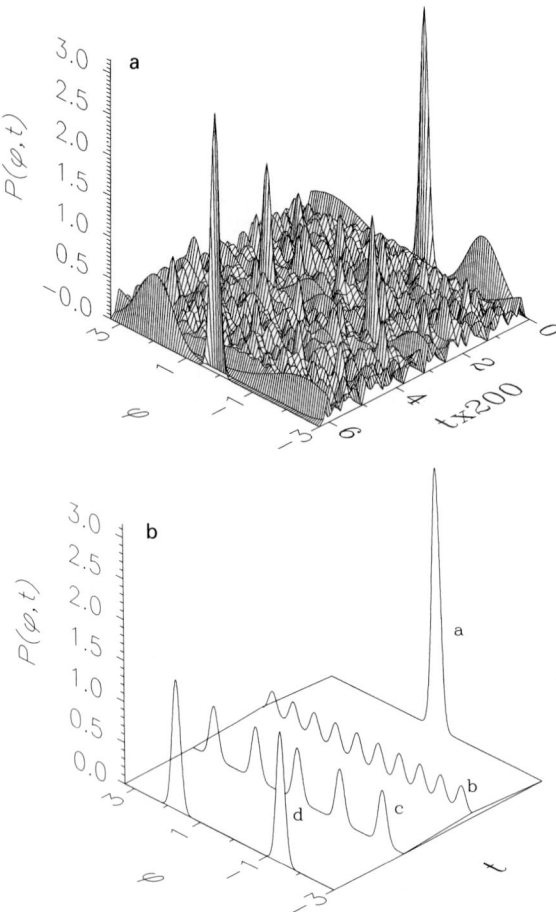

Fig. 9. Evolution of the phase distribution $P(\varphi, t)$ for $\kappa = 100$, $\bar{n}_d = 0$, $\gamma = 0$, $|\xi(0)| = 4$, and $\psi = 0$: (a) $t \in [0, \pi/\kappa]$; (b) $t = 0$ (curve a), $t = \pi/10\kappa$ (curve b), $t = \pi/5\kappa$ (curve c), $t = \pi/2\kappa$ (curve d).

$|\mu|^2$, $C(0) = -\mu^* v$, and $|\mu|^2 - |v|^2 = 1$. Equation (2.10) becomes:

$$f_{mn}(0) = \frac{1}{\pi(K(0))^{1/2}} \exp\left\{-\frac{B_{\mathscr{A}}(0)}{K(0)}|\bar{\eta}(0)|^2 - \left[\frac{C^*(0)}{2K(0)}\bar{\eta}^2(0) + \text{c.c.}\right]\right\}$$

$$\times \sum_{j=0}^{\min(m,n)} \frac{1}{j!(m-j)!(n-j)!} L^j(0) \left[-\frac{C^*(0)}{2K(0)}\right]^{(m-j)/2}$$

$$\times \left[-\frac{C(0)}{2K(0)}\right]^{(n-j)/2} H_{m-j}\left(\frac{\bar{\eta}^*(0)}{(-2C^*(0))^{1/2}}\right) H_{n-j}\left(\frac{\bar{\eta}(0)}{(-2C(0))^{1/2}}\right),$$

(2.162)

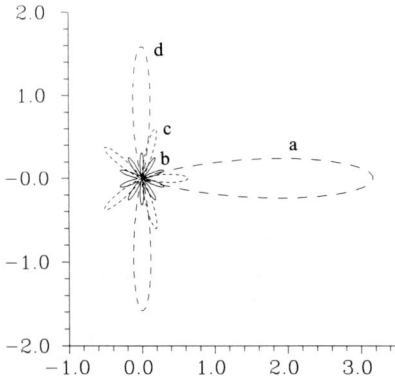

Fig. 10. Plot of the phase distributions $P(\varphi, t)$ in polar coordinates for $t = 0$ (curve a), $t = \pi/10\kappa$ (curve b), $t = \pi/5\kappa$ (curve c), $t = \pi/2\kappa$ (curve d), $\kappa = 100$, $\bar{n}_d = 0$, $\gamma = 0$, $|\xi(0)| = 4$, and $\psi = 0$.

where

$$L(0) = 1 - \frac{B_{\mathscr{A}}(0)}{K(0)}, \quad \bar{\eta}(0) = (K(0))^{-1/2}(B_{\mathscr{A}}(0)\xi(0) - C(0)\xi^*(0)), \tag{2.163}$$

and $H_l(x)$ are the Hermite polynomials

$$H_l(x) = \sum_{k=0}^{[l/2]} \frac{l!}{k!(l-2k)!}(-1)^k (2x)^{l-2k}. \tag{2.164}$$

In particular, for the Gaussian pure state the formula simplifies to:

$$f_{mn}(0) = \frac{1}{\pi|\mu|} \exp\left[-|\eta(0)|^2 + \frac{v^*}{2\mu}\eta^2(0) + \frac{v}{2\mu^*}\eta^{*2}(0)\right]\frac{1}{m!n!}$$

$$\times \left(\frac{v^*}{2\mu^*}\right)^{m/2}\left(\frac{v}{2\mu}\right)^{n/2} H_m\left(\frac{\eta^*(0)}{(2\mu^*v^*)^{1/2}}\right) H_n\left(\frac{\eta(0)}{(2\mu v)^{1/2}}\right), \tag{2.165}$$

where

$$\eta(0) = \mu\xi(0) + v\xi^*(0). \tag{2.166}$$

The statistical properties of this nonlinear oscillator and the system behavior are described completely by the functions $f_{mn}(t)$ given in eqs. (2.27) and (2.162). The system is periodic with period π/κ for $\bar{n}_d = 0$ and $\gamma = 0$. For $\bar{n}_d = 0$ and $\gamma \neq 0$, this property is preserved in part. The periodic behavior disappears for $\gamma \gg 0$ or $\bar{n}_d \gg 0$. A detailed analysis of the case $\bar{n}_d = 0$, $\gamma = 0$ would show not only that the initial states reproduce themselves at times

$(\pi/\kappa)k$ (k is an integer) during the evolution, but also that they form a superposition of two squeezed states at times $(\pi/2\kappa)(2k+1)$ (Tombesi and Mecozzi [1987]). The effect of damping on the generation of squeezed-state superpositions has been studied for the initial Gaussian pure state with the parameters $\mu = \cosh r$ and $\nu = \sinh r$ (Milburn, Mecozzi and Tombesi [1989]). Using the first-order perturbation contribution of a generic Markovian master equation and suitable separability assumptions, the effective damping rate of an arbitrary quantum superposition can be calculated (Braunstein [1992]). It has been shown that superpositions of squeezed states are less sensitive to damping than superpositions of coherent states.

The photon statistics of the third-order nonlinear oscillator with an initial Gaussian mixed state are related to the generalized superposition of coherent and chaotic fields with the quasidistribution (Peřinová, Křepelka and Peřina [1986]):

$$\phi_{\mathcal{A}}(\alpha, t) = \frac{1}{\pi(K(t))^{1/2}} \exp\left\{-\frac{B_{\mathcal{A}}(t)}{K(t)}|\alpha - \xi(t)|^2 + \frac{1}{2K(t)}\right.$$

$$\left. \times [C^*(t)(\alpha - \xi(t))^2 + \text{c.c.}]\right\}, \qquad (2.167)$$

where

$$\xi(t) = \xi(0)\exp\left(-\frac{\gamma}{2}t\right),$$

$$B_{\mathcal{A}}(t) = B_{\mathcal{A}}(0)\exp(-\gamma t) + (\bar{n}_d + 1)(1 - \exp(-\gamma t)), \qquad (2.168)$$

$$C(t) = C(0)\exp(-\gamma t), \qquad K(t) = B_{\mathcal{A}}^2(t) - |C(t)|^2.$$

The photon-number distribution and its factorial moments can also be expressed in terms of the Hermite polynomials (Peřinová, Lukš and Kárská [1990]). Moreover, the series in eq. (2.112) can be summed, and the variances [eqs. (2.74) and (2.78)] characterizing the standard and the principal squeezing may be expressed in a closed form. The dependence of the squeezing on the parameters of the initial quasidistribution, the relationship between the reproduction of the coherent part of the field and squeezing, and the effect of the dissipation on squeezing have been demonstrated numerically for an initial Gaussian pure state (Peřinová, Lukš and Kárská [1990]).

An analysis of the amplitude-squared squeezing (Mir and Razmi [1991]) and that of the effect of the photon statistics and the quadrature squeezing on the degree of coherence (Banerjee [1993]) have been presented for the lossless anharmonic oscillator and an initial Gaussian pure state.

2.6.4. The $SU(1,1)$ two-photon coherent state

This state (Perelomov [1986]),

$$|\xi(0)\rangle_{Pg} = (1 - |\xi(0)|^2)^{1/4} \sum_{n=0}^{\infty} \left[\frac{\Gamma(n+\frac{1}{2})}{n!\,\Gamma(\frac{1}{2})}\right]^{1/2} \xi^n(0)|2n\rangle, \quad |\xi(0)| \in [0, 1), \tag{2.169}$$

evolves in a lossless Kerr medium, into the state

$$|\psi_N(t)\rangle = (1 - |\xi(0)|^2)^{1/4} \sum_{n=0}^{\infty} \left[\frac{\Gamma(n+\frac{1}{2})}{n!\,\Gamma(\frac{1}{2})}\right]^{1/2} \xi^n(0)$$

$$\times \exp[-i\kappa 2n(2n-1)t]|2n\rangle. \tag{2.170}$$

The standard squeezing of the state [eq. (2.170)] has been studied by Bužek [1989c].

2.6.5. The k-photon coherent state

The conservative third-order nonlinear oscillator with the initial k-photon coherent state (D'Ariano, Rosetti and Vadacchino [1985]),

$$|\xi_k(0)\rangle = \left[\sum_{n=0}^{\infty} \frac{|\xi_k(0)|^{2n}}{(nk)!}\right]^{-1/2} \sum_{n=0}^{\infty} \frac{\xi_k^n(0)}{((nk)!)^{1/2}} |nk\rangle, \tag{2.171}$$

provides the resulting state (Bužek and Jex [1989])

$$|\psi_N(t)\rangle = \left[\sum_{n=0}^{\infty} \frac{|\xi_k(0)|^{2n}}{(nk)!}\right]^{-1/2} \sum_{n=0}^{\infty} \frac{\xi_k^n(0)}{((nk)!)^{1/2}} \exp[-i\kappa nk(nk-1)t]|nk\rangle. \tag{2.172}$$

This state has been analyzed from the principal squeezing viewpoint.

2.6.6. Squeezed and displaced number states

The squeezed and displaced number states $|\beta, M\rangle_g$, $\beta = \mu\xi(0) + \nu\xi^*(0)$, $|\mu|^2 - |\nu|^2 = 1$ (Kim, De Oliveira and Knight [1989a, b], Král [1990a]) have the coherent-state representation:

$$\langle\alpha|\beta, M\rangle_g = \frac{1}{(M!)^{1/2}} \xi^M H_M\left(\frac{\alpha^* - \xi^*(0)}{2\zeta\mu}\right) \langle\alpha|\beta\rangle_g, \tag{2.173}$$

where $\zeta = (-v^*/2\mu)^{1/2}$, $|\beta\rangle_g \equiv |\beta, 0\rangle_g$, the Gaussian pure state, and (Yuen [1976])

$$\langle\alpha|\beta\rangle_g = \mu^{-1/2} \exp\left(-\frac{1}{2}|\alpha|^2 - \frac{1}{2}|\beta|^2 - \frac{v}{2\mu}\alpha^{*2} + \frac{v^*}{2\mu}\beta^2 + \frac{1}{\mu}\alpha^*\beta\right). \quad (2.174)$$

The appropriate quasidistribution for the antinormal ordering of field operators reads:

$$\phi_{\mathcal{A}}(\alpha, 0) = \frac{1}{\pi}|\langle\alpha|\beta, M\rangle_g|^2. \quad (2.175)$$

The number-state representation of the initial state under study is given by:

$$\langle n|\beta, M\rangle_g = \left(\frac{n!M!}{\mu}\right)^{1/2} \exp\left(-\frac{1}{2}|\beta|^2 + \frac{v^*}{2\mu}\beta^2\right)$$

$$\times \sum_{j=0}^{\min(n,M)} \frac{1}{j!(n-j)!(M-j)!}\left(\frac{1}{\mu}\right)^j \chi^{n-j}\zeta^{M-j}$$

$$\times H_{n-j}\left(\frac{\beta}{2\chi\mu}\right) H_{M-j}\left(-\frac{\xi^*(0)}{2\zeta\mu}\right), \quad (2.176)$$

where $\chi = (v/2\mu)^{1/2}$. The initial values of $f_{mn}(t)$ are of the form:

$$f_{mn}(0) = \frac{1}{\pi}(n!m!)^{-1/2}\langle n|\beta, M\rangle_g {}_g\langle\beta, M|m\rangle. \quad (2.177)$$

The statistical properties of the squeezed and displaced number states propagating in a lossy Kerr medium have been studied by Král [1990b]. For a lossless medium the quasidistribution $\phi_{\mathcal{A}}(\alpha, L)$, where L is the propagation length, has been determined and the computation of the photon-number distribution $p(n, t)$ and its factorial moments $\langle W^k\rangle_{\mathcal{N}}$ has been outlined. The behavior of the phase quasidistribution:

$$Q(\varphi, L) = \int_0^\infty \phi_{\mathcal{A}}(r\exp(i\varphi), L) r\, dr \quad (2.178)$$

has been investigated in the dependence on the displacement $\xi(0)$. The effect of quantum fluctuations has been involved in the quasidistribution $\phi_{\mathcal{A}}(\alpha, L)$ only approximately.

The evolution of squeezed and displaced number states in the free and dissipative third-order nonlinear oscillator has been investigated by Peřinová and Křepelka [1993] from the point of view of nonclassical phenomena as the number squeezing in both the strong and weak sense, the principal squeezing of vacuum fluctuations, and the generation of superposition states. A destructive effect of losses on quantum coherence has been demonstrated.

The displaced number states $|\beta, M\rangle$ with the quasidistribution related to the antinormal ordering of field operators

$$\phi_{\mathscr{A}}(\alpha, 0) = \frac{1}{\pi} \exp(-|\alpha - \beta|^2) \frac{|\alpha - \beta|^{2M}}{M!} \qquad (2.179)$$

have been considered at the input of the third-order nonlinear oscillator with a quiet reservoir (Brisudová [1991]). The study of the standard quadrature variances has revealed that as few as one incoherent photon; i.e., the initial displaced number state $|\beta, 1\rangle$, makes the output field be not squeezed. This initial state evolves over the half-period into the superposition of two quantum states:

$$\left|\psi\left(t = \frac{\pi}{2\kappa}\right)\right\rangle = 2^{-1/2}\left[\exp\left(-i\frac{\pi}{4} + i\frac{\pi}{2}M\right)|i\beta, M\rangle \right.$$
$$\left. + \exp\left(i\frac{\pi}{4} - i\frac{\pi}{2}M\right)|-i\beta, M\rangle\right]. \qquad (2.180)$$

The damping prevents the rise of superposition states. In the conservative case, the quasidistribution $\phi_{\mathscr{A}}(\alpha, t = \pi/2\kappa)$ splits into component quasidistributions appropriate to two displaced number states and an interference non-quasidistribution. The damping smoothes out the shape of the quasidistribution and makes it rotationally symmetric.

2.7. OPTICALLY BISTABLE TWO-PHOTON MEDIUM

In light of the recent observations of squeezed electromagnetic radiation (Slusher, Hollberg, Yurke, Mertz and Valley [1985], Wu, Kimble, Hall and Wu [1986]), there is much interest in finding systems which exhibit a significant amount of squeezing. The degenerate parametric process with classical pumping is a prototype for the generation of squeezed states (Peřinová, Křepelka and Peřina [1986], Peřinová, Lukš and Szłachetka [1989], Peřina [1991]). Assuming the pump is so strong that it may be considered as classical, we write the Hamiltonian for the generation of the second subharmonic radiation in the form:

$$\hat{H} = \hbar\{\omega(\hat{a}^+\hat{a} + \tfrac{1}{2}) - \tfrac{1}{2}g[\hat{a}^2 \exp(i2\omega t - i\varepsilon) + \text{H.c.}]\}. \qquad (2.181)$$

Here the phase synchronization is assumed, so that the coupling constant g is positive real, 2ω is the pump frequency and the pump phase is ε; the pump amplitude is involved in the coupling constant.

The results of § 2.6 imply that the conservative third-order nonlinear oscillator produces periodically squeezed light from a coherent state at the input. The nonlinearity $\hat{a}^{+2}\hat{a}^2$ causes a phase shift of the complex amplitude of this oscillator, which depends on the input field energy as follows from the Heisenberg expression for the operator $\hat{a}(t)$: $\hat{a}(t) = \exp[-i\kappa t \hat{a}^+(0)\hat{a}(0)]\hat{a}(0)$. Due to a feedback, the output intensity of the nonlinear oscillator exhibits a bistable behavior depending on the input field intensity. The model for the optical bistability based on the dissipative third-order nonlinear oscillator is described by the Hamiltonian

$$\hat{H} = \hat{H}_F + \hat{H}_N + \hat{H}_R + \hat{H}_I + \hat{H}_D, \qquad (2.182)$$

where $\hat{H}_F$, $\hat{H}_N$, $\hat{H}_R$, $\hat{H}_I$ are defined in eqs. (2.2), (2.3), (2.4), and (2.5), respectively, and

$$\hat{H}_D = i\hbar[E \exp(-i\omega_L t)\hat{a}^+ - E^* \exp(i\omega_L t)\hat{a}] \qquad (2.183)$$

characterizes the coupling with the coherent driving field of frequency ω_L. Bistability requires that the detuning $\Delta = \omega - \omega_L$ exceed a critical value $\Delta^2 > 3(\gamma/2)^2$, and that the sign of the detuning be opposite to that of the anharmonicity, $\kappa\Delta < 0$. This model has been discussed by Drummond and Walls [1980]. This system has also been treated from the point of view of squeezing (Collett and Walls [1985]). On using the squeezing spectra, it has been shown that at critical points; e.g., the turning points for optical bistability, perfect standard squeezing is possible in principle.

The dynamics of the system (2.182) has been investigated with the aid of the quasidistributions $\phi_{(s)}(\alpha, t)$ for the normal ($\phi_{(1)} \equiv \phi_{\mathcal{N}}$), the antinormal ($\phi_{(-1)} \equiv \phi_{\mathcal{A}}$), and the symmetric ($\phi_{(0)} \equiv \phi_{\text{sym}}$) orderings. In particular, results for the Wigner quasidistribution are presented and compared with those for the quasidistribution $\phi_{\mathcal{A}}(\alpha, t)$. They enable us to calculate the transition rates between two stable states of the bistable system, which determine the ultimate stability of the two states.

The quantum dynamics of a coherently driven lossless nonlinear oscillator, especially the effect of increasingly strong driving on the mean photon number, quadrature squeezing, second-order correlation function, and on the behavior of the $\phi_{\mathcal{A}}$ quasidistribution have been investigated by McNeil [1993]. It has been found that while the dynamics do maintain some quantum aspects as the strength of the driving is increased, the spectacular quantum features of the undriven nonlinear oscillator are progressively lost.

A composition of the above systems described by the Hamiltonian

$$\hat{H} = \hbar\{\omega(\hat{a}^+ \hat{a} + \tfrac{1}{2}) + \tfrac{1}{2}g[\mathrm{i}\exp(\mathrm{i}2\omega t)\hat{a}^2 + \mathrm{H.c.}] + \kappa \hat{a}^{+2}\hat{a}^2\} \quad (2.184)$$

has been considered for an initial coherent state (Tombesi and Yuen [1984], Gerry and Rodriguez [1987b], Wielinga and Milburn [1993]) and for an initial SU(1,1) coherent state (Gerry and Vrscay [1988b]). The evolution of the input states in this optically bistable two-photon medium has been investigated for short interaction times only under the restriction to approximate methods. Assuming an initial coherent state, it has been found that an enhanced transient standard squeezing as well as photon antibunching may occur. The squeezing is sensitive to changes of the coupling constant κ. The enhancement of squeezing occurs for small values of $|\kappa|$. For larger values of $|\kappa|$, the squeezing is higher, but its duration is shorter. The behavior of the standard quadrature variance results from the interference of two-photon and anharmonic terms in eq. (2.182). With the initial SU(1,1) coherent state, the anharmonic term seems to revoke the squeezing, which is the faster the stronger is the coupling (Gerry and Vrscay [1988b]). The two-photon term has a similar effect on the squeezing. The behavior of this system is a consequence of a quantum interference and quantum fluctuations. An instance of the Hamiltonian (2.184) has been considered by Wielinga and Milburn [1993] to show that this quantum optical model can display coherent tunneling between near coherent states determined from parity eigenstates corresponding to the classical fixed points.

A parametric amplifier driven by a periodically pulsed pump field inside a cavity containing a Kerr medium has been considered by Milburn and Holmes [1991]. A coherent state has been used as the initial state. The dynamics of such a system are modelled as a kicked third-order nonlinear oscillator. The kick is constituted by the pulsed parametric amplifier. In between kicks, the dynamics are determined by the Kerr nonlinearity and damping. A classical description of the system in the absence of damping exhibits rich phase-space structures, including fixed points of multiple period and chaos. The damping can be interpreted as the effect of measurement on quantum features of the system under study. It has been concluded that the quantum and classical dynamics of the classically chaotic system converge proportionally to the losses.

The squeezing properties of a dissipative system appropriate to the Hamiltonian description [eq. (2.184)] have been analyzed using a solution of the Langevin equation (García Fernández, Colet, Toral, San Miguel and Bermejo [1991]). The steady states of the Langevin and Heisenberg equations have been obtained and their local stability has been investigated. Below threshold, a squeezed vacuum state is obtained and the nonlinear effects in the fluctuations have been taken into account by a Gaussian decoupling. In the case above threshold, the standard squeezing has been predicted.

2.8. TWO-LEVEL ATOM IN A KERR MEDIUM

Despite its mathematical simplicity, the Jaynes–Cummings model for the interaction of a single two-level atom with a quantized cavity mode of the radiation field (Jaynes and Cummings [1963], Barnett, Filipowicz, Javanainen, Knight and Meystre [1986]) also describes such pure quantum-mechanical phenomena as the Rabi oscillations, collapses and revivals of the atomic inversion, sub-Poissonian photon statistics and squeezing of the cavity field (for a review, see Bužek and Jex [1990a]). Some of these phenomena have found experimental evidence using highly excited Rydberg atoms (Rempe, Walther and Klein [1987], Gentile, Hughey and Kleppner [1989]). As a modification of this model, a cavity filled with a Kerr medium can be considered. A lossy medium can be treated as the dissipative third-order nonlinear oscillator; the total Hamiltonian of the system in the rotating-wave approximation has the form (Werner and Risken [1991a]):

$$\hat{H} = \hbar \left\{ \omega_c (\hat{a}^+ \hat{a} + \tfrac{1}{2}) + \kappa \hat{a}^{+2} \hat{a}^2 + \frac{\omega_a}{2} \hat{\sigma}_z + g(\hat{a}\hat{\sigma}^+ + \hat{a}^+ \hat{\sigma}^-) \right\} + \hat{H}_R + \hat{H}_I.$$

(2.185)

where ω_c and ω_a are the frequencies of the cavity mode and the atom, respectively, $\hat{\sigma}_z, \hat{\sigma}^+, \hat{\sigma}^-$ are the Pauli spin matrices and g is the coupling constant. The Hamiltonians $\hat{H}_R$ and $\hat{H}_I$ are given in eqs. (2.4) and (2.5), respectively. Werner and Risken [1991a] have derived the equation of motion for the density operator $\hat{\rho}$ in the presence of cavity damping. On the assumption that the atom is initially in its upper state and the cavity mode is in a coherent state, the partial differential equations for the quasi-distributions related to the symmetric ordering of field operators have been solved. The expansion coefficients of the quasidistributions are expressed in terms of the expectation values of the field operators. This makes it possible to calculate the second-order moments of the field operators and the atomic inversion. It has been shown that the addition of a weak Kerr medium leads to more pronounced collapses and revivals of the atomic inversion, and to the conservation of the degree of initial transient squeezing predicted for the detuned Jaynes–Cummings model in spite of the damping and thermal noise. Deb and Ray [1993] have shown that the intracavity Kerr nonlinearity may also be effective in squeezing a photon-number distribution in micromaser and subsequently facilitating the realization of a number state with a low number of photons.

For a lossless Kerr medium in the cavity, the evolution of the system under the above conditions is described by the state vector (Bužek and Jex [1990a], Werner and Risken [1991b], Dung and Shumovsky [1991]):

$$|\psi(t)\rangle = \exp(-\tfrac{1}{2}|\xi(0)|^2) \sum_{n=0}^{\infty} \frac{\xi^n(0)}{(n!)^{1/2}}$$
$$\times [A_n(t)|n\rangle|e\rangle + (n+1)^{1/2} B_n(t)|n+1\rangle|g\rangle], \qquad (2.186)$$

where

$$A_n(t) = \exp[-i\kappa n(n-1)t] \cos(\lambda_n t) + i\frac{(\Delta + 2\kappa n)}{2\lambda_n} \sin(\lambda_n t),$$

$$B_n(t) = \exp[-i\kappa n(n-1)t] \left[-i\frac{g}{\lambda_n} \sin(\lambda_n t) \right], \qquad (2.187)$$

with the detuning Δ and the generalized Rabi frequency λ_n given by the relations

$$\Delta = \omega_c - \omega_a, \qquad \lambda_n = \left[\left(\frac{\Delta}{2} + \kappa n \right)^2 + (n+1)g^2 \right]^{1/2}; \qquad (2.188)$$

$|e\rangle$ and $|g\rangle$ denote the excited and the ground states, respectively. The quasidistribution $\phi_{\mathcal{A}}(\alpha, t)$ for the electromagnetic field can be expressed in the form (Werner and Risken [1991a, b]):

$$\phi_{\mathcal{A}}(\alpha, t) = \frac{1}{\pi} \exp(-|\xi(0)|^2 - |\alpha|^2)$$
$$\times \left[\left| \sum_{n=0}^{\infty} \frac{(\xi(0)\alpha^*)^n}{n!} A_n(t) \right|^2 + \left| \sum_{n=0}^{\infty} \frac{(\xi(0)\alpha^*)^n}{n!} \alpha^* B_n(t) \right|^2 \right].$$
$$(2.189)$$

The dynamics of the quasidistribution $\phi_{\mathcal{A}}(\alpha, t)$ have been discussed in the resonance case ($\Delta = 0$) and for some values of the parameters $|\xi(0)|$, g, and κ. The results of these analyses have included the reduced revival time of the atomic inversion, less distinct revivals with increasing nonlinearity, and the rise of the states similar to superpositions of coherent states. A weak atomic coupling does not destroy the generation of the superposition-like states, but a strong coupling gives rise to other non-Gaussian symmetric structures. The photon-number distribution and standard squeezing have also been studied (Bužek and Jex [1990a]). The interaction of the cavity mode with the two-level atom leads to a deformation of the initial Poisson distribution so that the photon-number distribution exhibits several peaks for $t > 0$. The oscillatory behavior becomes more pronounced in the case of small $|\kappa|$. In the limit of a strong nonlinear coupling, the photon-number distribution remains almost constant. This corresponds to the decoupling of the atom from the resonator mode; the field evolution in the cavity is described by the Hamiltonian $\hat{H}_N = \hbar\kappa \hat{a}^{+2}\hat{a}^2$, which conserves the initial photon statistics (Peřinová and Lukš [1988]).

2.9. INTERACTION OF AN ELECTROMAGNETIC FIELD WITH A KERR MEDIUM

The interaction of a single mode of the frequency ω with a medium modelled as a dissipative third-order nonlinear oscillator with the frequency ω_0 and located in a single-mode cavity is described in the rotating-wave approximation by the Hamiltonian (Reid and Walls [1985]):

$$\hat{H} = \hbar\{\omega(\hat{a}^+ \hat{a} + \tfrac{1}{2}) + \omega_0(\hat{b}^+ \hat{b} + \tfrac{1}{2}) + q\hat{b}^{+2}\hat{b}^2 + g(\hat{a}^+ \hat{b} + \hat{a}\hat{b}^+)\} + \hat{H}_R + \hat{H}_I,$$

(2.190)

where q is a real constant standing for the dependence on the radiation intensity and g is the field–medium coupling constant. The Hamiltonians $\hat{H}_R$ and $\hat{H}_I$ are given in eqs. (2.4) and [(2.5), $\hat{a} \to \hat{b}$], respectively. The dynamics of the system are solvable exactly in the limits $q = 0$, $g \neq 0$ and $q \neq 0$, $g = 0$. The first case corresponds to the frequency conversion (Peřinová [1981], Peřinová and Peřina [1981]), or the energy exchange between two waveguides of a linear coupler (Peřinová, Lukš, Křepelka, Sibilia and Bertolotti [1991]).

When the coupling between the two oscillators is different from zero, the dynamics can be solved in a closed form in the adiabatic limit; i.e., when the oscillator frequencies differ significantly. The adiabatic approximation makes it possible to introduce the third-order susceptibility and to write the effective Hamiltonian, in the conservative case, in the form:

$$\hat{H} = \hbar[\bar{\omega}(\hat{a}^+ \hat{a} + \tfrac{1}{2}) + \kappa \hat{a}^{+2}\hat{a}^2],$$

(2.191)

where

$$\bar{\omega} = \omega - \frac{g^2}{\Delta}, \quad \kappa = \frac{qg^4}{\Delta^4}, \tag{2.192}$$

and $\Delta = \omega - \omega_0$ is the detuning (Agarwal and Puri [1989a]). The Hamiltonian (2.191) has been focused in §§ 2.1–2.6.

Let us note that with the Hamiltonian (2.190), a thorough analysis of the bistability and hysteresis has been performed (Selloni and Schwendimann [1979]). The anharmonic oscillator coupled to an electromagnetic field has been shown to describe the dispersive optical bistability occurring only off the resonance.

The adiabatic elimination of the medium variable supposes that the time changes in the medium are significantly slower than those of the field. Generally, the dynamics must include both the medium and the field. Such complex dynamics have been described for a lossy medium by the coupled Langevin equations for a degenerate four-wave mixing based on an anharmonic oscillator model for the medium (Reid and Walls [1985]). An approximate linearized solution has been used to study the standard squeezing for an initial coherent state. Conditions for the optimum generation of squeezed light have been determined. Independent of the detuning, good squeezing is predicted if the intensity of the pump field could be increased so that it would compensate for losses. In forward four-wave mixing, the detrimental effects of losses on the squeezing are minimized (Kumar and Shapiro [1984]).

This interaction has been considered without dissipation for the fields close to resonance (Agarwal and Puri [1989a, 1990a]). For a small enough coupling constant q, the Hamiltonian can be transformed so that the fourth-order term is diagonalized and the resulting off-resonant terms are neglected. On assuming the input field to be in a coherent or number state and the atomic oscillator to be in the ground state, the wave function has been determined. The mean photon number collapses and revives during a periodic exchange of energy between the atomic and field oscillators. For moderately large values of q, a numerical solution can be obtained. Photon antibunching is predicted. With strong nonlinearity and initial number states, the collapses and revivals of the mean photon number do not occur during the periodic exchange of energy between the oscillators. For $q \gg g$ and the initial number state $|N\rangle$, the resonant transitions of the frequency ω occur only between the states $|0, N\rangle$ and $|1, N-1\rangle$. In this case, the system for the atom initially in the ground state is equivalent to a two-level system with the states $|0, N\rangle$ and $|1, N-1\rangle$.

To include the effects connected with the propagation of the eliptically polarized light in a Kerr medium, the field is to be described using two modes (Ritze and Bandilla [1979], Tanaś and Kielich [1979, 1983, 1984, 1990a], Ritze [1980], Agarwal and Puri [1989b, 1990b], Gantsog and Tanaś [1991a, d], Tanaś and Gantsog [1992a,b]). In the quantum description of the electromagnetic field it is convenient to write the vector potential operator of the field component-wise with the aid of the positive- and negative-frequency parts:

$$\hat{A}_j(r, t) = \hat{A}_j^{(+)}(r, t) + \hat{A}_j^{(-)}(r, t), \tag{2.193}$$

where the subscript j indexes the polarization component. Then a mode decomposition of the field is possible, which for a plane wave of the free field propagating in the medium of the refractive index $n(\omega)$, is of the form:

$$\hat{A}_j^{(+)}(r, t) = (\hat{A}_j^{(-)}(r, t))^+ = \sum_{k,s} \left[\frac{\hbar}{2\varepsilon_0 ckn(\omega)V} \right]^{1/2} e_{kj}^{(s)} \hat{a}_{ks}$$

$$\times \exp[-i(ckt - \mathbf{k} \cdot \mathbf{r})]. \tag{2.194}$$

Here $k = |\mathbf{k}|$ and $e_{kj}^{(s)}$ is the jth component of the unit polarization vector related to the polarization state s and the propagation vector $\mathbf{k}$, and V is the quantization volume. The annihilation and creation operators of a photon with the momentum $\hbar\mathbf{k}$ and polarization s, $\hat{a}_{ks}$ and $\hat{a}_{ks}^+$, obey the commutation relations

$$[\hat{a}_{ks}, \hat{a}_{k's'}^+] = \hat{1}\delta_{kk'}\delta_{ss'}, \quad [\hat{a}_{ks}, \hat{a}_{k's'}] = [\hat{a}_{ks}^+, \hat{a}_{k's'}^+] = \hat{0}. \tag{2.195}$$

The polarization vectors satisfy the orthogonality conditions

$$\sum_j e_{kj}^{(s)*} e_{kj}^{(s')} = \delta_{ss'}, \quad \sum_j e_{kj}^{(s)} k_j = 0. \tag{2.196}$$

For the monochromatic field of frequency ω propagating in the z direction, the index $\mathbf{k}$ can be omitted and we can write

$$\hat{A}_j^{(+)}(z, t) = \left[\frac{\hbar}{2\varepsilon_0 ckn(\omega)V} \right]^{1/2} \exp[-i(\omega t - kz)] \sum_{s=1}^{2} e_j^{(s)} \hat{a}_s, \tag{2.197}$$

where $k = n(\omega)\omega/c$. Because in eq. (2.197) the sum over two orthogonal polarizations remains, we have a two-mode description of the field. The decomposition of one mode of elliptically polarized field into two orthogonal modes,

$$e_j \hat{a} = e_j^{(1)} \hat{a}_1 + e_j^{(2)} \hat{a}_2, \tag{2.198}$$

can be rewritten using the conditions of eqs. (2.196) in the form:

$$\hat{a} = e_1^* \hat{a}_1 + e_2^* \hat{a}_2, \tag{2.199}$$

where e_j^* are the transformed components of the polarization vector. The annihilation and creation operators, $\hat{a}_j$ and $\hat{a}_j^+$, obey the commutation relation

$$[\hat{a}_j, \hat{a}_k^+] = \hat{1}\delta_{jk}, \quad j, k = 1, 2. \tag{2.200}$$

So far, the decomposition (2.199) is quite general and can be further specified either for two modes with mutually perpendicular linear polarizations or for right- and left-circularly polarized modes. For the modes linearly polarized in the x and y directions, it holds that

$$\hat{a} = e_x^* \hat{a}_x + e_y^* \hat{a}_y. \tag{2.201}$$

Here,

$$e_x = \cos\eta \cos\theta - \mathrm{i}\sin\eta \sin\theta, \qquad e_y = \cos\eta \sin\theta + \mathrm{i}\sin\eta \cos\theta, \tag{2.202}$$

η is the ellipticity parameter, $\eta \in [-\pi/4, \pi/4]$, and θ is the azimuth of the polarization ellipse of the incident beam. Considering the circular basis related to the right-circularly polarized vector $e^{(1)} = 2^{-1/2}(x + \mathrm{i}y)$ and the left-circularly polarized vector $e^{(2)} = 2^{-1/2}(x - \mathrm{i}y)$, where x and y are the unit vectors in the x and y directions, respectively, we get in accordance with eqs. (2.198) and (2.199):

$$\hat{a}_1 = 2^{-1/2}(\hat{a}_x - \mathrm{i}\hat{a}_y), \qquad \hat{a}_2 = 2^{-1/2}(\hat{a}_x + \mathrm{i}\hat{a}_y). \tag{2.203}$$

The interaction of the monochromatic light of frequency ω with the optically isotropic medium, which consists of N atoms (molecules), has been considered by Tanaś and Kielich [1983]. This interaction for $N = 1$ molecule is described in the electric-dipole approximation by the Hamiltonian

$$\hat{H} = \alpha_{jk}(\omega)\hat{A}_j^{(-)}\hat{A}_k^{(+)} + \tfrac{1}{2}\gamma_{jklm}(\omega)\hat{A}_j^{(-)}\hat{A}_k^{(-)}\hat{A}_l^{(+)}\hat{A}_m^{(+)}, \tag{2.204}$$

where $\alpha_{jk}(\omega)$ and $\gamma_{jklm}(\omega)$ are the tensors of the polarization and hyperpolarization of the molecule, respectively (Kielich [1981]). This Hamiltonian can be expressed in two equivalent forms, depending on whether the light is linearly polarized,

$$\hat{H} = \bar{\alpha}(\omega)(\hat{a}_x^+ \hat{a}_x + \hat{a}_y^+ \hat{a}_y) + \tfrac{1}{2}\{\bar{\gamma}_1(\omega)(\hat{a}_x^{+2} + \hat{a}_y^{+2})(\hat{a}_x^2 + \hat{a}_y^2)$$
$$+ [\bar{\gamma}_2(\omega) + \bar{\gamma}_3(\omega)](\hat{a}_x^{+2}\hat{a}_x^2 + \hat{a}_y^{+2}\hat{a}_y^2 + 2\hat{a}_x^+ \hat{a}_y^+ \hat{a}_y \hat{a}_x)\}, \tag{2.205}$$

or circularly polarized,

$$\hat{H} = \bar{\alpha}(\omega)(\hat{a}_1^+ \hat{a}_1 + \hat{a}_2^+ \hat{a}_2) + \tfrac{1}{2}\{4\bar{\gamma}_1(\omega)\hat{a}_1^+ \hat{a}_2^+ \hat{a}_2 \hat{a}_1$$
$$+ [\bar{\gamma}_2(\omega) + \bar{\gamma}_3(\omega)](\hat{a}_1^{+2}\hat{a}_1^2 + \hat{a}_2^{+2}\hat{a}_2^2 + 2\hat{a}_1^+ \hat{a}_2^+ \hat{a}_2 \hat{a}_1)\}. \qquad (2.206)$$

The quantities $\bar{\alpha}(\omega)$, $\bar{\gamma}_j(\omega)$, $j = 1, 2, 3$, are related to the expectation values of the polarization and hyperpolarization tensors over all possible orientations of the molecule.

The Hamiltonians (2.205) and (2.206) have been used together with an initial coherent state to study photon statistics and antibunching (Ritze and Bandilla [1979], Tanaś and Kielich [1979], Ritze [1980]), and the standard squeezing of the vacuum fluctuations (Tanaś and Kielich [1983, 1984], Horák [1989]). It has been proved that a coherent light propagating in a Kerr medium exhibits squeezing of its own quantum fluctuations. The maximum amount of squeezing can be attained for an initial state of a large photon number. The photon antibunching effect requires that the input field be elliptically polarized, whereas squeezing is insensitive to the polarization state of the field and can occur for any polarization. Because the lossless medium conserves the photon-number distribution, squeezed states with Poissonian photon statistics can be obtained.

The above quantum model for propagation of the elliptically polarized light in a Kerr medium has been applied to partially polarized light for initial coherent and number states (Agarwal and Puri [1989b, 1990b]). States have been proved to be generated which are macroscopic superpositions of coherent states. It has been found that if the input field is completely polarized, the quantum effects result in partial polarization of the output field, which contradicts the prediction of the classical theory. The energies in separate modes, the correlation between two modes, higher-order correlations, and the mean photon number have been studied in relation to the input states. The input photon statistics are found to make a considerable difference to the dynamics.

The phase properties of the elliptically polarized light propagating in a Kerr medium have been investigated in the framework of the Pegg–Barnett formalism (Gantsog and Tanaś [1991a, d], Tanaś and Gantsog [1991]). The Hermitian phase operator concept has been generalized to two-mode fields, and the phase distribution of the two orthogonal modes describing the elliptically polarized field has been computed. In addition, marginal phase distributions as well as the expectation values and variances of the phases have been determined and their dynamics studied. It appears that in the course of the propagation, a correlation between the phases of the two modes comes into being. The degree of correlation depends substantially on the asymmetry parameter d of the medium, $d = f(\bar{\gamma}_1(\omega), \bar{\gamma}_2(\omega), \bar{\gamma}_3(\omega))$. The highest correlation occurs for $d = 1/2$. This correlation reduces the variance of the difference of the phase operators for the separate modes. If $d = 1/2$, neither the variance of the phase-difference operator nor the degree of the field polarization is affected during the propagation. However, for $d \neq 1/2$, this variance rises rapidly to $1/2$, which is the value for the randomly distributed phase difference. To the randomization of the phase difference there corresponds a degradation of the degree of light polarization. The phase properties of the elliptically polarized light in a Kerr medium confirm the results predicting this degradation (Agarwal and Puri [1989b], Tanaś and Kielich [1990a]).

The formation of discrete superpositions of coherent states has been considered for elliptically polarized light propagating in a nonlinear Kerr medium (Gantsog and Tanaś [1991a]). It has been shown that superpositions with any number of components can be obtained if the evolution time is taken as a fraction M/N of the period, where M and N are mutually prime integers. Exact analytical formulas for finding the superposition coefficients have been given. It has been shown that the coupling between the two circularly polarized components of the elliptically polarized light caused by the asymmetry of the nonlinear properties of the medium can suppress the number of components if the asymmetry parameter takes on appropriate values. The phase distribution for the two-mode field exhibits a well-resolved, multi-peak structure, clearly indicating the generation of the discrete superpositions of coherent states.

Quantum fluctuations in the Stokes parameters, which are the expectation values of the Hermitian Stokes operators, have been studied for strong light propagating in a lossless and damping Kerr medium (Tanaś and Kielich [1990a], Tanaś and Gantsog [1992a, b]). In the case without losses (Tanaś and Kielich [1990a], Tanaś and Gantsog [1992a]), the periodic behavior of quantum evolution of the light polarization is revealed explicitly. Although the variances of the Stokes operators are reminiscent of the field operator variances determining squeezing, they have been shown not to fall below the level for a coherent state. The signal-to-noise ratio for the measurement of the Stokes parameters is reduced by quantum field fluctuations. All polarization parameters depend crucially on the asymmetry parameter of the nonlinear medium. Quantitative assessment of the destructive role of linear damping in the quantum evolution of the field has been performed (Tanaś and Gantsog [1992b]).

§ 3. Higher-Order Nonlinear Oscillators

The third-order nonlinear oscillator model can be generalized to a model of the $(2k-1)$th-order nonlinear oscillator (k-photon anharmonic oscillator) described in the conservative case by the Hamiltonian (Gerry [1987b])

$$\hat{H}_{Nk} = \hbar[\omega(\hat{a}^+\hat{a} + \tfrac{1}{2}) + \kappa \hat{a}^{+k}\hat{a}^k], \tag{3.1}$$

where the normal ordering of field operators is respected. The higher-order nonlinear process is of importance in self-focusing (Piekara, Moore and Feld [1974]). The coherent state $|\xi(0)\rangle$ evolves in a medium modelled in this way according to the relation

$$|\psi(t)\rangle = \exp(-\tfrac{1}{2}|\xi(0)|^2) \sum_{n=0}^{\infty} \frac{\xi^n(0)}{(n!)^{1/2}} \exp\left[i\left(\omega n - \kappa \frac{n!}{(n-k)!}\right)t\right]|n\rangle. \tag{3.2}$$

The possibility of the standard squeezing occurrence has been proved for $k = 3, 4$. In this case, however, the standard quadrature variances cannot be expressed in a closed form, and the available expression in terms of the infinite series makes difficulties even in appropriately chosen numerical computations.

The amplitude-squared squeezing of the light in the nonlinear oscillator described in eq. (3.1) has been investigated for an initial coherent state (Yang and Zheng [1989]). The measure for this squeezing,

$$Q = \frac{\langle(\Delta\hat{Y}_1)^2\rangle - \langle\hat{n} + \tfrac{1}{2}\hat{1}\rangle}{\langle\hat{n} + \tfrac{1}{2}\hat{1}\rangle}, \tag{3.3}$$

has been analyzed in the dependence on the mean photon number and the coupling constant for different k. It has been shown that the maximum amount of squeezing (i.e., the minimum of Q) decreases with increasing k.

The formation of discrete superpositions of coherent states in the course of evolution in the k-photon nonlinear oscillator has also been studied (Paprzycka and Tanaś [1992]). Exact analytical formulas for the superposition coefficients have been obtained. In contrast to the third-order nonlinear oscillator, the superposition components enter the superposition with different amplitudes, which makes the superposition less symmetrical. The phase distributions for the resulting states illustrate the amount of symmetry and show the number of components.

The $(2k-1)$th-order nonlinear oscillator described by the Hamiltonian

$$\hat{H}_{Sk} = \hbar[\omega(\hat{a}^+\hat{a} + \tfrac{1}{2}) + \kappa(\hat{a}^+\hat{a})^k], \tag{3.4}$$

where the symmetric ordering of field operators is preferred, has been considered for the initial coherent state $|\xi(0)\rangle$ (Tombesi and Mecozzi [1987]). The coherent-state representation of the state vector $|\psi(t)\rangle$ has been determined in the form:

$$\psi(\alpha) = \pi^{-1/2}\langle\alpha|\psi(t)\rangle = \pi^{-1/2}\exp[-\tfrac{1}{2}(|\alpha|^2 + |\xi(0)|^2)]$$

$$\times \sum_{n=0}^{\infty} \frac{(\alpha^*\xi(0))^n}{n!} \exp[-in\omega t - (-1)^k i n^k \kappa t], \quad (3.5)$$

and used for discussing the generation of macroscopically distinguishable quantum states.

It is possible to add higher-order optical Kerr nonlinearities to the third-order one. A conservative system described by the Hamiltonian

$$\hat{H} = \hbar\left[\omega(\hat{a}^+\hat{a} + \tfrac{1}{2}) + \frac{\kappa_1}{2}\hat{a}^{+2}\hat{a}^2 + \frac{\kappa_2}{3}\hat{a}^{+3}\hat{a}^3 + \frac{\kappa_3}{4}\hat{a}^{+4}\hat{a}^4\right] \quad (3.6)$$

has been treated for an initial coherent state (Tanaś and Kielich [1990b]). The coupling constants $\kappa_1, \kappa_2, \kappa_3$ are related to the nonlinear susceptibilities of the medium $\chi^{(3)}, \chi^{(5)}, \chi^{(7)}$, respectively. The third-order nonlinearity is connected to the optical Kerr effect (Kielich [1981]), and the coupling constant κ_1 is always different from zero even if its value is medium-dependent; it is usually very small. The values of further coupling constants decrease by many orders of magnitude as the order of the nonlinearity increases. The physical effects of these nonlinearities manifest themselves only if the input field intensity is high enough. Numerical results demonstrate a larger amount of squeezing of coherent light when higher-order nonlinearities are included.

§ 4. Coupled Dissipative Third-Order Nonlinear Oscillators

The recent increase of interest in systems of nonlinear optical waveguides arises from their possible use as high-speed switching elements. From the viewpoint of purely quantum effects; e.g., squeezing of vacuum fluctuations, the interaction in such a system could serve for an effective generation of squeezed states.

A planar dielectric waveguide is a medium whose dielectric permittivity depends on one direction, which we suppose to be parallel with the x axis. If the medium does not damp and does not amplify and if its permittivity is weakly dependent on the field frequency, the electromagnetic field can be analyzed in the framework of the second quantization. The vector potential operator can be decomposed as:

$$\hat{A}(r,t) = \sum_j \left(\frac{\hbar}{2\varepsilon_0 \omega_j}\right)^{1/2} (u_j(r)\hat{a}_j(t) + \text{H.c.}), \qquad (4.1)$$

where $\hat{a}_j(t)$ ($\hat{a}_j^+(t)$) are the annihilation (creation) field operators and $u_j(r)$, the orthogonal mode functions, are assumed to have the form

$$u_j(r) = \left(\frac{2\varepsilon_0 \omega_j}{\hbar}\right)^{1/2} f_j(x) \exp(\mathrm{i} k_j \cdot r), \qquad (4.2)$$

where k_j is the wave vector with the components $k_{jx} = 0$, k_{jy}, and k_{jz} ($k_j \cdot r = k_{jy} y + k_{jz} z$), and $f_j(x)$ is a suitable function of the variable x only.

In the case of a nonlinear waveguide the polarization is nonlinearly dependent on the field; assuming the second quantization to be in the form of eq. (4.1), we write the radiation Hamiltonian:

$$\hat{H} = \frac{1}{2}\int_V \left[\varepsilon_0 \hat{E}^2(r,t) + \frac{1}{\mu_0}\hat{B}^2(r,t)\right] \mathrm{d}^3 r + \frac{1}{2}\int_V \hat{P}(r,t)\cdot\hat{E}(r,t)\,\mathrm{d}^3 r, \qquad (4.3)$$

where ε_0 is the vacuum dielectric constant and μ_0 is the vacuum permeability. Here $\hat{E}$ and $\hat{B}$ are the electric field operator and the magnetic field operator, respectively:

$$\hat{E} = -\frac{\partial \hat{A}}{\partial t}, \qquad \hat{B} = \operatorname{rot} \hat{A}, \qquad (4.4)$$

where V is the volume of the radiation field and $\hat{P}(r,t)$ is the polarization operator. For the Kerr nonlinearity, it holds that

$$\hat{P}(r,t) = \varepsilon_0 [\chi^{(1)}(x) + \chi^{(3)}(x)(\hat{E}(r,t)\cdot\hat{E}(r,t))]\hat{E}(r,t). \qquad (4.5)$$

Substituting eq. (4.5) into eq. (4.3), respecting eqs. (4.4) and (4.1), and using the rotating-wave approximation, we obtain:

$$\hat{H} = \hbar\left[\sum_j \omega_j(\hat{a}_j^+ \hat{a}_j + \tfrac{1}{2}) + \sum_{\substack{k,l \\ k\neq l}} U_{kl} \hat{a}_k^+ \hat{a}_l + \sum_{k,l,m,n} V_{klmn} \hat{a}_k^+ \hat{a}_l^+ \hat{a}_m \hat{a}_n\right], \qquad (4.6)$$

where

$$U_{kl} \approx \tfrac{1}{2}(\omega_k \omega_l)^{1/2} \int_V \chi^{(1)}(x)(u_k(r)\cdot u_l(r))\,\mathrm{d}^3 r,$$

$$V_{klmn} \approx \frac{\hbar}{4\varepsilon_0}(\omega_k \omega_l \omega_m \omega_n)^{1/2} \int_V \chi^{(3)}(x)[2(u_k^*(r)\cdot u_l(r))(u_m^*(r)\cdot u_n(r))$$

$$+ (u_k^*(r)\cdot u_l^*(r))(u_m(r)\cdot u_n(r))]\,\mathrm{d}^3 r. \qquad (4.7)$$

Including losses, we write the Hamiltonian (4.3) for two modes of different frequencies in the form:

$$\hat{H} = \hbar \left\{ \sum_{j=1}^{2} \omega_j(\hat{a}_j^+ \hat{a}_j + \tfrac{1}{2}) + (g\hat{a}_1^+ \hat{a}_2 + \text{H.c.}) + \kappa_1 \hat{a}_1^{+2} \hat{a}_1^2 \right.$$
$$\left. + \kappa_2 \hat{a}_2^{+2} \hat{a}_2^2 + \kappa_3 \hat{a}_1^+ \hat{a}_2^+ \hat{a}_1 \hat{a}_2 \right\} + \hat{H}_{R2} + \hat{H}_{12}, \qquad (4.8)$$

where

$$\hat{H}_{RN} = \hbar \sum_l \sum_{j=1}^{N} \psi_l^{(j)} (\hat{c}_l^{(j)+} \hat{c}_l^{(j)} + \tfrac{1}{2}), \qquad (4.9)$$

$$\hat{H}_{IN} = \hbar \sum_l \sum_{j=1}^{N} (\eta_l^{(j)} \hat{c}_l^{(j)} \hat{a}_j^+ + \text{H.c.}), \qquad (4.10)$$

and $g = U_{12}$, $\kappa_j = V_{jjjj}$, $j = 1, 2$, $\kappa_3 = 4V_{1212}$. Here $\hat{c}_l^{(j)}$ is the annihilation operator of the lth oscillator in the jth reservoir with the frequency $\psi_l^{(j)}$, and $\eta_l^{(j)}$ are the coupling constants of the reservoir oscillators.

The statistical properties of the coupled nonlinear oscillators described by the Hamiltonian (4.8) with $g = 0$ have been studied for an initial coherent state (Peřina, Horák, Hradil, Sibilia and Bertolotti [1989], Horák and Peřina [1989]). The system dynamics have been described with the generalized Fokker–Planck equation for the quasidistribution related to the antinormal ordering of field operators in the interaction picture:

$$\frac{\partial}{\partial t} \phi_{\mathscr{A}}(\alpha_1, \alpha_2, t) = \left\{ \left(\frac{\gamma_1}{2} + 2i\kappa_1 |\alpha_1|^2 + i\kappa_3 |\alpha_2|^2 \right) \alpha_1 \frac{\partial}{\partial \alpha_1} \right.$$
$$+ \left(\frac{\gamma_2}{2} + 2i\kappa_2 |\alpha_2|^2 + i\kappa_3 |\alpha_1|^2 \right) \alpha_2 \frac{\partial}{\partial \alpha_2}$$
$$+ \sum_{j=1}^{2} \left[\frac{\gamma_j}{2} + i\kappa_j \alpha_j^2 \frac{\partial^2}{\partial \alpha_j^2} + \frac{\gamma_j}{2} (\bar{n}_{d_j} + 1) \frac{\partial^2}{\partial \alpha_j \partial \alpha_j^*} \right]$$
$$\left. + i\kappa_3 \alpha_1 \alpha_2 \frac{\partial^2}{\partial \alpha_1 \partial \alpha_2} + \text{c.c.} \right\} \phi_{\mathscr{A}}(\alpha_1, \alpha_2, t),$$

$$(4.11)$$

where γ_j are the damping constants and $\bar{n}_{d_j} = \bar{n}_{d_l}^{(j)}$ are the mean quantum numbers of the lth reservoir oscillator in the jth mode

$(\bar{n}_{d_l}^{(j)} = \langle \hat{c}_l^{(j)+}(0)\hat{c}_l^{(j)}(0)\rangle$ independently of l). The solution of eq. (4.11) in the case without losses $(\gamma_j = 0, \bar{n}_{d_j} = 0, j = 1, 2)$ for the initial coherent state $|\xi_1(0)\rangle|\xi_2(0)\rangle$ is of the form:

$$\phi_{\mathscr{A}}(\alpha_1, \alpha_2, t) = \frac{1}{\pi^2} \exp\left[-\sum_{j=1}^{2}(|\alpha_j|^2 + |\xi_j(0)|^2)\right]$$

$$\times \sum_{k,l,m,n} \frac{1}{k!l!m!n!} (\alpha_1 \xi_1^*(0))^k (\alpha_1^* \xi_1(0))^l$$

$$\times (\alpha_2 \xi_2^*(0))^m (\alpha_2^* \xi_2(0))^n$$

$$\times \exp\{[i\kappa_1 k(k-1) - i\kappa_1 l(l-1) + i\kappa_2 m(m-1)$$

$$- i\kappa_2 n(n-1) + i\kappa_3(km - ln)]t\}. \tag{4.12}$$

The coupled equations for the expectation values of the antinormally ordered field operators derived from eq. (4.11) have been solved approximately in terms of the generalized superposition of coherent and chaotic fields described by the antinormal quantum characteristic function:

$$C_{\mathscr{A}}(\beta_1, \beta_2, t) = \exp\left\{\sum_{j=1}^{2}[-B_{j\mathscr{A}}|\beta_j|^2 + \tfrac{1}{2}(C_j^* \beta_j^2 + \text{c.c.}) + (\xi_j^* \beta_j - \text{c.c.})]\right.$$

$$\left.+ (-B_{12}\beta_1^* \beta_2 + C_{12}^* \beta_1 \beta_2 + \text{c.c.})\right\}, \tag{4.13}$$

where

$$\xi_j = \langle \hat{a}_j \rangle, \quad B_{j\mathscr{A}} = \langle \Delta\hat{a}_j \Delta\hat{a}_j^+ \rangle_{\mathscr{A}}, \quad C_j = \langle (\Delta\hat{a}_j)^2 \rangle, \quad j = 1, 2,$$
$$B_{12} = \langle \Delta\hat{a}_1^+ \Delta\hat{a}_2 \rangle, \quad C_{12} = \langle \Delta\hat{a}_1 \Delta\hat{a}_2 \rangle; \tag{4.14}$$

here, $\hat{a}_j \to \exp(i\omega_j t)\hat{a}_j$ for notational simplicity. For the lossless system closed formulas for the parameters (4.14) are presented. In terms of eqs. (4.14), the standard quadrature variances for the single-mode fields [using eq. (2.74)] are

$$\langle(\Delta\hat{Q}_{\hat{P}_j})^2\rangle = -1 + 2(\langle\Delta\hat{a}_j\Delta\hat{a}_j^+\rangle_{\mathscr{A}} \pm \text{Re}\langle(\Delta\hat{a}_j)^2\rangle), \tag{4.15}$$

and for two-mode fields are

$$\langle(\Delta\hat{Q}_{\hat{P}_{12}}^{12})^2\rangle = -1 + c_1 \pm \text{Re } c_2, \tag{4.16}$$

where the quadrature operators $\hat{Q}_{12}$, $\hat{P}_{12}$ and the moments c_1 and c_2 are

$$\hat{Q}_{12} = 2^{-1/2}(\hat{Q}_1 + \hat{Q}_2), \qquad \hat{P}_{12} = 2^{-1/2}(\hat{P}_1 + \hat{P}_2),$$

$$c_1 = \langle \Delta\hat{a}_1 \Delta\hat{a}_1^+ \rangle_{\mathscr{A}} + \langle \Delta\hat{a}_2 \Delta\hat{a}_2^+ \rangle_{\mathscr{A}} + 2\,\mathrm{Re}\langle \Delta\hat{a}_1 \Delta\hat{a}_2^+ \rangle, \qquad (4.17)$$

$$c_2 = \langle (\Delta\hat{a}_1)^2 \rangle + \langle (\Delta\hat{a}_2)^2 \rangle + 2\langle \Delta\hat{a}_1 \Delta\hat{a}_2 \rangle.$$

According to the standard definition of squeezing (Walls [1983]), the mode (field) studied occurs in a squeezed state if either of the above variances is less than unity; nevertheless, this mode (field) obeys the uncertainty relation:

$$\langle (\Delta\hat{Q}_j)^2 \rangle \langle (\Delta\hat{P}_j)^2 \rangle \geq 1 \qquad (\langle (\Delta\hat{Q}_{12})^2 \rangle \langle (\Delta\hat{P}_{12})^2 \rangle \geq 1). \qquad (4.18)$$

The photon statistics of the system are determined exactly by the superposition of coherent and chaotic fields (Peřinová and Lukš [1988], Peřinová and Kárská [1989], Peřina [1991]):

$$p(n, t) = \sum_{m_1 + m_2 = n} \prod_{j=1}^{2} \frac{(\bar{n}_j(t))^{m_j}}{(\bar{n}_j(t) + 1)^{m_j + 1}} \exp\left[-\frac{|\xi_j(t)|^2}{\bar{n}_j(t) + 1}\right]$$

$$\times \frac{1}{m_j!} L_{m_j}^0\left(-\frac{|\xi_j(t)|^2}{\bar{n}_j(t)(\bar{n}_j(t) + 1)}\right), \qquad (4.19)$$

and

$$\langle W^k \rangle_{\mathscr{N}} = k! \sum_{m_1 + m_2 = k} \prod_{j=1}^{2} \frac{(\bar{n}_j(t))^{m_j}}{m_j!} L_{m_j}^0\left(-\frac{|\xi_j(t)|^2}{\bar{n}_j(t)}\right), \qquad (4.20)$$

where the mean numbers of coherent and chaotic photons are given by the formulas

$$\xi_j(t) = \xi_j(0) \exp\left(-\frac{\gamma_j}{2} t\right), \qquad \bar{n}_j(t) = \bar{n}_{d_j}[1 - \exp(-\gamma_j t)], \qquad (4.21)$$

respectively. For the marginal photon-number distributions and their factorial moments, eqs. (2.144) and (2.145) hold.

Another approximate method has been used when determining the dynamics of the considered system [eq. (4.8), $g = 0$] based on coupled Heisenberg–Langevin equations (Horák and Peřina [1989]). The solution has been obtained in terms of the approximation of the evolution of the photon-number operator consisting in a neglect of the time dependence in the multiplicative noise for an initial coherent state. The marginal single-mode standard squeezing and the behavior of the mean complex amplitude of the field (particularly, its collapses and revivals) have been studied. The coherent amplitude revives at the cost of quantum fluctuations.

The conservative system described by the Hamiltonian (4.8) with $g = 0$ has been considered for an initial two-mode squeezed vacuum state (Bužek and Jex [1990b]). The solution of coupled Heisenberg equations in the form:

$$\hat{a}_1(t) = \exp[-it(\omega_1 + 2\kappa_1 \hat{n}_1(0) + \kappa_3 \hat{n}_2(0))]\hat{a}_1(0),$$
$$\hat{a}_2(t) = \exp[-it(\omega_2 + 2\kappa_2 \hat{n}_2(0) + \kappa_3 \hat{n}_1(0))]\hat{a}_2(0), \quad (4.22)$$

has been used for an analysis of the single- and two-mode standard squeezing. In particular, it has been shown that for an initial two-mode coherent state the interaction between the two modes leads to a high degree of squeezing. An analysis of the considered conservative system for initial coherent states using the Schrödinger picture has been performed by Meng, Chai and Zhang [1993]. The quadrature squeezing in the first (signal) mode appears to be considerably influenced by the coupling between the two fields. The phase dynamics of the second (probe) mode, such as the phase distribution, phase shift, and phase variance, are important in determining the signal-field photon number according to the phase shift of the probe light (Imoto, Haus and Yamamoto [1985]).

On coupled Heisenberg–Langevin equations the investigation of a coupler described by the Hamiltonian (4.8), where $\kappa_1 = \kappa_2 = \kappa$, $\kappa_3 = 2\kappa\varepsilon$, has been based on the supposition that the linear coupling is much stronger than the nonlinear one; i.e., $g \gg \kappa$ (Horák, Bertolotti, Sibilia and Peřina [1989]). The waveguides of a linear coupler exchange the energy periodically, with the period $\bar{t} = \pi/g$ (Peřinová, Lukš, Křepelka, Sibilia and Bertolotti [1991]). A nonlinear coupler behaves similarly when the input energy is well below a critical value. If the input energy is larger than the critical value, the energy exchange between the modes does not occur; the single-mode energies are only modulated slightly. A possibility of standard squeezing in separate waveguides and the relationship between the mean coherent amplitudes exhibiting revivals and the standard quadrature variances have been investigated.

The quantum statistics of coupled dissipative third-order nonlinear oscillators with the Hamiltonian

$$\hat{H} = \hbar \left\{ \sum_{j=1}^{4} \omega_j(\hat{a}_j^+ \hat{a}_j + \tfrac{1}{2}) + \kappa(\hat{a}_1^+ \hat{a}_1 + \hat{a}_2^+ \hat{a}_2)(\hat{a}_3^+ \hat{a}_3 + \hat{a}_4^+ \hat{a}_4) \right.$$
$$\left. + \bar{\kappa}(\hat{a}_3^{+2}\hat{a}_3^2 + \hat{a}_3^+ \hat{a}_3 \hat{a}_4^+ \hat{a}_4 + \hat{a}_4^{+2}\hat{a}_4^2) \right\} + \hat{H}_{R4} + \hat{H}_{14}, \quad (4.23)$$

where the Hamiltonians $\hat{H}_{R4}$ and $\hat{H}_{14}$, defined in eqs. (4.9) and (4.10), respectively, have been studied for an initial coherent state (Peřinová and Kárská [1989]). Here $\hat{a}_1$ and $\hat{a}_2$ ($\hat{a}_3$ and $\hat{a}_4$) are the annihilation operators of the signal (pump) modes with the frequencies ω_1 and ω_2 (ω_3 and ω_4); and κ and $\bar{\kappa}$ are real coupling constants. The frequency resonance condition for the wave mixing reads $\omega_1 + \omega_2 = \omega_3 + \omega_4$. The appropriate generalized Fokker–Planck equation for the quasidistribution related to the antinormal ordering of field operators reads:

$$\frac{\partial}{\partial t}\phi_{\mathscr{A}}(\{\alpha_j\},t) = \left\{\sum_{j=1}^4 \frac{\gamma_j}{2} + \left[\frac{\gamma_1}{2} + i\kappa(|\alpha_3|^2 + |\alpha_4|^2)\right]\alpha_1 \frac{\partial}{\partial \alpha_1}\right.$$

$$+ \left[\frac{\gamma_2}{2} + i\kappa(|\alpha_3|^2 + |\alpha_4|^2)\right]\alpha_2 \frac{\partial}{\partial \alpha_2}$$

$$+ \left[\frac{\gamma_3}{2} + i\kappa(|\alpha_1|^2 + |\alpha_2|^2) + i\bar{\kappa}(2|\alpha_3|^2 + |\alpha_4|^2)\right]\alpha_3 \frac{\partial}{\partial \alpha_3}$$

$$+ \left[\frac{\gamma_4}{2} + i\kappa(|\alpha_1|^2 + |\alpha_2|^2) + i\bar{\kappa}(|\alpha_3|^2 + 2|\alpha_4|^2)\right]\alpha_4 \frac{\partial}{\partial \alpha_4}$$

$$+ \sum_{j=1}^4 \frac{\gamma_j}{2}(\bar{n}_{d_j} + 1)\frac{\partial^2}{\partial \alpha_j \partial \alpha_j^*}$$

$$+ i\bar{\kappa}\alpha_3^2 \frac{\partial^2}{\partial \alpha_3^2} + i\bar{\kappa}\alpha_4^2 \frac{\partial^2}{\partial \alpha_4^2} + i\kappa\alpha_1\alpha_3 \frac{\partial^2}{\partial \alpha_1 \partial \alpha_3}$$

$$+ i\kappa\alpha_1\alpha_4 \frac{\partial^2}{\partial \alpha_1 \partial \alpha_4} + i\kappa\alpha_2\alpha_3 \frac{\partial^2}{\partial \alpha_2 \partial \alpha_3} + i\kappa\alpha_2\alpha_4 \frac{\partial^2}{\partial \alpha_2 \partial \alpha_4}$$

$$\left. + i\bar{\kappa}\alpha_3\alpha_4 \frac{\partial^2}{\partial \alpha_3 \partial \alpha_4} + \text{c.c.}\right\}\phi_{\mathscr{A}}(\{\alpha_j\},t). \quad (4.24)$$

In eq. (4.24), γ_j (γ_{j+2}), $j=1,2$, are the damping constants related to signal (pump) modes 1, 2 (3, 4), and $\bar{n}_{d_j}$ are the corresponding mean quantum numbers of the reservoir oscillators. Equation (4.24) has been solved exactly in the quiet reservoir case and approximately in the noisy reservoir case. The approximation used involves the quantum noise in the same manner as in the case of four coupled linear oscillators. Because the probability distribution of the integrated intensity related to the antinormal ordering of field operators does not depend on the coupling constants $\kappa, \bar{\kappa}$, the photon statistics are insensitive to the nonlinearities, and they are determined by the photon statistics of the field modelled as the coupled dissipative linear oscillators. The single-mode fields evolve from a coherent state at time $t = 0$ to a chaotic state for t tending to infinity. The photon statistics of the two-mode signal and pump fields use that of the two-mode superposition of coherent and chaotic fields [eq. (4.13)]. The approximation solution of the generalized Fokker–Planck equation [eq. (4.24)] serves for the determination of the standard quadrature variances [eqs. (4.15) and (4.16)] and the principal quadrature variances for single modes:

$$\langle (\Delta \hat{Q}_j^{(p)})^2 \rangle = -1 + 2[\langle \Delta \hat{a}_j \Delta \hat{a}_j^+ \rangle - |\langle (\Delta \hat{a}_j)^2 \rangle|], \quad j = 1, 2, 3, 4, \quad (4.25)$$

and those for couples of modes (Lukš, Peřinová and Hradil [1988]):

$$\langle (\Delta \hat{Q}_{j,j+1}^{(p)})^2 \rangle = -1 + c_j - |c_{j+1}|, \quad j = 1, 3, \quad (4.26)$$

where c_1, c_2 are defined in eqs. (4.17) and c_3, c_4 are introduced similarly. From the explicit formulas of the standard and principal quadrature variances, it follows that this phenomenon depends on the initial complex coherent amplitudes, the coupling constants, the damping constants, and on the mean quantum numbers of the reservoir oscillators. The squeezing is shown not to occur in single signal modes and the corresponding couples. Squeezing occurs in single pump modes and the two-mode pump field. Neglecting quantum noise, the squeezing appears periodically, with the period dependent on the coupling constants $\kappa, \bar{\kappa}$; this period is equal to $2\pi/\kappa$ for $\kappa = \bar{\kappa}$. The increase of the initial photon number shortens the interval of squeezing ultimately to its extinction. The squeezing is also smoothed out with increasing damping and thermal noise. A relationship is found between the squeezing and the revivals of the coherent-field amplitude, which collapses and revives during the interaction with this nonlinear Kerr medium. The period of the field amplitude is the same as that of the principal quadrature variance. The quantum noise smoothes out both the revivals and the squeezing.

The system of N coupled dissipative third-order nonlinear oscillators described by the Hamiltonian

$$\hat{H} = \hbar \sum_{j=1}^{N} \omega_j (\hat{a}_j^+ \hat{a}_j + \tfrac{1}{2}) + \hat{H}_S + \hat{H}_{RN} + \hat{H}_{IN}, \quad (4.27)$$

where

$$\hat{H}_S = \hbar \sum_{k,l=1}^{N} \kappa_{kl} \hat{a}_k^+ \hat{a}_k \hat{a}_l^+ \hat{a}_l, \quad (4.28)$$

the Hamiltonians $\hat{H}_{RN}, \hat{H}_{IN}$ are defined in eqs. (4.9) and (4.10), and κ_{kl} are real coupling constants, has been considered for an arbitrary beginning state (Chaturvedi and Srinivasan [1991b]). The appropriate master equation for the reduced state operator in the interaction picture reads:

$$\frac{\partial}{\partial t}\hat{\rho}_r = -i \sum_{k,l=1}^{N} \kappa_{kl}[\hat{a}_k^+ \hat{a}_k \hat{a}_l^+ \hat{a}_l, \hat{\rho}_r]$$

$$+ \frac{1}{2} \sum_{k=1}^{N} \gamma_k(\bar{n}_{d_k} + 1)(2\hat{a}_k \hat{\rho}_r \hat{a}_k^+ - 2\hat{a}_k^+ \hat{a}_k \hat{\rho}_r - \hat{\rho}_r \hat{a}_k^+ \hat{a}_k)$$

$$+ \frac{1}{2} \sum_{k=1}^{N} \gamma_k \bar{n}_{d_k}(2\hat{a}_k \hat{\rho}_r \hat{a}_k^+ - \hat{a}_k^+ \hat{a}_k \hat{\rho}_r - \hat{\rho}_r \hat{a}_k^+ \hat{a}_k), \tag{4.29}$$

where γ_k are the damping constants and $\bar{n}_{d_k}$ represents the noise in the ultimate steady state. On casting the master equation in the thermofield dynamics notation and using group-theoretical considerations, the quasi-distribution related to the antinormal ordering of field operators can be obtained in terms of the initial-state matrix elements in the number-state basis:

$$\phi_{\mathscr{A}}(\{\alpha_j\}, t) = \frac{1}{\pi^N} \sum_{\{m_j\}=0}^{\infty} \sum_{\{n_j\}=0}^{\infty} \prod_{j=1}^{N} \left[\exp\left\{ \left[i(m_j - n_j) \sum_{k=1}^{N} \kappa_{jk} + \frac{\gamma_j}{2} \right] t \right\} \right.$$

$$\times \frac{\alpha_j^{*m_j} \alpha_j^{n_j}}{m_j! \, n_j!} \exp[(g_{\{m_j - n_j\}}(t) - 1)|\alpha_j|^2](E_{\{m_j - n_j\}}(t))^{m_j + n_j + 1} \right]$$

$$\times \sum_{\{p_j\}=0}^{\infty} \prod_{j=1}^{N} \frac{1}{p_j!} \left[\frac{\bar{n}_{d_j} + 1}{\bar{n}_{d_j}} g_{\{m_j - n_j\}}(t) \right]^{p_j}$$

$$\times [(m_j + p_j)!(n_j + p_j)!]^{1/2} \rho_{\{m_j + p_j\}, \{n_j + p_j\}}(0), \tag{4.30}$$

where

$$g_{\{m_j \, n_j\}}(t) = \frac{2\bar{n}_{d_j}}{\Omega_j + \Delta_j \coth\left(\frac{\gamma_j}{2}\Delta_j t\right)}, \tag{4.31}$$

$$E_{\{m_j - n_j\}}(t) = \frac{\Delta_j}{\Omega_j \sinh\left(\frac{\gamma_j}{2}\Delta_j t\right) + \Delta_j \cosh\left(\frac{\gamma_j}{2}\Delta_j t\right)}, \tag{4.32}$$

and

$$\Omega_j = 1 + 2\bar{n}_{d_j} + i \frac{2}{\gamma_j} \sum_{k=1}^{N} \kappa_{jk}(m_k - n_k), \quad \Delta_j = [\Omega_j^2 - 4\bar{n}_{d_j}(\bar{n}_{d_j} + 1)]^{1/2}.$$
(4.33)

Let us remark that the explicit formula for the quasidistribution $\phi_{\mathscr{A}}(\{\alpha_j\}, t)$ could be used as a mathematical device for the ensuing physical computations, as for example concrete initial states, dynamics of squeezing, the phase distribution, and other phase statistics.

§ 5. Use of a Kerr Medium for Generation and Detection of Nonclassical States

The results reviewed in previous sections speak of the immense effort devoted to the study of the relation of a Kerr medium to nonclassical states of radiation. With the quasidistribution appropriate to the antinormal ordering of field operators, significant differences between the classical and the quantum behavior have been observed. An effect of dissipation on this medium has been considered. It has been shown that every initial state can evolve in the lossless Kerr medium into the state which is a superposition of phase-shifted versions of the input, which is a very nonclassical phenomenon. The conditions have been found under which a Kerr medium is capable of producing quadrature squeezed states and number-phase squeezed states. The quantum features have been confirmed by investigating the photon number and phase fluctuations.

With the results achieved, the theoretical study has motivated the use of a Kerr medium for the generation of nonclassical states. A scheme for generating states of number and phase minimum uncertainty has been proposed based on two successive unitary evolutions from a coherent state (Kitagawa and Yamamoto [1986], Yamamoto, Machida, Imoto, Kitagawa and Björk [1987]). To the interaction Hamiltonians

$$\hat{H}_N = \hbar\kappa \hat{a}^{+2} \hat{a}^2,$$
(5.1)

$$\hat{H}_T = \hbar(\lambda \hat{a}^+ + \lambda^* \hat{a}),$$
(5.2)

where the subscript T stands for translation, the following unitary developments belong:

$$\hat{U}_N = \exp\left[i\kappa \frac{L}{c} \hat{n}(\hat{n} - \hat{1}) \right] \quad (5.3)$$

and

$$\hat{U}_T = \exp(\xi \hat{a}^+ - \xi^* \hat{a}), \quad (5.4)$$

where $\xi = i\lambda L'/c$ and L and L' are the interaction lengths. The Kerr medium realizes the self-phase modulation [eq. (5.3)] and a high-reflection mirror performs approximately the unitary displacement operation [eq. (5.4)]. The output mode is related to the input one as follows:

$$\hat{a}_{out} = \hat{U}_T \hat{U}_N \hat{a}_{in} \hat{U}_N^+ \hat{U}_T^+ = \exp\left(i2\kappa \frac{L}{c} \hat{n} \right) \hat{a}_{in} + \xi \hat{1}. \quad (5.5)$$

The corresponding quasidistribution related to the antinormal ordering of field operators is deformed by the self-phase modulation; its topography is of crescent shape. After the interference in the beam splitter, the photon-number uncertainty $\langle (\Delta \hat{n})^2 \rangle$ can be reduced to $\langle \hat{n} \rangle^{1/3}$, which for $\langle \hat{n} \rangle > 1$ is below the limit $\langle \hat{n} \rangle^{2/3}$ for the two-photon coherent state. The resulting state exhibits strongly sub-Poissonian photon statistics. The reduced photon-number uncertainty and enhanced phase uncertainty obey the Heisenberg uncertainty relation $\langle (\Delta \hat{n})^2 \rangle \langle (\Delta \hat{\phi})^2 \rangle = \frac{1}{4}$.

An optimization of the system proposed by Kitagawa and Yamamoto [1986] has been performed with respect to the Kerr nonlinearity and the mirror parameters in order to obtain the minimum output photon-number uncertainty, with a fixed-input mean photon number in a coherent state (Grønbech-Jensen and Ramanujam [1990]). The optimized system produces an output state with the Fano factor $d \approx \langle \hat{n} \rangle^{-0.4}$, which is better than the best obtainable for quadrature squeezed states.

The quantum evolution associated with the nonlinear Kerr Hamiltonian (5.1) can lead to the one-photon number state in a cavity that is pumped by a periodic sequence of short pulses of small intensity, provided the system damping is low enough (Leoński and Tanaś [1994]).

Quantum nondemolition measurement has been suggested based on the optical Kerr effect (Imoto, Haus and Yamamoto [1985]). The signal wave of the frequency ω_S propagates together with the pump wave of the frequency ω_P through a Kerr medium, where the phase of the pump wave is modulated by the photon number of the signal wave. The photon number of the signal wave is determined in a nondemolition way by the homodyne detection of the phase of the pump wave. The interaction Hamiltonian for the cross-phase modulation in the Kerr medium is of the form:

$$\hat{H} = \hbar\kappa \hat{a}_S^+ \hat{a}_S \hat{a}_P^+ \hat{a}_P, \tag{5.6}$$

where the coupling constant κ is proportional to the third-order susceptibility $\chi^{(3)}(\omega_S, \omega_S, -\omega_P, \omega_P)$. The self-phase modulation is eliminated by using a resonant $\chi^{(3)}$ medium or by cancelling the effect by means of a negative $\chi^{(3)}$ medium. The unitary development operator is of the form:

$$\hat{U}(z) = \exp\left(i\kappa \frac{z}{v} \hat{n}_S \hat{n}_P\right), \tag{5.7}$$

where v is the light velocity in the medium. The corresponding quasidistribution related to the antinormal ordering of signal field operators has been analyzed (Kitagawa, Imoto and Yamamoto [1987]). The uncertainty product

$$\frac{\langle(\Delta\hat{n}_S)^2\rangle\langle(\Delta\widetilde{\sin}\varphi_S)^2\rangle}{\langle\widetilde{\cos}\varphi_S\rangle^2} \tag{5.8}$$

depends on the pump wave intensity and rests near its minimum (1/4). In the framework of these number-phase minimum uncertainty states, near-number states can be obtained, $\langle(\Delta\hat{n}_S)^2\rangle < 1$, and these states tend to number states with increasing pump-wave amplitude.

The influence of the nonlinear Kerr effect on the quantum nondemolition measurement has been investigated (Sanders and Milburn [1989]). The measurement does not destroy the field, but does disturb its phase. Such phase disturbances reduce fringe visibility and enforce a principle of complementarity. The appropriate parameters have been determined to optimize the observation of the two complementary quantities.

A theoretical study of the quantum nondemolition properties of a scheme that uses a Kerr medium in a double-faced cavity has been presented with both the signal and the probe inputs being coherent fields incident from the two sides (Chaba, Collett and Walls [1992]). It has been found that with a suitable choice of parameters this system satisfies almost perfectly the criteria for quantum nondemolition measurements.

A photon-number quantum nondemolition measurement theory has been proposed that takes a lossy Kerr medium into account (Imoto and Saito [1989]). Restrictions of losses and requirements for the nonlinearity and signal and probe powers necessary to observe the quantum nondemolition effect have been obtained. The conclusion is reached that such measurements are possible in existing media. The appropriate set-up of measurement using the Kerr media has been analyzed (Martens and De Muynck [1992]). This system, described effectively by the interaction Hamiltonian,

$$\hat{H} = \chi^{(3)} \hat{a}_S^+ \hat{a}_S \hat{a}_P^+ \hat{a}_P + \chi^{(3)} \hat{a}_L^+ \hat{a}_L \hat{a}_P^+ \hat{a}_P + \tfrac{1}{2}\chi^{(3)}(\hat{a}_P^+ \hat{a}_P)^2$$
$$+ \tfrac{1}{2}\chi^{(3)}(\hat{a}_L^+ \hat{a}_L)^2 + \tfrac{1}{4}\chi^{(3)}(\hat{a}_S^+ \hat{a}_S)^2, \tag{5.9}$$

leads to a complete compensation of self-phase modulation even in the presence of losses.

The nonlinear Mach–Zehnder interferometer composed of two beam splitters with a nonlinear Kerr medium in one arm has been presented as a device whereby a pair of coherent states can be transformed into an entangled superposition of coherent states for which the notion of entanglement is generalized to include nonorthogonal, but distinct, component states (Sanders [1992a]). In contrast to previous optical schemes (Tan, Walls and Collett [1991]), the initial state used here possesses a positive Glauber–Sudarshan representation and is therefore a semiclassical state. The Kerr nonlinearity $\kappa(\hat{a}^+ \hat{a})^2$ is itself responsible for generating the nonclassical state.

The nonlinear Mach–Zehnder interferometer with a Kerr medium in each arm can be treated as an all-optical switching device. Fundamental limits that apply to switching in this quantum device have been studied for input coherent states by Sanders and Milburn [1992]. A measure to quantify the reduction in the switching efficiency has been introduced.

A Kerr medium can be used for the detection of nonclassical states. It has been shown that the second-order interference (in the sense of complex amplitudes) can be used for testing the quantum character of light propagating through a single optical fiber with a Kerr nonlinearity (Agarwal [1989]). This propagation is described by the effective Hamiltonian $\hat{H}_N$ given in eq. (5.1). The quantum theory provides for initial coherent light the visibility of the interference fringes $v = |\gamma(\tau)| \leq 1$, where $\gamma(\tau)$ is the degree of coherence

$$\gamma(\tau) = \frac{\langle \hat{a}^+(t)\hat{a}(t+\tau) \rangle}{\langle \hat{a}^+(t)\hat{a}(t) \rangle} \tag{5.10}$$

and τ is the duration of the interaction with the fiber, whereas the classical theory provides $v = 1$. In the quantum theory, the visibility of the interference pattern can differ significantly from unity. A visibility index less than unity is a measure of nonclassical features of the field.

Another means of detecting certain nonclassical states consists of a measurement of the total noise of the output state from a Kerr medium (Hillery

[1991]). The measure of the total noise of the single-mode field state,

$$T = \langle (\Delta \hat{Q})^2 \rangle + \langle (\Delta \hat{P})^2 \rangle = 4 \langle \Delta \hat{a}^+ \Delta \hat{a} \rangle + 2, \tag{5.11}$$

provides a possibility of determining whether the input state is nonclassical. For a classical input state, the total noise at the output has a minimum value which depends on the photon number and the interaction time. If the total noise is less than this minimum, then the input state was nonclassical. In this way a certain class of generalized coherent states can be detected.

Alongside the intrinsic usefulness of squeezed light in carrying out fundamental experiments in optical physics, there are a number of general areas in which the use of quadrature-squeezed and/or photon-number-squeezed light produced by a Kerr medium may be advantageous. These include spectroscopy, interferometry, precision measurement, light-wave communications, and theory of vision. Quantum fluctuations can limit the sensitivity of certain experiments in all of these areas (Yamamoto, Machida, Saito, Imoto, Yanagawa, Kitagawa and Björk [1990]).

Acknowledgement

The authors would like to thank Ing. J. Křepelka for his assistance with the numerical calculations.

References

Agarwal, G.S., 1989, Opt. Commun. **72**, 253.
Agarwal, G.S., and R.R. Puri, 1989a, Phys. Rev. A **39**, 2969.
Agarwal, G.S., and R.R. Puri, 1989b, Phys. Rev. A **40**, 5179.
Agarwal, G.S., and R.R. Puri, 1990a, in: Coherence and Quantum Optics VI, eds J.H. Eberly, L. Mandel and E. Wolf (Plenum Press, New York) p. 13.
Agarwal, G.S., and R.R. Puri, 1990b, in: Coherence and Quantum Optics VI, eds J.H. Eberly, L. Mandel and E. Wolf (Plenum Press, New York) p. 15.
Agarwal, G.S., S. Chaturvedi, K. Tara and V. Srinivasan, 1992, Phys. Rev. A **45**, 4904.
Akhmanov, S.A., R.V. Khokhlov and A.P. Sukhorukov, 1972, in: Laser Handbook, Vol. 2, eds F.T. Arecchi and E.V. Schulz-Dubois (North-Holland, Amsterdam) p. 1151.
Aragone, C., E. Chalbaud and S. Salamo, 1976, J. Math. Phys. **17**, 1963.
Ban, M., 1992, J. Math. Phys. **33**, 3213.
Bandilla, A., and H. Paul, 1969, Ann. Phys. Leipzig **23**, 323.
Banerjee, A., 1993, Quant. Opt. **5**, 15.
Barnett, S.M., and D.T. Pegg, 1986, J. Phys. A **19**, 3849.
Barnett, S.M., and D.T. Pegg, 1989, J. Mod. Opt. **36**, 7.
Barnett, S.M., P. Filipowicz, J. Javanainen, P.L. Knight and P. Meystre, 1986, in: Frontiers in Quantum Optics, eds E.R. Pike and S. Sarkar (Hilger, Bristol) p. 485.

Barthelemy, A., S. Maneuf and C. Froehly, 1985, Opt. Commun. **55**, 193.
Białynicka-Birula, Z., 1968, Phys. Rev. **173**, 1207.
Blow, K.J., R. Loudon and S.J.D. Phoenix, 1993, J. Mod. Opt. **40**, 2515.
Braunstein, S.L., 1992, Phys. Rev. A **45**, 6803.
Brisudová, M., 1991, J. Mod. Opt. **38**, 2505.
Bužek, V., 1989a, Phys. Lett. A **136**, 188.
Bužek, V., 1989b, Phys. Rev. A **39**, 5432.
Bužek, V., 1989c, Acta Phys. Slov. **39**, 344.
Bužek, V., and I. Jex, 1989, Acta Phys. Slov. **39**, 351.
Bužek, V., and I. Jex, 1990a, Opt. Commun. **78**, 425.
Bužek, V., and I. Jex, 1990b, Int. J. Mod. Phys. B **4**, 659.
Bužek, V., and P.L. Knight, 1991, Opt. Commun. **81**, 331.
Bužek, V., P.L. Knight and A. Vidiella Barranco, 1992, in: Workshop on Squeezed States and Uncertainty Relations, eds D. Han, Y.S. Kim and W.W. Zachary (NASA, Maryland) p. 181.
Carruthers, P., and M.M. Nieto, 1968, Rev. Mod. Phys. **40**, 411.
Chaba, A.N., M.J. Collett and D.F. Walls, 1992, Phys. Rev. A **46**, 1499.
Chaturvedi, S., and V. Srinivasan, 1991a, J. Mod. Opt. **38**, 777.
Chaturvedi, S., and V. Srinivasan, 1991b, Phys. Rev. A **43**, 4054.
Collett, M.J., and D.F. Walls, 1985, Phys. Rev. A **32**, 2887.
Daniel, D.J., and G.J. Milburn, 1989, Phys. Rev. A **39**, 4628.
D'Ariano, G., M. Rosetti and M. Vadacchino, 1985, Phys. Rev. D **32**, 1034.
Deb, B., and D.S. Ray, 1993, Phys. Rev. A **48**, 3191.
Drummond, P.D., and D.F. Walls, 1980, J. Phys. A **13**, 725.
Dung, H.T., and A.S. Shumovsky, 1991, Phys. Lett. A **160**, 437.
Gantsog, Ts., and R. Tanaś, 1991a, Quant. Opt. **3**, 33.
Gantsog, Ts., and R. Tanaś, 1991b, J. Mod. Opt. **38**, 1021.
Gantsog, Ts., and R. Tanaś, 1991c, Phys. Rev. A **44**, 2086.
Gantsog, Ts., and R. Tanaś, 1991d, J. Mod. Opt. **38**, 1537.
García Fernández, P., P. Colet, R. Toral, M. San Miguel and F.J. Bermejo, 1991, Phys. Rev. A **43**, 4923.
Gentile, T.R., B.J. Hughey and D. Kleppner, 1989, Phys. Rev. A **40**, 5103.
Gerry, C.C., 1987a, Opt. Commun. **63**, 278.
Gerry, C.C., 1987b, Phys. Lett. A **124**, 237.
Gerry, C.C., 1987c, Phys. Rev. A **35**, 2146.
Gerry, C.C., 1990, Opt. Commun. **75**, 168.
Gerry, C.C., and S. Rodriguez, 1987a, Phys. Rev. A **35**, 4440.
Gerry, C.C., and S. Rodriguez, 1987b, Phys. Rev. A **36**, 5444.
Gerry, C.C., and E.R. Vrscay, 1988a, Phys. Rev. A **37**, 1779.
Gerry, C.C., and E.R. Vrscay, 1988b, Phys. Rev. A **37**, 4265.
Gibbs, H.M., 1985, Optical Bistability: Controlling Light with Light (Academic Press, New York).
Grønbech-Jensen, N., and P.S. Ramanujam, 1990, Phys. Rev. A **41**, 2906.
Haake, F., H. Risken, C. Savage and D.F. Walls, 1986, Phys. Rev. A **34**, 3969.
Hasegawa, A., 1990, Optical Solitons in Fibers, 2nd enlarged Ed. (Springer, Berlin).
Hillery, M., 1987a, Opt. Commun. **62**, 135.
Hillery, M., 1987b, Phys. Rev. A **36**, 3796.
Hillery, M., 1991, Phys. Rev. A **44**, 4578.
Hong, C.K., and L. Mandel, 1985, Phys. Rev. A **32**, 974.
Horák, R., 1989, Opt. Commun. **72**, 239.
Horák, R., and J. Peřina, 1989, J. Opt. Soc. Am. B **6**, 1239.
Horák, R., M. Bertolotti, C. Sibilia and J. Peřina, 1989, J. Opt. Soc. Am. B **6**, 199.

Hradil, Z., 1991, Phys. Rev. A **44**, 792.
Imoto, N., and S. Saito, 1989, Phys. Rev. A **39**, 675.
Imoto, N., H.A. Haus and Y. Yamamoto, 1985, Phys. Rev. A **32**, 2287.
Infeld, E., and G. Rowlands, 1990, Nonlinear Waves, Solitons and Chaos (Cambridge University Press, Cambridge).
Jackiw, R., 1968, J. Math. Phys. **9**, 339.
Jaynes, E.T., and F.W. Cummings, 1963, Proc. Inst. Electr. Eng. (IEEE) **51**, 89.
Kärtner, F.X., and A. Schenzle, 1993, Phys. Rev. A **48**, 1009.
Kielich, S., 1981, Nonlinear Molecular Optics (Science, Moscow). In Russian.
Kim, M.S., F.A.M. De Oliveira and P.L. Knight, 1989a, Opt. Commun. **72**, 99.
Kim, M.S., F.A.M. De Oliveira and P.L. Knight, 1989b, Phys. Rev. A **40**, 2494.
Kitagawa, M., and Y. Yamamoto, 1986, Phys. Rev. A **34**, 3974.
Kitagawa, M., N. Imoto and Y. Yamamoto, 1987, Phys. Rev. A **35**, 5270.
Král, P., 1990a, J. Mod. Opt. **37**, 889.
Král, P., 1990b, Phys. Rev. A **42**, 4177.
Kumar, P., and J.H. Shapiro, 1984, Phys. Rev. A **30**, 1568.
Leoński, W., and R. Tanaś, 1994, Phys. Rev. A **49**, R20.
Loudon, R., 1973, The Quantum Theory of Light (Clarendon Press, Oxford).
Lukš, A., and V. Peřinová, 1987, Czech. J. Phys. B **37**, 1224.
Lukš, A., and V. Peřinová, 1991, Czech. J. Phys. **41**, 1205.
Lukš, A., and V. Peřinová, 1992, Phys. Rev. A **45**, 6710.
Lukš, A., and V. Peřinová, 1993, Phys. Scripta T **48**, 94.
Lukš, A., V. Peřinová and Z. Hradil, 1988, Acta Phys. Pol. A **74**, 713.
Lukš, A., V. Peřinová and J. Křepelka, 1992, Czech. J. Phys. **42**, 59.
Lukš, A., V. Peřinová and J. Peřina, 1988, Opt. Commun. **67**, 149.
Lynch, R., 1988, Opt. Commun. **67**, 67.
Maneuf, S., R. Desailly and C. Froehly, 1988, Opt. Commun. **55**, 193.
Martens, H., and W.M. De Muynck, 1992, Quant. Opt. **4**, 303.
McNeil, K.J., 1993, J. Mod. Opt. **40**, 1957.
Meng, H.X., C.L. Chai and Z.M. Zhang, 1993, Phys. Rev. A **48**, 3219.
Messiah, A., 1961, Quantum Mechanics (McGraw-Hill, New York).
Milburn, G.J., 1986, Phys. Rev. A **33**, 674.
Milburn, G.J., and C.A. Holmes, 1986, Phys. Rev. Lett. **56**, 2237.
Milburn, G.J., and C.A. Holmes, 1991, Phys. Rev. A **44**, 4704.
Milburn, G.J., A. Mecozzi and P. Tombesi, 1989, J. Mod. Opt. **36**, 1607.
Mir, M.A., and M.S.K. Razmi, 1991, Phys. Rev. A **44**, 6077.
Miranowicz, A., R. Tanaś and S. Kielich, 1990, Quant. Opt. **2**, 253.
Newton, R.G., 1980, Ann. Phys. **124**, 327.
Paprzycka, M., and R. Tanaś, 1992, Quant. Opt. **4**, 331.
Pegg, D.T., and S.M. Barnett, 1988, Europhys. Lett. **6**, 483.
Pegg, D.T., and S.M. Barnett, 1989, Phys. Rev. A **39**, 1665.
Perelomov, A.M., 1986, Generalized Coherent States and Their Applications (Springer, Berlin).
Peřina, J., 1991, Quantum Statistics of Linear and Nonlinear Optical Phenomena (Reidel, Dordrecht).
Peřina, J., R. Horák, Z. Hradil, C. Sibilia and M. Bertolotti, 1989, J. Mod. Opt. **36**, 571.
Peřinová, V., 1981, Opt. Acta **28**, 747.
Peřinová, V., and M. Kárská, 1989, Phys. Rev. A **39**, 4056.
Peřinová, V., and J. Křepelka, 1993, Phys. Rev. A **48**, 3881.
Peřinová, V., and A. Lukš, 1988, J. Mod. Opt. **35**, 1513.
Peřinová, V., and A. Lukš, 1990, Phys. Rev. A **41**, 414.
Peřinová, V., and J. Peřina, 1981, Opt. Acta **28**, 769.

Peřinová, V., J. Křepelka and J. Peřina, 1986, Opt. Acta **33**, 1263.
Peřinová, V., A. Lukš and M. Kárská, 1990, J. Mod. Opt. **37**, 1055.
Peřinová, V., A. Lukš, J. Křepelka, C. Sibilia and M. Bertolotti, 1991, J. Mod. Opt. **38**, 2429.
Peřinová, V., A. Lukš and P. Szłachetka, 1989, J. Mod. Opt. **36**, 1435.
Piekara, A.A., J.S. Moore and N.S. Feld, 1974, Phys. Rev. A **9**, 1403.
Reid, M.D., and D.F. Walls, 1985, J. Opt. Soc. Am. B **2**, 1682.
Rempe, G., H. Walther and N. Klein, 1987, Phys. Rev. Lett. **58**, 353.
Ritze, H.-H., 1980, Z. Phys. B **39**, 353.
Ritze, H.-H., and A. Bandilla, 1979, Opt. Commun. **29**, 126.
Saleh, B.E.A., and M.C. Teich, 1991, Fundamentals of Photonics (Wiley, New York).
Sanders, B.C., 1992a, Phys. Rev. A **45**, 6811.
Sanders, B.C., 1992b, Phys. Rev. A **45**, 7746.
Sanders, B.C., and G.J. Milburn, 1989, Phys. Rev. A **39**, 694.
Sanders, B.C., and G.J. Milburn, 1992, J. Opt. Soc. Am. B **9**, 915.
Schleich, W., M. Pernigo and Fam Le Kim, 1991, Phys. Rev. A **44**, 2172.
Selloni, A., and P. Schwendimann, 1979, Opt. Acta **26**, 1541.
Shapiro, J.H., 1992, in: Workshop on Squeezed States and Uncertainty Relations, eds D. Han, Y.S. Kim and W.W. Zachary (NASA, Maryland) p. 107.
Shen, Y.R., 1976, Rev. Mod. Phys. **48**, 1.
Singer, F., M.J. Potasek, J.M. Fang and M.C. Teich, 1992, Phys. Rev. A **46**, 4192.
Slusher, R.E., L.W. Hollberg, B. Yurke, J.C. Mertz and J.F. Valley, 1985, Phys. Rev. Lett. **55**, 2409.
Stoler, D., 1971, Phys. Rev. D **4**, 2309.
Susskind, L., and J. Glogower, 1964, Phys. **1**, 49.
Svelto, O., 1974, in: Progress in Optics, Vol. XII, ed. E. Wolf (North-Holland, Amsterdam) p. 1.
Tan, S.M., D.F. Walls and M.J. Collett, 1991, Phys. Rev. Lett. **66**, 252.
Tanaś, R., 1984, in: Coherence and Quantum Optics V, eds L. Mandel and E. Wolf (Plenum Press, New York) p. 645.
Tanaś, R., 1988, Phys. Rev. A **38**, 1091.
Tanaś, R., 1989, Phys. Lett. A **141**, 217.
Tanaś, R., 1991, Opt. & Spektrosk. **70**, 637.
Tanaś, R., Ts. Gantsog, 1991, J. Opt. Soc. Am. B **8**, 2505.
Tanaś, R., and Ts. Gantsog, 1992a, J. Mod. Opt. **39**, 749.
Tanaś, R., and Ts. Gantsog, 1992b, Opt. Commun. **87**, 369.
Tanaś, R., and S. Kielich, 1979, Opt. Commun. **30**, 443.
Tanaś, R., and S. Kielich, 1983, Opt. Commun. **45**, 351.
Tanaś, R., and S. Kielich, 1984, Opt. Acta **31**, 81.
Tanaś, R., and S. Kielich, 1990a, J. Mod. Opt. **37**, 1935.
Tanaś, R., and S. Kielich, 1990b, Quant. Opt. **2**, 23.
Tanaś, R., Ts. Gantsog, A. Miranowicz and S. Kielich, 1991, J. Opt. Soc. Am. B **8**, 1576.
Tanaś, R., A. Miranowicz and S. Kielich, 1991, Phys. Rev. A **43**, 4014.
Teich, M.C., and B.E.A. Saleh, 1989, Quant. Opt. **1**, 153.
Thacker, H.B., 1981, Rev. Mod. Phys. **53**, 253.
Titulaer, U.M., and R.J. Glauber, 1966, Phys. Rev. **145**, 1041.
Tombesi, P., and A. Mecozzi, 1987, J. Opt. Soc. Am. B **4**, 1700.
Tombesi, P., and H.P. Yuen, 1984, in: Coherence and Quantum Optics V, eds L. Mandel and E. Wolf (Plenum Press, New York) p. 751.
Vourdas, A., and R.F. Bishop, 1989, Phys. Rev. A **39**, 214.
Walls, D.F., 1983, Nature **306**, 141.
Werner, M.J., and H. Risken, 1991a, Phys. Rev. A **44**, 4623.
Werner, M.J., and H. Risken, 1991b, Quant. Opt. **3**, 185.

Wielinga, B., and G.J. Milburn, 1993, Phys. Rev. A **48**, 2494.
Wilson-Gordon, A.D., V. Bužek and P.L. Knight, 1991, Phys. Rev. A **44**, 7647.
Wu, L.-A., H.J. Kimble, J.L. Hall and H. Wu, 1986, Phys. Rev. Lett. **57**, 2520.
Yamamoto, Y., N. Imoto and S. Machida, 1986, Phys. Rev A **33**, 3243.
Yamamoto, Y., S. Machida, N. Imoto, M. Kitagawa and G. Björk, 1987, J. Opt. Soc. Am. B **4**, 1645.
Yamamoto, Y., S. Machida, S. Saito, N. Imoto, T. Yanagawa, M. Kitagawa and G. Björk, 1990, in: Progress in Optics, Vol. XXVIII, ed. E. Wolf (North-Holland, Amsterdam) p. 89.
Yang, X., and X. Zheng, 1989, J. Mod. Opt. **36**, 607.
Yariv, A., 1967, Quantum Electronics and Nonlinear Optics, 1st Ed. (Wiley, New York).
Yuen, H.P., 1976, Phys. Rev. A **13**, 2226.
Yurke, B., and D. Stoler, 1986, Phys. Rev. Lett. **57**, 13.
Zakharov, Y.E., and A.B. Shabat, 1971, Zh. Eksp. Teor. Fiz. **61**, 118.

E. WOLF, PROGRESS IN OPTICS XXXIII
© 1994 ELSEVIER SCIENCE B.V.
ALL RIGHTS RESERVED

III

GAP SOLITONS

BY

C. Martijn de Sterke

*School of Physics,
University of Sydney,
NSW 2006, Australia,
and Optical Fibre Technology Centre,
University of Sydney,
NSW 2006, Australia*

J. E. Sipe

*Department of Physics,
University of Toronto,
Toronto,
Canada M5S 1A7
and Ontario Laser and Lightwave Research Centre,
Toronto,
Canada M5S 1A7*

CONTENTS

	PAGE
§ 1. INTRODUCTION	205
§ 2. QUALITATIVE DESCRIPTION	208
§ 3. COUPLED-MODE THEORY	212
§ 4. STATIONARY SOLUTIONS	217
§ 5. SOLITARY-WAVE SOLUTIONS	225
§ 6. MULTIPLE SCALES	230
§ 7. DISCUSSION OF ANALYTIC RESULTS	237
§ 8. NUMERICAL RESULTS	239
§ 9. THE COUPLING PROBLEM	244
§ 10. EXPERIMENTAL GEOMETRIES	253
§ 11. OUTLOOK	256
ACKNOWLEDGEMENTS	258
REFERENCES	259

§ 1. Introduction

Gap solitons are electromagnetic field structures that can exist in a nonlinear optical medium, if there is also a periodic variation in the linear optical properties over a length scale on the order of the wavelength of light (Mills [1991]). In simple one-dimensional geometries, such as those we will consider here, they are similar in many ways to the simpler electromagnetic field structures of a soliton in a uniform optical fiber (see, e.g., Agrawal [1989]). In both this "regular" soliton, and in a gap soliton, the field structure gains its stability through a counter-balancing of the effect of group-velocity dispersion, which tends to disperse the energy of the pulse, and the effect of the nonlinearity, which tends to concentrate it (Dodd, Eilbeck, Gibbon and Morris [1982], Drazin and Johnson [1989]). The difference is that in a "regular" soliton, the group-velocity dispersion is due primarily to the underlying dispersion of the uniform material, while in a gap soliton it is due to the photonic band structure that results from the periodic variation in the linear dielectric constant. Since this periodic variation can often be controlled and even engineered, such systems offer the promise of media designed to produce solitonic behavior of interest, whether for the purpose of studying fundamental effects in nonlinear dynamics, or for optical-device engineering.

The stationary properties of nonlinear periodic structures were analyzed by Winful, Marburger and Garmire [1979], who showed that such systems, similar to nonlinear Fabry–Pérot cavities, can exhibit optical bistability (Gibbs [1985]). Winful and Cooperman [1982] were the first to discuss aspects of the dynamics of nonlinear periodic structures. Nevertheless, the term "gap soliton" was first used by Chen and Mills [1987a], who inferred the solitonic nature of the electric field within such structures from results of their numerical studies of nonlinear periodic structures. This conjecture was put on firmer footing in subsequent research by Sipe and Winful [1988], and de Sterke and Sipe [1988], who showed that in one limit the envelope function of the electric field satisfies the nonlinear Schrödinger equation, which has soliton solutions, and by Mills and Trullinger [1987], Christodoulides and Joseph [1989] and Aceves and Wabnitz [1989], who showed that the envelope function has solitary-wave solutions in another limit.

Structures that have been proposed for the experimental study of gap solitons include thin-film stacks, planar waveguides with a surface corrugation, and optical fibers with a grating grown inside. The first of these, unlike the others, is not a guided wave structure, and thus it is difficult to establish and maintain high field densities over extended lengths in such structures. More importantly, these structures are grown layer by layer and, in practice, cannot be made very long. On the other hand, the constituents can be chosen from a large variety of materials, so that those with the largest nonlinearities can be selected. The choice of materials in waveguiding structures is much more limited, since not every material makes a good waveguide. A variation in the thickness of the guiding layer in such structures leads to a variation in the mode index, which mimics a variation in the linear dielectric constant. A grating structure can thus be obtained by corrugating the surface. Moreover, by using holographic techniques all rulings of the surface corrugation can be written simultaneously, and, even if the rulings are written one at the time (e.g., by electron-beam writing), a considerable number of periods can still be obtained. Indeed, bistable switching, originally predicted to occur by Winful, Marburger and Garmire [1979], has recently been observed by Sankey, Prelewitz and Brown [1992] in a Si–SiO_2 planar waveguide structure. A possible alternative is the use of optical fibers (Agrawal [1989]). Here one is, of course, limited to the use of glass, which is one of the least nonlinear materials known. However, optical-fiber geometries are attractive because their low losses allow very long interaction lengths, which reduce the required nonlinearity. Gratings in optical fibers can either be written directly in the core (Kawasaki, Hill, Johson and Fujii [1978], Meltz, Morey and Glenn [1989]) or can be obtained by a periodic corrugation of the core–cladding interface, similar to the corrugation of planar waveguides. Although for years the former method allowed one to write only weak gratings, recent advances in the field have shown that an index modulation of as large as 0.01 can now be obtained (Mizrahi and Sipe [1992]). Finally, Herbert and Malcuit [1992] recently reported the observation of switching in a colloidal crystal consisting of charged microspheres suspended in an aqueous solution of an absorptive dye. Their experiments made use of the thermal nonlinearity of the system, which therefore had a very sluggish response. While they observed the predicted switching behavior, their results suffered from the fairly wide size distribution of their microspheres. More detailed descriptions of the two waveguide geometries are given in § 10.

Although quite different physically, the propagation of light in all these structures can, under appropriate excitation conditions, be essentially one-dimensional. Thus, instead of concentrating on the details of any one of these geometries, we can present an introduction to gap solitons by considering the propagation of light in a thin-film stack (see, e.g., Macleod [1969]). By changing only the final parameters, such as the group velocity and the group-velocity dispersion, we can use the results of such an analysis to treat a guided wave geometry as well. Other conditions being satisfied, gap solitons can exist regardless of the magnitude of the periodic variation of the dielectric constant. However, in most of the geometries proposed for observing gap solitons, that magnitude is small compared to the average dielectric constant. In such cases, the coupled-mode equations – familiar from integrated optics and fiber optics – can be used as the underlying equations to model the optical properties of the medium (Kogelnik [1969]). We do this here for convenience, both to allow a reasonably simple introduction, and to establish contact with areas of guided-wave optics. Furthermore, there are some exact analytic solutions of the nonlinear coupled-mode equations, and some numerical calculations dealing with gap solitons. However, we stress that the approximations inherent in the coupled-mode equations are not essential for gap solitons; indeed, more general discussions of gap solitons have been given by de Sterke and Sipe [1988].

The contents of this review are as follows. In § 2, we will give a simple qualitative discussion of gap solitons, which we think will be useful when reading the more technical sections. As mentioned, the usual quantitative description of gap solitons employs coupled-mode theory, although more general descriptions can be given. In § 3, therefore, we will give an outline of coupled-mode theory, leading to the nonlinear coupled-mode equations. In the three following sections we will discuss three sets of solutions to the coupled-mode equations. The first of these, in § 4, are the stationary solutions. These solutions lead one to the conclusion that nonlinear periodic solutions exhibit bistability. This is followed in § 5 by a description of a class of solitary-wave solutions to the coupled-mode equations. We will demonstrate in § 6 that, in the appropriate limit, the envelope of the electric field satisfies the nonlinear Schrödinger equation. In this limit, therefore, the system is integrable, and supports soliton solutions. In § 7, we will investigate the relationship between these solitons and the solitary-wave solutions discussed in § 5, as well as the relationship with "regular" fiber solitons. After the various analytic approaches to the physics of nonlinear periodic media, we will turn in § 8 to numerical solutions. These solutions show that periodic nonlinear media exhibit self-pulsations which become chaotic at high intensities. This is followed in § 9 by a short discussion of the issues involved in trying to generate gap solitons. In § 10, we will discuss in some detail two experimental schemes to detect gap solitons in the laboratory. Finally, in § 11 we will give a brief summary, and also an overview of related work done in this area.

§ 2. Qualitative Description

Before embarking on the formal description of gap solitons, we aim in this section at giving the reader a qualitative understanding of the physics of gap solitons and of the general issues involved. To do so, we disregard initially the nonlinearity, and concentrate on the properties of *linear* periodic media, such as that shown in fig. 1. Recall that in the limit we are considering here, the modulation amplitude is assumed to be small, or $\delta n \ll \bar{n}$ (fig. 1). If the modulation is not sinusoidal, then δn is to be interpreted as the amplitude of the lowest Fourier component. It is natural to compare the properties of the medium in fig. 1 with those of a medium with a uniform index $\bar{n}$. In general, these two media behave very similarly, but there is a crucial exception which occurs when $\lambda \approx \lambda_0$, where

$$\lambda_0 = 2\bar{n}d. \tag{2.1}$$

This "Bragg condition" (see, e.g., Ashcroft and Mermin [1976]) specifies that at $\lambda = \lambda_0$, exactly half a wavelength fits into each period of the grating. As a consequence, light which is Fresnel-reflected off interfaces which are an integer number of periods apart (see fig. 1) are all in phase, leading to a

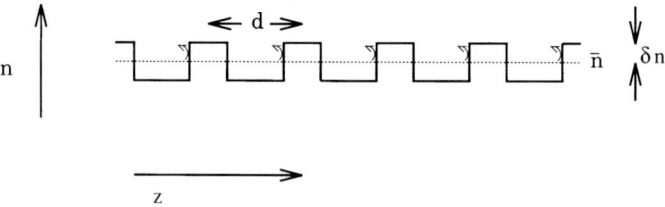

Fig. 1. Schematic of a linear periodic optical medium, showing the refractive index versus position. Note that δn is the amplitude of the lowest Fourier component of the index profile. Also indicated is the Fresnel-reflected field off each of the interfaces, which are an integer number of periods apart. At the Bragg resonance, all of these (small) contributions are in phase.

strong reflected wave. Clearly, at wavelengths far from this Bragg condition the light reflected off the various interfaces is mutually out of phase; as a consequence, the light then propagates through the structure essentially unimpeded.

In anticipation of the later sections, it is preferable to consider the properties of periodic structures as a function of the (angular) frequency ω, rather than as a function of the wavelength. Rewriting eq. (2.1) in this way, we find that

$$\omega_0 = \frac{\pi c}{\bar{n} d}, \tag{2.2}$$

where c is the speed of light in vacuum. As mentioned, it is not just light with the frequency ω_0 which is Bragg-reflected. Rather, light within a small range of frequencies $\Delta\omega$ around ω is Bragg-reflected. The extent of this range depends on the modulation depth. In the limit $\delta n \ll \bar{n}$ which we are considering here,

$$\frac{\Delta\omega}{\omega_0} \approx \frac{\delta n}{\bar{n}}, \tag{2.3}$$

as will be shown in §3. A sketch of the frequency dependence of the reflectivity R of a linear periodic medium is indicated by the solid line in fig. 2 (Macleod [1969], Kogelnik [1969]).

We now add the nonlinearity to the considerations. In particular, we consider an instantaneous Kerr response whereby the refractive index is

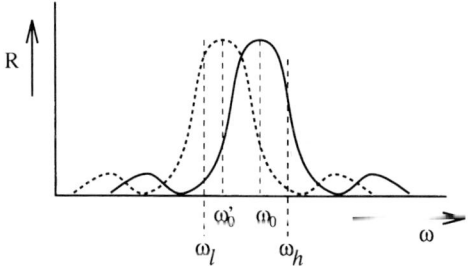

Fig. 2. Illustration of the intensity-dependent response of a nonlinear periodic structure. It shows that the Bragg wavelength at low intensities, ω_0, shifts to a lower frequency ω_0' at high intensity. As a consequence, the reflectivity R, which peaks around the Bragg frequency, also shifts to lower frequencies with increasing intensity. The reflectivity of light with frequency ω_l (initially) increases with intensity, while that of light at ω_h vanishes.

intensity-dependent (see, e.g., Agrawal [1989], and Shen [1984]):

$$n(I) = n_L + n^{(2)} I, \tag{2.4}$$

where I is the local intensity. In this review we take the nonlinearity to be positive, so that the refractive index *increases* with intensity, although negative nonlinearities give rise to very similar gap-soliton behavior. If, according to eq. (2.4), the refractive index is intensity-dependent, then by eq. (2.2), so is the Bragg frequency. What actually happens in a nonlinear periodic medium is discussed later in this review, but some of the main features can be understood from a "mean-field" approach in which one just follows the position of ω_0. From eq. (2.2) it is easy to see that if the nonlinearity is positive, then ω_0 shifts to lower frequencies. To first order, therefore, the highly reflective region illustrated in fig. 2 shifts down in frequency. Of course, eq. (2.3) would suggest that, apart from the position of ω, the size of the gap changes as well. However, this is a much smaller effect which can be neglected in the simple approach used here.

The shift of the Bragg frequency to lower frequencies, as illustrated in fig. 2, shows clearly how the structure's response to different frequencies changes when the intensity is increased. The low frequency ω_l is Bragg-reflected at low intensities, but is reflected even more strongly at high intensities. The higher frequency ω_h is also reflected at low intensities, but, in contrast, is scarcely reflected at high intensities. Because of the nonlinearity, light at this frequency detunes itself from the Bragg condition at high intensities; it then propagates through the system unimpeded. Wavelengths like ω_h are crucial in the formation of gap solitons. To see why, we refer to fig. 3 which, like fig. 1, shows the index as a function of position. Now assume that the intensity in the region indicated by the dashed line is high, while elsewhere it is low, and that the light has a frequency ω_h as in fig. 2.

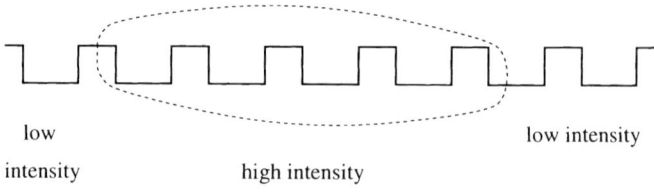

Fig. 3. Schematic of a nonlinear periodic medium (solid line) with a localized region with high field intensity indicated by the dashed line. As discussed in the text, such high-intensity regions – "gap solitons" – have a robustness which precludes them from dispersing.

In this context, the descriptor "high" means that the light is intense enough to render the system transparent for ω_h, as shown in fig. 2. In the indicated region in fig. 3, therefore, the light propagates freely. Now consider light traveling towards the right in this region. At the boundary of the high-intensity region, it "sees" the low-intensity region, which acts as a mirror, and the light is thus Bragg-reflected back in. The same, of course, happens on the other side. Thus, once one of these high-intensity regions has been created it tends to stay together since any light traveling away would be reflected back in! As explained in detail in subsequent sections, such high-intensity regions can have soliton properties. For this reason, and because they are clearly associated with the stop gap of the periodic medium, the term "gap soliton" introduced by Chen and Mills [1987a] for such regions is very appropriate. Apart from the stationary soliton just described, we shall see that gap solitons can have, in principle, any velocity between zero and the speed of light.

To conclude this section, we note that Bragg reflection of optical waves in linear, periodic optical media is similar to the Bragg reflection of electrons or X-rays in crystalline solids. In both cases, it leads to a range of frequencies (energies) over which the waves are strongly refected (Ashcroft and Mermin [1976], Yablonovitch [1987], John [1987]). In § 3 we shall see that for these frequencies (energies), no running-wave solutions of the wave equations (Schrödinger equation) exist for the electric field (wave function). In solids, the occurrence of Bragg reflection is known to lead to the opening of band gaps in the dispersion relation for the electrons, and to indicate the absence of these running waves (Ashcroft and Mermin [1976]). Optical waves in periodic media have a similar response; the periodicity opens photonic band gaps in the optical-dispersion relation, or "photonic band structure", of the medium (Yablonovitch [1987], John [1987]). Light with a frequency inside these gaps cannot propagate and is (Bragg-) reflected off the grating. This is illustrated in fig. 4, which shows the photonic band structure of a medium as in fig. 1. Indeed, it is seen to be a straight line with the slope $c/\bar{n}$, except near the edge of the Brillouin zone where $k = k_0 = \pi/d$, corresponding to the Bragg condition in eq. (2.2).

Figure 3 and the discussion above show clearly the inadequacy of the mean-field approach adopted earlier, as gap solitons occur when the intensity in the medium is non-uniform. This deficiency is corrected in subsequent sections in which we give a quantitative description of gap solitons using coupled-mode theory.

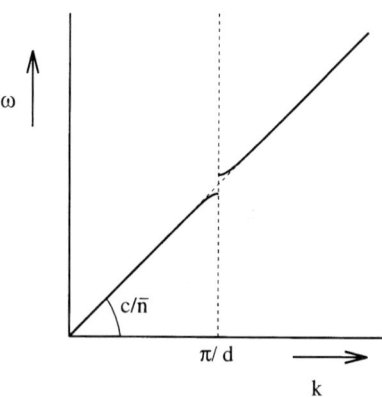

Fig. 4. Photonic band structure (dispersion relation) for a medium as shown in fig. 1, showing the frequency ω versus wavenumber k. Note the stop gap opening at the edge of the Brillouin zone, where $k = k_0 = \pi/d$. Light with frequencies inside the gap are Bragg-reflected.

Before finishing this qualitative description of gap solitons, we must qualify our earlier statement that frequencies like ω_h in fig. 2 are crucial in soliton formation, while frequencies like ω_l are not. In fact, subsequent sections will indicate that gap solitons can occur at any frequency. The significant point to remember, however, is that for positive nonlinearities, ω_h is much easier to tune out of the gap than ω_l, and therefore, as the light intensity increases, frequencies like ω_h will be the first ones to give rise to gap solitons.

§ 3. Coupled-Mode Theory

Consider a one-dimensional geometry where the electric field E and the polarization P are assumed to point in a direction perpendicular to the direction in which they vary (e.g., z). If there are no magnetic effects and we assume the light to be linearly polarized, then the Maxwell equations reduce to

$$\frac{\partial^2 E}{\partial z^2} - \frac{1}{c^2}\frac{\partial^2 E}{\partial t^2} = \mu_0 \frac{\partial^2 P}{\partial t^2}, \tag{3.1}$$

where μ_0 is the vacuum permeability. If for simplicity we restrict ourselves to a nondispersive linear medium, where $P(z,t) = \varepsilon_0 \chi(z) E(z,t)$, and where ε_0 is the vacuum permittivity, we can put $\varepsilon(z) \equiv (1 + \chi(z))$, and write eq. (3.1) in the form:

$$\frac{\partial^2 E}{\partial z^2} - \frac{\varepsilon(z)}{c^2}\frac{\partial^2 E}{\partial t^2} = 0. \tag{3.2}$$

Now suppose $\varepsilon(z) = \bar{n}^2 + \tilde{\varepsilon}(z)$, where $\bar{n}^2$ is the spatial average of $\varepsilon(z)$; then $\bar{n}$ is the average index of refraction of the medium, and in a uniform medium with this refractive index, light propagating with a frequency ω would be characterized by a wavenumber $k = \omega\bar{n}/c$. We consider an $\tilde{\varepsilon}(z)$ with a period d, and define $k_0 \equiv \pi/d$ as in § 2. Then we can expand $\tilde{\varepsilon}(z)$ in a Fourier series:

$$\tilde{\varepsilon}(z) = \sum_m \varepsilon_m e^{2ik_0 m z}, \tag{3.3}$$

where $m = 0$ is excluded from the sum, and the ε_m are complex coefficients with $\varepsilon_{-m} = \varepsilon_m^*$. We are interested in light at frequencies ω near the Bragg condition at $\omega_0 \equiv k_0 c/\bar{n}$. Under these circumstances, we shall argue shortly that only the $m = \pm 1$ terms in eq. (3.3) are important, so we take

$$\tilde{\varepsilon}(z) = 2\tilde{\varepsilon}\cos(2k_0 z), \tag{3.4}$$

where we have put $\varepsilon_1 = \varepsilon_{-1} = \tilde{\varepsilon}$, and have chosen our origin of z so that a peak value of $\varepsilon(z)$ is at $z = 0$. Now if $\tilde{\varepsilon}$ were zero, a solution of eq. (3.2) would be $E(z, t) = E_+ \exp[-i(\omega_0 t - k_0 z)] + E_- \exp[-i(\omega_0 t + k_0 z)] + \text{c.c.}$, where c.c. denotes the complex conjugate, and the constants E_+ and E_- label the amplitudes of the fields propagating to the right and left, respectively. The idea of a coupled-mode analysis (Kogelnik [1969]) is that if $\tilde{\varepsilon}$ is small, we can use basically this same form, but we must allow E_+ and E_- to vary weakly with z and t, since the perturbation can scatter a wave going to the right, causing it to go to the left (and vice versa). Therefore, we look for solutions of the form:

$$E(z, t) = E_+(z, t)e^{-i(\omega_0 t - k_0 z)} + E_-(z, t)e^{-i(\omega_0 t + k_0)} + \text{c.c.}. \tag{3.5}$$

We now insert eqs. (3.4) and (3.5) into eq. (3.2). In taking second derivatives, we find three kinds of terms: (i) those involving $k_0^2 E_\pm$ and $\omega_0^2 E_\pm$, (ii) those involving $k_0(\partial E_\pm/\partial z)$ and $\omega_0(\partial E_\pm/\partial t)$, and (iii) those involving $(\partial^2 E_\pm/\partial z^2)$ and $(\partial^2 E_\pm/\partial t^2)$. The third kind we neglect with respect to the second, since we suppose $E_\pm$ to be slowly varying with respect to ω_0^{-1} in time, and k_0^{-1} in space; viz.,

$$\left|\frac{\partial E_\pm}{\partial t}\right| \ll \omega_0 |E_\pm|, \qquad \left|\frac{\partial E_\pm}{\partial z}\right| \ll k_0 |E_\pm|. \tag{3.6}$$

Furthermore, since the terms of the first kind cancel because they are the solutions to the problem without a grating, we find that

$$E(z,t) = \left[+i\left(\frac{\partial E_+}{\partial z} + \frac{\bar{n}}{c}\frac{\partial E_+}{\partial t}\right) + \kappa E_- + \kappa E_+ e^{+2ik_0 z} \right] e^{-i(\omega_0 t - k_0 z)}$$

$$+ \left[-i\left(\frac{\partial E_-}{\partial z} - \frac{\bar{n}}{c}\frac{\partial E_-}{\partial t}\right) + \kappa E_+ + \kappa E_- e^{-2ik_0 z} \right] e^{-i(\omega_0 t + k_0 z)} + \text{c.c.},$$

(3.7)

where

$$\kappa \equiv \frac{\omega_0 \tilde{\varepsilon}}{2\bar{n}c}. \tag{3.8}$$

The remaining problematic terms in eq. (3.7) are the "non-phase-matched" terms involving $\exp(\pm 2ik_0 z)$. It is often claimed that one can demonstrate that these have little effect on $E_\pm$ by integrating eq. (3.7), but this approach is suspect since such an integration would introduce terms in $E_\pm$ which did not satisfy eqs. (3.6), on which the derivation of eq. (3.7) is based! A more correct approach is to note that the non-phase-matched terms in eq. (3.7), and similar terms that would appear if the full eq. (3.3) were used instead of eq. (3.4), indicate that there is simply more to the electric field than eq. (3.5) and eqs. (3.6) would suggest. However, such components of higher spatial frequency, which would couple to $E_\pm$ through the non-phase-matched terms, would only couple *back* to $E_\pm$ involving terms of order κ^2. In the limit of small $\tilde{\varepsilon}/\bar{n}^2$, we expect such effects to be negligible. Thus, for $E_\pm$, we may neglect the non-phase-matched terms in eq. (3.7), and take our equations to be

$$+i\frac{\partial E_+}{\partial z} + i\frac{\bar{n}}{c}\frac{\partial E_+}{\partial t} + \kappa E_- = 0, \qquad -i\frac{\partial E_-}{\partial z} + i\frac{\bar{n}}{c}\frac{\partial E_-}{\partial t} + \kappa E_+ = 0.$$

(3.9)

These are the linear coupled-mode equations. A more thorough discussion of some of the approximations made in obtaining the coupled-mode equations have been discussed by Kogelnik [1969] and by Sipe and Stegeman [1979]. We note that the solutions to the full eqs. (3.2) and (3.3) are well investigated; by comparing those solutions with the solution of the coupled-mode equations, the validity of the approximations used in the derivation of eqs. (3.9) may be confirmed (Ashcroft and Mermin [1976]). We mention in passing that if the dispersion of the constituent material is included, then

$c/\bar{n}$ in the coupled-mode equations should be replaced by the group velocity of the uniform medium at ω_0.

Before generalizing the mode equations to include nonlinear effects, we will discuss briefly the dispersion relation associated with the linear coupled-mode equations. To obtain the dispersion relations, we search for envelope functions of the form:

$$E_\pm(z, t) = A_\pm e^{-i(\Omega t - Qz)}, \tag{3.10}$$

leading to a set of agebraic equations in the constants $A_\pm$

$$\begin{bmatrix} c\Omega/\bar{n} - Q & \kappa \\ \kappa & c\Omega/\bar{n} + Q \end{bmatrix} \begin{bmatrix} A_+ \\ A_- \end{bmatrix} = 0. \tag{3.11}$$

We obtain from this the dispersion relation

$$c\Omega/\bar{n} = \pm \sqrt{\kappa^2 + Q^2}. \tag{3.12}$$

This dispersion relation, which is shown in fig. 5, confirms the discussion in §2 regarding the opening of a photonic stop gap in the dispersion relation to periodic media. Equation (3.12) shows that no running-wave solutions are allowed for $|\Omega| < \kappa c/\bar{n}$, so that the gap has a width $\Delta\omega$ equal to:

$$\Delta\omega = \Delta\Omega = 2\kappa c/\bar{n}, \tag{3.13}$$

which, with eq. (3.8), is easily seen to be consistent with eq. (2.3). It should be recalled at this point that in using the coupled-mode equations, we are dealing with slowly varying amplitudes, rather than with the field itself, and the center of the gap is thus located at $\omega = \omega_0$, $k = k_0$ [eq. (3.5)]. In §6, we

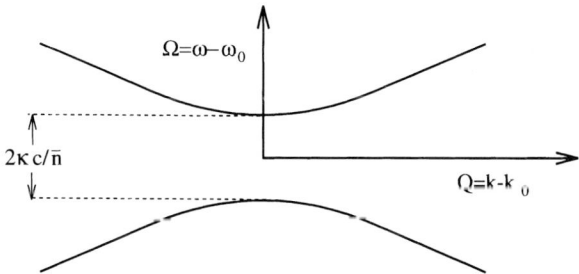

Fig. 5. Dispersion relation associated with the linear coupled-mode equations for light with wavenumbers around the edge of the Brillouin zone, where $k = k_0 = \pi/d$ (cf. fig. 4). For light with a frequency close to the Bragg condition ($|\omega - \omega_0| < \kappa c/\bar{n}$), no running-wave solutions are allowed.

will consider the linear dispersion relation in somewhat more detail, and also discuss the eigenvectors of eq. (3.11).

We now return to eq. (3.1), and include a nonlinear polarization $P_{NL}(z,t)$. Then, instead of eq. (3.2), we have

$$\frac{\partial^2 E}{\partial z^2} - \frac{\varepsilon(z)}{c^2}\frac{\partial^2 E}{\partial t^2} = \mu_0 \frac{\partial^2 P_{NL}}{\partial t^2}. \tag{3.14}$$

For simplicity, we treat the nonlinear properties of the medium as uniform, although this is not necessary. Introducing vector notation for a moment, we recall that for an electric field of the form

$$\boldsymbol{E}(\boldsymbol{r},t) = \boldsymbol{\xi}(\boldsymbol{r})e^{-i\omega t} + \text{c.c.}, \tag{3.15}$$

the lowest-order nonlinear response in an isotropic medium is of the form (see, e.g., Shen [1984]):

$$\boldsymbol{P}_{NL}(\boldsymbol{r},t) = \varepsilon_0(A(\boldsymbol{\xi}\cdot\boldsymbol{\xi}^*)\boldsymbol{\xi}(\boldsymbol{r})e^{-i\omega t} + \tfrac{1}{2}B(\boldsymbol{\xi}\cdot\boldsymbol{\xi})\boldsymbol{\xi}^*(\boldsymbol{r})e^{-i\omega t}) + \text{c.c.}, \tag{3.16}$$

plus, of course, terms responsible for third-harmonic generation and frequency mixing, which do not concern us here. The constants A and B in eq. (3.16) characterize the nonlinear response; far below the resonant frequencies of the electronic transitions, their contributions are equal. We assume that A and B are real, treating any two-photon absorption as being negligible. It can be shown that such a nonlinear response is equivalent to that in eq. (2.4).

Applying eq. (3.16) to the field described by eq. (3.5), we calculate the right-hand side of eq. (3.14) neglecting the time derivatives of $E_\pm$, and taking the values of A and B to be those at ω_0; since the nonlinear term is assumed to be small, we only consider the dominating part. Carrying on the analysis, we find instead of eqs. (3.9) that (Winful and Cooperman [1982], de Sterke and Sipe [1990a]):

$$+i\frac{\partial E_+}{\partial z} + i\frac{\bar{n}}{c}\frac{\partial E_+}{\partial t} + \kappa E_- + \Gamma_s|E_+|^2 E_+ + 2\Gamma_\times|E_-|^2 E_+ = 0,$$

$$-i\frac{\partial E_-}{\partial z} + i\frac{\bar{n}}{c}\frac{\partial E_-}{\partial t} + \kappa E_+ + \Gamma_s|E_-|^2 E_- + 2\Gamma_\times|E_+|^2 E_- = 0, \tag{3.17}$$

where in fact,

$$\Gamma_\times = \Gamma_s = \frac{1}{2}\frac{\omega^2}{\bar{n}c}\left(A + \frac{B}{2}\right). \tag{3.18}$$

We label Γ_s and $\Gamma_\times$ separately, since in some mode-coupling problems where the mode overlap is more complicated than the plane waves considered here, the self-phase mode modulation term Γ_s and cross-plane modulation term $\Gamma_\times$ can be different.

Before closing this section, we point out that the coupled-mode equations have important scaling properties which reduce the range of parameters that must be investigated. It is straightforward to check that the coupled-mode equations are unchanged under the substitutions $\kappa = \beta_1 \hat{\kappa}$, $\Gamma_{\times,s} = \beta_2^2 \hat{\Gamma}_{\times,s}$, $E_\pm = \sqrt{\beta_1} \hat{E}_\pm / \beta_2$, $z = \hat{z}/\beta_1$, and $t = \hat{t}/\beta_1$, where $\beta_{1,2}$ are arbitrary numbers. This means that one may fix the Γ's, and either κ, or L, where L is the system's length, without loss of generality. Note, however, that the dimensionless product κL, which represents the grating's strength, is unchanged under the transformation.

The nonlinear coupled-mode equations play a central role in this review. We show that they have traveling-wave solutions for wavelengths which would be Bragg-reflected at low intensities; these are exactly the gap solitons which were described on an intuitive level in § 2. In subsequent sections, we will investigate various exact and approximate solutions to the coupled-mode equations. The first of these are the stationary solutions to coupled-mode equations, which are discussed in the next section.

§4. Stationary Solutions

In this section, we will discuss the stationary solutions to the nonlinear coupled-mode equations first derived by Winful, Marburger and Garmire [1979]. These solutions are of the form:

$$F_\pm(z, t) = e_\pm(z) e^{-i\hat{\delta}ct/\bar{n}}, \tag{4.1}$$

and are subject to the standard boundary conditions (Kogelnik [1969]); i.e.,

$$e_-(0, t) = 0, \qquad e_+(-L, t) = A, \tag{4.2}$$

where the coordinate system as shown in fig. 6 was used, and $\hat{\delta}$ is a detuning. We note that, as implied by the notation, the detuning as defined in eq. (4.1) differs from that used in other sections. The system has a length L, and according to fig. 6, the origin is at the back of the system. In eq. (4.2), A is the amplitude of the incoming field, and $|A|^2$ is proportional to the incoming intensity [see eq. (8.1)]. The reflectivity R and the transmissivity T of the system are calculated from $|e_-(-L)/A|^2$ and $|e_+(0)/A|^2$, respectively.

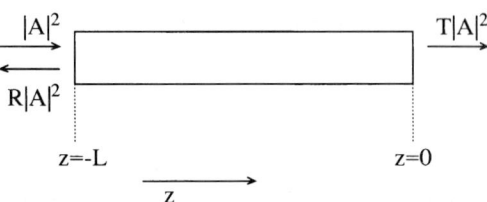

Fig. 6. Schematic of the setup used for the calculation of the stationary solutions to the coupled-mode equations. The length of the system is L, which is positive.

Assuming solutions as in eq. (4.1), the coupled-mode equations turn into ordinary differential equations with the detuning as a parameter. On the basis of the (stationary) solutions to these equations, which were first reported by Winful, Marburger and Garmire [1979], one can predict that periodic structures exhibit bistability. However, to describe dynamical effects like propagation, one needs to go beyond the stationary solutions; this will be done in later sections. In general, the stationary solutions are written in terms of Jacobi elliptic functions (Abramowitz and Stegun [1970], Byrd and Friedman [1954]), with complicated arguments and parameters. The main reason for this complexity is that the solutions involve the roots of a cubic equation; although such equations can be solved exactly, the resulting expressions for the roots usually do not lend themselves to being written in any simple way. The exception, as we will see, occurs in the middle of the gap where $\hat{\delta} = 0$. In that case, the roots of the cubic equations can be found very simply, and the solutions then reduce to a fairly simple form, although they are still written in terms of Jacobi elliptic functions.

One way to solve eqs. (3.17) with the *ansatz* eq. (4.1) is to rewrite them in terms of the four (real) parameters defined by

$$A_0 = |e_+|^2 + |e_-|^2, \quad A_1 = e_+ e_-^* + e_+^* e_-,$$
$$A_2 = i(e_+ e_-^* - e_+^* e_-), \quad A_3 = |e_+|^2 - |e_-|^2. \tag{4.3}$$

These parameters have the property:

$$A_0^2 = A_1^2 + A_2^2 + A_3^2. \tag{4.4}$$

Though this may not seem very helpful in itself, it has the advantage of reducing the number of parameters from four to three [by eq. (4.4)], because by using these parameters the absolute phase of field is taken out. Our definitions in eqs. (4.3) follow from

$$A_i \equiv (e_+^*, e_-^*) \sigma^i \begin{pmatrix} e_+ \\ e_- \end{pmatrix}, \tag{4.5}$$

where the σ^i indicate the Pauli matrices. Since the A parameters are similar to those first introduced by Stokes (see, e.g., Born and Wolf [1980]) to describe the state of polarization of a single beam of light, we henceforth refer to them as "Stokes parameters". Rewriting the coupled-mode equations in terms of the Stokes parameters leads to:

$$\frac{dA_0}{dz} = -2\kappa A_2, \quad \frac{dA_1}{dz} = 2\hat{\delta} A_2 + 3\Gamma A_0 A_2,$$

$$\frac{dA_2}{dz} = -2\hat{\delta} A_1 - 2\kappa A_0 - 3\Gamma A_0 A_1, \quad \frac{dA_3}{dz} = 0, \tag{4.6}$$

where $3\Gamma = 2\Gamma_\times + \Gamma_s$, as is consistent with eq. (4.4). The distinction between the self- and cross-phase modulation can be dropped, because within the formalism of the Stokes parameters, this is the only combination of the Γ's which appears. Note further that, as expected for a lossless system, the net energy flow,

$$|e_+|^2 - |e_-|^2 \equiv S, \tag{4.7}$$

is uniform according to the second of eqs. (4.6).

It is perhaps instructive to remark here that S is the natural parameter to label the states of the system – every value of S uniquely describes such a state. However, in experimental situations one cannot control S directly; rather, the control parameter is the intensity of the incoming radiation $|A|^2$ [eqs. (4.2)]. These parameters are related by

$$S = T|A|^2, \tag{4.8}$$

where T is intensity-dependent. Thus, although the state of the system is determined uniquely by the energy flow S, as a function of the incoming intensity $|A|^2$ the system's state may no longer be single-valued, leading to bistable behavior as shown below (Winful, Marburger and Garmire [1979], Chen and Mills [1987a, b]).

Returning to eqs. (4.6), we see that this set has another conserved quantity; viz.,

$$\hat{\delta} A_0 + \kappa A_1 + \tfrac{3}{4}\Gamma A_0^2. \tag{4.9}$$

Using the two conserved quantities [eqs. (4.7) and (4.9)], and the mutual relation between the A_i's, [eq. (4.4)], we can eliminate three of the Stokes parameters from eqs. (4.6) and find an equation in the energy A_0 only:

$$\frac{dA_0}{dz} = \pm 2\sqrt{(A_0 - S)[\kappa^2(A_0 + S) - (A_0 - S)(\hat{\delta} + \tfrac{3}{4}\Gamma(A_0 + S))^2]}, \quad (4.10)$$

where, as follows from the boundary conditions [eqs. (4.2)], we require that $A_0 = S$ at the back of the structure. Note the different footing of the front and back interfaces of the structure: the field inside the stack is determined uniquely by the fact that at the rear interface, one has only an outgoing wave. The position of the front interface determines only how far one has to integrate forward, and in this way determines the transmissivity, but does not influence the field shape inside.

Equation (4.10) is known to be solvable in terms of Jacobi elliptic functions (Winful, Marburger and Garmire [1979], Abramowitz and Stegun [1970], Byrd and Friedman [1954]). In order to write these equations, it is necessary to know the roots of the quartic equation under the radical in eq. (4.10). The root $A_0 = S$ is readily identified, but the others are nontrivial and must be found by solving the cubic equation in square brackets in eq. (4.10). The solution to this cubic equation leads, of course, to three roots which we denote by $a > S > c > d$. We note that in most practical situations, the three roots are all found to be real. Using standard methods (Byrd and Friedman [1954]), one finds that the energy density A_0 is given by

$$A_0 = S + \frac{(a - S)(S - c)sn^2}{(a - c) - (a - S)sn^2}, \quad (4.11)$$

where sn represents:

$$sn = sn\left[\sqrt{(a - c)(S - d)}\tfrac{3}{4}\Gamma z \left| \frac{(a - S)(c - d)}{(a - c)(S - d)} \right.\right]; \quad (4.12)$$

sn is one of the Jacobi elliptic functions, and the coordinate system was taken according to fig. 6. Note that we have adopted the convention used in Abramowitz and Stegun [1970]; viz., that the sum of the parameter and complementary parameter of a Jacobi elliptic function equal unity. Now the sn function is periodic, and varies between -1 and $+1$; we thus see from eq. (4.11) that the energy density A_0 is also periodic, and that it varies between the two largest roots, a and S, of the quartic equation under the radical in eq. (4.10). In the limit in which $S \to 0$, the period diverges, and the elliptic function reduces to hyperbolic functions (Abramowitz and Stegun [1970]). This is consistent with the well-known result for linear periodic functions; viz., that the field consists of solutions which grow and decay exponentially. With increasing S the period decreases monotonically. Once we have calculated A_0 in the most general case, we can find A_1 and A_2 as well [eqs. (4.4) and (4.9)], and thus find the electric field, apart from an overall phase factor. Note from the discussion above that in the general nonlinear case we did not really have to specify whether frequencies are inside or outside the photonic gap at low intensities. Evidently, the nonlinear system puts all frequencies on an equal footing. The reason, which was also discussed at the end of § 2, can be seen from fig. 2; although ω_h is relatively easily tuned out of the gap, ω_l can also be tuned out of the gap, albeit at much higher intensities. Similarly, frequencies which were originally below the photonic stop gap, for instance, will be tuned into the gap at very high intensities, and tuned out again at even higher ones. According to the coupled-mode equations, all frequencies exhibit qualitatively the same behavior, although at dramatically different intensities.

The periodicity of the energy density can be explained qualitatively as follows. If the frequency of the radiation is chosen to be in the stop gap, the solutions grow and decay exponentially. However, if the intensity of the radiation inside the structure is high enough, it can be locally tuned out of the gap as also mentioned in § 2. Solutions growing and decaying exponentially to the left and right of this region can then be connected in the high-intensity region to form a self-consistent solution as shown schematically in fig. 7. The result that the energy density through the medium varies periodically leads one to conclude that the electric field is also periodic, apart from

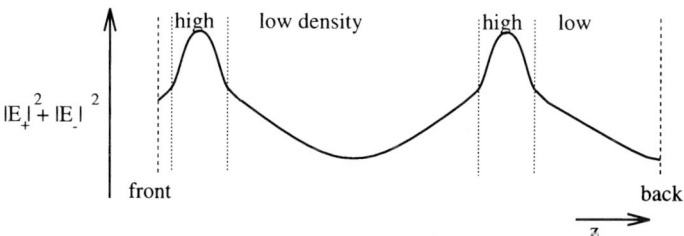

Fig. 7. Schematic of the energy density $A_0 = |E_+|^2 + |E_-|^2$ as a function of position z. The energy density (solid line) is a periodic function consisting of low-energy parts where the field is tuned in the gap, leading to solutions which grow and decay exponentially, and high-energy parts where the field is strong enough to tune itself out of the gap, leading to sinusoidal solutions. These borders between these regions are indicated by the dotted lines, while the dashed lines indicate the front and the back of the structure. As derived in the text, the energy density is actually given by a Jacobi elliptic function.

an overall phase factor. However, if the period equals exactly the system's length, then the fields at the front and at the back of the system are identical, implying that it is perfectly transparent. In fact, the period decreases monotonically with S, and every time an integer number of periods fits inside the structure, the transmissivity is unity. This behavior is illustrated in fig. 8, which shows a typical example of the transmissivity as a function of the energy flow S. As mentioned above, the control parameter is not S, but rather the incoming intensity $|A|^2$ at the front of the system, which is related to S by eq. (4.8). As a function of this parameter, the transmissivity is no longer single-valued. This is shown in fig. 9, which clearly predicts bistable behavior. Physically, this bistable behavior occurs for light which is Bragg-reflected at low intensities, but is tuned out of the gap at higher intensities, thus leading to a transmissivity of order unity. The size of the bistable region is strongly dependent on the detuning $\hat{\delta}$. This is not unexpected, because frequencies such as ω_h in fig. 2 are easily "lifted" out of the gap, while frequencies such as ω_l require much higher fields. Consequently, for a positive nonlinearity the bistable region is found to be largest when $\hat{\delta} \approx -\kappa$ and smallest for $\hat{\delta} \approx \kappa$, while the bistable behavior vanishes for detunings somewhat larger than κ.

We note here that the dotted parts of the curves in figs. 8 and 9 are

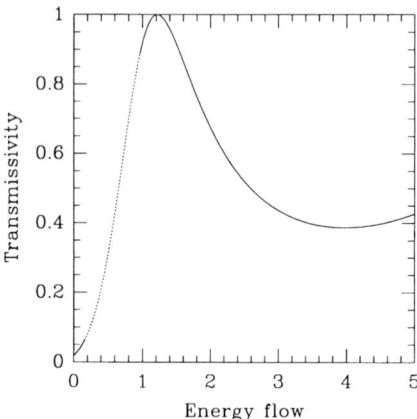

Fig. 8. Typical example of the dependence of the transmissivity T on the energy flow S, for a nonlinear periodic system. The parameters taken here are: $\kappa = 5$, $L = 1$, $\hat{\delta} = 4.75$, and $\Gamma_x = \Gamma_s = 0.1$. The dotted line indicates unstable solutions, while the solid line indicate stable solutions according to the simple argument in the text. However, numerical simulations and a stability analysis show that these solutions may be unstable against the formation of side bands.

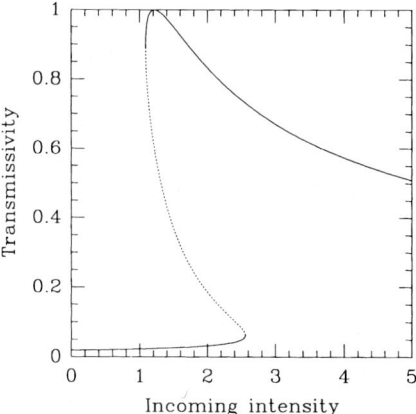

Fig. 9. Typical example of the dependence of the transmissivity T on the incoming intensity $|A|^2$, for a nonlinear periodic system, indicating bistability. The parameters taken here are: $\kappa = 5$, $L = 1$, $\hat{\delta} = 4.75$, and $\Gamma_x = \Gamma_s = 0.1$. The dotted line indicates unstable solutions, while the solid line indicates stable solutions according to the simple argument in the text. However, numerical simulations and a stability analysis show that these solutions may be unstable against the formation of side bands.

unstable against amplitude fluctuations. A general proof of this property is given in Appendix E of Gibbs [1985]. However, the following simple argument gives the general idea. Recall eq. (4.8), and now assume that by some fluctuation the energy flow through the system carries, say, an extra photon with energy $\hbar\omega$. By eq. (4.8), this would imply that the transmissivity changes by an amount $\hbar\omega(\partial T/\partial S)$. For a given incoming intensity $|A|^2$, this would therefore lead to an energy flow $|A|^2(T + \hbar\omega(\partial T/\partial S))$, the second term describing the system's initial response to the fluctuation $\hbar\omega$ in the energy flow. In turn, this response would lead to yet another response and so forth. The total response of the system becomes the sum of all of these, which can be written as a geometrical series. This leads one to conclude that the system is *unstable* against the original functuation if $|A|^2 \hbar\omega(\partial T/\partial S) > \hbar\omega$, or

$$\frac{S}{T}\frac{\partial T}{\partial S} > 1, \qquad (4.13)$$

where eq. (4.8) was used. Note that, as required, this result does not depend on the size of initial perturbation. Rewriting this condition in terms of T and $|A|^2$ then shows that the dotted parts of the curves in figs. 8 and 9 are unstable. This is not to say that the rest of the curve is necessarily stable against all conceivable fluctuations – to ascertain this requires either a linear stability analysis or the study of full time-dependent solutions to the coupled-mode equations. In fact, parts of the upper branch in fig. 9 are often unstable to side-band formation, leading to self-pulsations, and ultimately to chaotic behavior (Winful and Cooperman [1982], de Sterke and Sipe [1990c], Winful, Zamir and Feldman [1991].) We will explore this topic further in § 7.

We now return briefly to fig. 9, which shows that nonlinear periodic media may exhibit bistability. It is useful to appreciate that the states of the system on the lower and upper branches are quite different (Chen and Mills [1987b]). With reference to fig. 9, it is easy to see that on the lower branch the period of the electric field distribution exceeds the system's length; the field maxima seen in fig. 7 thus do not occur. In such a state, the system behaves essentially as if it were linear; recall that for a linear system the field is roughly a negative exponential function when going from the front of the system to the back. In contrast, on the upper branch the field density exhibits at least one maximum, and the system is thus in a profoundly nonlinear state. Indeed, the self-pulsations referred to in the paragraph above are only observed to occur for states on the upper branch, and not for those on the lower branch where the nonlinearity plays a minor role.

To finish this section, we will show briefly how the general expression for A_0 in eq. (4.11) simplifies when the frequency is tuned to the middle of the gap ($\hat{\delta} = 0$). In that case, finding the roots of the cubic equation [eq. (4.10)] is straightforward. The resulting expression can be rewritten after some algebra, and the application of an "ascending Landen transformation" (Abramowitz and Stegun [1970]) of the elliptic function. Consistent with Winful, Marburger and Garmire [1979], it is then found that

$$A_0/S = \text{nd}[2\cosh \tilde{\gamma} \kappa z \,|\, 1/\cosh^2 \tilde{\gamma}], \tag{4.14}$$

where nd is another of the Jacobi elliptic functions, and $\tilde{\gamma}$ is defined through

$$\sinh \tilde{\gamma} \equiv \frac{3\Gamma S}{4\kappa}. \tag{4.15}$$

In a similar way, the expressions for other Stokes parameters, and in fact the entire solution, take very simple forms in this limit.

As mentioned earlier, the Stokes formalism in itself gives the electric field up to an absolute phase factor. Although this phase is often not required, sometimes it is; e.g., when the stationary solutions are the subject of a stability analysis as performed by de Sterke [1992b]. One must then resort to the coupled-mode equations to obtain this final piece of information on the electric field inside the structure. We remark that the absolute phase of the fields depends upon the phase-modulation parameters Γ_s and Γ_x separately, and not, like the Stokes parameters, on only the linear combination $\Gamma_s + 2\Gamma_x$.

This finishes our discussion of the stationary solutions for the electric field inside nonlinear periodic structures. It should be mentioned that other methods have been used. Chen and Mills [1987b], for example, solved directly the wave equation without reference to coupled-mode theory. To do so, they noted that in a homogeneous nonlinear layer, the wave equation can be solved exactly in terms of Jacobi elliptic functions (Chen and Mills [1987c]). Starting at the back, just as in a linear thin-film stack, they then solved for the fields in each of the layers, connecting the fields in each of the layers using the Maxwell *saltus* conditions. In this way they found the exact electric field inside the structure, and thus the reflectivity and transmissivity. However, such a method is not easily generalized to include time-dependent fields. Mills and Trullinger [1987] showed, starting from the coupled-mode equations, that in the same time-independent limit and for a medium of infinite extent, the electric field satisfies the *double sine-Gordon equation*.

§ 5. Solitary-Wave Solutions

The next set of solutions to the coupled-mode equations we will discuss are those found by Aceves and Wabnitz [1989]. As we shall see, these are solitary-wave solutions with frequencies inside the stop gap, and can travel through the structure at any velocity between zero and the speed of light. They thus have the characteristics of gap solitons, though they are not, strictly speaking, "proper" solitons. A special case of the solutions of Aceves and Wabnitz [1989] was reported somewhat earlier by Christodoulides and Joseph [1989]. However, because the method of Aceves and Wabnitz [1989] is both general and well laid out, we follow it here. Their starting point is the massive Thirring model (Kaup and Newell [1977], Kuznetsov and Mikhailov [1977]) of field theory, which is very similar to eqs. (3.17). In fact, although the interpretation of the various terms is completely different, the only difference is that in the massive Thirring model the self-phase modulation terms are absent. Recognizing that the massive Thirring model is integrable, Aceves and Wabnitz [1989] found solitary-wave solutions to the coupled-mode equations by reworking suitably the one-soliton solutions to the massive Thirring model. This section reviews how this is done.

The massive Thirring model (Kaup and Newell [1977], Kuznetsov and Mikhailov [1977]) is obtained from the coupled-mode equations in the limit in which $\Gamma_S \to 0$. While in many practical geometries one finds that $\Gamma_S = \Gamma_\times$, and situations in which $\Gamma_\times = 0$ is conceivable, we are not aware of any realistic geometries in which $\Gamma_S = 0$ and $\Gamma_\times \neq 0$. Thus, the massive Thirring model is not relevant directly in optical applications; but, it is a useful starting point. The one-soliton solutions to the massive Thirring model, which we denote by $\tilde{E}_\pm$, read (Kaup and Newell [1977], Kuznetsov and Mikhailov [1977]):

$$\tilde{E}_+ = \pm \sqrt{\pm \frac{\kappa}{2\Gamma_\times} \frac{1}{\Delta}} \sin \tilde{\delta} e^{\pm i\sigma} \operatorname{sech}(\theta \mp i\hat{\delta}/2),$$

$$\tilde{E}_- = -\sqrt{\pm \frac{\kappa}{2\Gamma_\times} \Delta} \sin \tilde{\delta} e^{\pm i\sigma} \operatorname{sech}(\theta \pm i\hat{\delta}/2),$$

(5.1)

where the choice of signs is determined by that of the linear and nonlinear coupling coefficients. Furthermore,

$$\theta = \gamma\kappa \sin \tilde{\delta}(z - vct/\bar{n}), \qquad \sigma = \gamma\kappa \cos \tilde{\delta}(vz - ct/\bar{n}), \qquad (5.2)$$

where

$$v = (1 - \Delta^4)/(1 + \Delta^4), \tag{5.3}$$

and

$$\gamma = 1/\sqrt{1 - v^2}, \tag{5.4}$$

is the Lorentz factor. These solutions are thus characterized by two parameters, $\tilde{\delta}$ ($0 \leq \tilde{\delta} \leq \pi$) and v ($-1 < v < 1$). Equations (5.1) and (5.2) show that $\tilde{\delta}$ determines the soliton's width, height, and spectrum, and thus plays the role of a detuning parameter (but is defined differently than the $\hat{\delta}$ in § 4). For every value of the parameters, eqs. (5.1) represent a two-component solution with a single maximum. Note that for $\tilde{\delta} \to 0$, the soliton is wide and has a low amplitude, and we thus refer to this as the low-intensity limit. Since the frequency content is then concentrated about the upper (lower) edge (fig. 2) of the gap if the nonlinearity is positive (negative), this limit corresponds to that of ω_h in fig. 2. As $\tilde{\delta} \to \pi$ (the "high-intensity" limit), the soliton is narrow and the spectrum is concentrated about the lower (upper) edge, while when $\tilde{\delta} = \pi/2$, it peaks in the middle of the gap. Equations (5.2) show that the parameter v determines the soliton's velocity. It can take any value between 0 and ± 1, implying that the soliton can have any velocity between zero and the speed of light in the medium.

To find the solutions to the coupled-mode equations ($\Gamma_\times \neq 0$) in eqs. (3.17), we follow Aceves and Wabnitz [1989] and try solutions of the form:

$$E_\pm(z, t) = \alpha \tilde{E}_\pm(z, t) e^{i\eta(\theta)}. \tag{5.5}$$

We know from § 3 that in the static limit ($v = 0$), and if $\Gamma_\times = \Gamma_S$, we must have $\alpha^2 = 2/3$ and $\eta = 0$, since in this limit $\Gamma_\times$ and Γ_S play identical roles – the difference between the coupled-mode equations and the massive Thirring model in this limit is just the size of the nonlinear coefficient.

We now substitute Aceves and Wabnitz's *ansatz* for the electric field [eq. (5.5)] into the coupled-mode equations [eq. (3.17)]. Since the $\tilde{E}_\pm$'s are solutions to the massive Thirring model, many terms cancel and one is left with the two coupled equations:

$$\gamma\kappa \sin\tilde{\delta}\frac{d\eta}{d\theta} = \frac{1}{1+v}(-2(\alpha^2 - 1)\Gamma_\times |\tilde{E}_+|^2 - \alpha^2 \Gamma_S |\tilde{E}_-|^2),$$

$$\gamma\kappa \sin\tilde{\delta}\frac{d\eta}{d\theta} = \frac{1}{1-v}(+2(\alpha^2 - 1)\Gamma_\times |\tilde{E}_-|^2 + \alpha^2 \Gamma_S |\tilde{E}_+|^2). \tag{5.6}$$

Amazingly, perhaps, this set of equations is found (Aceves and Wabnitz [1989]) to be mutually consistent if

$$\frac{1}{\alpha^2} = 1 + \frac{\Gamma_S}{2\Gamma_\times}\frac{1+v^2}{1-v^2}, \tag{5.7}$$

which is in agreement with our remarks just below eq. (5.5). It is now easy to solve for the phase function η. Integrating either of eqs. (5.6) using eq. (5.7), we find for the phase function:

$$e^{i\eta} = \left(-\frac{e^{2\theta} + e^{\mp i\tilde{\delta}}}{e^{2\theta} + e^{\pm i\tilde{\delta}}}\right)^{2\Gamma_S v/[2\Gamma_\times(1-v^2)+\Gamma_S(1+v^2)]}, \tag{5.8}$$

which, as expected, vanishes when either $v = 0$ or $\Gamma_S = 0$. Like eqs. (5.1), eqs. (5.5) also represent solutions which can travel with any velocity between zero and the speed of light and which have a single maximum. Equation (5.7) exhibits an interesting dynamic effect in that for large velocities the amplitude is much smaller than that of the massive Thirring model soliton. The reason for this is that at high velocities, almost all of the soliton's energy is in E_+ if the soliton propagates in the forward direction, and in E_- if it propagates backwards. The self-phase modulation is now very strong, of course, and completely dominates the cross-phase modulation. The required intensity for a gap soliton is therefore much less than that required for a massive Thirring model soliton.

We will discuss some of the properties of these solutions. First, while the massive Thirring model is integrable, the coupled-mode equations appear not to be, and eqs. (5.5) thus only specify a solitary-wave solution of the coupled-mode equations. Next note that, strictly speaking, the solitary-wave solutions require an infinite domain, in contrast with the stationary solutions of § 4 which were specified for a finite interval.

In § 9, we will make use of quantities which do not depend on time as the field evolves according to the coupled-mode equations. While three such conserved quantities are known, we use only the fact that the energy content of a gap soliton is time-independent, as is to be expected for wave propagation in a lossless medium; the other conserved quantities correspond to momentum, and free energy (cf. Kaup and Newell [1977], Kuznetsov and Mikhailov [1977]). Indeed, by using the coupled-mode equations [eqs. (3.17)], it can be shown that

$$Q \equiv \int_{-\infty}^{+\infty} (|E_+|^2 + |E_-|^2) \, dz, \tag{5.9}$$

is conserved. Substituting the expression for the envelope functions, it is found that (cf. Kaup and Newell [1977], Kuznetsov and Mikhailov [1977]):

$$Q = \frac{2\tilde{\delta}}{\Gamma} \alpha^2. \tag{5.10}$$

Equations (5.1) and (5.5) show that the solitary-wave solution contains both forward- and backward-propagating components, which are "bound" together by the nonlinearity. Recall that in linear systems without a grating, these components would travel in opposite directions at the speed of light! Apart from an overall multiplicative factor, the two components have the same intensity profile. In addition, when $v = 0$ [and thus $\Delta = 1$ by eq. (5.3)], E_+ and E_- have equal strength. In general, however,

$$\frac{\int_{-\infty}^{+\infty} |E_+|^2 \, dz}{\int_{-\infty}^{+\infty} |E_-|^2 \, dz} = \Delta^{-4} = \frac{(1+v)}{(1-v)}, \tag{5.11}$$

so that the asymmetry between the components increases with the velocity.

An interesting feature of gap solitons is found when they are written in terms of their modulus and phase:

$$E_\pm(x, t) = |E_\pm(x, t)| e^{i\phi_\pm(x,t)}. \tag{5.12}$$

One often expresses the phase in terms of the *instantaneous frequency* ω_i through the relation

$$\omega_i(t) \equiv -\frac{\partial \phi}{\partial t}, \tag{5.13}$$

where a variation of ω_i over the duration of a pulse indicates that it is chirped. It is well known, for example, that a pulse propagating through a medium with a Kerr nonlinearity attains a phase ϕ which is proportional to the instantaneous intensity of the pulse (see, e.g., Agrawal [1989]), so that by eq. (5.13)

$$\omega_i(t) \propto \frac{\partial |E(t)|^2}{\partial t}. \tag{5.14}$$

An initially unchirped pulse, therefore, attains an instantaneous frequency which is proportional to the *gradient* of the intensity; for a pulse with a symmetric intensity profile, the instantaneous frequency is thus an odd function of time. In a gap soliton, in contrast, the instantaneous frequency is found to be proportional to the intensity itself. Because the intensity profile is symmetric, the instantaneous frequency is symmetric as well. For example, for E_+:

$$\omega_i(t) = \Gamma f(v) \frac{c}{n} |E_+(t)|^2, \tag{5.15}$$

where

$$f(v) = \frac{1}{2} \frac{v(3 + 4v - v^2)}{(1+v)^2(1-v)}, \tag{5.16}$$

is only a function of the soliton velocity.

The solitary-wave solutions of Aceves and Wabnitz are most useful for determining the external field required to excite a gap soliton with specified parameters. Consider, for example, a semi-infinite nonlinear periodic structure with average index $\bar{n}$ and modulation as in eq. (3.4) and illustrated in fig. 1. The incident medium is taken to be linear and uniform, with refractive index $n = \bar{n}$. We now wish to excite a gap soliton with prescribed $\tilde{\delta}$ and v in the nonlinear periodic medium. Since the average indices in two half-spaces are identical, no Fresnel reflection occurs, and the required incoming field at the front of the system is described uniquely by $E_+(-L, t)$ from eqs. (5.1). By the same argument one may conclude that, apart from exciting the prescribed gap soliton, the incoming field sets up a reflected field $E_-(-L, t)$

at the front of the system. We see immediately from eq. (5.11) that if we wish to excite a slowly propagating soliton, the reflected wave carries almost as much energy as the incident wave, so that a very small fraction actually enters the nonlinear periodic medium. In fact, for $v \ll 1$ the fraction of the energy entering the gap soliton equals $2v$ from eq. (5.11). We see here a specific example of the general problem encountered when exciting gap solitons. In order to set up a field profile which is clearly a gap soliton, we need $|v| \ll 1$. However, in that limit the coupling is inefficient, so that very large external fields are required. We will return to this issue in some more detail in § 9.

§ 6. Multiple Scales

In this section we show that in the limit in which $v, \tilde{\delta} \to 0$ the coupled-mode equation reduce to the nonlinear Schrödinger equation, using a derivation which closely follows that of de Sterke and Sipe [1990a]. In this limit, therefore, the coupled-mode equations are integrable (Dodd, Eilbeck, Gibbon and Morris [1982], Drazin and Johnson [1989]), and are independent of the value of Γ_s. Before doing so, we stress that the nonlinear Schrödinger equation is a valid description of gap solitons in this limit even if the coupled-mode equations are not. In fact, its applicability can be shown based directly on the wave equation [eq. (3.14)], and thus does not require that the periodic component is weak (Sipe and Winful [1988], de Sterke and Sipe [1988]). We return to this point below.

Before tackling the full nonlinear coupled equations, we must find explicit expressions for the electric field in the linear limit. In doing so, we introduce a notation which will be particularly useful in this context. Let us now reconsider the linear coupled-mode equations [eqs. (3.9)]. Following de Sterke and Sipe [1990a], we introduce the column vector

$$F(z, t) \equiv \begin{bmatrix} E_+(z, t) \\ E_-(z, t) \end{bmatrix}. \tag{6.1}$$

We then write eqs. (3.9) in the form:

$$[O + \sigma^1 \kappa] F(z, t) = 0, \tag{6.2}$$

where we have introduced the matrix operator

$$O \equiv i\sigma^3 \frac{\partial}{\partial z} + i \frac{\bar{n}}{c} \sigma^0 \frac{\partial}{\partial t}, \tag{6.3}$$

and we use two of the Pauli spin matrices,

$$\sigma^1 \equiv \begin{bmatrix} 0 & 1 \\ 1 & 0 \end{bmatrix}, \quad \sigma^3 \equiv \begin{bmatrix} 1 & 0 \\ 0 & -1 \end{bmatrix}, \tag{6.4}$$

as well as the unit matrix σ^0. We now seek solutions of the form

$$F(z,t) = f e^{-i(\Omega t - Qz)}. \tag{6.5}$$

Of course, this is identical to eq. (3.10), leading to an eigenvalue problem as in eq. (3.11), with a solution which in turn leads to the dispersion relation as in eq. (3.12). The eigenvectors can similarly be found to be:

$$f^{(+)} = \frac{1}{\sqrt{2\sqrt{\kappa^2 + Q^2}}} \begin{bmatrix} \frac{\kappa}{\sqrt{\sqrt{\kappa^2 + Q^2} - Q}} \\ -\sqrt{\sqrt{\kappa^2 + Q^2} - Q} \end{bmatrix} \equiv \begin{bmatrix} f_+^{(+)} \\ f_-^{(+)} \end{bmatrix},$$

$$f^{(-)} = \frac{1}{\sqrt{2\sqrt{\kappa^2 + Q^2}}} \begin{bmatrix} \sqrt{\sqrt{\kappa^2 + Q^2} - Q} \\ \frac{\kappa}{\sqrt{\sqrt{\kappa^2 + Q^2} - Q}} \end{bmatrix} \equiv \begin{bmatrix} f_+^{(-)} \\ f_-^{(-)} \end{bmatrix}, \tag{6.6}$$

and have been chosen to be normalized. Note that as $Q \to \infty$, we have $(f_+^{(+)}, f_-^{(+)}) \to (1, 0)$ and $(f_+^{(-)}, f_-^{(-)}) \to (0, 1)$. In this limit, far detuned from the Bragg condition, the solutions are essentially plane waves of a field propagating to the left or the right. On the other hand, for $Q = 0$, $(f_+^{(+)}, f_-^{(+)}) = 2^{-1/2}(1, -1)$ and $(f_+^{(-)}, f_-^{(-)}) = 2^{-1/2}(1, 1)$. Here, right at the Bragg resonance, the solutions are standing waves, such that the solution at the lower frequency concentrates the energy where the dielectric constant is highest [eq. (3.4)]. Finally, note from eqs. (3.5) and (6.5) that the electric field $E(z,t)$ may be written as $u_{k+}(z) \exp[i(kz - \omega t)] + \text{c.c.}$ for the Ω_+ solution, where $k = k_0 + Q$ and $\omega = \omega_0 + \Omega_+$, and as $u_{k-}(z) \exp[i(kz - \omega t)] + \text{c.c.}$ for the Ω_- solution, where $k = k_0 + Q$ and $\omega = \omega_0 + \Omega_-$, with

$$u_{k\pm}(z) = [f_+^{(\pm)} + f_-^{(\pm)} e^{-2ik_0 z}]. \tag{6.7}$$

Since $u_{k\pm}(z+d) = u_{k\pm}(z)$, these take the form of Bloch solutions for the photonic bands of eq. (3.12); their existence is guaranteed by the form of eqs. (3.2) and (3.3), just as in the corresponding electron problem in solid-state physics (Ashcroft and Mermin [1976]). What we have done is to find their explicit form within the approximation of coupled-mode theory. We stress that these Bloch functions are exact solutions to the linear problem –

they fully account for the scattering between E_+ and E_- (Ashcroft and Mermin [1976]).

We now return to the fully nonlinear coupled-mode equations [eqs. (3.17)], which we write by analogy with eq. (6.2) as

$$[O + \sigma^1 \kappa] F + V = 0, \tag{6.8}$$

where the nonlinear operator V is

$$V = \begin{bmatrix} \Gamma_S |E_+|^2 E_+ + 2\Gamma_\times |E_-|^2 E_+ \\ \Gamma_S |E_-|^2 E_- + 2\Gamma_\times |E_+|^2 E_- \end{bmatrix}. \tag{6.9}$$

Our Bloch solutions satisfy eq. (6.2), but clearly do not satisfy eq. (6.8). However, we seize upon the idea that, if the nonlinearity is weak, its overall effect should be to modulate a Bloch function, an exact solution to the linear problem, on length and time scales much longer than k_0^{-1} and ω_0^{-1}, respectively. This leads to the idea of searching for solutions which vary on several different length and time scales. To implement this so-called *method of multiple scales* (Nayfeh [1973], Dodd, Eilbeck, Gibbon and Morris [1982], de Sterke and Sipe [1988, 1990a]), we first introduce a new set of spatial and temporal variables:

$$z_n = \mu^n z, \qquad t_n = \mu^n t, \tag{6.10}$$

where $\mu \ll 1$, for all non-negative integers n. We treat both the z_n and t_n as independent variables; this then separates the different scales. In terms of these variables, derivations are written in the form:

$$\frac{\partial}{\partial z} = \frac{\partial}{\partial z_0} + \mu \frac{\partial}{\partial z_1} + \mu^2 \frac{\partial}{\partial z_2} + \cdots, \tag{6.11}$$

and similarly for t, showing clearly that for increasing n, the z_n describe variations on longer and longer length scales. We now search for solutions of eq. (6.8) of the form (de Sterke and Sipe [1990a]):

$$F(z, t) = [\mu a(z_i; t_i) f^{(+)} + \mu^2 b_2(z_i; t_i) f^{(-)} \\ + \mu^3 b_3(z_i; t_i) f^{(-)} + \cdots] e^{-i(\Omega_+ t_0 - Qz_0)}, \tag{6.12}$$

with $i = 1, 2, 3, \ldots$, and where the functions $a(z_i; t_i)$ and $b_j(z_i; t_i)$ are yet to be determined. Since the nonlinearity is assumed to be weak, these functions are expected to be slowly varying. We therefore refer to them as "envelope functions". Since $\mu \ll 1$, we assume that the soliton solutions consist mainly of a Bloch function on the upper Ω_+ branch, with Q and Ω_+ being fixed (in a similar fashion, we could have looked for a solution consisting mainly of a Bloch function on the lower branch). This Bloch function is modulated over long time scales by the nonlinearity, and a small amount of the other Bloch function mixed in as well. We can write eq. (6.12) as

$$F(z, t) = \mu F_1 + \mu^2 F_2 + \mu^3 F_3 + \cdots, \qquad (6.13)$$

where

$$F_1 = a(z_i; t_i) F^{(+)}(z_0, t_0), \qquad F_2 = b_2(z_i; t_i) F^{(-)}(z_0, t_0) e^{-i(\Omega_+ - \Omega_-)t_0},$$
$$F_3 = b_3(z_i; t_i) F^{(-)}(z_0, t_0) e^{-i(\Omega_+ - \Omega_-)t_0}. \qquad (6.14)$$

After writing

$$O = O_0 + \mu O_1 + \mu^2 O_2 + \cdots, \qquad (6.15)$$

where

$$O_n = i\sigma^3 \frac{\partial}{\partial z_n} + i \frac{\bar{n}}{c} \sigma^0 \frac{\partial}{\partial t_n}, \qquad (6.16)$$

[cf. eqs. (6.3) and (6.11)], we then substitute eqs. (6.13) and (6.15) into the coupled-mode equation [eq. (6.8)], and identify terms with corresponding powers of μ. Because the lowest order appearing in the nonlinear operator V is μ^3 [eq. (6.9)], $V = \mu^3 V_3 + \cdots$, for $i = 1, 2, 3$ we find:

$$[\sigma^1 \kappa + O_0] F_1 = 0, \qquad [\sigma^1 \kappa + O_0] F_2 + O_1 F_1 = 0,$$
$$[\sigma^1 \kappa + O_0] F_3 + O_1 F_2 + O_2 F_1 + V_3 = 0. \qquad (6.17)$$

The first of these equations is identically satisfied [see eq. (6.2)]. For the second equation, we must evaluate

$$[\sigma^1 \kappa + O_0] F_2 = b(z_i; t_i) [\sigma^1 \kappa + O_0] e^{-i(\Omega_+ - \Omega_-)t_0} F^{(-)}(z_0, t_0)$$
$$= b(z_i; t_i) \frac{\bar{n}}{c} [\Omega_+ - \Omega_-] \sigma^0 e^{-i(\Omega_+ - \Omega_-)t_0} F^{(-)}(z_0, t_0), \qquad (6.18)$$

where we have used eq. (6.2). The second of eqs. (6.17) then yields the matrix equation

$$b(z_i; t_i) \frac{\bar{n}}{c} [\Omega_+ - \Omega_-] \sigma^0 f^{(-)} + O_1 a(z_i; t_i) f^{(+)} = 0. \qquad (6.19)$$

To obtain two scalar equations, we dot eq. (6.19) into the vectors $f^{(+)}$ and

$f^{(-)}$. For this and for later operations, we shall need the identities:

$$f^{(-)} \cdot f^{(+)} = f^{(+)} \cdot f^{(-)} = 0, \qquad f^{(-)} \cdot f^{(-)} = f^{(+)} \cdot f^{(+)} = 1,$$

$$f^{(+)} \cdot \sigma^3 \cdot f^{(+)} = -f^{(-)} \cdot \sigma^3 \cdot f^{(-)} = \frac{Q}{\sqrt{\kappa^2 + Q^2}},$$

$$f^{(+)} \cdot \sigma^3 \cdot f^{(-)} = f^{(-)} \cdot \sigma^3 \cdot f^{(+)} = \frac{\kappa}{\sqrt{\kappa^2 + Q^2}}, \tag{6.20}$$

which follow immediately from the definitions. Dotting eq. (6.19) into $f^{(+)}$, and using eqs. (6.20), we find that

$$\Omega'_+ \frac{\partial a}{\partial z_1} + \frac{\partial a}{\partial t_1} = 0, \tag{6.21}$$

where we have put

$$\Omega'_+ = \frac{d\Omega}{dQ} = \frac{c}{\bar{n}} \frac{Q}{\sqrt{\kappa^2 + Q^2}}; \tag{6.22}$$

Ω'_+ is the group velocity of a Bloch function on the upper branch with wavenumber Q. Equation (6.21) shows that the envelope function a cannot depend independently on z_1 and t_1. Indeed, defining the two new variables,

$$\zeta_1 = z_1 - \Omega'_+ t_1, \qquad \tau_1 = t_1, \tag{6.23}$$

we see from eq. (6.21) that the function a does not depend on τ_1, so $a = a(\zeta_1; z_2, ...; t_2, ...)$. We thus obtain the reasonable result that, to this order, the evelope function a can take on any shape, but must travel at the group velocity associated with the Bloch function it modulates. We now dot eq. (6.19) into $f^{(-)}$ and find that $b_2 = b_2(\zeta_1; z_2, ...; t_2, ...)$ as well, and that

$$b_2(\zeta_1; z_2, ...; t_2, ...) = -\frac{1}{2} i \frac{\kappa}{\kappa^2 + Q^2} \frac{\partial a}{\partial \zeta_1}. \tag{6.24}$$

Turning now to the third of eqs. (6.17), we must first evaluate V_3. Using eqs. (6.14) in eq. (6.9), we can write:

$$V_3 = \tfrac{1}{2}|a|^2 a \left[(\Gamma_s + 2\Gamma_\times)\sigma^0 + (\Gamma_s - 2\Gamma_\times) \frac{Q}{\sqrt{\kappa^2 + Q^2}} \sigma^3 \right] f^{(+)}. \tag{6.25}$$

Dotting our final equation into $f^{(-)}$ leads to an equation for b_3 which does not concern us here; dotting it into $f^{(+)}$, on the other hand, leads to:

$$i\frac{\partial a}{\partial \tau_2} + \frac{1}{2}\Omega''_+ \frac{\partial^2 a}{\partial \zeta_1^2} + \alpha|a|^2 a = 0, \tag{6.26}$$

where we introduced ζ_2, τ_2 as in eqs. (6.23), and where

$$\Omega''_+ = \frac{d^2\Omega_+}{dQ^2} = \frac{c}{\bar{n}} \frac{\kappa^2}{(\kappa^2 + Q^2)^{3/2}} \tag{6.27}$$

is the group-velocity dispersion of the upper-branch Bloch function at Q. Furthermore, in eq. (6.26) the coefficient α, which is given by

$$\alpha = \frac{1}{2}\frac{c}{\bar{n}}\left[(\Gamma_s + 2\Gamma_\times) + (\Gamma_s - 2\Gamma_\times)\left(\frac{\Omega'_+}{c/\bar{n}}\right)\right], \tag{6.28}$$

describes the effect of the nonlinearity at this level.

We stop at this point, and let $\mu \to 1$ in the usual spirit of perturbation analyses. Then eq. (6.26) becomes simply the nonlinear Schrödinger equation,

$$i\frac{\partial a}{\partial \tau} + \frac{1}{2}\Omega''_+ \frac{\partial^2 a}{\partial \zeta^2} + \alpha|a|^2 a = 0, \tag{6.29}$$

where in terms of the laboratory coordinates we have:

$$\zeta = z - \Omega'_+ t, \qquad \tau = t. \tag{6.30}$$

This result [eq. (6.29)] was first obtained by Sipe and Winful [1988] and by de Sterke and Sipe [1988, 1990a].

From eq. (6.29), we know that for the positive nonlinearity ($\alpha > 0$) we are considering here, a soliton solution can be constructed by modulating any Bloch function on the upper branch where $\Omega_+ > 0$. Similarly, for a negative nonlinearity ($\alpha < 0$), soliton solutions can be constructed by modulating Bloch functions on the lower branch of the dispersion curve. This asymmetry between states on the upper and lower branches was pointed out earlier in fig. 2 where frequencies above ω_0 (like ω_h) are tuned easily out of the gap, while for frequencies like ω_l this is much more difficult.

At this point in the discussion, it is useful to give a qualitative description of the approximations leading to eq. (6.29) (de Sterke and Sipe [1988, 1990a]). The key to this can be found in the *ansatz* in eq. (6.12): since $\mu \ll 1$, we assume here that a Bloch function on the upper branch dominates the electric field. This restriction to the use of the nonlinear Schrödinger equation is best illustrated in the stationary limit in which a is assumed to be varying harmonically (cf. § 2). Clearly, eq. (6.29) cannot be used when the frequency of the field is, for instance, in the middle of the photonic band gap, since at this frequency the Bloch functions of the upper and the lower branches are equally important. For eq. (6.29) to be valid, therefore, the frequency must be close to the upper edge of the gap. Considering now electric fields with a more general time dependence, we see that for eq. (6.29) to be valid, the frequency content must be concentrated around the upper edge of the gap. This precludes not only fields with frequency elsewhere in the gap, but also, for example, short optical pulses with a spectral width of the same order of magnitude as the size of the gap.

We now pause briefly and take a step back to compare the quite different "philosophies" between the approach taken in the present section (Sipe and Winful [1988], de Sterke and Sipe [1988, 1990a]), and that of Aceves and Wabnitz [1989] described in § 5. It is important to recall that we are giving a quantitive description of media with a periodic component in the linear refractive index, as well as a nonlinearity in the refractive index. The two philosophies differ in the way these two effects are taken into account. The approach taken in the present section could be described as a "solid-state physics" approach. After all, use is made of Bloch functions, which are the eigenfunctions of periodic linear media, and the nonlinearity is accounted for by modulating these functions. Clearly, in this approach the nonlinearity and the periodicity are on a different footing: one could say that the periodicity is taken into account exactly, while the nonlinearity is treated as a perturbation. This should be contrasted with the "optics" approach in § 5. There, the electric field is written as a superposition of forward- and backward-propagating modes (which are the eigenmodes for *uniform* media), and both the periodicity and the nonlinearity are accounted for by modulations of the eigenfunctions. The nonlinearity and the periodicity are thus treated on an equal (perturbative) footing.

We have thus shown in this section that, for the range of frequencies discussed above, the description of the electric field in terms of forward- and backward-propagating waves is not the most efficient. Instead, it is better to write the electric field in terms of Bloch functions. This has the advantage of reducing the coupled-mode equation – actually, a set of two equations – to just a single equation, the nonlinear Schrödinger equation. Since in this approach the periodicity is accounted for exactly, it must be able to deal with systems with arbitrarily deep index modulations (de Sterke and Sipe [1988], Ashcroft and Mermin [1976]). In this case, one would not use the coupled-mode equations at all (since these require the modulation to be weak), but start directly from the wave equation for the electric field. As shown by de Sterke and Sipe [1988], this approach also leads to the conclusion that the electric-field envelope satisfies the nonlinear Schrödinger equation.

In the discussion above, we mentioned that the multiple-scales method we used represents a "solid-state" approach to the problem at hand. This is a notion which can be put more concretely: de Sterke and Sipe [1989b] have shown that in the limit of harmonic time-dependence and vanishing group velocity, the results of the multiple-scale approach are identical to those which would follow from a standard "effective mass" treatment (see, e.g., Callaway [1974]), a method used widely in solid-state physics. However, the derivation of the effective mass formalism is quite different than the method of multiple scales used here. It thus sheds a different light on the approximation made. In fact, the standard effective mass derivation quantifies the "slowness" at which the envelope function must vary for the approximation to be valid; viz., its Fourier components should be negligible at a third of the width of the Brillouin zone, or, in other words, the envelope functions must vary slowly on the scale of three lattice constants (de Sterke and Sipe [1989b]).

In the next section we will discuss the connection between the results above and those in § 5. In particular, we will show that the one-soliton solutions to the nonlinear Shrödinger equations can be found as special cases of the solitary-wave solutions to the coupled-mode equations.

§ 7. Discussion of Analytic Results

In § 6, we showed that under suitable conditions, the envelope function of the electric field in a nonlinear periodic medium satisfies the nonlinear Schrödinger equation, and the envelope function thus has soliton solutions. We next investigate the relationship between these solitons and the solutions of the coupled-mode equations discussed in § 5. For convenience, we henceforth refer to all of these solutions as "gap solitons", although the latter, strictly speaking, are solitary waves.

The one-soliton solution to the nonlinear Schrödinger equation [eq. (6.29)] reads (Dodd, Eilbeck, Gibbon and Morris [1982], Drazin and Johnson [1989], de Sterke and Sipe [1988, 1990a]):

$$a(\zeta, \tau) = \sqrt{\frac{C_1^2}{\Omega''_+ \alpha}} \exp\left[i\left(C_2\zeta - \frac{C_2^2\tau}{2} + \frac{C_1^2\tau}{2}\right)\bigg/\Omega''_+\right] \operatorname{sech}\left[\frac{C_1}{\Omega''_+}(\zeta - C_2\tau)\right],$$
(7.1)

where the velocities C_1 and C_2 can in principle be chosen freely. Clearly, C_2 determines the soliton's velocity, while C_1 determines its amplitude and width. However, since a is an envelope function which is assumed to vary slowly, we must require that $C_1, C_2 \ll c/\bar{n}$. In the simple case of $C_2 = 0$, the soliton travels with velocity Ω'_+. If, in contrast, $Q = 0$, then $\Omega'_+ = 0$, and the soliton moves with velocity C_2. A soliton with given velocity can thus be constructed from a range of suitable combinations of Ω'_+ and C_2, each corresponding to a different combination of Bloch function and envelope function, chosen such that the actual electric field (i.e., the product of these functions) is unchanged. Here we choose to modulate the Bloch function at $Q = 0$, where $\Omega'_+ = 0$, and the coordinates ζ, τ reduce to the laboratory coordinates z, t. Furthermore, $\Omega_+ = c\kappa/\bar{n}$, $\Omega''_+ = c/(\bar{n}\kappa)$, and, from eqs. (6.6), $f_\pm^T = 2^{-1/2}(1, \mp 1)$. If we further set $C_1 = cp/\bar{n}$ and $C_2 = cq/\bar{n}$, so that use of the envelope-function approach requires that $p, q \ll 1$, we find for the one-soliton solutions to eq. (6.29) that

$$F(z,t) = \sqrt{\frac{\pm \kappa}{2\Gamma_\times + \Gamma_s}} \, e^{\pm i\phi_1} \, \text{sech}(\phi_2)$$

$$\times \left\{ \begin{bmatrix} 1 \\ \pm 1 \end{bmatrix} + \frac{1}{2}(q + ip \tanh(\phi_2)) \begin{bmatrix} 1 \\ \mp 1 \end{bmatrix} \right\}, \quad (7.2)$$

where we used eq. (6.12), and where we defined

$$\phi_1 = q\kappa z + (c/\bar{n})\kappa t(-1 + \tfrac{1}{2}(p^2 - q^2)), \qquad \phi_2 = p\kappa(z - (c/\bar{n})qt). \quad (7.3)$$

It is straightforward to show that if $p, q \ll 1$, these expressions are identical to the solutions found in § 5, as required (Aceves and Wabnitz [1989], de Sterke and Sipe [1990a]). Of course, the nonlinear Schrödinger equation also has higher-order soliton solutions (Dodd, Eilbeck, Gibbon and Morris [1982], Drazin and Johnson [1989]), and thus in the appropriate limit the coupled-mode equations must allow such solutions as well. Furthermore, the massive Thirring model discussed in § 5 also has higher-order soliton solutions (David, Harnad and Shnider [1984]). Although higher-order soliton solutions are known to exist in two limiting cases, it is at present unknown whether they have a general counterpart among the solutions to the coupled-mode equations.

To finish this section, we will discuss briefly the relationship between gap solitons and the "regular" optical-fiber solitons (see, e.g., Agrawal [1989]). The common characteristic of these two classes of solitons is that both are envelope functions which satisfy the nonlinear Schrödinger equation in appropriate limits. The first difference is that for gap solitons the envelope function modulates Bloch functions, while for regular solitons it modulates unidirectional modes. The most important *practical* difference, however, is that the group-velocity dispersion Ω''_+ is determined mainly by the periodicity in the case of gap solitons, while for regular solitons it is determined mainly by the dispersion of glass. The former is much larger than the latter – indeed, the group velocity Ω'_+ in a periodic structure varies between $\pm c/\bar{n}$, while in an optical fiber the group velocity is always dominated by the properties of the glass. As a consequence, gap solitons can in principle attain any velocity between zero and $\pm c/\bar{n}$, while regular solitons always travel at velocities around $c/1.5$. More generally, however, because of the strong curvature of the dipersion relation of periodic media, gap solitons are very sensitive to perturbations. While this may not be a desirable property if one wants to transfer information between different points, it may be highly desirable if those perturbations are applied intentionally for the purpose of, say, studying nonlinear dynamics, or possibly for device design.

We have thus come to the conclusion that one can use either of two methods to describe the properties of nonlinear periodic media. The first of these are the coupled-mode equations. These are valid anywhere in the photonic stop gap, but require that the refractive-index modulation is weak. In the other approach, one applies the method of multiple scales directly to the wave equation. In this way one can deal with systems with arbitrarily deep index modulations, but one is restricted to electric fields with frequency content close to the upper edge of the stop gap (if the nonlinearity is positive). Our experience thus far has been that the coupled-mode approach appears to be the most practical approach in many applications. The reason for this is not related to the ranges of validity. Actually, the fact that coupled-mode theory makes direct use of the forward- and backward-propagating mode makes it the preferred approach for relating electric fields in embedding media to those in the periodic nonlinear medium; the embedding media are uniform in general, and in such media the forward- and backward-propagating modes do not mix.

§ 8. Numerical Results

In this section, we will discuss numerical simulations of the coupled-mode equations, following Winful and Cooperman [1982], de Sterke and Sipe [1990b], and Winful, Zamir and Feldman [1991]. In our particular numerical simulations we make use of a fourth-order collocation method to integrate over the characteristics of eqs. (3.17). This results in an efficient and numerically very robust procedure. For details, the reader is referred to de Sterke, Jackson and Robert [1991]. It should be noted that this procedure represents a "genuine" simulation, in that the input variable is the amplitude of the incoming radiation and not the energy flow through the system. Although the latter was the central quantity in calculating the stationary solutions, it plays no particular role when the full time dependence is included.

One of the aims of numerical work is to simulate realistic experimental geometries. In doing so, the scaling relationships given at the end of § 3 (de Sterke and Sipe [1990b]) are most useful. They imply that an experimental geometry can be simulated with any system which has the same grating strength κL; the response for the actual experimental system can then be found by choosing suitable values for β_1 and β_2. Once one knows the field strengths in the actual system, one can calculate the incoming intensity, I, from the well-known relation

$$I = \frac{2\bar{n}}{Z} |E_+|^2, \tag{8.1}$$

where the factor 2 must be included because [by eq. (3.5)] the actual field is taken to be twice the real part of the complex field.

Most of the simulations pertain to the response of the nonlinear periodic medium to an external electric field, either (quasi) stationary or pulsed. An important part of these efforts is to ascertain the correctness and the stability of the analytic solutions found in § 4, § 5, and § 6. Aceves and Wabnitz [1989] found that the soliton/solitary-wave solutions from § 5 and § 6 are stable, a result that has been confirmed by other workers. Indeed, these solutions can be seen to propagate through nonlinear periodic media without changing shape. Recently, Aceves, De Angelis and Wabnitz [1992] have also investigated the modulational instability of the nonlinear coupled-mode equations and its possible relation to the formation of gap solitons.

We investigated the switching behavior of nonlinear periodic media which can be expected on the basis of the stationary solutions to the coupled-mode equations (de Sterke and Sipe [1990b]), following earlier work by Winful and Cooperman [1982]. We now give a brief description of our own simulations of that switching. To start, recall fig. 9, which shows the transmissivity as a function of the incoming intensity. It shows a low- and a high-transmissivity branch, connected by a branch which is known to be unstable. In our fully time-dependent calculations, we first let the system settle in a state on the lower branch with a certain value of the amplitude of the incoming radiation A. Such a dynamic equilibrium is well described by the analysis in § 4. Once the system settles, we raise the incoming amplitude slightly, and again we let the system come to a dynamic equilibrium, so that we can follow the branch "adiabatically". In fact, the small increment is smeared out and thus not even applied instantaneously. Recall now that according to fig. 9, the system "jumps" off the lower branch for a certain value of the incoming intensity. *If* the system then settles on the upper branch, the incoming intensity is decreased slowly so that the system follows adiabatically the upper branch, until it jumps back to the lower branch, essentially confirming the stationary results. In this case, the time-dependent solutions only add information regarding switching times. However, this is found to be an atypical situation! In fact, the work of Winful and Cooperman [1982], as well as our own (de Sterke and Sipe [1990b]), indicates that the system often does not settle on the upper branch at all. Rather, the system exhibits self-pulsations which are either periodic, or, at high intensities, chaotic. Winful, Zamir and Feldman [1991] have attributed this side-band instability to a four-photon parametric oscillation taking place inside the structure.

Figure 10a shows the transmitted amplitude $\sqrt{T}|A|^2$ as a function of time (in round-trip times $T_r = 2\bar{n}L/c$) when the state of the system jumps from the lower branch and does settle on the upper branch; the system now does not exhibit instability. Similarly, fig. 10b shows the transmitted amplitude versus time when the system jumps back from the upper branch to the lower branch. For these examples, the parameters were taken to be identical to those in figs. 8 and 9 ($\kappa = 5$, $L = 1$, $\hat{\delta} = 4.75$, and $\Gamma = 0.1$). In both figs. 10a and 10b, the switching is seen to take many round-trip times, even though the response of the material is assumed to be instantaneous. To understand this sluggish response, recall that the field distribution inside the structure changes profoundly when switching occurs. As discussed in § 4, on the lower branch the system behaves essentially as if it were linear and has an approximately exponential field distribution, while on the upper branch the nonlinearity dominates, leading to a field distribution which exhibits a maximum (see figs. 7 and 9). The sluggish behavior now occurs simply because a large amount of energy at a frequency close to the Bragg condition must be relocated in the system.

We now give an example of a situation in which the sysem does *not* settle in a state on the upper branch. This can be accomplished by choosing a frequency which is somewhat deeper inside the stop gap of the photonic structure. Figures 11a and 11b show examples of the transmitted amplitude

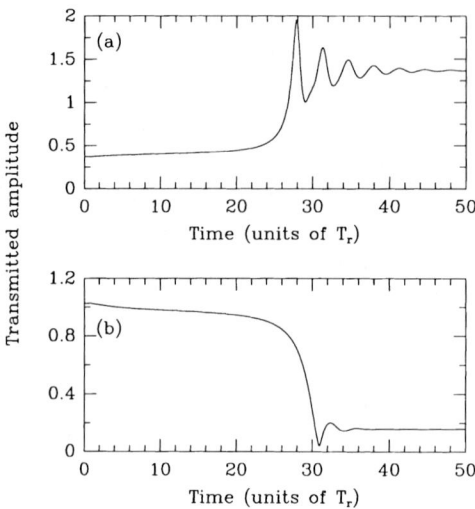

Fig. 10. (a) Time dependence of the transmitted amplitude during the transition from the low-transmission to the high-transmission state as in fig. 9, when the incoming intensity is about 2.5. The origin of time in this figure is determined by the intensity change which forces the system to jump. (b) Time dependence of the transmitted amplitude during the transition from the high-transmission to the low-transmission state as in fig. 9, when the incoming intensity is about unity.

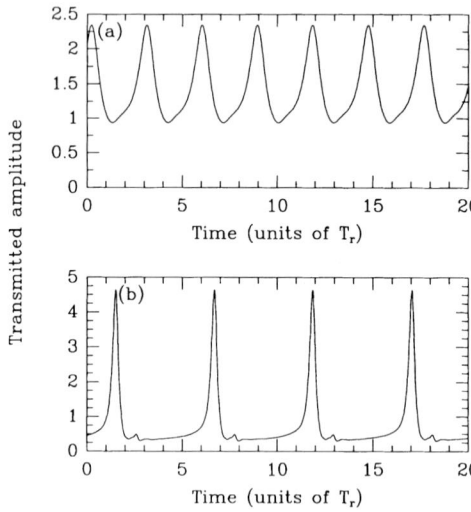

Fig. 11. Time dependence of the transmitted amplitude during the transition from the low-transmission to the high-transmission state for a system with $\kappa = 5$, $L = 1$, $\Gamma_x = \Gamma_s = 0.1$, and $\hat{\delta} = 4.5$ (a), $\hat{\delta} = 4.0$ (b), clearly showing that the system exhibits self-pulsations.

which results when the system with parameters as in figs. 10 jumps off the lower branch, for a frequency such that $\hat{\delta} = 4.5$ (fig. 11a), and $\hat{\delta} = 4.0$ (fig. 11b). The figures show clearly that the system is unstable and exhibits self-pulsations. In fact, de Sterke and Sipe [1990b] have argued that the instability leads to a periodic train of gap solitons which are created near the front end of the system. They subsequently move towards the back, where they disintegrate upon encountering the back surface of the system. They then release their energy in a single burst, which gives rise to the structure in the transmitted signal shown in fig. 11. Work by Winful and co-workers, and by ourselves, has shown that if the detuning from the upper edge of the band gap is increased further, the transmitted signal remains "spiky", but undergoes a period-doubling, ultimately leading to chaos (Winful and Cooperman [1982], de Sterke and Sipe [1990b]). Since this has been discussed sufficiently in the references cited above, we will not do so here.

We have seen that the system may fail to settle in a state on the high-transmission branch. In order to find out when to expect this behavior, and to find the associated pulsation frequency, a linear stability analysis was applied by de Sterke [1992b]. In such an analysis, one investigates whether small deviations from the stationary solutions grow or decay in time; growth of these deviations would indicate instability. Since in a linear stability analysis these deviations are assumed to be small; thus giving rise to a linear problem, such analyses cannot indicate chaotic behavior. In fig. 12 we show the result of such an analysis for a system with parameters as in figs. 10 ($\kappa = 5$, $L = 1$, and $\Gamma = 0.1$). The dotted lines in the figure indicate the extent of the bistable region as a function of frequency. They show that the bistability vanishes for frequencies close to the upper edge of the photonic gap at $\hat{\delta} \approx 1.06\kappa$, and that the bistable region increases monotonically for frequencies deeper inside the gap. The solid line indicates the border between stable states on the upper branch (clear) and those states on the upper branch which are unstable against the band formation (dotted). It shows that for $\hat{\delta} < 0.75\kappa$, the entire upper branch is unstable, preventing bistable switching for these frequencies. One way to avoid this may be the use of a material with a sluggish nonlinear response; in the example given in their paper, Winful and Cooperman [1982] have shown that the self-pulsation vanishes if the system's response time exceeds about a single round-trip time of the structure. Winful, Zamir and Feldman [1991] have also suggested that the instability may be turned to advantage by using it in a periodic optical pulse generator. The pulse frequency in such a structure would be

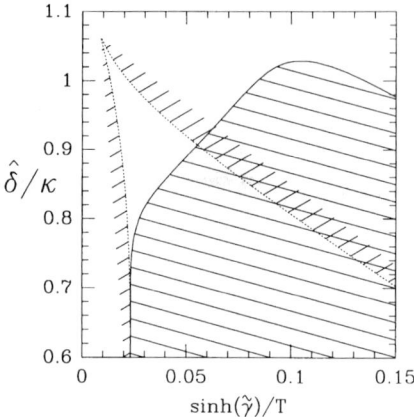

Fig. 12. Results of a stability analysis for a nonlinear stack with $\kappa L = 5$ and $\Gamma_x = \Gamma_s = 0.1$; $\hat{\delta}/\kappa$ is a normalized detuning, while $\sinh \hat{\gamma}/T$ is a normalized incoming intensity. Shown below the solid curve as a function of the detuning of the incoming light, is the region where the system is unstable against side-band formation. The bistable region is indicated by the two dotted lines. Bistable switching can thus be accomplished in the clear, triangularly shaped region.

tunable by the frequency of the incoming radiation, as well as by its frequency.

§ 9. The Coupling Problem

In this section, we will dicuss briefly some of the issues involved in launching gap solitons in the laboratory. Since this requires efficient coupling of light from the incoming medium into the nonlinear periodic structure, we refer to this as the "coupling problem". The main difficulty was alluded to at the end of § 5; viz., for the wavelengths of interest, the structure would normally (i.e., at low intensities) be highly reflective. Computer simulations have shown that the coupling of light into a nonlinear periodic medium can happen in either of two ways. In the first of these, one considers a very powerful pulse to be incident on a nonlinear periodic medium; if the pulse is strong enough, then an instability occurs inside the medium, similar to that discussed in § 8. While this instability occurs, energy transfer into the medium is very efficient, and a large fraction of the incoming energy then enters the nonlinear periodic medium. This energy propagates towards the other end of the medium, and while it propagates it attains the properties of a gap soliton, shedding energy which is not required. The main characteristic of this rather complicated process is that while the system is in a stable state the reflection is high, but when the instability occurs the reflection diminishes. This leads to the appearance of low-intensity "bites" taken out of the reflected intensity. For a semi-infinite geometry, this result is illustrated in figs. 13. Figure 13a shows the intensity of the incoming light (dotted line), and of the reflected light (solid line), as functions of time for the geometry which will be discussed in detail in § 10. Note the two "bites" taken out of

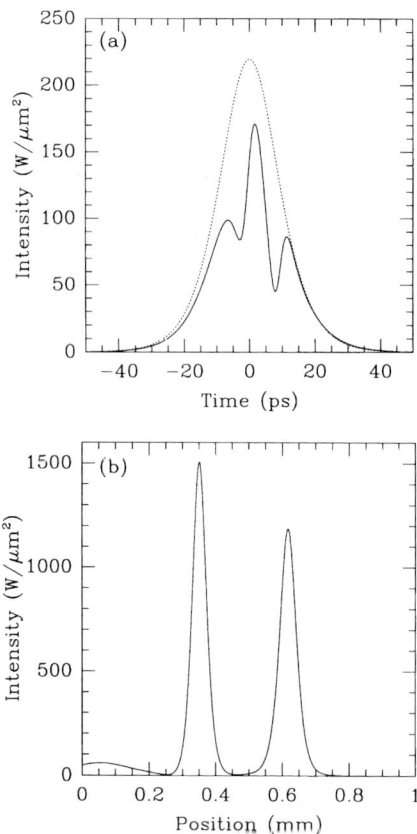

Fig. 13. (a) Typical example of the instability-mediated formation of a gap soliton. Shown are the intensities of the incoming radiation (dotted line) and reflected radiation (solid line) as functions of time. Note the "bites" taken out of the reflected signal. These indicate the occurrence of an instability which allows energy to enter the structure, thus leading to a temporary drop in the reflection. The energy that has entered the structure leads to gap solitons, as shown by the intensity inside the structure (b) at $t = 21$ ps. The interface is taken to be at the origin.

the reflected signal, indicating efficient energy coupling into the system. This is borne out by fig. 13b, in which the field intensity inside the structure at $t = 21$ ps indicates that two gap solitons have been formed. The peak intensity of the gap solitons is much larger than that of the incoming pulse; this is due to the instability which mediates the formation of the gap solitons. In spite of their high peak intensity, the gap solitons carry only a small fraction of the energy of the incoming pulse, because they are (spatially) much narrower. Since the instability plays a central role in this process, the parameters of the gap solitons which are formed cannot, at present, be predicted.

The other method which can be used to launch a gap soliton, and the method discussed in more detail in the remainder of this section, was mentioned briefly in § 6. Because the electric field associated with a gap soliton is known, one can determine exactly the external field required to launch it. In this case, a gap soliton is formed immediately, and no propagation is required to shed the undesired energy. Similar to figs. 13, figs. 14 show the intensity of the incoming and reflected energy when a gap soliton is formed in this way. Note especially that the intensity of the reflected light does not exhibit the behavior observed in fig. 13a; viz., it is a smooth function with the same shape as that of the incoming radiation. Note that the peak intensity of the gap soliton which is formed (fig. 14a) is about twice as large as that of the incoming pulse, consistent with de Sterke and Sipe [1989a]. In spite of this, only a small fraction of the incoming energy enters the medium (recall that the gap soliton is much narrower than the incoming pulse). The low coupling efficiency is consistent with eq. (5.11) and the discussion following it; for low-velocity gap solitons, the fraction of the incoming energy entering the medium is approximately $2v$. Finally, in contrast to the method described above, for the method illustrated in figs. 14 the parameters of the incoming pulse allow one to launch a gap soliton with prescribed properties, in this case $v = 0.04$ and $\delta = 0.10\pi$. In the remainder of this section, we will discuss additional features of this launching process.

As discussed in the previous sections, gap solitons have two free parameters; viz., the detuning parameters $\tilde{\delta}$, and a velocity v. We now first argue that the entire parameter space described by $\tilde{\delta}$ and v is not of primary interest. As a criterion, we use the condition that the soliton be "manifestly nonlinear". By this we mean if the same field were incident on the same medium *without* the nonlinearity, then the results would be clearly different. This requirement limits the frequency content of the soliton, for if there is too much energy at frequencies outside the photonic stop gap, then the

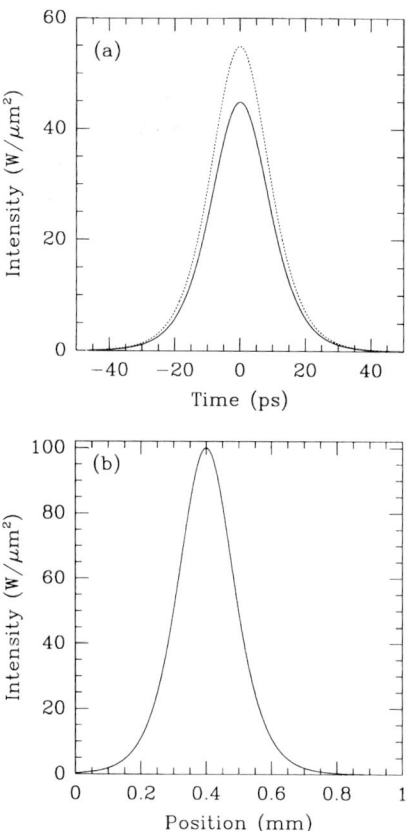

Fig. 14. Typical example of the formation of a gap soliton by a specially chosen incident pulse, as discussed in the text. Note that, in contrast to the situation in fig. 13, the reflected intensity (solid curve) in (a) follows the incident intensity (dotted curve) closely, apart from an overall multiplicative factor. The meaning of the curve in (b) is the same as in fig. 13b.

soliton is not manifestly nonlinear. In searching for manifestly nonlinear gap solitons, we use two approaches. The first of these is an approximation leading to analytic results, while the second is purely numerical and is exact.

The requirement formulated in the previous paragraph leads to two obvious conditions for the frequency content of the gap soliton: (i) the center frequency of the solitary wave should be inside the gap, and (ii) the width of the frequency spectrum should be less than the width of the gap. Of course, these requirements are necesary, but they are not sufficient. They do, however, lead easily to analytic results. An exact argument with sufficient conditions follows below, and gives qualitatively similar results. The main difference between the two arguments occurs at large values for $\tilde{\delta}$. We now consider separately the requirements formulated under (i) and (ii) below.

(i) Equations (5.1), (5.2), and (5.5) show that the dominating time dependence is

$$e^{i\sigma} = e^{i\gamma\kappa\cos\tilde{\delta}(vz-ct/\bar{n})}, \tag{9.1}$$

so that the center frequency Ω, of the gap soliton is given by

$$\Omega = \frac{c}{\bar{n}} \frac{\kappa\cos\tilde{\delta}}{\sqrt{1-v^2}}. \tag{9.2}$$

We now require that this center frequency be inside the gap. Since the gap extends from $-c\kappa/\bar{n}$ to $+c\kappa/n$,

$$\left| \frac{\kappa\cos\tilde{\delta}}{\sqrt{(1-v^2)}} \right| < \kappa. \tag{9.3}$$

It is easy to show that this is equivalent to

$$v < \sin\tilde{\delta}. \tag{9.4}$$

(ii) To obtain the extent of the spectrum of the gap soliton, we need the Fourier transform of

$$\text{sech}(\theta \pm i\tilde{\delta}/2). \tag{9.5}$$

Using contour integration, it is found that the Fourier transform is proportional to

$$\text{sech}\left(\frac{\pi\bar{\omega}}{2}\right) e^{-\bar{\omega}\tilde{\delta}/2}, \tag{9.6}$$

where

$$\bar{\omega} = \frac{\bar{n}}{c} \frac{\omega - \Omega}{\kappa\gamma v \sin\tilde{\delta}}. \tag{9.7}$$

As mentioned above, the width of the frequency spread should be less than the width of the gap, which equals $2\kappa c/n$, or approximately

$$\kappa\gamma v \sin\tilde{\delta} < \kappa. \tag{9.8}$$

This is easily found to be equivalent to

$$\frac{v}{\sqrt{1-v^2}} \sin\tilde{\delta} < 1. \tag{9.9}$$

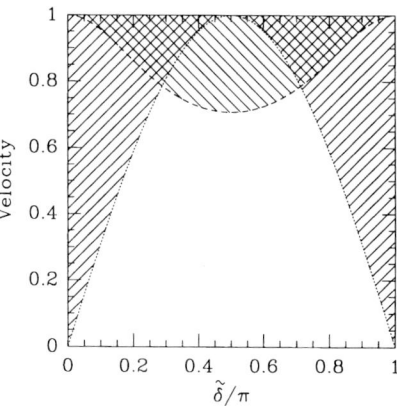

Fig. 15. Graphical representations of eq. (9.10), showing approximately the range of parameters $\tilde{\delta}$ and v for which gap solitons are manifestly nonlinear (clear region).

Combining now the two results obtained above [eqs. (9.4) and (9.9)] leads to:

$$v < \sin \tilde{\delta} < \frac{\sqrt{1-v^2}}{v}. \qquad (9.10)$$

These conditions are shown in fig. 15, the dotted line representing the equality in eq. (9.4), and the dashed line representing the equality in eq. (9.9). The permissible region is clearly indicated. The main conclusion from this argument is that the region of very high velocities is excluded. From eq. (9.10) it is found that, independent of the parameter $\tilde{\delta}$, solitary waves with velocities higher than $\sqrt{\frac{1}{2}(1+\sqrt{5})} \approx 0.79$ are not manifestly nonlinear. The reason, of course, is that a large fraction of its frequency content lies outside the photonic stop gap, and the nonlinear nature of the response of the medium is then not sufficiently prominent.

We will now discuss briefly some numerical results. To obtain these, the Fourier transform of gap solitons was taken, and the fraction of the energy within the stop gap was determined as a function of the parameters $\tilde{\delta}$ and v. The curves in fig. 16 show for which choice of parameters a fixed fraction (viz., 50% and 90%) of the energy is inside the gap. These figures lead to the same conclusion as the simple qualitative discussion above; the region of high velocities must be excluded, since too much energy is at frequencies outside the gap. It is satisfying that the high velocities are excluded, since the most characteristic property of gap solitons is that they can travel slowly!

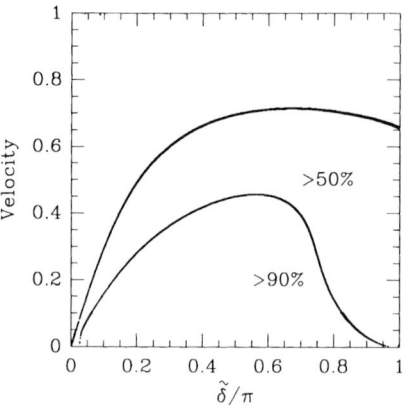

Fig. 16. Range of parameters δ and v for which 50% and 90% of the gap soliton's energy falls inside the photonic band gap. These exact results are in qualitative agreement with the approximate results in fig. 15 in that gap solitons with high velocities are not manifestly nonlinear.

We now use the fact that gap-soliton solutions can be traced back in time to give the external field which would be ideally necessary to generate it. We again consider a semi-infinite geometry; the field is incident from the left through a uniform linear medium with refractive index $\bar{n}$. The interface between this medium and the nonlinear periodic medium, which is taken to be semi-infinite, is taken to be at $z=0$. Neglecting index jumps between the incoming medium and the periodic nonlinear structure, it is seen easily that the required incoming field at the interface (e.g., $z=0$) is given by $E_+(0,t)$, while the reflected field at the interface equals $E_-(0,t)$. Since the incident field travels at a velocity $c/\bar{n}$, the amplitude of the incoming field at an arbitrary position in the uniform medium must equal

$$E_+(0, t - z\bar{n}/c)\mathrm{e}^{-\mathrm{i}(\omega_0 t - k_0 z)} + \text{c.c.}, \tag{9.11}$$

where E_+ was defined in eq. (5.5). At an arbitrary position in the uniform medium, the generated reflected field is given by:

$$E_-(0, t + z\bar{n}/c)\mathrm{e}^{-\mathrm{i}(\omega_0 t - k_0 z)} + \text{c.c.} \tag{9.12}$$

(recall that the incident medium is semi-infinite and extends to the left of the interface which is at $z=0$), where E_- was defined in eq. (5.5) as well. Expressions (9.11) and (9.12) can, of course, also be used to confirm eq. (5.11), which gives the fraction of the incoming energy which is reflected when an incident pulse is incident on a nonlinear periodic medium. As also discussed in the paragraph below eq. (5.11), this equation allows one to define a transmission coefficient $\mathcal{T}$ given by

$$\mathcal{T} = \frac{2v}{1+v}. \tag{9.13}$$

We now combine the results obtained thus far. Recall that according to the inequality described by eq. (9.10), high-velocity gap solitons are not acceptable. On the other hand, it is advantageous to have efficient energy transfer into the material; the transmission $\mathcal{T}$ should thus be as large as possible. From eqs. (9.4) and (9.13), one therefore requires that

$$v = \sin \tilde{\delta} \approx \tilde{\delta}, \tag{9.14}$$

where the last relation holds because high velocities have been rejected.

Next we include energy considerations. Of course, the energy content of an external pulse should exceed the energy of the desired solitary wave. The latter is evaluated easily, since it is proportional to the conserved quantity Q. With our definitions of the envelope functions [eq. (3.5)] the electromagnetic energy, W_r, required to generate a solitary wave with prescribed values of the parametes $\tilde{\delta}$ and v is

$$W_r = 2\varepsilon Q, \tag{9.15}$$

where ε is the (average) permittivity of the structure, or with eqs. (5.10) and (3.18),

$$W_r = \frac{2\bar{n}\tilde{\delta}}{n^{(2)}\omega} \frac{1}{1 + \frac{1}{2}\frac{1+v^2}{1-v^2}}. \tag{9.16}$$

Note that the energy content of a gap soliton does not depend on κ; the energy contained by a gap soliton with given parameters thus does not depend on the modulation depth of the periodic structure. Now consider the second multiplying factor on the right-hand side of eq. (9.16). It shows that the energy content of a gap soliton drops at high velocities. However, since by the inequality of eq. (9.10) high velocities are unacceptable, this factor is always of order unity for the velocities we are interested in. For this reason, we take this factor to have the value 2/3, which it attains at vanishing velocity. Making this substitution in eq. (9.16) leads to:

$$W_r = \frac{4\bar{n}\tilde{\delta}}{3n^{(2)}\omega}, \tag{9.17}$$

showing that in the limit under consideration, the gap soliton's energy content depends only on the detuning.

Now we finally combine all previous results to find a criterion for the external energy required to launch a gap soliton. Taking the ratio of eq. (9.17), the energy content of a gap soliton, and eq. (9.13), the efficiency of energy transfer over the interface, and keeping in mind that for the gap solitons of interest $v = \tilde{\delta}$ [eq. (9.14)], it is found that the energy per unit area, W_i, of the incoming field should satisfy the surprisingly simple criterion:

$$W_i \approx \frac{2n}{3n^{(2)}\omega}, \qquad (9.18)$$

independent of the modulation depth of the grating, and independent of v and $\tilde{\delta}$, as long as v and $\sin\tilde{\delta}$ are substantially below unity. This criterion expresses the fact that for desirable gap solitons, the parameter dependence of the energy content and of the energy transfer over the interface just cancel, so that the external energy is independent of the gap soliton's parameters.

Finally, following earlier work (de Sterke and Sipe [1989b]) we have also considered the product $\Delta n_L \Delta n_{NL} \mathcal{N}^2$, where $\mathcal{N}$ is the number of periods in the structure, Δn_L is the modulation depth of the grating, and Δn_{NL} is the induced nonlinear index change. Using the results derived above, it is straightforward to show that a necessary condition for the launching of gap solitons is that

$$\Delta n_L \Delta n_{NL} \mathcal{N}^2 > 1, \qquad (9.19)$$

which is consistent with the result of de Sterke and Sipe [1989b].

Recall that the requirements formulated above are necessary to observe gap solitons which are manifestly nonlinear. This condition leads to the rejection of gap solitons which travel too rapidly. However, in order to launch gap solitons which truly travel much slower than the speed of light – indeed, this characteristic is one of the hallmarks of gap solitons – the ">" signs above should be replaced by "≫". It should be noted that the results above are valid if one wants to minimize the peak intensity of the incoming light. However, for very strong gratings, this would lead to unrealistically long incoming pulses. Under these conditions, therefore, one might want to use another, more severe, criterion to optimize the launching conditions.

We thus see that the coupling of light with a wavelength close to the Bragg wavelength is quite inefficient between a uniform linear medium and a periodic nonlinear medium. In a rather different context, such issues have been discussed in the design of thin-film stacks (Macleod [1969]). There it is considered to be associated with a mismatch of the impendances of the two media involved. A remedy suggested to alleviate this problem is to make use of materials with a refractive index which differs from $\bar{n}$ (Macleod [1969]), leading to Fresnel reflections at the interfaces. It has been shown that with large contrast ratios, the coupling efficiency can be improved considerably if one chooses carefully the thickness and the index of the first layer of the nonlinear periodic structure. Following standard thin-film design procedures, it is even possible to improve considerably the coupling efficiency by adding an "anti-reflection" layer of suitable thickness between the two media. The reader is referred to de Sterke [1992a] for details.

Haus [1992] has recently discussed the use of a second grating to accomplish impedance matching, and thus to facilitate the launching of gap solitons. His approach appears to be most suitable to solitons with a fequency content outside the photonic band gap, and may thus be less appropriate to launch manifestly nonlinear gap solitons. More recently, it was discovered that the launching of gap solitons can also be facilitated by using non-uniform gratings. In particular, it has been found in numerical simulations that by tapering the front of the grating (i.e., by letting the modulation depth of the gratings increase gradually to its maximum value), or by chirping it (i.e., letting the period of the grating gradually attain its final value), the power requirements of the incoming radiation can be reduced considerably (de Sterke and Sipe [1993]). This work is in its initial stages at the time of this writing, and we will therefore not discuss it here in more detail.

As a final point, it should be mentioned that this "tracing back" in time to find the external field required to generate a gap soliton, was applied first by de Sterke and Sipe [1989a] using the results of the multiple-scales method described in § 6.

§ 10. Experimental Geometries

As mentioned in § 1, Sankey, Prelewitz and Brown [1992] observed bistable switching in a Si–SiO$_2$ waveguide structure, which was first predicted to occur by Winful, Marburger and Garmire [1979]. Their samples are strongly nonlinear due to free-carrier effects. In fact, their effective $n^{(2)} \approx 10^{-10}$ cm^2/W, which is more than three orders of magnitude larger than the electronic nonlinearity in GaAs, and more than five orders of magnitude larger than that of silica glass. However, optical nonlinearities associated with the creation of free carriers, although large, have a sluggish response which is determined by the carrier lifetime. In the samples used by Sankey, Prelewitz and Brown [1992], this lifetime is between 1 and 5 ns, only slightly smaller than the length of their laser pulses.

The geometry used by Sankey, Prelewitz and Brown [1992] to observe switching in nonlinear periodic media is schematically shown in fig. 17. The light is guided in the silicon layer. Since the structure has a silicon substrate the modes, strictly speaking, are leaky. However, the leakage may be made arbitrarily small by taking the SiO_2 buffer layer to be sufficiently thick. The middle grating in fig. 17 is where the gap-soliton effects are expected to occur, while the auxiliary gratings on the left and right are used to couple the light in and out of the guide; they allow for the simultaneous observation of the light-reflected and transmitted by the middle grating. They make use of a Q-switched Nd:YAG laser, emitting at 1.064 μm. Their pulses have a length of 25–50 ns, and the pulse energy is roughly between 1 and 20 μJ. The reader is referred to the original paper for more details.

In fig. 18 we show one of the experimental results obtained by Sankey, Prelewitz and Brown [1992] with the structure described above. The upper trace in the figure represents the incoming pulse, which has a width of 25 ns and an energy of 5 μJ. The solid lower trace gives the reflected intensity, and the dotted lower curve gives the transmitted intensity. It is clear that,

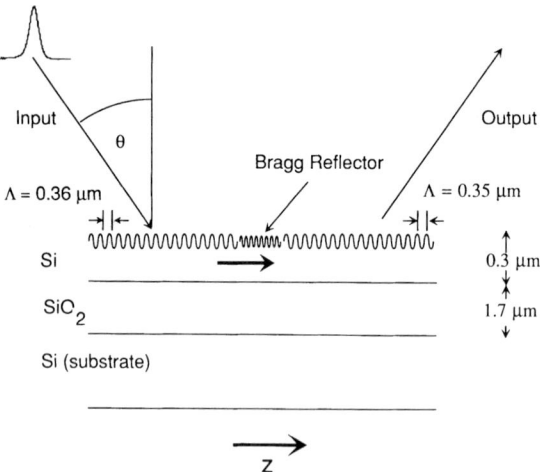

Fig. 17. Schematic of the waveguide geometry used by Sankey, Prelewitz and Brown [1992] to observe bistable switching.

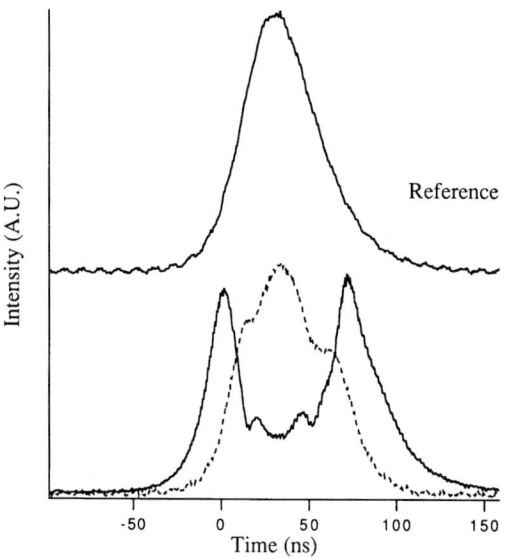

Fig. 18. Experimental result obtained by Sankey, Prelewitz and Brown [1992] showing nonlinear switching in the geometry from fig. 17. The upper trace gives the incoming intensity as a function of time, while the solid and dotted lower traces give the reflected and transmitted intensities, respectively, all in arbitrary units.

as expected, at low intensities the periodic structure acts as a mirror, while at higher intensities the nonlinearity shifts the photonic band gap, leading to strong transmission. When shorter pulse are used, Sankey et al. observe additional effects, which are thought to be due to sample heating and carrier recombination effects. However, the detailed interpretation has not yet been worked out and we will not comment further on it here.

As mentioned above, the experimental results of Sankey, Prelewitz and Brown [1992] are obscured somewhat by auxiliary effects such as carrier recombination and sample heating. A possible alternative geometry, in which such effects would play less of a role, is shown in fig. 19. Such structures, developed by AT&T Bell laboratories (Slusher, Islam, Soccolich, Hobson, Pearton, Tai, Sipe and de Sterke [1991]), are made of GaAs and AlGaAs, and the nonlinearity is of electronic origin. As a consequence, it is only about 2×10^{-14} cm^2/W, much smaller than that of Sankey et al., but the response time is of the order of tens of femtoseconds. It therefore acts essentially instantaneously if laser pulses longer than roughly 1 ps are used. The structure shown in fig. 19 has a period of 2700 Å. The relative width of the gap $\Delta\lambda/\lambda_0 \approx \delta n/n \approx 0.03/3.1 \approx 0.01$. It has a 1 μm guiding layer of

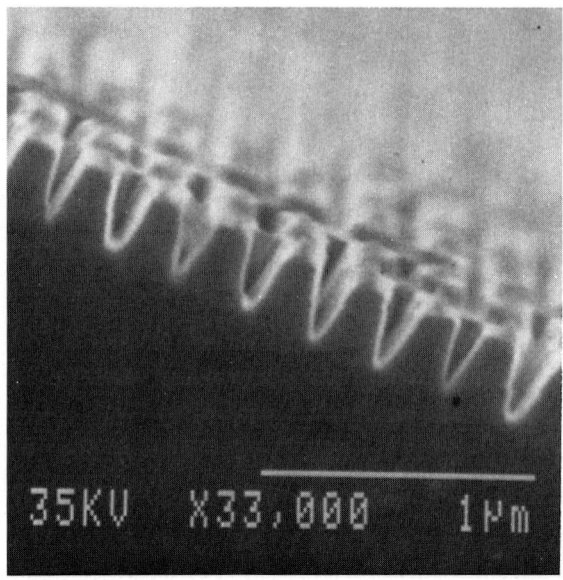

Fig. 19. Scanning electron microscope image of GaAs–AlGaAs structure which has been proposed for gap-soliton experiments. The parameters of the structure are discussed in the text.

$Al_{0.2}Ga_{0.8}As$ ($n = 3.27$ at $\lambda = 0.5$ μm) and $Al_{0.4}Ga_{0.6}As$ ($n = 3.075$ at $\lambda = 0.5$ μm) over a 2.5 μm $Al_{0.8}Ga_{0.2}As$ ($n = 2.98$) layer, all grown by molecular beam epitaxy on a GaAs wafer. Note in fig. 19 the large modulation depth, which is approximately equal to the period, and which results in the relatively large index modulation. The gratings were produced by etching patterns in photo-resist exposed to interfering laser beams. To avoid two-photon absorption in such structures, one must use light with wavelengths below half the band gap of the constituents, as discussed by Ho, Soccolich, Islam, Levi and Slusher [1991]. For this reason it is proposed that a color center laser be used, which emits at wavelengths around 1.60 μm. Although gap solitons have not been observed in such structures, attempts to detect them are expected to commence in the near future.

§ 11. Outlook

In this review we have tried to give an overview of the main issues concerning theoretical and experimental aspects of gap solitons. While the efforts in the last few years have been mostly theoretical, the recent observation by Sankey, Prelewitz and Brown [1992] of all-optical switching in nonlinear periodic media, first predicted to occur by Winful, Marburger and Garmire [1979], has shown that experiments in this area are possible. We would expect that in the near future other predicted effects will also be verified experimentally. Of particular interest would be the observation of low-velocity propagation of light through such media, first predicted by Aceves and Wabnitz [1989]. It appears, at present, that the geometry of Sankey, Prelewitz and Brown [1992] is not well suited for those types of experiments because of the sluggishness of their nonlinearity; systems with an electronic nonlinearity, which has a much faster response time, are better suited for this purpose.

Finally, we briefly mention some work on closely related topics. Winful [1985] proposed that the pulse reshaping which occurs when light propagates through nonlinear periodic media can be used for pulse compression. Trutschel and Lederer [1988] considered stationary solutions to the coupled-mode equations with more general nonlinearities than a simple Kerr nonlinearity. Dutta Gupta [1989] also studied stationary solutions, but at non-normal incidence. Wabnitz [1989] has pointed out that the coupling of modes propagating in the *same* direction in periodic structures (e.g., coupling between orthogonally polarized modes) gives rise to the same equations [eqs. (3.17)], but with space and time interchanged. He therefore pointed out that gap-soliton-like solutions are also expected to occur in structures exhibiting this forward mode coupling. Coste and Peyraud [1989a, b], and Peyraud and Coste [1989] also considered the properties of a nonlinear periodic medium, but from a dynamical system perspective. In their first paper, Coste and Peyraud [1989a] used a simple model in which the medium was thought of as consisting of δ-function-like nonlinear sheets, separated by vacuum. This allowed them to simplify the analysis considerably, but it is not clear whether the observed behavior is general, or a particular feature of the model adopted. In this and their other papers, these authors showed that wave propagation in nonlinear periodic media exhibits a multitude of instabilities. Our own numerical simulations of the coupled-mode equations, in contrast, have shown a remarkable degree of stability of gap solitons. At this point, it is not clear whether this is due to the investigation of different parts of parameter space, or to the particular description employed. Danckaert, Fobelets, Veretennicoff, Vitrant and Reinisch [1991] described a very general, exact method to determine stationary solutions of nonlinear periodic structures, and compared their results to those from envelope-function approaches. Further, Bilbault and Remoissenet [1991] have shown that nonlinear electrical superlattices comprised of nonlinear capacitors sandwiched between linear electrical transmission lines exhibit very similar behavior to that described in this review. In particular, they observed hyperbolic secant-shaped field profiles, in agreement with the analysis described here. However, they also observed a substantial amount of third-harmonic generation; we would expect third-harmonic generation to be less prominent in the optical domain, as different dispersion characteristics make phase matching less likely. Cada, Acklin, Proctor, He and Dupertuis [1991] and He and Cada [1991] have also considered switching behavior in nonlinear periodic media. However, these authors did not make use of the shifting of the photonic band gap, but of the Fabry–Pérot-like fringes which occur just outside the gap. The switching behavior which Cada, Acklin, Proctor, He and Dupertuis [1991] observed, and discussed theoretically (He and Cada [1991]), is thus not associated with gap-soliton formation. De Sterke and Sipe [1991] and Steel and de Sterke [1993] studied the propagation of gap solitons through media with gain as well as ordinary and two-photon loss. We should also mention the work of Herbert and Malcuit [1992] and of Herbert, Capinsky and Malcuit [1992], who used colloidal crystals to obtain an optical limiting effect. To do so, the light is tuned to be just outside the photonic band gap at low intensities; however, due to a thermal nonlinearity, it is shifted into the gap at higher intensities, leading to an increased reflection.

In conclusion, we hope to have shown that the study of the properties of nonlinear periodic media is fruitful and interesting. Moreover, while in the past most efforts have been theoretical, it is also our hope that the injection of results of recent experiments and the anticipated results of future experiments will bring additional vigor to the field.

Acknowledgements

Jointly, we wish to thank: A. B. Aceves, T. G. Brown, V. Mizrahi, N. D. Sankey, R. E. Slusher, M. J. Steel, G. Town, and H. G. Winful for many useful discussions. We are grateful to T. G. Brown for providing us with figs. 17 and 18, and to R. E. Slusher for providing us with fig. 19. This work was supported by the Australian Research Council, a University of Sydney Research Grant, the Natural Sciences and Engineering Research Council of Canada, and by the Ontario Laser and Lightwave Research Centre.

References

Abramowitz, M., and I.A. Stegun, 1970, Handbook of Mathematical Functions (Dover, New York) ch. 16.
Aceves, A.B., and S. Wabnitz, 1989, Phys. Lett. A **141**, 37.
Aceves, A.B., C. De Angelis and S. Wabnitz, 1992, Opt. Lett. **17**, 1566.
Agrawal, G.P., 1989, Nonlinear Fiber Optics (Academic Press, San Diego).
Ashcroft, N.W., and N.D. Mermin, 1976, Solid State Physics (Holt, Rinehard and Winston, New York).
Bilbault, J.M., and M. Remoissenet, 1991, J. Appl. Phys. **70**, 4544.
Born, M., and E. Wolf, 1980, Principles of Optics, 6th Ed. (Pergamon Press, Oxford).
Byrd, P.F., and M.D. Friedman, 1954, Handbook of Elliptic Integrals for Physicists and Engineers (Springer, Berlin).
Cada, M., B. Acklin, M. Proctor, J. He and M.A. Dupertuis, 1991, Investigation of a nonlinear periodic structure: the first step towards a new type of optical bistable devices, in: Integrated Photonics Research, Monterey, CA, April 9–11, 1991, postdeadline papers, p. 47.
Callaway, J., 1974, Quantum Theory of the Solid State (Academic Press, New York) section 5.3.1.
Chen, W., and D.L. Mills, 1987a, Phys. Rev. Lett. **58**, 160.
Chen, W., and D.L. Mills, 1987b, Phys. Rev. B **36**, 6269.
Chen, W., and D.L. Mills, 1987c, Phys. Rev. B **35**, 524.
Christodoulides, D.N., and R.I. Joseph, 1989, Phys. Rev. Lett. **62**, 1746.
Coste, J., and J. Peyraud, 1989a, Phys. Rev. B **39**, 13086.
Coste, J., and J. Peyraud, 1989b, Phys. Rev. B **39**, 13096.
Danckaert, J., K. Fobelets, I. Veretennicoff, G. Vitrant and R. Reinisch, 1991, Phys. Rev. B **44**, 8215.
David, D., J. Harnad and S. Shnider, 1984, Lett. Math. Phys. **8**, 27.
de Sterke, C.M., 1992a, Phys. Rev. A **45**, 2012.
de Sterke, C.M., 1992b, Phys. Rev. A **45**, 8252.
de Sterke, C.M., and J.E. Sipe, 1988, Phys. Rev. A **38**, 5149.
de Sterke, C.M., and J.E. Sipe, 1989a, Opt. Lett. **14**, 871.
de Sterke, C.M., and J.E. Sipe, 1989b, Phys. Rev. A **39**, 5163.
de Sterke, C.M., and J.E. Sipe, 1990a, Phys. Rev. A **42**, 550.
de Sterke, C.M., and J.E. Sipe, 1990b, Phys. Rev. A **42**, 2858.
de Sterke, C.M., and J.E. Sipe, 1991, Phys. Rev. A **43**, 2467.
de Sterke, C.M., and J.E. Sipe, 1993, Opt. Lett. **18**, 269.
de Sterke, C.M., K.R. Jackson and B.D. Robert, 1991, J. Opt. Soc. Am. B **8**, 403.
Dodd, R.K., J.C. Eilbeck, J.D. Gibbon and H.C. Morris, 1982, Solitons and Nonlinear Wave Equations (Academic Press, London).
Drazin, P.G., and R.S. Johnson, 1989, Solitons: An Introduction (Cambridge University Press, Cambridge).
Dutta Gupta, S., 1989, J. Opt. Soc. Am. B **6**, 1927.
Gibbs, H.M., 1985, Optical Bistability: Controlling Light with Light (Academic Press, Orlando, FL).
Haus, H.A., 1992, Opt. Lett. **17**, 1134.
He, J., and M. Cada, 1991, IEEE J. Quantum Electron. **QE-27**, 1182.
Herbert, C.J., and M.S. Malcuit, 1992, Optical bistability in nonlinear periodic structures, Optical Society of America Annual Meeting, paper WQ1.
Herbert, C.J., W.S. Capinsky and M.S. Malcuit, 1992, Opt. Lett. **17**, 1037.
Ho, S.T., C.E. Soccolich, M.N. Islam, A.F.J. Levi and R.E. Slusher, 1991, Appl. Phys. Lett. **59**, 2558.

John, S., 1987, Phys. Rev. Lett. **65**, 2486.
Kaup, D.J., and A.C. Newell, 1977, Lett. Nuovo Cim. **20**, 325.
Kawasaki, B.S., K.O. Hill, D.C. Johnson and Y. Fujii, 1978, Opt. Lett. **3**, 66.
Kogelnik, H., 1969, Bell Sys. Techn. J. **48**, 2909.
Kuznetsov, E.A., and A.V. Mikhailov, 1977, Teor. Mat. Fiz. **30**, 193.
Macleod, H.A., 1969, Thin Film Optical Filters (Hilger, London).
Meltz, G., W.W. Morey and W.H. Glenn, 1989, Opt. Lett. **14**, 823.
Mills, D.L., 1991, Nonlinear Optics (Springer, Berlin).
Mills, D.L., and S.E. Trullinger, 1987, Phys. Rev. B **36**, 947.
Mizrahi, V., and J.E. Sipe, 1992, Strong photosensitive phase gratings in optical fibres, Optical Society of America Annual Meeting, paper ThWW3.
Nayfeh, A., 1973, Perturbation Methods (Wiley, New York).
Peyraud, J., and J. Coste, 1989, Phys. Rev. B **40**, 12201.
Sankey, N.D., D.F. Prelewitz and T.G. Brown, 1992, Appl. Phys. Lett. **60**, 1427.
Shen, Y.R., 1984, The Principles of Nonlinear Optics (Wiley, New York).
Sipe, J.E., and G.I. Stegeman, 1979, J. Opt. Soc. Am. **69**, 1676.
Sipe, J.E., and H.G. Winful, 1988, Opt. Lett. **13**, 132.
Slusher, R.E., M.N. Islam, C.E. Soccolich, W. Hobson, S.J. Pearton, K. Tai, J.E. Sipe and C.M. de Sterke, 1991, Gap solitons in buried Bragg grating AlGaAs waveguides, Optical Society of America Annual Meeting 1991, paper TuBB2.
Steel, M.J., and C.M. de Sterke, 1993, Phys. Rev. A **48**, 1625.
Trutschel, U., and F. Lederer, 1988, J. Opt. Soc. Am. B **5**, 2530.
Wabnitz, S., 1989, Opt. Lett. **14**, 1071.
Winful, H.G., 1985, Appl. Phys. Lett. **46**, 527.
Winful, H.G., and G.D. Cooperman, 1982, Appl. Phys. Lett. **40**, 298.
Winful, H.G., J.H. Marburger and E. Garmire, 1979, Appl. Phys. Lett. **35**, 379.
Winful, H.G., R. Zamir and S. Feldman, 1991, Appl. Phys. Lett. **58**, 1001.
Yablonovitch, E., 1987, Phys. Rev. Lett. **65**, 2059.

E. WOLF, PROGRESS IN OPTICS XXXIII
© 1994 ELSEVIER SCIENCE B.V.
ALL RIGHTS RESERVED

IV

DIRECT SPATIAL RECONSTRUCTION OF OPTICAL PHASE FROM PHASE-MODULATED IMAGES

BY

VALENTIN I. VLAD* and DANIEL MALACARA

Centro de Investigaciones en Óptica, A.C.,
Apdo. Post. 948,
37000 León, Gto., México

* Permanent address: The Institute of Atomic Physics, P.O. Box MG-6, R-76900 Bucharest, Romania.

CONTENTS

	PAGE
§ 1. INTRODUCTION	263
§ 2. DIRECT SPATIAL RECONSTRUCTION OF OPTICAL PHASE (DSROP) FROM FRINGE PATTERNS	265
§ 3. DIRECT SPATIAL RECONSTRUCTION OF OPTICAL PHASE (DSROP) FROM MORE GENERAL PHASE-MODULATED IMAGES	281
§ 4. DIRECT SPATIAL RECONSTRUCTION OF OPTICAL PHASE (DSROP) FOR TIME- AND SPACE-PHASE-MODULATED IMAGES	294
§ 5. AN OVERVIEW OF DSROP METHODS. RANGE OF MEASUREMENT, ACCURACY AND SPEED	298
§ 6. APPLICATIONS OF THE METHODS OF DSROP	302
§ 7. CONCLUDING REMARKS	311
ACKNOWLEDGEMENTS	313
REFERENCES	313

§ 1. Introduction

The spatial variations of the phase play a key role in the optical recording of wavefronts, optical testing and metrology, as well as optical information processing.

Spatial variations of the phase have been observed directly, in some approximation limits, since the nineteenth century within classical interferometric and schlieren work. Closer to our time, F. Zernike [1942] discovered the phase-contrast method, and D. Gabor [1949] the holographic method; both are recognized as outstanding scientific discoveries. Leith and Upatnieks [1962] have shown that in holography, the optical phase is recorded efficiently by the modulation of a spatial carrier frequency (tilted reference wavefront) and is reconstructed by a product demodulation of the phase-modulated pattern with the same spatial carrier frequency. Holography was not only a strongly developing research field, but also stimulated further research and analogies in the classical interferometry, moiré and speckle-pattern methods. Thus, encoding spatial phase-modulation in fringe patterns and analyzing, more or less automatically, these fringe patterns to reconstruct the phase information has been a useful and interesting exercise in the last twenty years, as is shown by the large number of papers in the field.

A proposed classification scheme for the methods of *fringe pattern* (FP) analysis, in order to reconstruct the spatial phase-modulation, involves at least four groups (Malacara [1992]):

Single pattern analysis
  (i) Fringe sampling with global and local interpolation.
  (ii) Space heterodyne demodulation of fringe patterns (direct-measuring interferometry).
  (iii) Fourier analysis of fringe patterns.

Multiple pattern analysis
  (iv) Phase-shifting methods.

The first group of methods (fringe sampling with global and local interpolation) uses the most interesting and sophisticated numerical analysis to process the FP as an ordinary image (Reid [1986, 1988], Womack, Jonas, Koliopoulos, Underwood, Wyant, Loomis and Hayslett [1979], Loomis [1979], Nakadate, Magome, Honda and Tsujiuchi [1981], Yatagai, Nakadate, Idesawa, Yamashi and Susuki [1982], Hayslett and Swantner [1980], Becker, Maier and Wegner [1982], Malacara, Carpio-Valadez and Sánchez-Mondragón [1990], Spik and Salbut [1992]). The fourth class of methods use several phase-shifted FPs to obtain the phase distribution directly (Carré [1966], Bruning, Herriott, Gallagher, Rosenfeld, White and Brangaccio [1974], Wyant [1975], Yatagai, Nakadate, Idesawa and Saito [1982], Hariharan, Oreb and Brown [1983], Willemin and Dandliker [1983], Greivenkamp [1984], Reid, Rixon and Messer [1984], Yatagai [1984], Wyant, Oreb and Hariharan [1984], Nakadate and Saito [1985], Thalmann and Dandliker [1985], Creath [1985, 1988], Nakadate [1986], Robinson and Williams [1986], Kujawinska, Spik and Wojciak [1989], Massie [1987], Rodriguez-Vera, Kerr and Mendoza-Santoyo [1991], Vrooman and Maas [1991], Greivenkamp and Bruning [1991], Joenathan [1991], Kwon [1992], Schmidt, Creath and Kujawinska [1992], Brown and Pryputniewicz [1992], Kujawinska [1992]).

The second and third classes of methods – also called spatial-carrier FP analysis – go back to the holographic type of phase-modulation and demodulation, implement the entire demodulation procedure in computer image processors and yield in a direct (fast) way the two-dimensional phase distribution which modulated the spatial carrier (reference FP).

The analogy between a phase-modulated FP and a hologram was recognized by Ichioka and Inuiya [1972], who also developed the first demodulation procedure in the space domain. Later on, this analogy was followed by many authors, who used their analyses in the space domain (Mertz [1983], Macy [1983], Womack [1984], Toyooka and Tominaga [1984], Ransom and Kokal [1986], Frankowski, Stobbe, Tischer and Schillke [1989]) and in the spatial-frequency (Fourier) domain (Takeda, Ina and Kobayashi [1982], Womack [1984], Nugent [1985], Kreis [1986], Bone, Bachor and Sandeman [1986], Ru, Honda, Tsujiuchi and Ohyama [1988], Roddier and Roddier [1987, 1991]). Recently, these groups of methods have been used in such important applications as optical testing, optical profilometry, plasma diagnostics and flow characterization (Takeda and Mutoh [1983], Dzubur and Vukicevic [1984], Takeda and Tung [1985], Takeda [1989], Kalal, Nugent and Luther-Davies [1987], Snyder and Hesselink [1988], Kujawinska, Spik and Wojciak [1989], Toyooka and Iwaasa [1986], Vlad, Popa and Apostol [1991], Malacara [1991], Suematsu and Takeda [1991],

Suganuma and Yoshizawa [1991], Parker [1991], Praeter and Swain [1992]) as well as in the design of new instruments such as the Carl Zeiss Interferometer Direct 100 (Freischland, Kuechel, Wiedman, Keiser and Mayer [1990], Kuechel [1990], Doerband, Wiedman, Wegmann, Kuechel and Freischland [1990], Freischland [1992]).

Considering the generality of holographic concepts, the increasing importance of fringe image analysis and the increasing interest in spatial-carrier FP computer processing, this work aims to:

(i) provide a complete analogy between the phase reconstruction in holography and in digitized phase-modulated images (outputs of interferometric, moiré and speckle-pattern instruments), generating new processing methods,

(ii) demonstrate the phase reconstruction for time- and space-phase-modulated images by real-time and double-exposure computer processing,

(iii) illustrate the concepts with computer experiments, and

(iv) show the importance of these concepts for a large number of applications.

We remark upon introducing the concept of arbitrary phase reference in the computer processing of phase-modulated images that we describe all these methods [from the classes (ii) and (iii)] as *direct spatial reconstruction of optical phase* (DSROP). By "direct" and "spatial", we are to understand that the phase-modulated image is detected directly by a high-resolution TV camera from an optical setup (i.e., without scanning or phase-shifting devices). By "reconstruction", we define the entire process of phase demodulation, unwrapping and representation by computer algorithms.

§ 2. Direct Spatial Reconstruction of Optical Phase (DSROP) from Fringe Patterns

A phase-modulated image can be defined as a bi-dimensional function of the form:

$$I(x, y) = A(x, y) + B(x, y) \cos \phi(x, y), \tag{2.1}$$

where $A(x, y)$ is a positive, slowly varying background, $B(x, y)$ is the positive, slowly varying contrast, and $\phi(x, y)$ is the phase-spatial distribution.

In the case of optical fringe patterns, there is often a spatial carrier frequency modulated by the object phase (Vest [1979], Ostrovsky, Butusov and Ostrovskaya [1982], Hariharan [1984], Jones and Wykes [1983]):

$$\phi(x, y) = 2\pi u_0 x + \phi_m(x, y), \tag{2.2}$$

where $u_0 = 1/\Lambda$ is the spatial carrier frequency and $\phi_m(x, y)$ is the phase modulation. Thus, many interferograms, holographic interferograms and moiré patterns can be described by the (detected) intensity pattern:

$$I(x, y) = A(x, y) + B(x, y) \cos[2\pi u_0 x + \phi_m(x, y)]. \tag{2.3}$$

All optical FPs strongly resemble the holographic recorded patterns. Actually, we can consider that the fringe pattern from eq. (2.3) is the result of the interference of an arbitrary optical wavefront (object) with a tilted plane reference wavefront with a tilt θ given by $\sin \theta/\lambda = u_0$, like in holography (fig. 1). The difference between holographic and DSROP methods consists of the reconstruction of a 3D image in the former and the reconstruction of the 3D phase map in the latter. In both cases, the phase demodulation can be analyzed in space and in spatial-frequency (Fourier) domains.

2.1. PHASE DEMODULATION IN THE SPACE DOMAIN

Following the holographic analogy, we can accomplish the phase demodulation by the multiplication of FP by the reference (fig. 1b):

$$R(x, y) = \exp(2i\pi u_0 x). \tag{2.4}$$

This reference can be obtained from (i) the external optical setup (two-plane wavefront interference, or reference grating image), (ii) the recorded FP with a zone without phase modulation, or (iii) a computer-generated pattern, with an a priori knowledge of the spatial carrier frequency, u_0. The result of the multiplication can be written in a simple way:

$$\begin{aligned} I(x, y) \cdot R(x, y) &= [A(x, y) + B(x, y) \cos(2\pi u_0 x + \phi_m)] \\ &\quad \times [\cos 2\pi u_0 x + i \sin 2\pi u_0 x] \\ &= A \exp(2i\pi u_0 x) + \tfrac{1}{2} B \exp[i(4\pi u_0 x + \phi_m)] \\ &\quad + \tfrac{1}{2} B (\cos \phi_m - i \sin \phi_m). \end{aligned} \tag{2.5}$$

The first two terms of eq. (2.5) are rapidly varying functions (if $u_0 \gg (\partial \phi_m/\partial x)_{\max}$). Thus, it is possible to obtain only the slowly varying third term by smoothing (integrating) this pattern over a space interval with the dimension $\Lambda_0 = 1/u_0$.

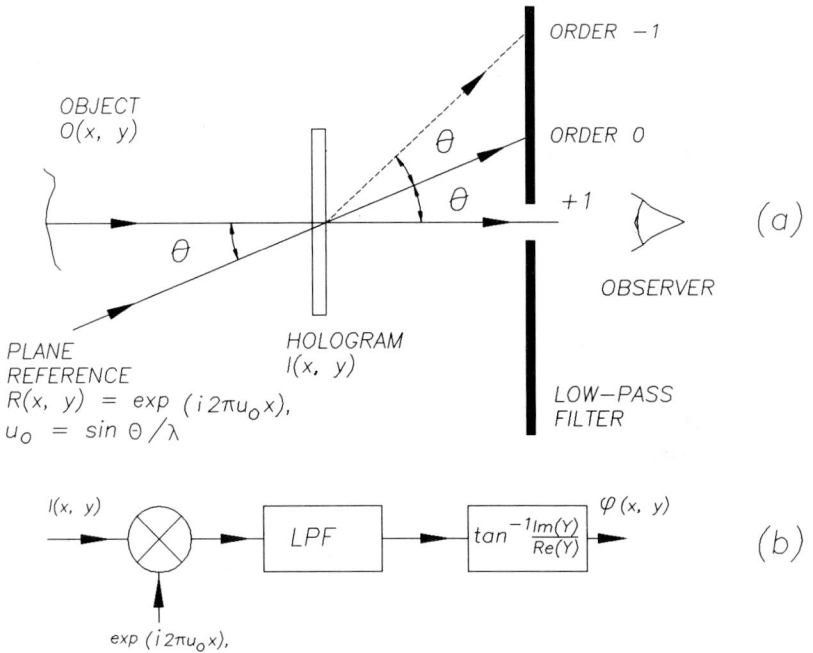

Fig. 1. (a) Holographic recording and reconstruction of the object $O(x, y) = O_0(x, y) \times \exp[i\phi(x, y)]$ with a tilted plane reference wave. The recording step yields a complex transmittance with spatial carrier:

$$I(x, y) = |O_0(x, y)|^2 + 1 + 2O_0(x, y) \cos[2\pi u_0 x - \phi(x, y)]$$
$$= A(x, y) + B(x, y) \cos[2\pi u_0 x + \phi_m(x, y)]$$

The reconstruction process is done by the multiplication of $I(x, y)$ by $R(x, y)$ and low-pass filtering of the result. (b) The steps of DSROP: the smoothing method in the space domain.

The smoothing method starts with the assumptions that the carrier frequency is taken as high as the numerical resolution in the image allows for; i.e., $u_0 = \frac{1}{3}$, and A, B and ϕ_m are approximately constant over the fringe period (Ichioka and Inuiya [1972], Toyooka and Tominaga [1984]). Then, integrating eq. (2.5) over an interval equal to the fringe spacing $\Lambda_0 = 1/u_0$, it is clear that all harmonic functions with frequencies u_0 and $2u_0$ vanish, and the results are

$$Y(x, y) = \frac{1}{\Lambda_0} \int_0^{\Lambda_0} I(x, y) \cdot R(x, y) \, dx$$

$$= \tfrac{1}{2} B[\cos \phi_m(x, y) - i \sin \phi_m(x, y)], \qquad (2.6)$$

and

$$\phi_m(x,y) = -\tan^{-1}\frac{\text{Im}[Y(x,y)]}{\text{Re}[Y(x,y)]}. \tag{2.7}$$

In this method we have used digital image smoothing procedures equivalent to the integration; i.e., average and median with masks 3×1 or larger, which are simple, fast and relatively accurate with mean errors of about 5% (fig. 2).

It is interesting to develop a smoothing method when the local phase is allowed to vary linearly over one fringe period:

$$\phi_{mi} = a_i + b_i x; \quad a_i, b_i = \text{const.}; \quad b_i \ll u_0. \tag{2.8}$$

In this case, by integration, the first two terms of eq. (2.5) vanish again (the second one vanishes due to the condition $b_i \ll u_0$), and it is possible to calculate:

$$\begin{aligned} C_{Ai} &= \frac{B_i u_0}{2}\int_0^{\Lambda_0}\cos(a_i + b_i x)\,\mathrm{d}x = \frac{B_i u_0}{2b_i}\int_{a_i}^{a_i+b_i\Lambda_0}\cos\theta\,\mathrm{d}\theta \\ &= \frac{B_i u_0}{2b_i}[\sin(a_i+b_i\Lambda_0)-\sin a_i] = \frac{B_i}{2b_i\Lambda_0}[\sin(a_i+b_i\Lambda_0)-\sin a_i] \\ &\approx \tfrac{1}{2}B_i\cos\phi_{mi}. \end{aligned} \tag{2.9}$$

By the same arguments,

$$S_{Ai} = -\frac{B_i u_0}{2}\int_0^{\Lambda_0}\sin(a_i+b_i x)\,\mathrm{d}x \approx \tfrac{1}{2}B_i\sin\phi_{mi}, \tag{2.10}$$

and again

$$\phi_m(x,y) = \tan^{-1}\frac{\text{Im}[Y(x,y)]}{\text{Re}[Y(x,y)]}. \tag{2.11}$$

Thus, the smoothing method remains accurate even when the local phase varies linearly and slowly over one fringe period. A similar attempt was reported by Toyooka and Tominaga [1984], but the rough approximation they introduced in the final step of phase computation was equivalent to setting $b_i = 0$, i.e., constant phase in the fringe period.

The sinusoidal-fitting method (Mertz [1983], Macy [1983]) is a space version of the phase demodulation in the Fourier domain. One again assumes that the carrier frequency is as high as the numerical resolution as the image allows for, i.e., 3 pixels/fringe period, and that the functions $A(x,y)$, $B(x,y)$ and $\phi(x,y)$ vary slowly over this period. Thus, $u_0 = \tfrac{1}{3}$, and the fitting of the

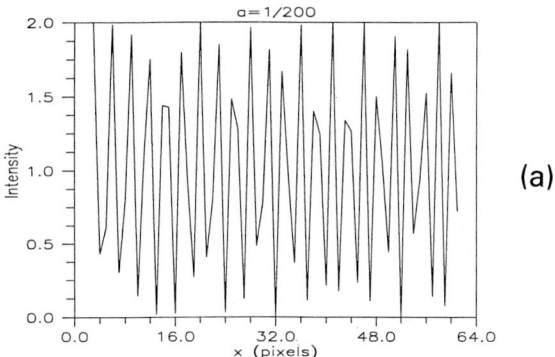

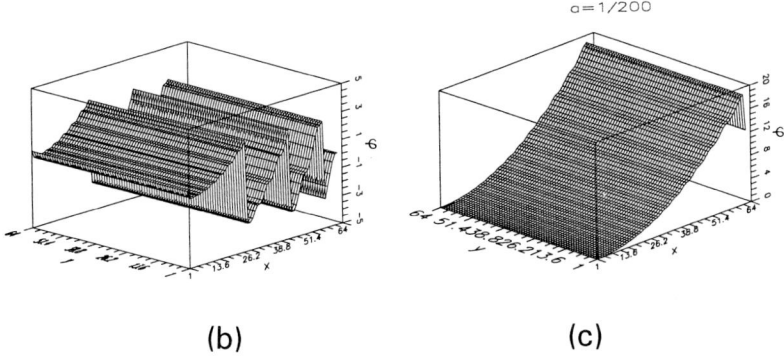

Fig. 2. Some experimental results of DSROP – the smoothing method with an average window 3×3 in space domain: (a) the intensity distribution of the phase-modulated image;

$$I(x) = 1 + \cos[(2\pi/3)x + ax^2] \quad \text{for } a = \tfrac{1}{200};$$

(b) the 3D calculated (wrapped, modulo 2π) phase distribution PD; (c) the 3D unwrapped PD; (d) 2D sections through the calculated wrapped PD, the dashed curve represents the orginal PD; (e) 2D sections through the calculated unwrapped (continuous line) and the original (dashed line) PD; (f) the intensity distribution $I(x)$ for $a = \tfrac{1}{2000}$; (g) 2D sections through the calculated (continuous line) and original (dashed line) PD ($a = \tfrac{1}{2000}$); (h) 3D plot of the calculated PD corresponding to the phase-modulated image $I(x, y) = 1 + \cos[(2\pi/3)x + a(x^2 + y^2)]$ for $a = \tfrac{1}{2000}$. One can observe in the upper corner the beginning of a phase wrapping.

FP, $I(x, y)$, with the first three terms of the Fourier series in each 3-pixel window,

$$f(x, y) = f_0 + f_1 \cos(2\pi x/3) + f_2 \sin(2\pi x/3), \tag{2.12}$$

leads to the following results:

$$f_1 = \cos\phi_m(x, y), \qquad f_2 = -\sin\phi_m(x, y), \tag{2.13}$$

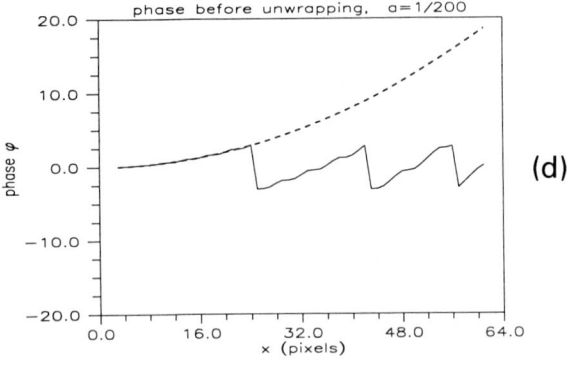

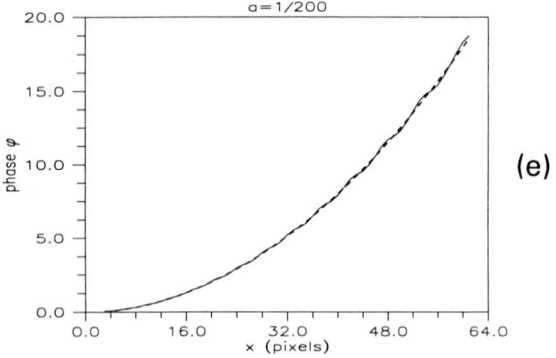

Fig. 2. Continued.

so that

$$\phi_m(x, y) = -\tan^{-1}\left(\frac{f_2}{f_1}\right). \tag{2.14}$$

The numerical evaluation of the Fourier coefficients in the 3-pixel window (with equally spaced points at $-2\pi/3, 0, 2\pi/3$) gives:

$$f_1(j, k) = -\tfrac{1}{2}I(j-1, k) + I(j, k) - \tfrac{1}{2}I(j+1, k)$$

$$f_2(j, k) = -\frac{\sqrt{3}}{2}I(j-1, k) - \frac{\sqrt{3}}{2}I(j+1, k) \tag{2.15}$$

for all $j = 1, ..., N-1$ and $k = 1, ..., N-1$ (the image format is $(N+1) \times (N+1)$ pixels with $j, k = 0, ..., N$). Taking into account the constant (and unimportant) contribution of the carrier (tilt) to the phase, one

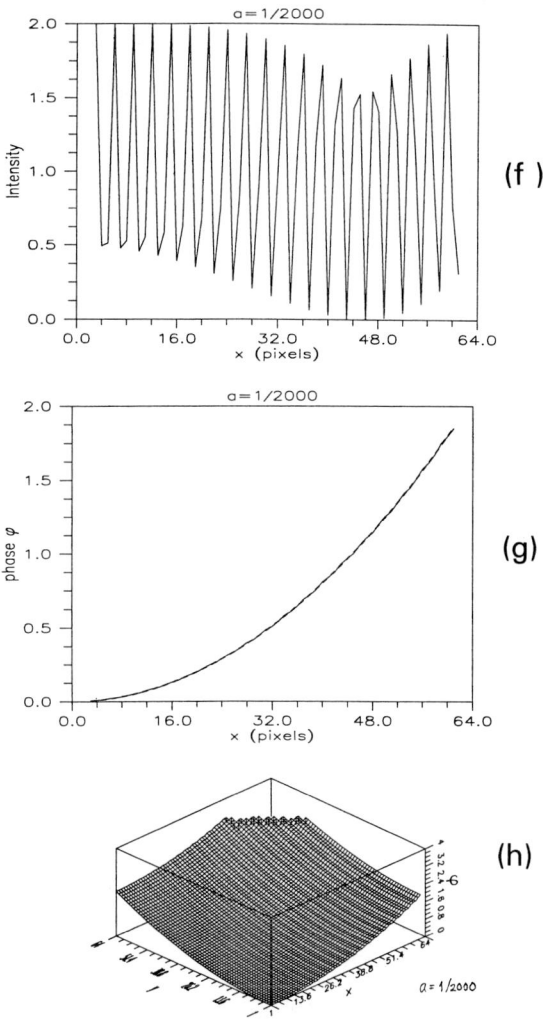

Fig. 2. Continued.

can write the pure phase modulation as

$$\phi_m(j,k) = -\tan^{-1}\left(\frac{f_2}{f_1}\right) - \left(\frac{2\pi j}{3}\right) \pmod{2\pi}. \tag{2.16}$$

A modified sinusoidal-fitting method, in which the fringe period is not 3 pixels and the local phase is allowed to vary slowly, was developed by Ransom and Kokal [1986] and led to a simple change in eq. (2.14), namely the multiplication of f_1 by the correction factor:

$$K = \sqrt{2}\,\frac{\sin \Omega h}{1 - \cos \Omega h}, \qquad (2.17)$$

where

$$\Omega = \left[\frac{2\pi}{3} + \phi'_{mi} \cdot \frac{1}{3u_0}\right] 3u_0,$$

and h is the sampling interval. In general, a large number of pixels per fringe does not improve the accuracy due to the violation of the assumption that A, B and ϕ_m are constant over one fringe.

Recently, Frankowski, Stobbe, Tischer and Schillke [1989] proposed the parameter estimation method for spatial phase reconstruction from FP, in which the matching of a theoretical intensity model [e.g., eq. (2.2)] to the experimentally recorded intensity pattern is done by minimizing a quadratic loss function. With the same approximations as in the sinusoidal-fitting method, the analytical result is almost identical to that of eq. (2.10).

Another method for phase demodulation in the space domain, which is equivalent to spatial band-phase filtering in the Fourier domain, will be described at the end of the next section.

2.2. PHASE DEMODULATION IN THE SPATIAL FREQUENCY (FOURIER) DOMAIN

Phase demodulation in the Fourier domain relies on the fundamental description given by Leith and Upatnieks [1962] to the wavefront recording and reconstruction in holography (fig. 3). In the case of optical FP, the operations are analogous to the holograms recorded with a plane reference. Thus, eq. (2.3) can be written as

$$I(x, y) = A(x, y) + \tfrac{1}{2}B(x, y)\exp(i[2\pi u_0 x + \phi_m(x, y)])$$
$$+ \tfrac{1}{2}B(x, y)\exp(-i[2\pi u_0 x + \phi_m(x, y)])$$
$$= A(x, y) + C(x, y)\exp(2i\pi u_0 x) + C^*(x, y)\exp(-2i\pi u_0 x), \qquad (2.18)$$

where

$$C(x, y) = \tfrac{1}{2}B(x, y)\exp[i\phi_m(x, y)]. \qquad (2.19)$$

Taking the Fourier transform of eq. (2.18) and denoting the spectra with the same letters but covered by tilde(s), we obtain

$$\tilde{I}(u, v) = \tilde{A}(u, v) + \tilde{C}(u - u_0, v) + \tilde{C}^*(u + u_0, v). \qquad (2.20)$$

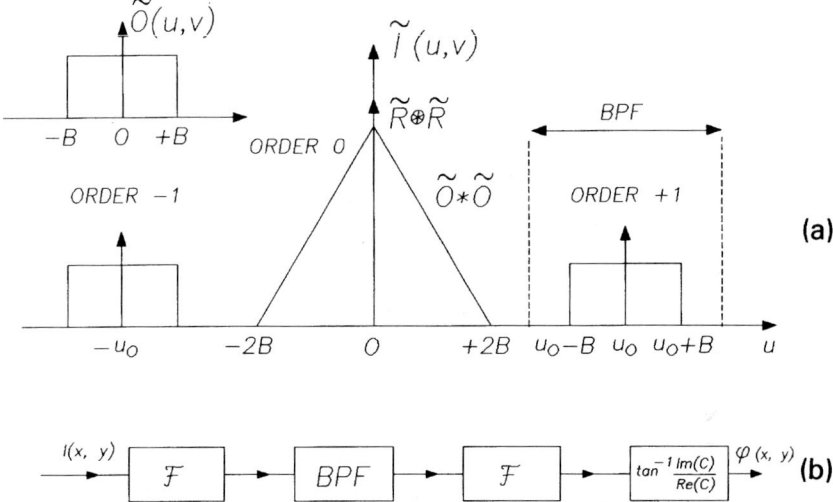

Fig. 3. (a) The spectrum of the hologram transmittance, $I(x, y)$, and the demodulation by the band-pass filtering of the order $+1$ (the origin of the Fourier plane is taken on the axis of the reference beam). The sideband separability condition is $u_0 \geqslant 3B$ ($u_0 \gg |\partial\phi_0/\partial x|_{\max}$). (b) The steps of DSROP: the band-pass filtering method in the Fourier domain.

If the carrier frequency is larger than the half-sum of the bandwidth of $\tilde{A}(u, v)$ and $\tilde{C}(u, v)$ [u and v are the spatial frequencies corresponding to x and y],* the spectrum consists of three well-defined bands (fig. 3). A bandpass filtering of the term $\tilde{C}(u, v)$ can be done by multiplication of the spectrum from eq. (2.22) with the transfer function $H(u - u_0, v)$. This function can be (i) the rectangular window (with the drawbacks of sharp truncation), (ii) the Hamming window:

$$H(u, v) = 0.54 + 0.46 \cos(2\pi u/\Lambda_0) \quad \text{for } u < \Lambda_0/2,$$

$$= 0 \quad \text{elsewhere} \quad (2.21)$$

(with smoother truncation and smaller oscillations in the result), or (iii) special polynomial windows (vide Takeda, Ina and Kobayashi [1982], Takeda [1989], Womack [1984], Pratt [1978], Frankowski, Stobbe, Tischer and Schillke [1989]).

After the band-pass filtering and computation of the inverse Fourier transform of the selected term, $\tilde{C}(u, v)$, the phase modulation can be obtained

---

* A more physical meaning of this condition is that the fringes in the image must be open and never cross any line parallel to the x-axis more than once. This is met if u_0 is large enough.

directly:

$$\phi_m(x, y) = -\tan^{-1}\frac{\operatorname{Im}[C(x, y)]}{\operatorname{Re}[C(x, y)]}. \qquad (2.22)$$

Some experimental results obtained with this method for the same FP used with the smoothing method in the space domain are shown in fig. 4. The Fourier domain approach is equivalent completely to the spatial approach, as the entire filtering process of the spectrum can be described by a convolution of the initial FP with an weighting function (impulse response):

$$\Im(x', y) = \int_{-\infty}^{\infty} I(x, y) h(x' - x, y) \, dx, \qquad (2.23)$$

where

$$h(x, y) = \mathfrak{F}[H(x, y)], \qquad (2.24)$$

and $\mathfrak{F}$ means the Fourier transformation (Womack [1984]).

We have solved the equivalent space problem for the band-pass filtering in a convenient manner using the differential Gaussian filter with the transfer function (fig. 5):

$$H_1(u, v) = [\exp(-\rho^2/2\alpha_1^2) - \exp(-\rho^2/2\alpha_2^2)]\sigma(u) = H(\rho)\sigma(u), \qquad (2.25)$$

where $\rho^2 = u^2 + v^2$, and $\sigma(u)$ is the unity step function. The central frequency of this band-pass filter is

$$\rho_c = 2\alpha_1 \alpha_2 \sqrt{\frac{\ln \alpha_2 - \ln \alpha_1}{\alpha_2^2 - \alpha_1^2}} \sim \frac{\sqrt{2}}{2}(\alpha_1 + \alpha_2) \qquad (2.26)$$

(when $\alpha_1 \sim \alpha_2$), and its bandwidth is

$$\Delta\rho \approx \Delta\rho(\alpha_1, \alpha_2) \sim \rho_c \cdot q(\alpha_1/\alpha_2), \qquad (2.27)$$

where $q(\alpha_1/\alpha_2)$ is a function of the ratio α_1/α_2 only. It follows that for $\alpha_1/\alpha_2 =$ constant, the ratio $\Delta\rho/\rho_c$ is constant and, moreover, the maximum value of $H(\rho)$ is constant.

The impulse response of the differential Gaussian filter has another

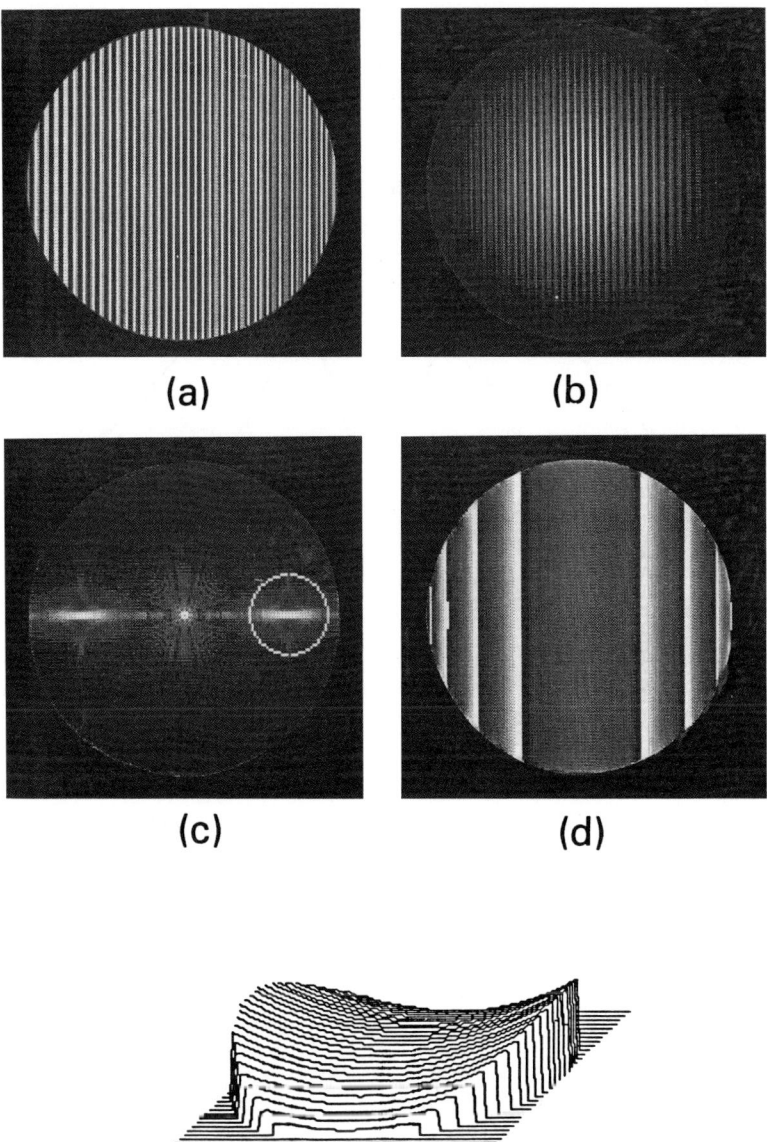

Fig. 4. Some experimental results of DSROP in the Fourier domain by the band-pass method with a Hamming window: (a) the phase-modulated image, $I(x, y) = 1 + \cos[(2\pi/3)x + ax^2)]$ for $a = \frac{1}{200}$, (b) Hamming filtering of $I(x)$, (c) Fourier spectrum of $I(x)$, (d) wrapped phase distribution calculated by the band-pass filtering method in the Fourier domain, (e) 3D plot of the calculated unwrapped phase distribution corresponding to $I(x)$.

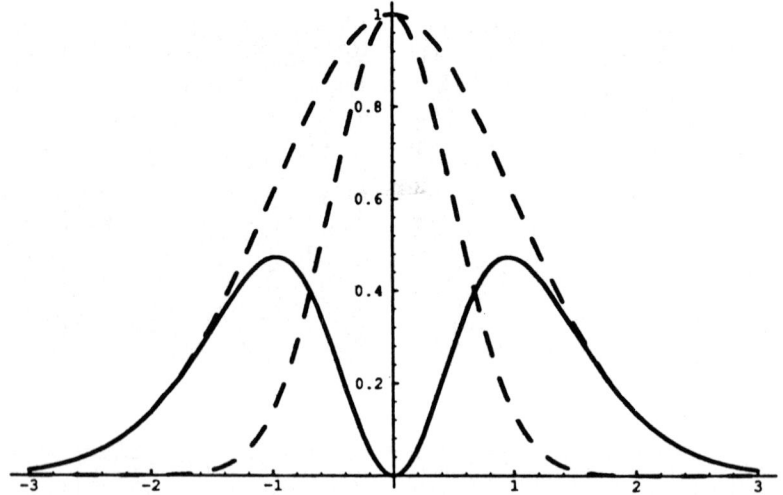

Fig. 5. The transfer function of a differential Gaussian filter, $H(\rho)$ (continuous line). The original Gaussian functions are plotted with dashed lines.

differential Gaussian form:

$$h_1(r) = \left[\frac{1}{\sqrt{2\pi\sigma_1^2}}\exp\left(-\frac{r^2}{2\sigma_1^2}\right) - \frac{1}{\sqrt{2\pi\sigma_2^2}}\exp\left(-\frac{r^2}{2\sigma_2^2}\right)\right] * \frac{1}{2}\left[\delta(x) + i\frac{1}{\pi x}\right], \quad (2.28)$$

where

$$r^2 = x^2 + y^2, \quad \sigma_1 = (2\pi\alpha_1)^{-1}, \quad \text{and} \quad \sigma_2 = (2\pi\alpha_2)^{-1}. \quad (2.29)$$

Equation (2.28) can further be written as:

$$h_1(r) = \frac{\sqrt{2}}{2}\pi[\alpha_1 \exp(-2\pi^2\alpha_1^2 r^2) - \alpha_2 \exp(-2\pi^2\alpha_2^2 r^2)] + \tfrac{1}{2}\mathcal{H}_x[h(r)], \quad (2.30)$$

where $\mathcal{H}_x[.]$ is the Hilbert transform on the x coordinate. Thus, once we have selected the convenient central frequency and the bandwidth of the filter, as in eqs. (2.26) and (2.27), one can calculate α_1, α_2 and the impulse response of the band-pass filter is determined completely. The numerical convolution of the FP with this impulse response can be written as

$$I(x, y) = \sum_{m=0}^{M-1}\sum_{n=0}^{N-1} I(m, n)h(x - m, y - n). \quad (2.31)$$

An approximative solution for rapid processing can be further obtained by developing the differential Gaussian transfer function in power series and retaining only the two terms:

$$H(\rho) = \left(1 - \frac{\rho^2}{2\alpha_1^2} + \ldots\right) - \left(1 - \frac{\rho^2}{2\alpha_2^2} + \ldots\right) \approx \frac{1}{2}\left(\frac{1}{\alpha_2^2} - \frac{1}{\alpha_1^2}\right)\rho^2, \qquad (2.32)$$

and

$$H_1(u, v) = \frac{1}{2}\left[\frac{1}{\alpha_2^2} - \frac{1}{\alpha_1^2}\right]\rho^2 \sigma(u). \qquad (2.33)$$

The corresponding impulse response is

$$h_1(r) = \tfrac{1}{2}h(r) + \tfrac{1}{2}h(r) * i\frac{1}{\pi x} = \tfrac{1}{2}h(r) + i\tfrac{1}{2}\mathcal{H}_x[h(r)], \qquad (2.34)$$

and the convolution with the FP takes the form

$$\begin{aligned}\Im(x, y) &= I(x, y) * \left[h(x, y) * \frac{1}{2}\left[\delta(x) + i\frac{1}{\pi x}\right]\right] \\ &= \tfrac{1}{2}\nabla^2[I(x, y) + i\mathcal{H}_x[I(x, y)]] \\ &= \tfrac{1}{2}[\nabla^2[I(x, y)] + i\mathcal{H}_x[\nabla^2[I(x, y)]]], \qquad (2.35)\end{aligned}$$

where

$$\nabla^2[I] = \frac{\partial^2 I}{\partial x^2} + \frac{\partial^2 I}{\partial y^2} \qquad (2.36)$$

is the Laplace operator.

One can remark immediately that there are two solutions for performing the band-pass filtering: (i) for the first, construct the analytical signal corresponding to the FP, $I(x, y)$, using the 1D Hilbert transform, and filter this signal with a weighted Laplacian window, and (ii) for the second, filter the FP with a weighted Laplacian window and construct the analytical signal corresponding to the filtered image.

Following the second variant and assuming that $A(x, y)$, $B(x, y)$ and the phase modulation, $\phi_m(x, y)$ are slowly varying functions such that all $\phi'_m \approx$ ct and $\phi''_m \approx 0$ in the Laplacian window, we can calculate:

$$\nabla^2 I(x, y) \approx -B[(2\pi u_0 + \phi'_{mx})^2 + \phi'^2_{my}]\cos[2\pi u_0 x + \phi_m(x, y)]. \qquad (2.37)$$

In the same conditions, the Hilbert transform yields:

$$\mathcal{H}_x[\nabla^2 I(x, y)] \approx B[(2\pi u_0 + \phi'_{mx})^2 + \phi'^2_{my}]\sin[2\pi u_0 x + \phi_m(x, y)]. \qquad (2.38)$$

Thus, the filtered FP takes the form:

$$\Im(x, y) = \frac{B}{4}\left(\frac{1}{\alpha_2^2} - \frac{1}{\alpha_1^2}\right)[(2\pi u_0 + \phi'_{mx})^2 + \phi'^2_{my}]$$

$$\times [\cos[2\pi u_0 x + \phi_m(x, y)] + i \sin[2\pi u_0 x + \phi_m(x, y)]], \quad (2.39)$$

and the phase can again be obtained as

$$\phi_m(x, y) = -\tan^{-1}\frac{\operatorname{Im}[\Im(x, y)]}{\operatorname{Re}[\Im(x, y)]} - 2\pi u_0 x. \quad (2.40)$$

The strong similarity between this method of phase demodulation and the demodulation of the phase-modulated signals in practical communication systems (radio, TV) is noteworthy.

2.3. PHASE UNWRAPPING AND REPRESENTATION

The phases obtained with the method just described are indeterminate by an additive term $2m\pi$, where m is any constant, because the arctangent is defined in the limited interval from $-\pi/2$ to $\pi/2$. Thus the correct phase is more correctly written as

$$\phi_{mc}(x, y) = \phi_m(x, y) + 2\pi m. \quad (2.41)$$

On the other hand, the sampling theorem requires that each fringe contains at least two pixels, so that at the highest frequency, the change between two pixels is always smaller than π. The phase unwrapping procedure consists of adding multiples of 2π to the phase whenever discontinuities in the phase appear, but these large discontinuities may be avoided by having the fringe-pattern spatial frequency within the limits of the sampling theorem. Then, the unwrapping may be performed if it is assumed that the phase increases monotonically (Takeda, Ina and Kobayashi [1982], Kreis [1986]), and by the use of the following relations, beginning with the condition $m(x_1, y) = 0$,

$$m(x_i, y) = \begin{cases} m(x_{i-1}, y) & \text{if } |\phi(x_i, y) - \phi_{-1}(x_{i-1}, y)| < \pi, \\ m(x_{i-1}, y) + 1 & \text{if } |\phi(x_i, y) - \phi_{-1}(x_{i-1}, y)| \geq \pi, \end{cases} \quad (2.42)$$

with $i = 1, 2, 3, \ldots$.

The procedure to unwrap the phase in two directions and to determine if the value of m decreases instead of increases when the phase does not increase monotonically has also been described by Takeda, Ina and Kobayashi [1982], Macy [1983], Kreis [1986], Roddier and Roddier

[1987], Barr, Coudé du Foresto, Fox, Paczulp, Richardson, Roddier and Roddier [1991], and Bone [1991]. This procedure is applied in two perpendicular directions (x, y) in an iterative manner. The following relations are used, beginning wth $m(x_1, y_1) = 0$

$$m(x_1, y_j) = \begin{cases} m(x_1, y_{j-1}) & \text{if } |\phi(x_1, y_j) - \phi_{-1}(x_1, y_{j-1})| < \pi, \\ m(x_1, y_{j-1}) + 1 & \text{if } |\phi(x_1, y_j) - \phi_{-1}(x_1, y_{j-1})| \leqslant -\pi, \\ m(x_1, y_{j-1}) - 1 & \text{if } |\phi(x_1, y_j) - \phi_{-1}(x_1, y_{j-1})| \geqslant \pi, \end{cases}$$

(2.43)

with $j = 2, 3, \ldots$, and

$$m(x_i, y_j) = \begin{cases} m(x_{i-1}, y_j) & \text{if } |\phi(x_i, y_j) - \phi_{-1}(x_{i-1}, y_j)| < \pi, \\ m(x_{i-1}, y_j) + 1 & \text{if } |\phi(x_i, y_j) - \phi_{-1}(x_{i-1}, y_j)| \leqslant -\pi, \\ m(x_{i-1}, y_j) - 1 & \text{if } |\phi(x_i, y_j) - \phi_{-1}(x_{i-1}, y_j)| \geqslant \pi, \end{cases}$$

(2.44)

with $i = 2, 3, \ldots$.

Spatial frequencies higher than those permitted by the sampling theorem may be allowed only if some special assumptions are made about the nature of the phase, namely that the phase is smooth and has continuous derivatives, as shown by Greivenkamp [1987].

The phase-modulation distribution can be displayed using different modern graphics representations; e.g., topograms, projections, or perspectives. In many cases, numerical features (e.g., aberration coefficients, first and second derivatives of the phase, etc.) are necessary for optical scientists and designers. In these cases, the polynomial fitting of the phase distribution is imposed as an analytic representation step.

The phase-modulation distribution may be represented analytically by means of many types of functions, but the Zernike polynomials are the most commonly used due to their unique and desirable properties. Zernike polynomials $U(\rho, \theta)$ are represented in polar coordinates and are orthogonal in the unit circle (pupil area), satisfying the condition

$$\int_0^1 \int_0^{2\pi} U_n^1 U_{n'}^{1'} \rho \, d\rho \, d\theta = \frac{\pi}{2(n+1)} \delta_{nn'} \delta_{11'}.$$

(2.45)

In this expression, the Zernike polynomials are represented with an index n, but often two indices n and m are used, since they are dependent on two coordinates. Thus, these polynomials can be separated into two functions, one depending only on the radius ρ and the other depending only on the angle θ, as follows:

$$U_{nm} = R_n^{n-2m} \begin{bmatrix} \sin \\ \cos \end{bmatrix} (n-2m)\theta, \qquad (2.46)$$

where sin is used for $n - 2m > 0$, cos is used for $n - 2m \leq 0$, and n is the degree of the radial polynomial $R_n(\rho)$, and $0 \leq m \leq n$. The maximum exponent for the radial polynomials is $n - 2m$. Thus it is easy to see that there are $\frac{1}{2}(n+1)(n+2)$ linearly independent polynomials U_{nm} of degree $\leq n$, one for each allowed pair of numbers n and $n - 2m$. The radial polynomial is given by

$$R_n^{n-2m}(\rho) = \sum_{s=0}^{m} (-1)^s \frac{(n-s)!}{s!(m-s)!(n-m-s)!} \rho^{n-2s}. \qquad (2.47)$$

As in the orthogonality condition, a single index r may be used, if defined by

$$r = \frac{n(n+1)}{2} + m + 1. \qquad (2.48)$$

Thus, a phase distribution $\phi(\rho, \theta)$ over a wavefront may be represented by a linear combination of Zernike polynomials as follows:

$$\phi(\rho, \theta) = \sum_{n=0}^{k} \sum_{m=0}^{n} A_{nm} U_{nm} = \sum_{r=0}^{L} A_r U_r. \qquad (2.49)$$

The coefficients A_r for this linear combination of polynomials U_r may be found by means of a least squares procedure. At this point, it must be pointed out that Zernike polynomials are orthogonal on the continuous unit circle. The polynomials which are orthogonal over the set of measured data points have to satisfy the condition

$$\sum_{i=1}^{N} V_r(\rho_i, \theta_i) V_p(\rho_i, \theta_i) = \delta_{rp}. \qquad (2.50)$$

The phase distribution is then represented in terms of these orthogonal polynomials as

$$\phi(\rho, \theta) = \sum_{r=1}^{L} B_r V_r(\rho, \theta). \qquad (2.51)$$

By use of the orthogonality condition, it may be shown that the coefficients B_r are given by

$$B_p = \frac{\sum_{i=1}^{N} \phi(\rho_i, \theta_i) V_p}{\sum_{i=1}^{N} V_p^2}, \qquad (2.52)$$

where $\phi(\rho_i, \theta_i)$ are the measured values of the phase distribution. These polynomials V_p are found by means of a Gram–Schmidt orthogonalization as a linear combination of the Zernike polynomials U_i. We begin by writing

$$V_r = U_r + \sum_{s=1}^{r-1} D_{rs} V_s. \tag{2.53}$$

Using the orthogonality condition, we find

$$D_{rp} = \frac{\sum_{i=1}^{N} U_r V_p}{\sum_{i=1}^{N} V_p^2}. \tag{2.54}$$

Then, the coefficients C_{rs} that define the orthogonal polynomials V_r as a linear combination of the Zernike polynomials U_r are

$$V_r = U_r + \sum_{i=1}^{r-1} C_{ri} U_i. \tag{2.55}$$

The desired coefficients A_r may be found by

$$A_r = B_r + \sum_{i=r+1}^{L} B_i C_{ir}. \tag{2.56}$$

With these relations, the phase fitting may be performed in terms of Zernike polynomials. For details on this procedure, see Malacara and De Vore [1992].

§ 3. Direct Spatial Reconstruction of Optical Phase (DSROP) from more General Phase-Modulated Images

A key parameter in the analysis of phase-modulated images, the product demodulation and the separation of the sidebands is the spatial carrier, or in more general terms, the reference image. As in interferometry, holography, moiré methods and speckle pattern (SP) interferometry, the reference image can be (i) a one-dimensional periodic (sinusoidal) pattern (corresponding to a tilted plane wave, as was shown in § 2), (ii) a chirped radially symmetric pattern (corresponding to a spherical wave), (iii) a sheared periodic pattern (like in shearing interferometry, moiré and SP methods), (iv) an arbitrary deterministic reference image (like in holographic associative memory), or (v) an arbitrary random reference image (like in holographic interferometry and digital speckle-pattern interferometry).

In this section, we shall describe DSROP for the last four types of reference images, providing this field of computer image processing with the great generality met in existing optical methods. Obviously, other particular types of references can be imagined and used in the reconstruction of the optical phase, but the ones mentioned above cover the most important practical cases.

3.1. DSROP WITH A CHIRPED RADIALLY SYMMETRIC PATTERN

In this case, the phase-modulated image takes the form:

$$I(x, y) = A(x, y) + B(x, y) \cos[D_0(x^2 + y^2) + \phi_m(x, y)]. \tag{3.1}$$

The name "chirped signal" is given in communication theory to signals with linearly varying carrier frequency, and is related strongly to pulse compression. In optics, it is related to spherical waves, lens transmittances and Fresnel transformations (fig. 6).

The chirped radially symmetric FP from eq. (3.1) can be easily demodulated in the space domain with the computer-generated reference:

$$R(x, y) = \exp[iD_0(x^2 + y^2)], \tag{3.2}$$

with $2D_0 x \gg (\partial \phi_m/\partial x)_{max}$ and $2D_0 y \gg (\partial \phi_m/\partial y)_{max}$. The product of the FP

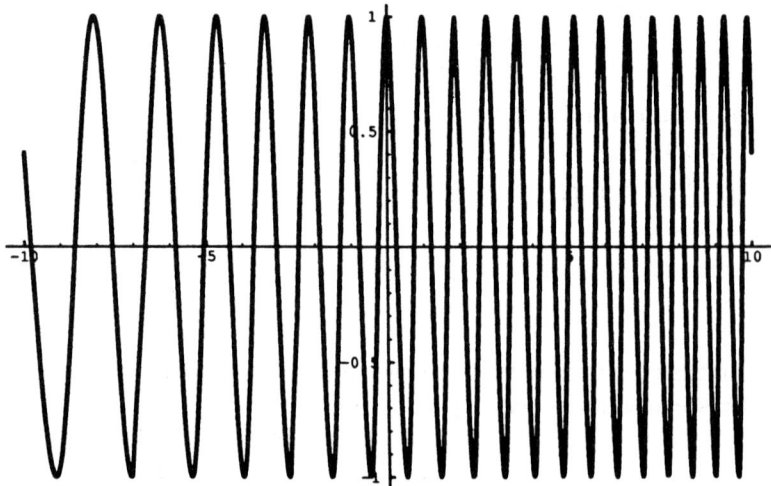

Fig. 6. The 2D section of the chirped signal $I(x) = \cos[x^2/5 + 2\pi x]$.

and the reference is:
$$I \cdot R = A \cdot R + \tfrac{1}{2}B \exp[i2D_0(x^2+y^2) + i\phi_m(x,y)] + \tfrac{1}{2}B \exp[-i\phi_m(x,y)]. \tag{3.3}$$

Then, using digital low-pass filtering (e.g., with a 3×3 average window), we can obtain
$$Y(x,y) = \tfrac{1}{2}B \exp(-i\phi_m(x,y)) = \frac{B}{2}[\cos\phi_m(x,y) - i\sin\phi_m(x,y)], \tag{3.4}$$

and the phase distribution
$$\phi_m(x,y) = -\tan^{-1}\frac{\operatorname{Im}[Y(x,y)]}{\operatorname{Re}[Y(x,y)]}. \tag{3.5}$$

Ru, Honda, Tsujiuchi and Ohyama [1988] have analyzed the pattern from eq. (3.1) arising from single-path interferometry, where defocusing (measured by D_0) rather than tilt can be more easily accomplished by Fresnel transformation. If the Fresnel transformation (denoted by Fr{.}) in $z = 1/D_0$, is applied to the chirped pattern from eq. (3.1) the result can be written as

$$\operatorname{Fr}[I(x,y)] = \tilde{A}(u,v) * \frac{\pi}{D_0} \exp\left[i\pi^2 \frac{u^2+v^2}{D_0} - i\frac{\pi}{2}\right]$$
$$+ \tilde{C}(u,v) + \tilde{C}^*(u,v) * \frac{\pi}{2D_0} \exp\left[i\pi^2 \frac{u^2+v^2}{2D_0} - i\frac{\pi}{2}\right], \tag{3.6}$$

where all previous notations were preserved and * signifies the 2D convolution integral. The separability conditions for the three terms from eq. (2.54) are $D_0 \gg (\partial^2\phi_m/\partial^2 x)$ and $D_0 \gg (\partial^2\phi_m/\partial^2 y)$. Moreover, if D_0 is large, the first and third terms can be eliminated easily by a digital low-pass filtering in the Fresnel space, at $z = 1/D_0$. It follows that

$$C(x,y) = \mathfrak{F}^{-1}[\tilde{C}(u,v)] = \tfrac{1}{2}B(x,y)\exp(i\phi_m(x,y))] \tag{3.7}$$

and
$$\phi_m(x,y) = \tan^{-1}\frac{\operatorname{Im}[C(x,y)]}{\operatorname{Re}[C(x,y)]}. \tag{3.8}$$

It is interesting to note that in image holographic interferometry (Brandt [1969], Boone and De Backer [1973]) and in Fourier holographic interferometry (Roychoudhuri and Machorro [1978], Vlad and Popa [1982]), the object displacements yield only two types of interference patterns (localized always on the holographic plate). These are: (i) the in-plane displacements which produce straight-parallel (sinusoidal) fringes, and (ii) the out-of-plane displacements which produce circular concentric fringes of the form shown in eq. (3.1).

The 3D deformations produce circular concentric fringes with the center displaced from the optical axis. Both of the two types of patterns can be processed with DSROP algorithms to obtain rapidly and accurately the full information about any 3D deformation of a diffuse object by comparison with a reference of the same form defined by the computer. Thus, DSROP algorithms can provide a modern alternative to those proposed earlier (Vlad, Popa and Solomon [1986]). However, we have to mention that an external scanning of the hologram with a small aperture is still necessary if we want to reconstruct 3D deformations of the object.

3.2. DSROP WITH A SHEARED PERIODIC PATTERN

The reference pattern which multiplies the detected phase-modulated image in the computer can be transformed geometrically in various ways by simple programs. For example, one can translate, rotate, invert or change the scale of the reference pattern as it is used in shearing interferometry, the moiré methods and in SP interferometry. However, we shall develop here only the case of a rotated sinusoidal carrier investigated previously as moiré deflectometry (Kafri [1980], Kafri and Livnat [1981], Vlad, Popa and Apostol [1991]). In this case, the tested (transparent or reflecting) object is "seen" by the TV camera through a grating which, for convenience, is rotated in-plane and around the x-axis by a small angle, $\theta/2$, like in fig. 7. The object yields small deflections of the coherent illumination beam, and grating-line shifts along the x-axis by $x_0 = x_0(x, y)$ and along the y-axis by $y_0 = y_0(x, y)$, so the detected intensity pattern can be written:

$$I(x, y) = I_0 + I_1 \cos^2 2\pi u_0 \left[-(x - x_0) \sin \frac{\theta}{2} - (y - y_0) \cos \frac{\theta}{2} \right]$$

$$= A(x, y) + B(x, y) \cos 4\pi u_0 \left[-(x - x_0) \sin \frac{\theta}{2} - (y - y_0) \cos \frac{\theta}{2} \right]. \tag{3.9}$$

If the reference pattern is a grating rotated around the x-axis by the angle $(-\theta/2)$ modulating a 1D carrier along the x-axis; e.g.,

$$R_c(x, y) = \left[\cos 4\pi u_0 \left(x \sin \frac{\theta}{2} + y \sin \frac{\theta}{2} \right) \right] \cos \left(8\pi u_0 x \sin \frac{\theta}{2} \right), \tag{3.10}$$

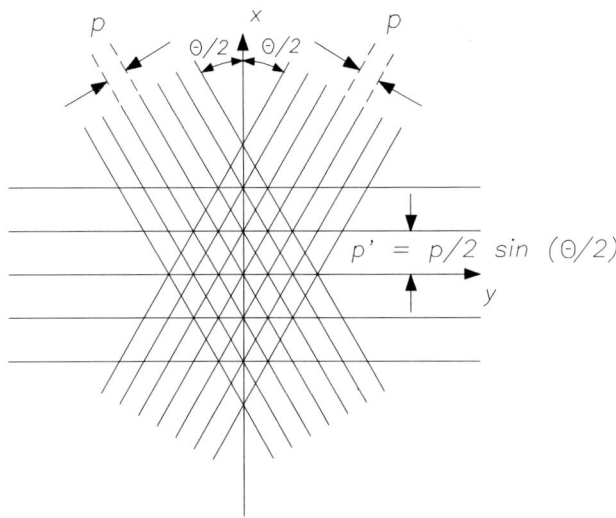

Fig. 7. Rotated sinusoidal carrier (grating) and the corresponding sheared reference used in DSROP (for computer moiré deflectometry).

then

$$I \cdot R_c \approx A \cdot R_c + \tfrac{1}{2}B \cos\left[8\pi u_0 \left(y - \frac{y_0}{2}\right) \cos\frac{\theta}{2}\right] \cdot \cos\left(8\pi u_0 x \sin\frac{\theta}{2}\right)$$

$$+ \tfrac{1}{2}B \cos\left[8\pi u_0 \left(x + \frac{y_0}{2\tan\dfrac{\theta}{2}}\right) \sin\frac{\theta}{2}\right] \cdot \cos\left(8\pi u_0 x \sin\frac{\theta}{2}\right)$$

$$= A \cdot R_c + \tfrac{1}{4}B \cos\left[8\pi u_0 \left(x \sin\frac{\theta}{2} + \left(y - \frac{y_0}{2}\right)\right) \cos\frac{\theta}{2}\right]$$

$$+ \tfrac{1}{4}B \cos\left[8\pi u_0 \left(x \sin\frac{\theta}{2} - \left(y - \frac{y_0}{2}\right)\right) \cos\frac{\theta}{2}\right]$$

$$+ \tfrac{1}{4}B \cos\left[8\pi u_0 \left(2x + \frac{y_0}{2\tan\dfrac{\theta}{2}}\right) \sin\frac{\theta}{2}\right] + \tfrac{1}{4}B \cos\left(4\pi u_0 y_0 \cos\frac{\theta}{2}\right).$$

$$(3.11)$$

The phase shift of the product moiré pattern is significant only along the x-axis:

$$\Delta\phi = 4\pi u_0 \frac{y_0}{\tan\left(\frac{\theta}{2}\right)}. \tag{3.12}$$

This phase shift can be obtained from eq. (3.11) by a low-pass filtering which will eliminate all terms except one; viz.,

$$Y_c = \cos\left(4\pi u_0 y_0 \cos\frac{\theta}{2}\right). \tag{3.13}$$

By multiplication with a quadrature reference:

$$R_s(x, y) = \left[\cos 4\pi u_0 \left(x \sin\frac{\theta}{2} + y \sin\frac{\theta}{2}\right)\right] \sin\left(8\pi u_0 x \sin\frac{\theta}{2}\right), \tag{3.14}$$

and by the same filtering, one can obtain

$$Y_s = \sin\left(4\pi u_0 y_0 \cos\frac{\theta}{2}\right), \tag{3.15}$$

and finally,

$$y_0 = \frac{1}{4\pi u_0 \cos\left(\frac{\theta}{2}\right)} \tan^{-1}\frac{Y_s}{Y_c}. \tag{3.16}$$

Similarly, by rotating the external grating and the reference pattern (grating) in the orthogonal position, one can obtain x_0. As shown previously, the deflection angles, which are related to the phase gradients, can be determined by dividing x_0 and y_0 by $(nz_T)/\sin(\theta/2)$, the distance between the external grating and the surface of the TV sensor (an integer multiple n, of the Talbot distance, $z_T = 2\Lambda_0/\lambda$, where λ is the light wavelength).

It was proved that the sensitivity of the *computer moiré deflectometry* (CMD), a one-path system, is comparable with that obtained in interferometry. Consequently, the fast detection of the phase gradients is very important. Moreover, it was shown that CMD without grating tilting can bring information on the second derivatives of the optical phase (Vlad, Popa and Apostol [1991]).

3.3. DSROP WITH AN ARBITRARY DETERMINISTIC REFERENCE IMAGE

In § 2, we defined in general form the phase-modulated image in eq. (2.1). The phase in eq. (2.1) can be thought of as a phase difference between a reference wavefront and an object wavefront; i.e.,

$$\phi_m(x, y) = \phi_R(x, y) - \phi_0(x, y). \tag{3.17}$$

The intensity $I(x, y)$ can then be interpreted as the interference pattern in the associative holographic memory (Gabor [1969]), in which an object wavefront is recorded with a complex reference (etalon) one (fig. 8a).

In the spatial domain, the object phase demodulation can be accomplished by multiplication by the reference pattern stored in the computer (fig. 8b):

$$R(x, y) = \exp[i\phi_R(x, y)]; \tag{3.18}$$

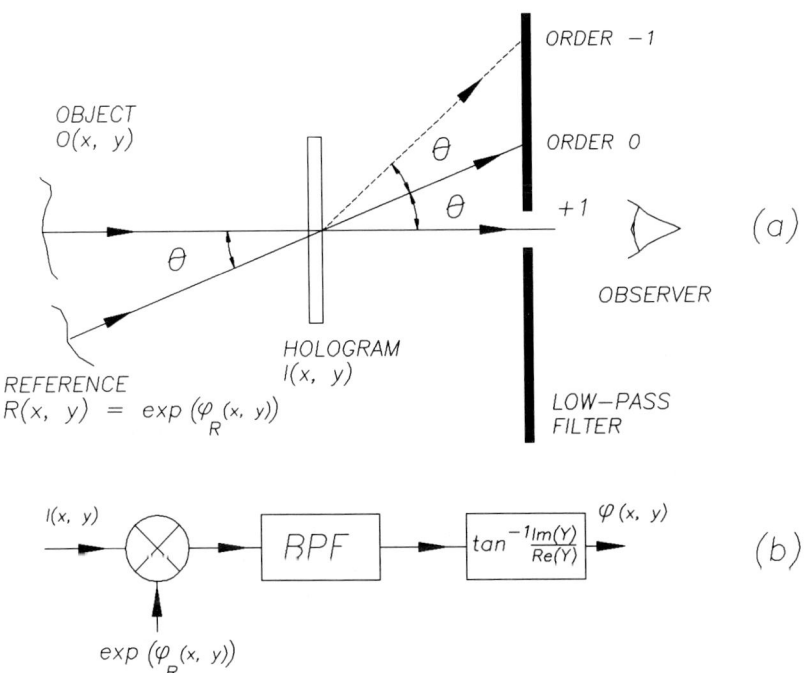

Fig. 8. DSROP with arbitrary deterministic reference: (a) associative holographic recording of object $O(x, y)$ with an arbitrary deterministic reference $R(x, y)$. The reconstruction process consists of the multiplication of the hologram transmittance $I(x, y)$ by the reference $R(x, y)$ and the optical low-pass filtering of the result. (b) The steps of DSROP, the smoothing method in the space domain.

that is,

$$I \cdot R = A \cdot R + \tfrac{1}{2} B \exp[i(2\phi_R - \phi_0)] + \tfrac{1}{2} B \exp(i\phi_0). \qquad (3.19)$$

If we assume that the mean spatial frequency of the reference is much higher than that of the object (i.e., ϕ_R varies more rapidly than ϕ_0), the first two terms of eq. (3.13) vary much more rapidly than the third one. By a low-pass filtering of $(I \cdot R)$, we can obtain

$$\phi_0(x, y) = \tan^{-1} \frac{[\text{Im LPF}(I \cdot R)]}{[\text{Re LPF}(I \cdot R)]}. \qquad (3.20)$$

Alternatively, $\phi(x, y)$ can be thought of as a deviation of the external object wavefront from a computer-generated (almost perfect) reference. Let us consider this case in the Fourier domain (Kreis [1986]). The phase-modulated image can be written as

$$\begin{aligned} I(x, y) &= A(x, y) + \tfrac{1}{2} B(x, y) \exp[i[\phi_R(x, y) - \phi_0(x, y)]] \\ &\quad + \tfrac{1}{2} B(x, y) \exp[-i[\phi_R(x, y) - \phi_0(x, y)]] \\ &= A(x, y) + C(x, y) + C^*(x, y). \end{aligned} \qquad (3.21)$$

By (discrete) Fourier transform, eq. (3.21) becomes

$$\tilde{I}(u, v) = \tilde{A}(u, v) + \tilde{C}(u, v) + \tilde{C}^*(u, v). \qquad (3.22)$$

There are at least two cases in which we can define separability conditions for the terms from the spectrum $I(u, v)$. In the first of these, the reference pattern is a narrow-bandwidth signal; one can define its mean spatial frequency: $u_{Rm} = (1/2\pi)(\partial \phi_R/\partial x)_m$. Considering a band-limited object with the half-bandwidth $u_{0\,\text{max}}$ and

$$u_{R\,\text{max}} - u_{Rm} \ll u_{0\,\text{max}}, \qquad (3.23)$$

this situation is close to that of a sinusoidal spatial carrier (fig. 9a), and the separability condition can be written as

$$u_{Rm} > 3 u_{0\,\text{max}}. \qquad (3.24)$$

In the second case, the reference pattern is band-limited with the half-bandwidth close to the object one; i.e., $u_{Rm} \approx u_{0\,\text{max}}$ and with a mean spatial frequency, u_{Rm} (fig. 9b). In this case, an approximate separability condition is

$$u_{Rm} > 4 u_{0\,\text{max}}. \qquad (3.25)$$

In these cases, the spectrum of $I(x, y)$ is trimodal and, by a band-pass filtering (multiplication of the spectrum with a suitable transfer function), it is possible to select $\tilde{C}(u, v)$. Then, computing the (discrete) inverse Fourier

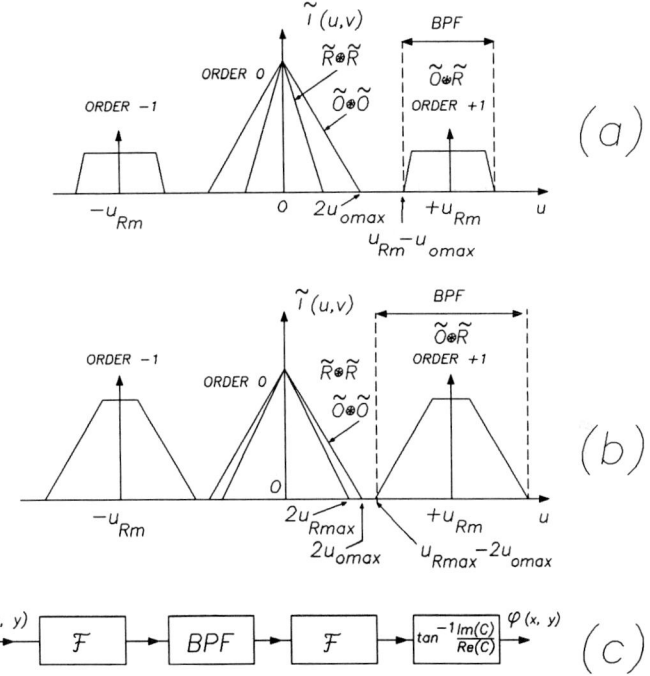

Fig. 9. The spectrum of the associative hologram transmittance $I(x, y)$: (a) for a narrow-band reference spectrum (separability condition: $u_{Rm} \geqslant 3u_{0\max}$), (b) for a reference spectrum with the width close to the object spectrum bandwidth (separability condition: $u_{Rm} \geqslant 4u_{0\max}$), (c) the steps of DSROP, the band-pass filtering method in the Fourier domain.

transform of $\tilde{C}(u, v)$, one obtains $\tilde{C}(x, y)$, and finally

$$\phi_0(x, y) = \phi_R(x, y) - \tan^{-1} \frac{\operatorname{Im}[C(x, y)]}{\operatorname{Re}[C(x, y)]}. \tag{3.26}$$

The band-pass filtering approach can also be implemented in the space domain. If we know the mean and maximum slopes of the phase, $(\partial \phi/\partial x)_m$ and $|\partial \phi/\partial x|_{\max}$, respectively, it is possible to define the central frequency

$$u_c = \frac{1}{2\pi} \left[\frac{\partial \phi}{\partial x} \right]_m, \tag{3.27}$$

and the bandwidth

$$\Delta u = \frac{1}{\pi} \left[\left| \frac{\partial \phi}{\partial x} \right|_{\max} - \left[\frac{\partial \phi}{\partial x} \right]_m \right] \tag{3.28}$$

of the band-pass filter.

Assuming that we again use the differential Gaussian filter with the transfer function from eq. (2.25), with the central frequency

$$\rho_c = 2\alpha_1\alpha_2 \sqrt{\frac{\ln\alpha_2 - \ln\alpha_1}{\alpha_2^2 - \alpha_1^2}} \sim \frac{\sqrt{2}}{2}(\alpha_1 + \alpha_2) \qquad (3.29)$$

(when $\alpha_1 \sim \alpha_2$), and that its bandwidth is

$$\Delta\rho \approx \Delta\rho(\alpha_1, \alpha_2) \sim \rho_c \cdot q(\alpha_1/\alpha_2), \qquad (3.30)$$

one can completely define the filter by calculating the Gaussian widths α_1, α_2 from ρ_c and $\Delta\rho$ imposed by the phase-modulated image (mean and maximum phase slopes of the investigated object or deformation).

If we take only the first term from the power series of the transfer function of the differential Gaussian filter

$$H(\rho) \approx \frac{1}{2}\left[\frac{1}{\alpha_2^2} - \frac{1}{\alpha_1^2}\right]\rho^2, \qquad (3.31)$$

it is easy to form the single-sideband filter

$$H_1(\rho) = H(\rho) \cdot \sigma(u), \qquad (3.32)$$

and to obtain the corresponding impulse response

$$h_1(r) = \tfrac{1}{2}h(r) + i\tfrac{1}{2}\mathcal{H}_x[h(r)], \qquad (3.33)$$

where $\sigma(u)$ is the unit step function and $\mathcal{H}_x[.]$ is the Hilbert transform with respect to the variable x.

Thus, in the space domain, the FP must be convolved with the impulse response $h(x, y)$; i.e.,

$$\mathfrak{I}(x, y) = I(x, y) * h_1(x, y)$$

$$\approx \frac{1}{4}\left(\frac{1}{\alpha_2^2} - \frac{1}{\alpha_1^2}\right)[\nabla^2 I(x, y) + i\mathcal{H}_x[\nabla^2 I(x, y)]]. \qquad (3.34)$$

Assuming that the background, the contrast and the phase modulation are slowly varying functions, as all $\phi'_m \approx ct$ and $\phi''_m \approx 0$, we can calculate

$$\nabla^2 I(x, y) \approx -B[(\phi'_{mx})^2 + (\phi'_{my})^2]\cos\phi_m(x, y), \qquad (3.35)$$

$$\mathcal{H}_x[\nabla^2 I(x, y)] \approx B[(\phi'_{mx})^2 + (\phi'_{my})^2]\sin\phi_m(x, y), \qquad (3.36)$$

and the filtered FP takes the form:

$$\mathfrak{I}(x, y) \approx \frac{B}{4}\left(\frac{1}{\alpha_2^2} - \frac{1}{\alpha_1^2}\right)[(\phi'_{mx})^2 + (\phi'_{my})^2]$$

$$\times [-\cos[\phi_m(x,y)] + i\sin[\phi_m(x,y)]]. \qquad (3.37)$$

Finally, one can obtain the spatial phase distribution:

$$\phi_m(x, y) = -\tan^{-1} \frac{\operatorname{Im}[\Im(x, y)]}{\operatorname{Re}[\Im(x, y)]}. \tag{3.38}$$

Summarizing the procedure, the steps to obtain the phase distribution are
  (i) smoothing of the intensity data,
  (ii) processing of the smoothed intensity pattern by the Laplacian window (with the weight $\frac{1}{4}(1/\alpha_2^2 - 1/\alpha_1^2)$),
  (iii) convolution of the result of (ii) with $(-1/\pi x)$; i.e., 1D Hilbert transform,
  (iv) division of the resulted images: (iii) by (ii),
  (v) calculation of the phase distribution by the arctangent of the result of (iv), as in eq. (3.38) (modulo 2π),
  (vi) unwrapping of the phase distribution, and
  (vii) smoothing of the phase distribution by a large-size median filter to eliminate the residual noise and oscillations.

In the case of speckle interferograms recorded with a specular reference, the intensity detected by the TV sensor can be written as:

$$I(x, y) = I_0(x, y) + I_R(x, y) + 2\sqrt{I_0(x, y)I_R(x, y)} \cos \phi(x, y), \tag{3.39}$$

where I_0 and I_R are the intensities of the object and reference beams, respectively, and $\phi(x, y)$ is the phase of the speckle pattern. Creath [1985] has used phase-shifting interferometry to determine the phase $\phi(x, y)$. In the same conditions, it is possible to obtain the phase from only one image using the band-pass filtering approach demonstrated previously. Before applying the Laplacian window filtering, the speckle pattern must be smoothed digitally. After the calculation of the phase distribution, the result must be enhanced against the residual noise by several passes of a median window with large sizes (5×5, 11×11; Creath [1985], Nakadate and Saito [1985]).

3.4. DSROP WITH AN ARBITRARY RANDOM REFERENCE IMAGE

This type of phase reconstruction is related mostly to electronic speckle-pattern interferometry [ESPI] (Butters and Leendertz [1971], Butters, Jones and Wykes [1978], Da Costa [1978], Stetson [1978], Slettemoen [1980], Løkberg [1980], Jones and Wykes [1983], Nakadate and Saito [1985], Løkberg and Slettemoen [1987]), although in conventional holography a diffuse reference beam was sometimes found to be advantageous (Waters [1972], Biedermann and Ek [1975]).

ESPI is a holographic method with the recording accomplished using low resolution devices (TV cameras). This particularity imposes some special features on the image recording and reconstruction; viz.,
   (i) the recording of an almost in-axis image hologram (with a small separation of the sidebands),
   (ii) the imaging with a small aperture system (f/20, ..., f/70) to yield a speckle size compatible with the resolution of the TV camera,
   (iii) the position of the pinhole of the reference beam is usually placed in the center of the aperture to ensure an approximate constant (low) spatial frequency across the hologram,
   (iv) the electronic processing of the video signal is equivalent to the image reconstruction in the holography, and
   (v) the speckle is a random carrier, but also a noise with special statistical properties, noise that must be suppressed in the final step of phase reconstruction.

Assuming that the intensity on the TV monitor screen is averaged over many speckles, the r.m.s. brightness of the image over that area is proportional to $\langle I_m^2 \rangle^{1/2}$, with

$$\langle I_m^2(x, y) \rangle = (g\gamma_0)^2 \langle I_0^2 \rangle + (g\gamma_R)^2 \langle I_R^2 \rangle + 2(g\gamma_{OR})^2 \langle I_0 \rangle \langle I_R \rangle \cos^2 \phi(x, y)$$

$$= \sum_{i=0}^{3} T_i(x, y), \qquad (3.40)$$

where $\langle I_0 \rangle$ and $\langle I_R \rangle$ are the respective average object and reference intensities at the TV camera; g is the gradient of the TV sensor characteristic, γ_0 and γ_R are the respective r.m.s. contrasts of the smoothed object and reference auto-interferences, γ_{OR} is the r.m.s. contrast of the object and reference cross-interference, and $\phi(x, y)$ is the phase of the speckle pattern. Equation (3.40) is valid in both cases of specular and diffuse reference with some changes of γ_R and γ_{OR} (Slettemoen [1980]).

Thus, we are strongly interested in extracting the cross-interference term from the speckle interference pattern and in smoothing as much as possible the speckle noise and its consequences from the phase distribution. To do that, the separability conditions in the Fourier domain must be studied, as in the case of conventional holography. If the power (Wiener) spectra of the square root of the terms of eq. (3.40) are denoted by S_i, it is possible to write the spectrum of the average brightness of the image as:

$$I_m(u, v) = \sum_{i=1}^{3} S_i(u, v). \tag{3.41}$$

Figure 10 illustrates the spectra corresponding to three types of apertures (i) circular aperture (Butters, Jones and Wykes [1978]), (ii) two-slit aperture (Biedermann and Ek [1975]), and (iii) periodic strip aperture (transparent–opaque–mirror; Slettemoen [1980]). In case (i), taking into account that usually $\gamma_R \ll \gamma_0$, it is possible to suppress the object auto-interference with respect to the cross-interference term simply by increasing the reference-to-object intensity ratio. In cases (ii) and (iii), the cross-interference term can be selected by band-pass filtering. This filtering can be accomplished in spatial or Fourier domains, as we have done previously.

The advantages of the use of a speckle reference are the easier adjustments,

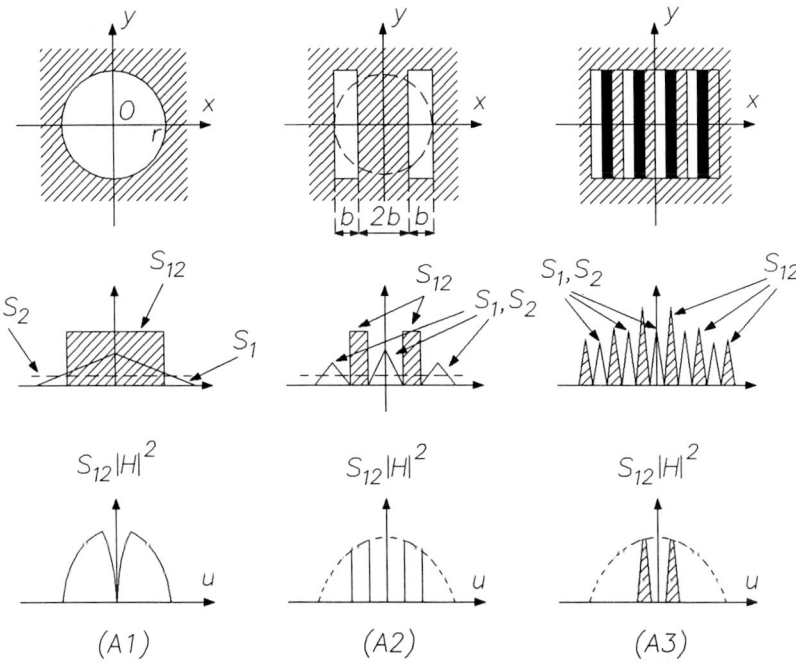

Fig. 10. Apertures used in ESPI with specular and speckle references and the separation of the sidebands in the signal spectra (first row): (A1) circular aperture; (A2) double-slit aperture, (A3) periodic strip aperture. In the second row, the corresponding signal spectra are shown, with the hatching of the useful cross-interference spectra, S_{12}. In the third row, the corresponding electronic filtered specta are presented.

the compactness of the setup, the elimination of the wedge from the surface of the TV sensor, and the easier equalization of the optical paths.

By using the TV image storage, the unwanted auto-interference term can be eliminated digitally by subtraction. This method, related to the Butters and Leendertz [1971] double exposure ESPI, will be studied in more detail in the next section.

In any case, to obtain precise results, the speckle interference patterns described by eq. (3.40) can be smoothed digitally before any DSROP procedure. In the final step of image processing, the phase distribution must be enhanced against the residual noise by several passes of a median window with large sizes (5×5, 11×11; Nakadate and Saito [1985], Creath [1985], Robinson and Williams [1986]).

§ 4. Direct Spatial Reconstruction of Optical Phase (DSROP) for Time- and Space-Phase-Modulated Images

In the previous analysis, some analogies in the reconstruction of phase from phase-modulated images (intensity distributions) by computers and by holograms were exploited largely to develop different DSROP methods. Now, we shall analyze the possibilities of reconstructing, by computer processing, the phase changes in time in a manner analogous to holographic interferometry. Thus, there are at least two methods (modes) of optical phase reconstruction. These are: (i) real-time DSROP; i.e., phase reconstruction at time intervals smaller than the TV frame period ($1/25$ s for European standard), and representation as an animated 3D map [if the object evolves slowly enough and the computer processing is fast enough, or if the object evolution was recorded on a video-cassette (or film) and is then processed frame by frame], and (ii) double-exposure DSROP; i.e., phase reconstruction from pairs of phase-modulated images corresponding to two "frozen" states of the time-evolving object. Obviously, the time evolution of the object can be produced by the time evolution of another physical parameter which changes the phase modulation.

4.1. REAL-TIME METHOD

In this method, the time-varying fringe pattern

$$I(x, y, t_i) = A(x, y) + B(x, y) \cos[2\pi u_0 x - \phi_m(x, y, t_i)], \tag{4.1}$$

is processed in each frame interval by a phase-demodulation procedure in the space domain (which is usually faster). One can multiply $I(x, y, t)$ by a carrier reference:

$$R(x, y) = \exp(i2\pi u_0 x) \tag{4.2}$$

to obtain, as in eq. (2.7),

$$\phi_m(x, y, t_i) = \tan^{-1} \frac{\text{Im}[Y(x, y, t_i)]}{\text{Re}[Y(x, y, t_i)]}, \tag{4.3}$$

and to animate the object time-varying phase (profile or refractive index).

Another possibility is to use as a reference a pre-recorded fringe pattern arising from an almost-perfect object (etalon) or computer-generated etalon pattern of the form:

$$R(x, y) = \exp(i[2\pi u_0 x - \phi_E(x, y)]) \tag{4.4}$$

to obtain

$$\phi_0(x, y, t_i) = \phi_E(x, y) - \tan^{-1} \frac{\text{Im}[Y(x, y, t_i)]}{\text{Re}[Y(x, y, t_i)]}, \tag{4.5}$$

which in manufacturing and testing is usually a smoothly and slowly varying function. The same result would be obtained from phase-modulated images without a definite spatial carrier; e.g., those described in § 3.3.

An interesting particular case is that of a low-frequency ($\omega < 2\pi \cdot 10$ rad/s) sinusoidal vibration of the optical phase modulation:

$$I(x, y, t) = A(x, y) + B(x, y) \cos[2\pi u_0 x - \phi_1(x, y) \sin \omega t]. \tag{4.6}$$

Multiplying by $R_c(x, y) = \cos 2\pi u_0 x$ and spatially low-pass filtering the result, one obtains

$$(I \cdot R_c)_3 = \frac{B}{2} \cos[\phi_1(x, y) \sin \omega t]$$

$$= \frac{B}{2} \left[J_0(\phi_1) + 2 \sum_{n=1}^{\infty} J_{2n}(n\phi_1) \sin 2n\omega t \right]. \tag{4.7}$$

Upon multiplying further by $R_s(x, y) = \sin 2\pi u_0 x$ and spatially low-pass filtering the result, one obtains

$$(I \cdot R_s)_3 = \frac{B}{2} \sin[\phi_1(x, y) \sin \omega t]$$

$$= \frac{B}{2} \left[2 \sum_{n=0}^{\infty} J_{2n+1}(n\phi_1) \sin(2n+1)\omega t \right]. \tag{4.8}$$

If the time signal $(I \cdot R_c)_3$ is temporally low-pass filtered, the result is

$$\text{LPF}(I \cdot R_c)_3 = \frac{B}{2} \cdot 2J_0[\phi_1(x, y)]; \tag{4.9}$$

if $(I \cdot R_s)$ is temporally band-pass filtered around the frequency ω and then (envelope) rectified, the result is

$$\text{BFP}(I \cdot R_s)_3 = \frac{B}{2} \cdot 2J_1[\phi_1(x, y)]. \tag{4.10}$$

Dividing eq. (2.54) by eq. (2.53), one obtains

$$\frac{\text{BPF}(I \cdot R_s)_3}{\text{LPF}(I \cdot R_c)_3} = \frac{2J_1[\phi_1(x, y)]}{J_0[\phi_1(x, y)]} \approx \phi_1(x, y), \quad \phi_1 < 0.3. \tag{4.11}$$

In this manner, it is possible to visualize in (nearly) real time the distribution of the low-frequency vibration amplitudes of holographic tables, large mechanical structures, buildings, infrasound devices, etc. For large structures, assuming stationary vibrations, one can assemble a large pattern of phase as a mosaic of subimages which met the format and the resolution of the TV camera.

Band-pass filtering in the space domain, demonstrated in § 2.2 and § 3.3, is another procedure which can be implemented easily for real-time DSROP, and is the fastest variant.

Certainly, the real-time DSROP is determined strongly by the power of the image-processing equipment, but state-of-the art interferometers, like the Direct Interferometer 100 (Carl Zeiss), contain powerful host systems and pipeline image processors, which will be able to implement such real-time processing (Doerband, Wiedman, Wegmann, Kuechel and Freischland [1990], Freischland, Kuechel, Wiedman, Keiser and Mayer [1990]). For slower equipment, frame-by-frame processing remains a possibility.

4.2. DOUBLE-EXPOSURE METHOD

The observation of optical phase evolution by the double-exposure method consists of recording two phase-modulated images corresponding to two (close) states of the investigated object:

$$I_1(x, y) = A(x, y) + B(x, y) \cos[2\pi u_0 x - \phi_1(x, y)], \tag{4.12}$$

and

$$I_2(x, y) = A(x, y) + B(x, y) \cos[2\pi u_0 x - \phi_2(x, y)]. \tag{4.13}$$

We can demodulate independently the fringe patterns from eqs. (4.12) and (4.13) by multiplication with the computer-generated reference

$$R(x, y) = \exp(i2\pi u_0 x) \qquad (4.14)$$

as in § 2.1, to obtain the phase difference, corresponding to the change of the investigated object between exposures:

$$\phi_1(x, y) - \phi_2(x, y) = -\tan^{-1}\frac{\text{Im}[Y_1(x, y)]}{\text{Re}[Y_1(x, y)]} + \tan^{-1}\frac{\text{Im}[Y_2(x, y)]}{\text{Re}[Y_2(x, y)]}. \qquad (4.15)$$

Actually, if $\phi_1(x, y) - \phi_2(x, y)$ is small and we know the variation of the parameter which led to this phase variation (e.g., time, pressure, temperature, etc.), we can display the phase gradient with respect to that parameter. Moreover, an animation following the successive phase differences, when the free parameter takes different (say, increasing) values, is also possible.

Alternatively, the computer can generate the quadrature signal of $I(x, y)$ and then, after a spatial band-pass filtering,

$$R_2(x, y) = B \exp[i(2\pi u_0 x - \phi(x, y))]. \qquad (4.16)$$

Multiplying eq. (4.12) by eq. (4.16), the result is

$$\begin{aligned}I_1 \cdot R_2 &= A \cdot R_2 + B^2[\cos(2\pi u_0 x - \phi_1)] \\ &\quad \times [\cos(2\pi u_0 x - \phi_2) + i \sin(2\pi u_0 x - \phi_2)] \\ &= A \cdot R_2 + \tfrac{1}{2}B^2[\cos(4\pi u_0 x - \phi_1 - \phi_2)] + \tfrac{1}{2}B^2[\cos(\phi_1 - \phi_2)] \\ &\quad + i\tfrac{1}{2}B^2[\sin(2\pi u_0 x - \phi_1 - \phi_2)] - i\tfrac{1}{2}B^2[\sin(\phi_1 - \phi_2)]. \end{aligned} \qquad (4.17)$$

After a spatial low-pass filtering, we get:

$$\phi_1(x, y) - \phi_2(x, y) = -\tan^{-1}\frac{\text{Im}[\text{LPF}(I_1 \cdot R_2)]}{\text{Re}[\text{LPF}(I_1 \cdot R_2)]}. \qquad (4.18)$$

In the particular case where the second phase-modulated image is a sheared version of the first one, one can display different derivatives of the phase, some of them related to the transverse aberrations of the tested object (wavefront) (Malacara [1992]) or to some mechanical parameters such as bending moments, strains and stresses (Nakadate and Saito [1985]).

In the double-exposure method, the FP can be demodulated by band-pass filtering in the space domain (described in § 2.2 and § 3.3), and in the final step the phase difference can be calculated by subtracting the derived phase distributions as in eq. (4.15). A similar method was used by Creath [1985] in phase-shifting ESPI.

An interesting double-exposure method, developed by Nakadate and Saito [1985] in phase-shifting ESPI (using five different images), can also be implemented by DSROP methods (using two images only). In our terms, this method implies the recording of two speckle interferograms, with the object in slightly different states:

$$I_1(x, y) = I_0(x, y) + I_R(x, y) + 2\sqrt{I_0(x, y)I_R(x, y)} \cos \phi(x, y), \quad (4.19)$$

and

$$I_2(x, y) = I_0(x, y) + I_R(x, y)$$
$$+ 2\sqrt{I_0(x, y)I_R(x, y)} \cos [\phi(x, y) - \Delta\phi(x, y)]. \quad (4.20)$$

The square difference of the patterns from eqs. (4.19) and (4.20) leads to:

$$|I_1(x, y) - I_2(x, y)|^2 = 4I_0 I_R \{\cos \phi - \cos[\phi(x, y) - \Delta\phi(x, y)]\}$$
$$= 8I_0 I_R \sin^2 \phi \sin^2(\Delta\phi/2). \quad (4.21)$$

When the ensemble average is taken over many realizations of these secondary fringes, the resulting image is

$$\langle |I_1(x, y) - I_2(x, y)|^2 \rangle = 8I_0 I_R \sin^2(\Delta\phi/2)$$
$$= 4I_0 I_R [1 - \cos(\Delta\phi(x, y))]. \quad (4.22)$$

After smoothing this speckle interferogram, it is possible to apply any described DSROP method to determine $\Delta\phi(x, y)$.

§ 5. An Overview of DSROP Methods. Range of Measurement, Accuracy and Speed

5.1. A SYSTEMATIC CLASSIFICATION OF DSROP METHODS

The sources of optical phase-modulated images (PMI) are (i) interferometry (I), (ii) holographic interferometry (HI), (iii) moiré methods (M), and (iv) speckle-pattern methods (SP). In table 1, we show a systematic classification of DSROP methods with respect to the type of reference pattern used in the computer demodulation and to the type of phase-modulated image introduced in the computer (stationary/time-varying). In the upper part of each entry is written the source of PMI in which holographic, real-time and double-exposure methods were used with a specific type of reference. In the lower part, we have shown whether the correspond-

TABLE 1
Overview of DSROP methods.

Mode Reference	Holographic (Stationary PMI)	Real-time (Time-varying PMI)	Double-exposure (Time-varying PMI)
1 Sinusoidal	I,HI,M,SP	HI,M	HI,M,SP
	DSROP: ex SD, FD	DSROP: new	DSROP: new
2 Chirped (Rot.	I,HI,M	HI,M	HI,M
symmetric)	DSROP: ex FD, new SD	DSROP: new	DSROP: new
3 Sheared	I,HI,M,SP	I,M	I,M,HI,SP
sinusoidal	DSROP: new	DSROP: new	DSROP: new
4 Arbitrary	HI	HI	HI
deterministic	DSROP: ex FD, new SD	DSROP: new	DSROP: new
5 Arbitrary	SP	SP	SP
random	DSROP: ex FD, new SD	DSROP: new	DSROP: new

ing DSROP method exists (denoted by "ex") or is new (denoted by "new"). Each DSROP method can be implemented in the space domain (SD) and the Fourier domain (FD).

5.2. THE RANGE OF MEASUREMENT, ACCURACY AND SPEED OF DSROP METHODS

The range of measurement in DSROP is limited by the basic relations developed in the phase modulation and demodulation in communication theory. Thus, to avoid overmodulation, the phase derivative is upper-limited by the carrier frequency. On the other hand, in DSROP methods the spatial carrier frequency was taken as high as possible, at the resolution limit of the TV sensor, u_{RTV}; i.e.,

$$\frac{1}{2\pi}\frac{\partial \phi_m}{\partial x} \ll u_0 = u_{RTV}. \tag{5.1}$$

To be more precise, taking into account again the holographic constraints (viz., the maximum frequency in the object spectrum must be smaller than one-third of the carrier frequency; fig. 1), we can write:

$$\frac{1}{2\pi}\left|\frac{\partial \phi_m}{\partial x}\right|_{max} \leqslant \frac{u_0}{3}, \tag{5.2}$$

or

$$\left|\frac{\partial \phi_m}{\partial x}\right|_{max} \leqslant \frac{2\pi}{3}u_0. \tag{5.3}$$

A similar relation was obtained previously by Takeda and Mutoh [1983], in a slightly different approach. As far as $u_0 = u_{RTV}$, eq. (5.3) can be generalized to the case of phase modulation without a carrier (or with an arbitrary deterministic carrier).

Assuming that the dimension of the image detected by the TV camera is N_x points, one can obtain the maximum range of monotonically varying phase that it is possible to measure by DSROP:

$$|\phi_m|_{max} \leqslant \frac{2\pi}{3} u_0 N_x. \tag{5.4}$$

For $N_x = 512$ pixels and $u_0 = \frac{1}{3}$, eq. (5.4) yields:

$$|\phi_m|_{max} \leqslant 57 \times (2\pi). \tag{5.5}$$

If the phase is varying slowly in a nonmonotonic way, we may assume as its first harmonic:

$$\phi_m(x) = a \cos(2\pi u_m x). \tag{5.6}$$

Equation (5.3) leads to:

$$2\pi a u_m \leqslant \frac{2\pi}{3} u_0, \tag{5.7}$$

or

$$a u_m \leqslant \tfrac{1}{3} u_0. \tag{5.8}$$

Equation (5.8) shows that in this case the amplitude of the phase modulation is limited by the ratio between the carrier frequency and the modulation frequency. One can calculate, for example, that for measuring $a_{max} = \pi$, one needs $u_m < u_0/9$; for $a_{max} = 2\pi$, one needs $u_m < u_0/18$, and for $a_{max} = 10 \times 2\pi$, one must impose that $u_m < u_0/190$ (which means a little bit less than one oscillation for $N_x = 512$ pixels).

The relation between the signal bandwidth B, the phase amplitude (modulation index), and the modulation frequency is known in communication theory as the Carson relation, and for large amplitudes takes the form (Stremler [1982]):

$$B \approx 2u_m(a + 1). \tag{5.9}$$

Imposing $B < u_0$, one can obtain a slightly different relation than that of eq. (5.8), but with the same qualitative features.

In any case, the phase-slope variations will be limited by the resolution of the TV sensor. It is now clear that if we want to measure accurately a large phase variation with a limited resolution in the TV image, we have to record only parts of the objects with slow phase variations. Thus, it is sometimes not possible to characterize large objects all at once, and a mosaic of FPs must be carefully designed with proper phase continuity at the borders.

Our experiments with the smoothing method in the conditions demonstrated above have shown relative errors of the phase of less than 5% (due to the approximations introduced in the calculations and the replacement of the integral by the simple and fast average-window algorithm). The demodulation in the Fourier domain, with the same conditions and FP, led in general to smaller errors. The errors appear, in this case, due mainly to the subjective selection of the central frequency and bandwidth of the band-pass filter. The computing time was essentially shorter for demodulation in the space domain.

The signal-to-noise ratio is also the largest possible in the case of the spatial heterodyne methods for phase demodulation as used in DSROP.

For the demodulation in the Fourier domain, Takeda [1989] derived the following r.m.s. phase error equation:

$$\sqrt{\langle [\Delta\phi(x, y)]^2 \rangle} = (n/N)^{1/2}/S(x, y), \tag{5.10}$$

where n is the numer of sample points inside the band-pass filter in the Fourier domain, N is the total number of sample points, and $S(x, y)$ is the signal-to-noise ratio, defined as

$$S(x, y) = \frac{B(x, y)}{\sqrt{2}\sigma}, \tag{5.11}$$

with $B(x, y)$ being the fringe contrast and σ the r.m.s. value of the noise.

The accuracy of different variants of DSROP are further determined by the numerical algorithms which must run on properly sampled data and which decrease the noise. An initial smoothing of the FP is always indicated and, for speckle interferograms, is imposed. For this reason and for its high processing speed, the smoothing method is particulary suitable in any first attempt of processing an FP with a carrier frequency. When the phase-modulated image has no carrier frequency (arbitrary reference), the band-pass filtering (BPF) method in the space domain takes the place of the smoothing method. The accuracy of the BPF method is determined strongly by the knowledge of the central frequency and the bandwidth of the filter used; i.e., of the mean and maximum phase slopes of the investigated object or deformation.

Another source of errors in both space and Fourier domains is the sharp cut of the fringe patterns at the boundaries, leading to the spread of the spectrum sidebands and to the difficulties in finding the cut-off frequency (bandwidth) of the filter. These errors were studied and minimized by Roddier and Roddier [1987], Green, Walker and Robinson [1988], and Kujawinska, Spik and Robinson [1989]. Some possible errors due to TV sensor nonlinearity and their correction algorithm were studied by Nugent [1985]. Additional errors can be introduced by the unwrapping algorithms, which yield in some cases phase dislocations (Takeda, Ina and Kobayashi [1982], Macy [1983], Barr, Coudé du Foresto, Fox, Paczulp, Richardson, Roddier and Roddier [1991], Bone [1991]).

§ 6. Applications of the Methods of DSROP

DSROP methods can be applied to any type of interferometry (e.g., conventional, holographic, or ESPI) as well as to the moiré methods. Applications cover at least three important areas: optical testing, optomechanical testing and profilometry, and optical diagnosis of inhomogeneous transparent materials. Most phase-shifting methods can be converted into DSROP methods, with the advantages of the acquisition and processing of only one image.

6.1. OPTICAL TESTING

Various interferometric and moiré methods are used in modern optical testing (Malacara [1992]) for the measurement of shape and different features of the surface shape (e.g., Zernike coefficients, Seidel coefficients, etc.). The phase map of optical components was obtained using the sinusoidal fitting method (Macy [1983]), the smoothing method (Tooyoka and Tominaga [1984]) in the space domain, and the band-pass filtering method in the Fourier domain (Takeda and Kobayashi [1984], Doerband and Tiziani [1985], Kujawinska, Spik and Wojciak [1989], Frankowsky, Stobbe, Tischer and Schillke [1989], Vlad and Petris [1994]).

Recently, DSROP algorithms have been implemented in the new Zeiss interferometer, DIRECT 100, in an advanced hardware and software design (Kuechel [1990], Doerband, Wiedman, Wegmann, Kuechel and Freischland

[1990]). The high-precision of interferometric testing of spherical mirrors with long radii of curvature was proved with this interferometer (Freischland, Kuechel, Wiedman, Keiser and Mayer [1990], Freischland [1992]). The Zeiss interferometer uses a spatial heterodyne algorithm implemented in the modern digital video-processing hardware. The high processing speed makes this instrument less sensitive to instabilities, especially with large testing arrangements. Another important feature of this instrument is the real-time display of the calculated wavefronts. The accuracy in the measurement of spherical mirrors is in the angstrom range and is highly reproducible, even for long beam paths in air.

The phase demodulation in the Fourier plane is also used in the evaluation of large optical surfaces by an instrument called ECAPS (a product of the Lockheed Co.) and in the testing of large-diameter telescope mirrors (3.5 m honeycomb mirrors) with accuracy of 0.01 µm (Barr, Coudé du Foresto, Fox, Paczulp, Richardson, Roddier and Roddier [1991]). Some of our results in interferometric testing of wavefronts are shown in figs. 11 and 12.

It is noteworthy that DSROP with an arbitrary deterministic reference can be applied to the testing of aspherical optical components without null compensators (Malacara and Vlad [1992], Servin, Malacara, Malacara and Vlad [1994]).

6.2. OPTO-MECHANICAL TESTING AND PROFILOMETRY

Object deformations and vibrations have been investigated largely by holographic interferometry, ESPI and moiré methods. Since the first paper by Takeda, Ina and Kobayashi [1982], DSROP methods have been used frequently to obtain phase maps of object deformations and profiles.

In holographic interferometry, Dzubur and Vukicevic [1984] have used a modified method of sandwich holographic interferometry, providing inherently a spatial carrier frequency, and the band-pass filtering method in the Fourier domain to yield ultra-high resolution of the surface deformation. The method was tested for a bent cantilever beam and for a clamped, supported beam with precise results. The rigid-body motion, an adversary of quantitative holographic interferometry, was turned into a useful tool for phase demodulation. For double-exposure holographic interferometry, Takeda and Tung [1985] have shown experimentally that the phase demodulation in the Fourier domain can attain a subfringe sensitivity in the measurement of object deformations. Recently, we have demonstrated the

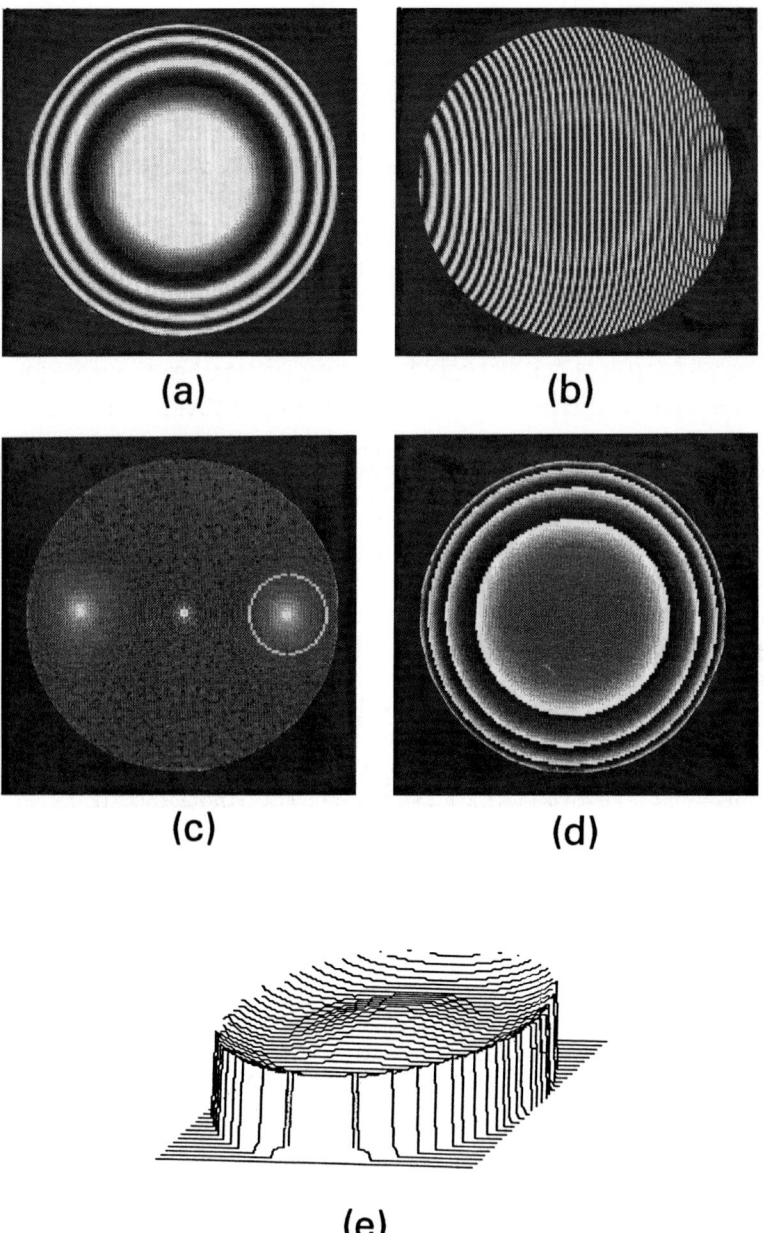

Fig. 11. Application of the DSROP, the band-pass filtering method in the Fourier domain, in optical testing: (a) the interferogram of a wavefront with spherical aberration $W_S = 3.3\lambda$, (b) the interferogram of the same wavefront with a carrier frequency (tilt, $u_0 = \frac{1}{3}$), (c) the Fourier spectrum of the interferogram from (b), (d) the wrapped phase distribution calculated by the band-pass filtering method in the Fourier domain, (e) 3D plot of the unwrapped phase distribution corresponding to the spherical aberration ($W_S = 3.3\lambda$).

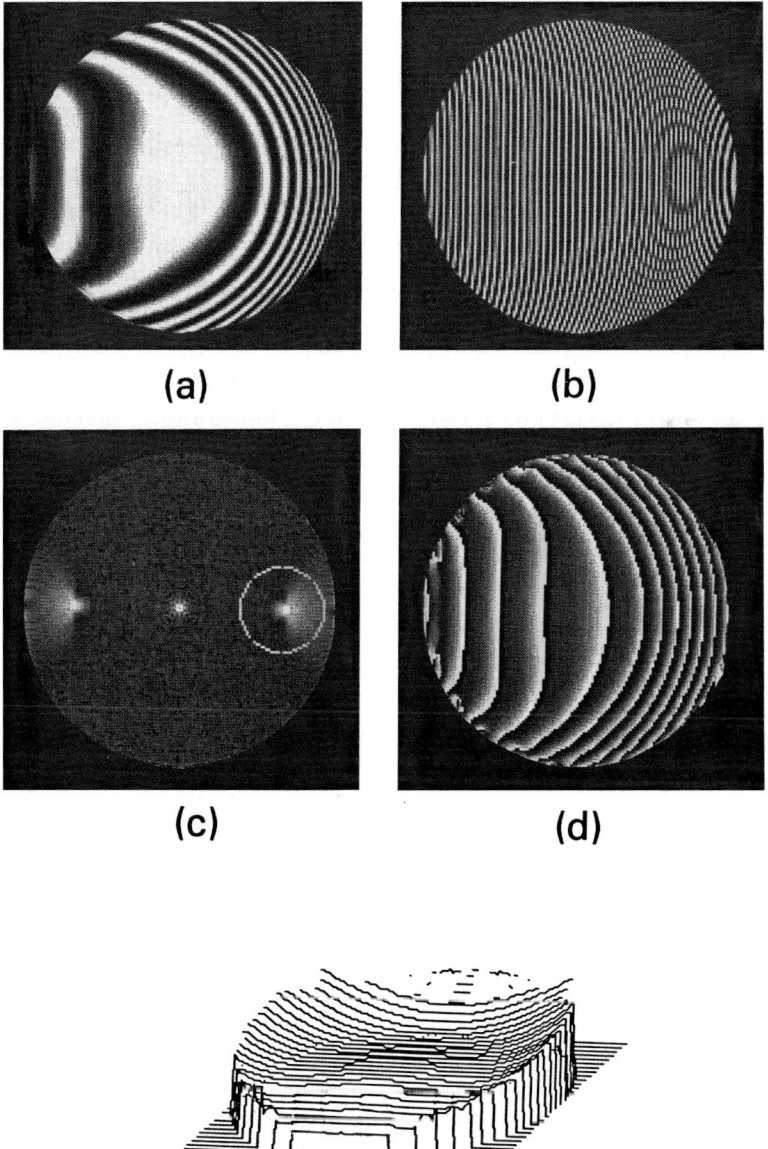

Fig. 12. Application of the DSROP, the band-pass filtering method in the Fourier domain, in optical testing: (a) the interferogram of a wavefront with spherical aberration $W_S = 3.3\lambda$ and coma aberration ($W_C = 5\lambda$), (b) the interferogram of the same wavefront with a carrier frequency (tilt, $u_0 = \frac{1}{3}$), (c) the Fourier spectrum of the interferogram from (b), (d) the wrapped phase distribution calculated by the band-pass filtering method in the Fourier domain, (e) 3D plot of the unwrapped phase distribution corresponding to the spherical and coma aberrations ($W_S = 3.3\lambda$ and $W_C = 5\lambda$).

same features by using a simpler and faster smoothing method (Solano, Vlad, Malacara and Voicu [1994]).

In optical profilometry, Takeda and Mutoh [1983] demonstrated a fast-Fourier-transform method for phase demodulation. They used an optical setup similar to that of projection moiré topography, but the grating image projected on the object was detected directly by the TV camera without the second grating (to generate the moiré fringes). The DSROP method has shown a much higher sensitivity than the conventional moiré techniques, and was capable of fully automatic discrimination between depressions and elevations on the object surface. Other advantages of this method were the elimination of the fringe interpolation and of the errors yielded by the spurious moiré fringes generated by the higher harmonic components of the grating. Improved results were obtained recently by Suganuma and Yoshizawa [1991] with the enhancement of visibility and multiplication of the moiré fringes, and by Simova and Stoev [1993] with the holographic moiré.

The smoothing method in the space domain was used by Toyooka and Tominaga [1984] and Toyooka and Iwaasa [1986] for the automatic profilometry of 3D reflecting and diffusing objects, respectively. In comparison with the Fourier-transform method, it was shown to simplify strongly the calculations, with a corresponding decrease of the computing time, but with an additional source of errors due to neglect of the phase gradient; the latter remains in an acceptable range if the object slope is not very large.

Some of our results using the smoothing method with the 3×3 averaging window are shown in figs. 13, 14 and 15. The good accuracy of the method in a spatial phase demodulation of the periodic PMI corresponding to a periodic corrugated object (figs. 13b–e; $a = 1$; $u_m = u_0/20$). The results are improved further by an additional smoothing with a 3×3 average window (figs. 13c and e). Upon increasing the profile height, a, the accuracy remains acceptable (figs. 14 and 15), but if we want to preserve the accuracy in a larger measurement range, the modulation frequency (phase slope variations) must be decreased acording to eq. (5.8). For large phase variations, we have to introduce a local "magnification" of the PMI; i.e., for a fixed resolution, to decrease the size of the analyzed pattern. The same problem also appears in holographic interferometry; when increasing the deformation, the fringes become so close that a local optical magnification must be introduced to make fringe counting possible (with the corresponding decrease of the size of the analyzed area). The computing time for our algorithm on an IBM (Risk) 6000 computer was less than 5 s, and ensured a mean error of less than 5%.

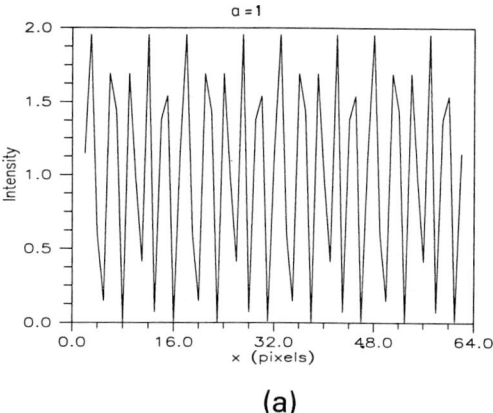

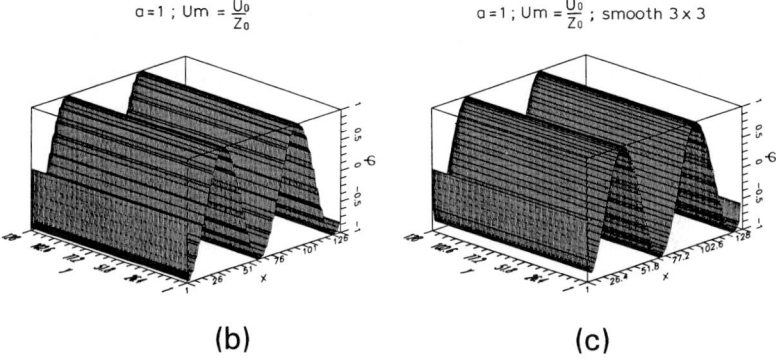

Fig. 13. Application of the DSROP, the smoothing method in the space domain, in profilometry: (a) the intensity distribution in the periodic PM image $I(x) = 1 + \cos[(2\pi/3)x + a \cos 2\pi u_m x]$; $a = 1$; $u_m = (2\pi/\Lambda) = u_0/20$; $u_0 = \frac{1}{3}$, (b) 3D plot of the object phase (profile) calculated from $I(x)$, (c) 3D plot of the object phase after an additional smoothing by 3×3 averaging, (d) 2D plot of the object phase (profile) calculated from $I(x)$ (solid line). The original object phase is shown with dashed line. The mean error is 6.49%. (e) 2D plot of the object phase calculated from $I(x)$ and smoothed by 3×3 averaging. The mean error is 5.98%.

Recently, phase demodulation in the Fourier domain was introduced in ESPI (Preater and Swain [1992]). We have used the smoothing method with the average window (and additional median filterings) in the phase-map calculation in ESPI for different in-plane movements (Moore, Vlad, Mendoza and Mendoza [1994]). In this case, the errors are larger than 5% and the computing time was in the range of 20 s (depending on the number

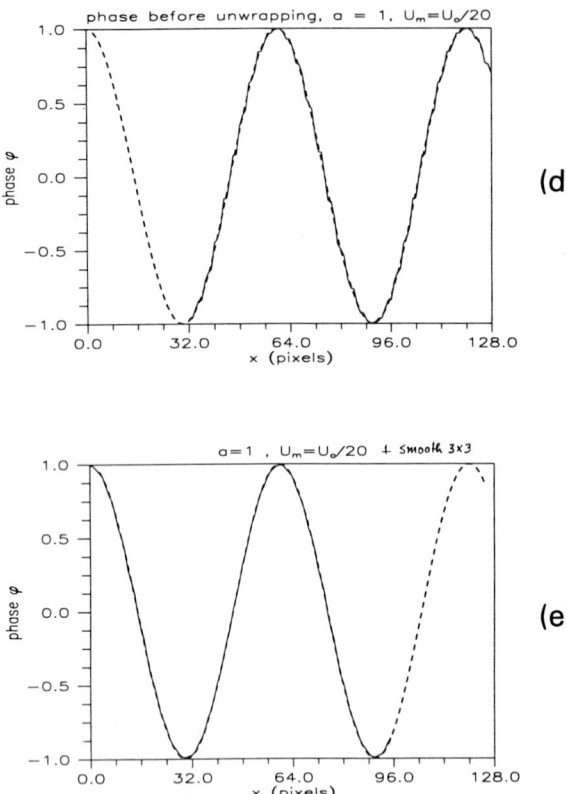

Fig. 13. Continued

of filterings taken for the smoothing of speckles). The DSROP methods can be applied in ESPI setups and in algorithms for measuring in-plane, out-of-plane and 3D deformations (Nakadate and Saito [1985]).

6.3. OPTICAL DIAGNOSIS OF INHOMOGENEOUS TRANSPARENT MATERIALS

The optical diagnosis of inhomogeneous transparent media is a powerful tool in the measurement of plasma, fluid flow, heat transfer and doping (controlling the gradient index) phenomena.

The first application of the band-pass filtering method in the Fourier domain for measurement of the phase (index) distribution in heat transfer from a heat source to air was made by Takeda, Ina and Kobayashi [1982].

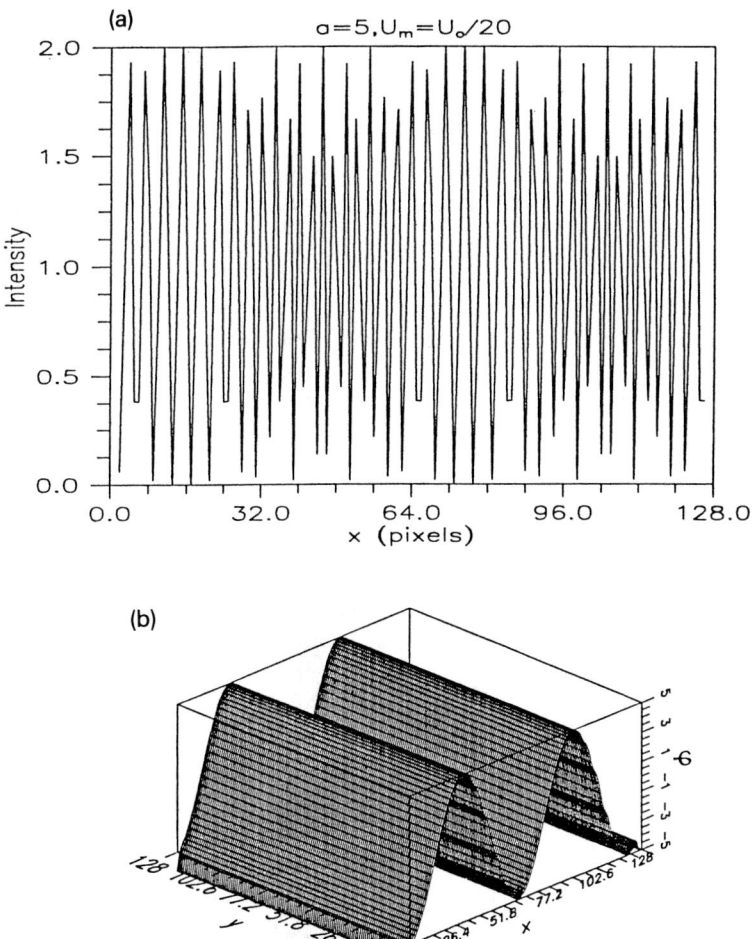

Fig. 14. Application of the DSROP, the smoothing method in the space domain, in profilometry ($a = 5$; $u_m = u_0/20$; $u_0 = \frac{1}{3}$): (a) the intensity distribution in the periodic PM image, $I(x)$, for $a = 5$, $u_m = u_0/20$, $u_0 = \frac{1}{3}$, (b) 3D plot of the object phase calculated from $I(x)$ ($a = 5$), (c) 2D plot of the wrapped object phase calculated from $I(x)$ ($a = 5$), (d) 2D plot of the unwrapped object phase calculated from $I(x)$ ($a = 5$), and smoothed by 3×3 averaging. The mean error between the original (dashed line) and calculated (solid line) phases is 7.17%; before the additional smoothing, it was 15.4%.

The phase profile was determined uniquely, precisely, and without sign ambiguity.

Automated analysis of the magnetic fields in laser-induced plasmas by the Fourier-transform method was accomplished by Kalal, Nugent and Luther-Davies [1987]. They used a modified Nomarski interferometer which

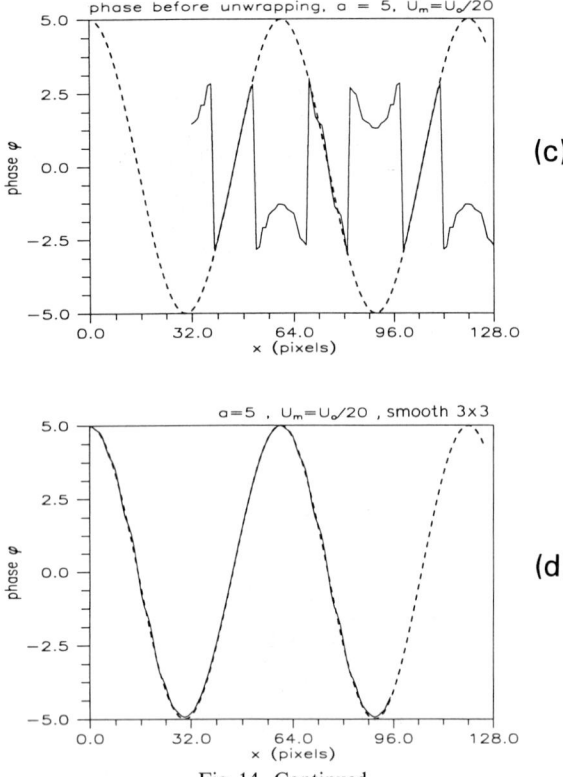

Fig. 14. Continued.

encodes both the strength of the magnetic field and the plasma density into the amplitude and the phase of the interferogram, respectively, and demonstrated experimentally that both can be reconstructed correctly by the use of band-pass filtering in the Fourier domain.

Snyder and Hesselink [1988] demonstrated the first "instantaneous" optical tomographic reconstruction of a nonstationary fluid flow. The experimental architecture consisted of a holographic setup, in which 18 double-exposure holographic interferograms of the flow were recorded from uniformly separated directions. All holographic interferograms were recorded simultaneously in 300 µs, so as to "freeze" jets of helium (inner) and air (outer) co-flowing with a speed of 6 m/s. Between the exposures, the reference wave was tilted by a small amount, so that finite fringe holographic interferograms were recorded. The holographic interferograms were then processed by the bandpass filtering method in the Fourier domain. The phase distributions were

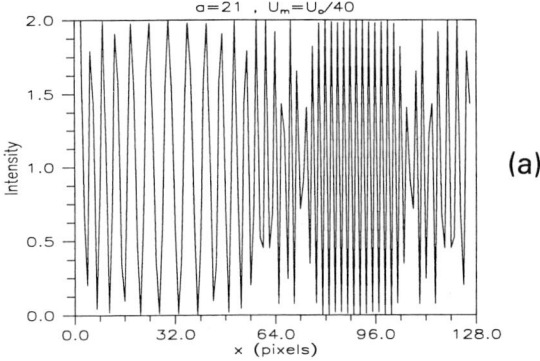

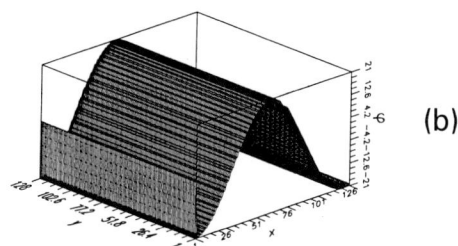

Fig. 15. Application of the DSROP, the smoothing method in the space domain, in profilometry ($a = 21$, $u_m = u_0/40$, $u_0 = \frac{1}{3}$): (a) the intensity distribution in the periodic PM image, $I(x)$, for $a = 21$, $u_m = u_0/40$, $u_0 = \frac{1}{3}$, (b) 3D plot of the object phase calculated from $I(x)$ ($a = 21$), (c) 2D plot of the wrapped object phase calculated from $I(x)$ ($a = 21$), (d) 2D plot of the unwrapped object phase calculated from $I(x)$ ($a = 21$), and smoothed by 3×3 averaging. The mean error between the original (dashed line) and calculated (solid line) phases is 15.1%; before the additional smoothing, it was $\sim 30\%$.

used for the convolution-back-projection processing, which reconstructed the spatial concentration of the co-flowing gases at different distances from the nozzle. The spatial resolution was 3 mm and the concentration errors were less than 3%. By using absorption techniques and a tunable laser, the species concentration may also be measured by the same methods.

§ 7. Concluding Remarks

The direct spatial reconstruction of optical phase (DSROP) and its applications in various interferometric and moiré methods can generate a large

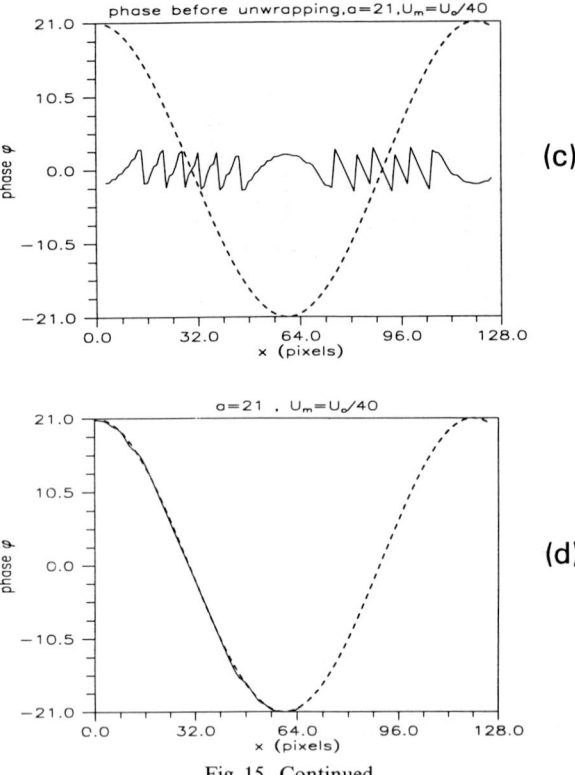

Fig. 15. Continued.

number of new methods in which optics and computer image processing can find optimal interconnections. Many phase-shifting interferometric methods can be "shifted" to DSROP equivalents, with the advantages of single (or double) image processing, greater insensitivity to perturbations, and the elimination of phase-shifting devices; the resolution limitation is introduced by the real TV sensors.

In this paper, we have improved some DSROP methods, such as the smoothing method (with the averaging window) and the band-pass filtering method (with the differential Gaussian filter) in the space domain. Some new methods are also proposed, including DSROP in the space domain for patterns with chirped radially symmetric (Fresnel and Gabor zone plates) carriers, DSROP for phase-modulated images with an arbitrary random reference, and some DSROP methods for space- and time-PMI. All of these contributions are introduced in a general review of the field, which can provide the frame for new developments.

Acknowledgements

One of the authors (V.V.) expresses his gratitude to Dr. Cristina Solano for the invitation to C.I.O. México, and for the excellent conditions created for the research work. Both of us are grateful to Dr. Bernardo Mendoza for his help in the implementation of the smoothing algorithm and the use of all resources of the IBM (Risk) 6000 computer. Thanks are addressed to Dr. Andrew Moore for useful discussions regarding phase-shifting ESPI, the range of measurement and 2D unwrapping. Our thanks are addressed also to Abundio Dávila and Carlos Treviño for their efforts in developing PC computer programs for our DSROP methods. We also thank the Consejo Nacional de Ciencia y Tecnología, México, for financial support.

References

Barr, L.D., V. Coudé du Foresto, J. Fox, G.A. Paczulp, J. Richardson, C. Roddier and F. Roddier, 1991, Large-mirror testing facility at the National Optical Astronomy Observatories, Opt. Eng. **30**, 1405.

Becker, F., G.E.A. Maier and H. Wegner, 1982, Automatic evaluation of interferograms, Proc. SPIE **359**, 386.

Biedermann, K., and L. Ek, 1975, A recording and display system for hologram interferometry with low resolution imaging devices, J. Phys. E **8**, 571.

Bone, P.M., 1991, Fourier fringe analysis: the two-dimensional phase unwrapping problem, Appl. Opt. **30**, 3627.

Bone, D.J., H.A. Bachor and R.J. Sandeman, 1986, Fringe pattern analysis using 2-D Fourier transforms, Appl. Opt. **25**, 1653.

Boone, P.M., and L.C. De Backer, 1973, Determination of three orthogonal displacement components from one double exposure hologram, Optik **37**, 61.

Brandt, G.B., 1969, Image plane holography, Appl. Opt. **8**, 1421.

Brown, G.M., and R.J. Pryputniewicz, 1992, Nanomeasurements by heterodyne holographic interferometry, Proc. SPIE **1755**, Paper 1756-31.

Bruning, J.H., D.R. Herriott, J.E. Gallagher, D.P. Rosenfeld, A.D. White and D.H. Brangaccio, 1974, Digital wavefront measurement interferometer, Appl. Opt. **13**, 2693.

Butters, J.N., and J.A. Leendertz, 1971, Speckle pattern and holographic techniques in engineering metrology, Opt. Laser Technol. **3**, 26.

Butters, J.N., R. Jones and C. Wykes, 1978, Electronic speckle pattern analysis, in: Speckle Metrology, ed. R.K. Erf (Academic Press, New York) p. 267.

Carre, P., 1966, Installation et utilisation du comparateur photoelectrique et interferentiel du Bureau International des Poids et Measures, Metrologia **2**, 13.

Creath, K., 1985, Phase shifting speckle interferometry, Appl. Opt. **24**, 3053.

Creath, K., 1988, Phase measurement interferometry techniques, in: Progress in Optics, Vol. XXVI, ed. E. Wolf (North-Holland, Amsterdam) p. 349.

Da Costa, G., 1978, Transient phenomena analysis, in: Speckle Metrology, ed. R.K. Erf (Academic Press, New York) p. 267.

Doerband, B., and H.J. Tiziani, 1985, Testing aspheric surfaces with computer-generated holograms: analysis of adjustment and shape errors, Appl. Opt. **24**, 2604.
Doerband, B., W. Wiedman, U. Wegmann, W. Kuechel and K.R. Freischland, 1990, Software concept for the new Zeiss interferometer, Proc. SPIE **1332**, 664.
Dzubur, A., and D. Vukicevic, 1984, Ultrahigh resolution sandwich holography, Appl. Opt. **23**, 1474.
Frankowski, G., I. Stobbe, W. Tischer and F. Schillke, 1989, Investigation of surface shapes using a carrier frequency based analysis system, Proc. SPIE **1121**, 89.
Freischland, K.R., 1992, Wavefront integration from difference data, Proc. SPIE **1755**, Paper 1755-28.
Freischland, K., M. Kuechel, W. Wiedman, W. Keiser and M. Mayer, 1990, High precision interferometric testing of spherical mirrors with long radius of curvature, Proc. SPIE **1332**, 8.
Gabor, D., 1949, Microscopy by reconstructed wavefronts, Proc. R. Soc. (London) Ser. A **197**, 454.
Gabor, D., 1969, Associative holographic memory, IBM J. Res. Dev. **13**, 156.
Green, R.J., J.G. Walker and D.W. Robinson, 1988, Investigation of the Fourier transform method of fringe pattern analysis, Opt. Laser Eng. **8**, 29.
Greivenkamp, J.E., 1984, Generalized data reduction for heterodyne interferometry, Opt. Eng. **23**, 350.
Greivenkamp, J.E., 1987, Sub Nyquist interferometry, Appl. Opt. **26**, 5245.
Greivenkamp, J.E., and J.H. Bruning, 1992, in: Optical Shop Testing, 2nd Ed., ed. D. Malacara (Wiley, New York).
Hariharan, P., 1984, Optical Holography (Cambridge University Press, London).
Hariharan, P., B.F. Oreb and N. Brown, 1983, Real time holographic interferometry: a microcomputer system for measurement of vector displacements, Appl. Opt. **22**, 876.
Hayslett, C.R., and W. Swanter, 1980, Wave front derivation from interferograms by three computer programs, Appl. Opt. **19**, 3401.
Ichioka, Y., and M. Inuiya, 1972, Direct phase detecting system, Appl. Opt. **11**, 1507.
Joenathan, C., and R. Torroba, 1991, Modified electronic speckle pattern interferometer employing off-axis reference beam, Appl. Opt. **30**, 1169.
Jones, R., and C. Wykes, 1983, Holographic and Speckle Interferometry (Cambridge University Press, London).
Kafri, O., 1980, Non coherent method for mapping phase objects, Opt. Lett. **5**, 555.
Kafri, O., and A. Livnat, 1981, Reflective surface analysis using moiré deflectometry, Appl. Opt. **20**, 3098.
Kalal, M., K.A. Nugent and B. Luther-Davies, 1987, Phase amplitude imaging: its applications to fully automated analysis of magnetic field measurements in laser induced plasmas, Appl. Opt. **26**, 1674.
Kreis, T., 1986, Digital holographic interference-phase measurement using the Fourier transform method. J. Opt. Soc. Am. A **3**, 847.
Kuechel, M., 1990, The new Zeiss interferometer, Proc. SPIE **1332**, 665.
Kujawinska, M., 1992, Expert system for analysis of complicated fringe pattern, Proc. SPIE **1755**, Paper 1755-33.
Kujawinska, M., A. Spik and D.W. Robinson, 1989, Quantitative analysis of transient events by ESPI, Proc. SPIE **1121**, 416.
Kujawinska, M., A. Spik and J. Wojciak, 1989, Fringe pattern analysis using Fourier transform techniques, Proc. SPIE **1121**, 130.
Kwon, O.Y., 1992, Advances in phase shifting interferometric technology, Proc. SPIE **1755**, Paper 1755-01.
Leith, E.N., and J. Upatnieks, 1962, Reconstructed wavefronts and communication theory, J. Opt. Soc. Am. **52**, 1123.

Løkberg, O.J., 1980, Advances and applications of electronic speckle pattern interferometry (ESPI), Proc. SPIE **215**, 92.
Løkberg, O.J., and G.Å. Slettemoen, 1987, Basic electronic speckle-pattern interferometry, in: Applied Optics and Optical Engineering, Vol. X, eds R.R. Shannon and J.C. Wyant (Academic Press, San Diego) p. 455.
Loomis, J.S., 1979, Analysis of interferograms from axicons, Proc. SPIE **171**, 64.
Macy Jr, W.W., 1983, Two-dimensional fringe pattern analysis, Appl. Opt. **22**, 3898.
Malacara, D., 1992, Optical Shop Testing, 2nd Ed. (Wiley, New York).
Malacara, D., and S.L. De Vore, 1992, in: Optical Shop Testing, 2nd Ed., ed. D. Malacara (Wiley, New York).
Malacara, D., and V.I. Vlad, 1993, Spatial phase reconstruction by computer demodulation of the fringe images, 16th Congress ICO, Budapest, paper POP 4.11.
Malacara, D., J.M. Carpio-Valadez and J.J. Sánchez-Mondragón, 1990, Wavefront fitting with discrete orthogonal polynomials in unit radius circle, Opt. Eng. **29**, 672.
Massie, N.A., 1987, Digital-heterodyne interferometry, Proc. SPIE **816**, 40.
Mertz, L.N., 1983, Real time fringe pattern analysis, Appl. Opt. **22**, 1535.
Moore, A., V.I. Vlad, F. Mendoza and B. Mendoza, 1994, Spatial phase reconstruction in ESPI by smoothing of the fringe pattern, Appl. Opt., submitted.
Nakadate, S., 1986, Vibration measurement using phase shifting time average holographic interferometry, Appl. Opt. **25**, 4162.
Nakadate, S., and H. Saito, 1985, Fringe scanning speckle pattern interferometry, Appl. Opt. **24**, 2172.
Nakadate, S., N. Magome, T. Honda and J. Tsujiuchi, 1981, Hybrid holographic interferometer for measuring three-dimensional deformations, Opt. Eng. **20**, 246.
Nakadate, S., T. Yatagai and H. Saito, 1980, Digital speckle pattern shearing interferometry, Appl. Opt., **19**, 4241.
Nugent, K.A., 1985, Interferogram analysis using an accurate fully automated algorithm, Appl. Opt. **24**, 3101.
Ostrovsky, Y.I., N. Butusov and G.V. Ostrovskaya, 1982, Interferometry by Holography (Springer, Berlin).
Parker, D.H., 1991, Moiré patterns in three dimensional Fourier space, Opt. Eng. **30**, 1534.
Pratt, W.K., 1978, Digital Image Processing (Wiley, New York).
Preater, R.W., and R.C. Swain, 1992, Fourier transform fringe analysis of ESPI fringes from rotating components, Proc. SPIE **1821**, Paper 1821-55.
Ransom, P.L., and J.V. Kokal, 1986, Interferogram analysis by a modified sinusoid fitting technique, Appl. Opt. **25**, 4199.
Reid, G.T., 1986, Automatic fringe pattern analysis: A review, Opt. & Lasers in Eng. **7**, 37.
Reid, G.T., 1988, Image processing techniques for fringe pattern analysis, Proc. SPIE **954**, 468.
Reid, G.T., R.C. Rixon and H.I. Messer, 1984, Absolute and comparative measurements of three dimensional shape by phase-measuring moiré topography, Opt. Laser Technol. **16**, 315.
Robinson, D.W., and D.C. Williams, 1986, Digital phase stepping speckle interferometry, Opt. Commun. **57**, 26.
Roddier, C., and F. Roddier, 1987, Interferogram analysis using Fourier transform techniques, Appl. Opt. **26**, 1668.
Roddier, C., and F. Roddier, 1991, Wavefront reconstruction using iterative Fourier transforms, Appl. Opt. **30**, 1325.
Rodriguez-Vera, R., D. Kerr and F. Mendoza-Santoyo, 1991, Three-dimensional contouring of diffuse objects by Talbot-projected fringes, J. Mod. Optics **38**, 1935.
Roychoudhuri, C., and R. Machorro, 1978, Holographic nondestructive testing at the Fourier plane, Appl. Opt. **17**, 848.
Ru, Q.-S., T. Honda, J. Tsujiuchi and N. Ohyama, 1988, Fringe analysis by using 2-D Fresnel transform, Opt. Commun. **66**, 21.

Schmidt, J., K. Creath and M. Kujawinska, 1992, Spatial and temporal phase measurement techniques: A comparison, Proc. SPIE **1755**, Paper 1755-27.
Servin, M., D. Malacara, Z. Malacara and V.I. Vlad, 1994, Sub-Nyquist null aspheric testing using a computer stored compensator, Opt. Eng., submitted.
Simova, E.S., and K.N. Stoev, 1993, Automated Fourier transform fringe-pattern analysis in holographic moiré, Opt. Eng. **32**, 2286.
Slettemoen, G.Å., 1980, Electronic speckle pattern interferometric system based on a speckle reference beam, Appl. Opt. **19**, 616.
Snyder, R., and L. Hesselink, 1988, Measurements of mixing fluid flows with optical tomography, Opt. Lett. **13**, 87.
Solano, C., V.I. Vlad, D. Malacara and L. Voicu, 1994, Fringe pattern smoothing for automatic spatial phase reconstruction in holographic interferometry, in: Proc. ROMOPTO'94, September 5–8 1994, Bucharest, Proc. SPIE, to be published.
Spik, A., and L.A. Salbut, 1992, Software package for interferometric, refractive index distribution measurement, Proc. SPIE **1755**, Paper 1755-34.
Stetson, K., 1978, Miscellaneous topics in speckle metrology, in: Speckle Metrology, ed. R.K. Erf (Academic Press, New York) p. 295.
Suematsu, M., and M. Takeda, 1991, Wavelength shift interferometry for distance measurements using the Fourier transform technique for fringe analysis, Appl. Opt. **30**, 4046.
Suganuma, M., and T. Yoshizawa, 1991, Three dimensional shape analysis by use of a projected grating image, Opt. Eng. **30**, 1529.
Takeda, M., 1989, Spatial carrier heterodyne techniques for precision interferometry and profilometry: An overview, Proc. SPIE **1121**, 73.
Takeda, M., and S. Kobayashi, 1984, Lateral aberration measurements with a digital Talbot interferometer, Appl. Opt. **23**, 1760.
Takeda, M., and K. Mutoh, 1983, Fourier transform profilometry for the automatic measurement of 3-D object shapes, Appl. Opt. **22**, 3977.
Takeda, M., and Z. Tung, 1985, Subfringe holographic interferometry by computer-based spatial-carrier fringe-pattern analysis, J. Opt. (Paris) **16**, 127.
Takeda, M., H. Ina and S. Kobayashi, 1982, Fourier-transform method of fringe pattern analysis, J. Opt. Soc. Am. **72**, 156.
Thalmann, R., and R. Dandliker, 1985, Holographic contouring using electronic phase measurement, Opt. Eng. **24**, 930.
Toyooka, S., and Y. Iwaasa, 1986, Automatic profilometry of 3-D diffuse objects by spatial phase detection, Appl. Opt. **25**, 1630.
Toyooka, S., and M. Tominaga, 1984, Spatial fringe scanning for optical phase measurement, Opt. Commun. **51**, 68.
Vest, C.M., 1979, Holographic Interferometry (Wiley, New York).
Vlad, V.I., and A. Petris, 1994, Fidelity evaluation of the phase conjugation in photorefractive crystals using the spatial heterodyne demodulation of optical interferograms, Proc. ICO Topical Meeting, Frontiers in Information Optics, April 4–8 1994, Kyoto, Japan.
Vlad, V.I., and D. Popa, 1982, Time-average holographic interferometry in Fourier plane, Proc SPIE **370**, 92.
Vlad, V.I., D. Popa and I. Apostol, 1991, Computer moiré deflectometry using the Talbot effect, Opt. Eng. **30**, 300.
Vlad, V.I., D. Popa and S. Solomon, 1986, Hybrid holographic digital processing system for three dimensional displacement measurement, Proc. SPIE **700**, 344.
Vrooman, H.A., and A.M. Maas, 1991, Image Processing Algorithms for the analysis of phase shifted speckle interference patterns, Appl. Opt. **30**, 1636.
Waters, J.P., 1972, Object motion compensation by speckle reference beam holography, Appl. Opt. **11**, 630.

Willemin, J.-F., and R. Dandliker, 1983, Measuring amplitude and phase of microvibrations by heterodyne speckle interfermetry, Opt. Lett. **8**, 102.

Womack, K.H., 1984, Frequency domain description of interferogram analysis, Opt. Eng. **23**, 396.

Womack, K.H., J.A. Jonas, C.L. Koliopoulos, K.L. Underwood, J.C. Wyant, J.S. Loomis and C.R. Hayslett, 1979, Microprocessor-based instrument for analysis of video interferograms, Proc. SPIE **192**, 134.

Wyant, J.C., 1975, Use of an ac heterodyne lateral shear interferometer with real time wavefront correction systems, Appl. Opt. **14**, 2622.

Wyant, J.C., B.F. Oreb and P. Hariharan, 1984, Testing aspherics using two wavelength holography: Use of digital electronic techniques, Appl. Opt. **23**, 4020.

Yatagai, T., 1984, Fringe scanning Ronchi test for aspherical surfaces, Appl. Opt. **23**, 3676.

Yatagai, T., M. Idesawa, Y. Yamashi and M. Suzuki, 1982, Interactive fringe analysis system: Applications to moiré contourogram and interferogram, Opt. Eng. **21**, 901.

Yatagai, T., S. Nakadate, M. Idesawa and H. Saito, 1982, Automatic fringe analysis using digital image processing techniques, Appl. Opt. **21**, 432.

Zernike, F., 1942, Phase contrast, new method for the microscopic observation of transparent objects, Physica **9**, 686.

E. WOLF, PROGRESS IN OPTICS XXXIII
© 1994 ELSEVIER SCIENCE B.V.
ALL RIGHTS RESERVED

V

IMAGING THROUGH TURBULENCE IN THE ATMOSPHERE

BY

MARK J. BERAN and JASMIN OZ-VOGT

Faculty of Engineering,
Tel Aviv University,
Ramat Aviv, Israel

CONTENTS

		PAGE
§ 1.	INTRODUCTION	321
§ 2.	PROPAGATION OF LIGHT THROUGH THE ATMOSPHERE	330
§ 3.	IMAGING OF OBJECTS IN THE TURBULENT ATMOSPHERE	358
REFERENCES		383

§ 1. Introduction

Imaging through turbulent media is a subject of extensive research. It is of importance in optical and radio astronomy, remote sensing, target identification and the investigation of the properties of the medium itself. Turbulent velocity fluctuations in the propagation medium cause image distortion by generating a statistical temperature field, which in turn gives rise to random inhomogeneities in the index of refraction. Although variations in the index of refraction cause distortion of electromagnetic and acoustic fields at all wavelengths, in this review we shall restrict our attention to the visible portion of the EM spectrum. In addition, we shall consider only light propagation in the atmosphere and shall not discuss the effects of aerosols. Much of the research that we review here, however, has direct application in other problems of wave propagation in random media.

The most obvious effects of turbulence on an image are a wandering of the image center and a general blurring of the image details. These effects are due mainly to phase changes introduced into the optical field as the field propagates through turbulence in the atmosphere. In optical astronomy for example, this causes a serious problem, since without correction the resolving power of an earth-based telescope is reduced drastically. Typically, the turbulent atmosphere reduces the effective telescope size to somewhere between 5 and 20 cm, depending upon the atmospheric conditions. It is thus essential to find effective methods for removing the effects of turbulence and improving the resolving power of the telescope.

In this review we shall discuss the current state of understanding of the propagation of light through the atmosphere. We shall summarize the available theory and experiments that may be used to analyze image distortion and to compensate for it. We shall then discuss the most important methods that are presently used to compensate for image distortion caused by atmospheric turbulence. We shall make a special point of emphasizing the assumptions that are used in both the underlying propagation theory and the compensation methods. For example, we shall review our knowledge of the statistical properties of the index-of-refraction field at both large and small scales, discuss the importance of optical intensity fluctuations in addition to phase fluctuations, and note the far-reaching effects of the assumption of isoplanicity in compensation techniques.

To set the stage for the future technical discussions, we will begin by discussing the difference between long- and short-time exposures in imaging. We shall assume that we are dealing with filtered light, so that $\Delta v/\bar{v} \ll 1$, where Δv is the frequency spread and $\bar{v}$ is the mean frequency. The minimum characteristic time in which the turbulent field changes, t_T, is about a millisecond. A short exposure of an image is assumed to take place over a time interval t_S, where $t_S \ll t_T$. A long exposure of an image takes place over a time interval t_L, where $t_L \gg t_T$. For incoherent imaging, we assume $\Delta v t_S \gg 1$.

1.1. LONG EXPOSURES

In the absence of turbulence, the quality of the image is limited principally by the finite diameter, d_L, of the lens. If the distance from the object to the lens is z_0, and mean wavenumber of the light is $\bar{k}$, then any object detail smaller than a characteristic size $d_0 = \pi z_0/\bar{k}d_L$ appears blurred in the image. We shall refer to this as the diffraction-limited resolution of the object. The main goal of image processing is to restore the image up to this limit. In fig. 1A, we show a point object imaged by a lens of diameter d_L. The image in the focal plane has a characteristic spread of order $l_D = f(1/\bar{k}d_L)$. When a turbulent atmosphere is present and the distortion resulting from the effects of the turbulence is strong, we obtain an image like that pictured in fig. 1B, where $l_T \gg l_D$.

In order to predict the pattern of a long-exposure image, we shall make

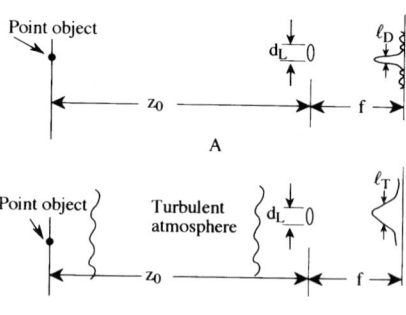

Fig. 1. Imaging of a point object with and without turbulence.

INTRODUCTION

two assumptions: (i) the object is self-luminous or illuminated by incoherent radiation such as sunlight, and (ii) light coming from all points of the object encounters turbulence with the same statistical properties. If these conditions are satisfied, the image pattern of the object intensity $\langle I(x) \rangle$ is obtained mathematically as a convolution of the form:

$$\langle I(x) \rangle = \int S(x - x') I_o(x') \, dx'. \tag{1.1}$$

Here $I_o(x)$ is the intensity pattern we would obtain in the absence of turbulence, and $S(x)$ is the turbulence-affected pattern of a non-resolvable object (such as a distant point object) in the image plane.

Equation (1.1) may be solved formally by taking the Fourier transform of both sides of the equation, and dividing the Fourier transform of $\langle I(x) \rangle$ by the Fourier transform of $S(x)$, $\tilde{S}(\kappa)$. The function $S(x)$ is determined by measuring or calculating the image of a point source located in the vicinity of the object. Since many sources of noise are always present and the function $S(x)$ is known imperfectly, the inversion procedure is of necessity much more complex. In § 3.2 we shall discuss imaging procedures that are used to solve eq. (1.1) under various conditions.

From an experimental point of view, it is convenient to obtain images using long exposures. Unfortunately, in high-resolution astrophysical applications it has not been possible to process the long exposure of a badly distorted image with sufficient accuracy to remove the degradation. For many reasons, the image of the point source needed for the image-processing procedure cannot be measured with sufficient accuracy to be used in inverting eq. (1.1). However, under less demanding conditions, the long-exposure image may be used successfully in image processing.

In the past decades, considerable attention has been given to developing image-processing techniques for improving the resolving power of ground-based telescopes that image astronomical or satellite objects. These techniques make use of short-exposure images. It is important to note that most of the theory developed to predict the statistical properties of the optical field requires time averages of the order of t_L. In the discussion of short exposures, the theory is not directly applicable to each short exposure. However, many of the techniques depend on averages of processed information from a sequence of short exposures. Here the theory is directly applicable. Moreover, knowing the long-time averages of the statistical properties of the optical field allows us to estimate the characteristic spatial properties of the short exposures; e.g., we cannot predict the exact image distortion occurring in a particular short exposure. Nevertheless, we can predict with some confidence the overall properties of the images which will be distorted.

To solve eq. (1.1) it is not necessary to assume isoplanicity in each short exposure; i.e., it is not necessary to assume that each point on the object experiences essentially the same turbulent fluctuations as the optical field propagates from the object to the image plane. It is only necessary to assume that the statistical properties of the turbulent fluctuations are homogeneous in any plane perpendicular to the mean propagation direction. However, the assumption of isoplanicity is so important in the entire field of short-exposure image processing that we give a brief discussion of the concept before proceeding.

1.2. ISOPLANICITY

The assumption of isoplanicity is used most commonly in the imaging of objects that are located far above the portion of the atmosphere that introduces measurable turbulence-induced distortion. Figure 2A depicts a typical imaging configuration. The imaging is performed with a lens of diameter d_L, the characteristic length of the object is d_0, z is the distance from the object to the earth, and ψ is the zenith angle of the center of the object relative to the z-axis. Assuming $d_0 \ll z$, we denote the angular separation by $\theta = d_0/z$. The distance H is the altitude at which the turbulence in the atmosphere ceases to have a significant influence on the light propagation.

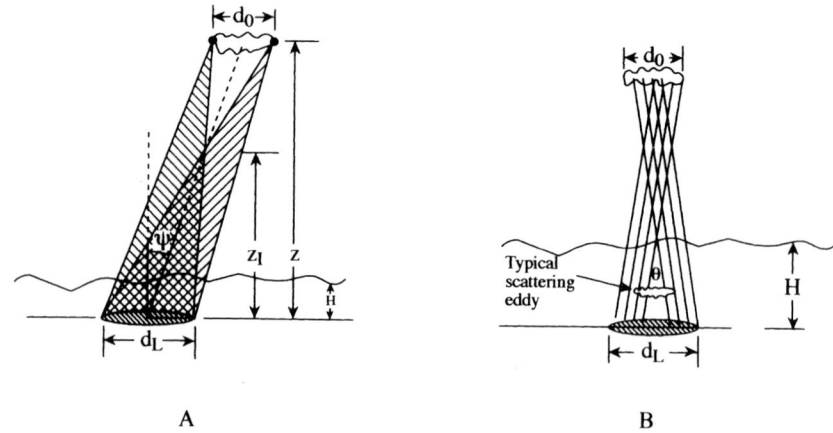

Fig. 2. Conditions for isoplanicity.

The distance H is of the order of 20–30 km, and we assume that $z \gg H$. In order to resolve an object with characteristic size d_0 the lens size must be of order $d_L = \pi z/(\bar{k} d_0)$. From fig. 2A, we see that if $z_1 \gg H \sec \psi$ the fields from even the most remote points of the object overlap almost completely in the turbulent region. Thus, essentially the same fluctuations are encountered along all propagation paths. The condition $z_1 \gg H \sec \psi$ then becomes $\theta = d_0/z \ll [\pi/(\bar{k} H \sec \psi)]^{1/2}$. As we shall see in § 2.4, this condition is necessary regardless of whether or not all points on the object and lens are coplanar. Therefore, it suffices to consider only the coplanar case.

In addition to the above overlap condition on θ, we require a condition on the nature of the essentially plane wave propagation within the turbulent region (see fig. 2B). We require that two rays, originating from the same point at height H and separated in angular direction by θ, remain within a horizontal distance that is much less than the characteristic turbulent eddy sizes that contribute to the turbulent scattering. If this condition is satisfied, the two rays impinging on the lens have experienced the same phase changes. Unfortunately, in order to determine which eddy sizes are important in the scattering process we require a solution of the scattering problem. As we shall see in the next section, the eddy sizes of importance are of the same order as the characteristic coherence length of the light impinging on the lens. Fried [1966] has termed this length r_0, and measurements show that typically it is about 5–20 cm for visible light. For the two rays to remain within a horizontal distance much less than r_0 for propagation from the height H to the lens, we must have $\theta \ll r_0/H \sec \psi$. In any imaging problem, the more stringent of the two conditions on θ must be satisfied.

For adaptive optics (see § 1.3.1), a reference point source must be placed close enough to the object that the point source and all points on the object lie within an angle θ that satisfies the conditions given above.

The above discussion is very qualitative. In § 2.4 we shall summarize the quantitative analyses that appear in the literature (see, e.g., Fried [1982]). However, as the reader will see, the assumptions necessary to obtain the quantitative results are quite strong. It is not clear if the detailed calculations are sufficiently accurate to determine the degradation in images when θ begins to violate the conditions given above.

We have just treated the case of viewing objects that lie far beyond the turbulent region of the atmosphere. For horizontal propagation in the atmosphere, the situation with respect to satisfying the condition of isoplanicity is much less satisfactory. First we note that now H is equal to z, and we can no longer assume that the light impinging on the atmosphere is a plane wave. For all the rays on the object to experience the same turbulent field, we must satisfy both the conditions $d_0 \ll l_c$ and $d_0 \ll (\pi z/\bar{k})^{1/2}$, where l_c is the characteristic size of the eddies contributing to the eddy distortion. Assuming that $l_c \sim O(r_o)$, this puts an upper limit of r_o on the size of the object that may be imaged. In addition, as we shall show in § 2.4, there is another condition that may be used if $d_L \ll d_0$.

1.3. SHORT EXPOSURES

A short exposure is taken in a time interval t_S that is very short compared to the time interval t_T over which the turbulent field changes from one realization to another. The short-exposure intensity $I_S(x)$ is related to $I_o(x)$ by the relation

$$I_S(x) = \int S_S(x, x')I_o(x')\,dx'. \tag{1.2}$$

Here the short-exposure point spread function $S_S(x, x')$ depends on both x and x'. Note that $S(x - x') = \langle S_S(x, x')\rangle$ depends on the difference $x - x'$.

Any short exposure of the image-intensity pattern of a point source looks very little like the image-intensity pattern of a long exposure. Each pattern looks quite random, and it is only upon averaging a large sequence of short exposures that one obtains the long-exposure pattern (see fig. 3). Looked at from the perspective of an angular spectrum of plane waves impinging upon the lens, a short exposure reflects the fact that the intensity of a plane wave in a particular direction varies randomly from exposure to exposure. In each exposure, the light traverses a different realization of the stochastic index-of-refraction field.

The angular spectrum of plane waves that results from the interaction of the optical field and the turbulence is often separated artificially into two segments. An angular-tilt segment of the spectrum is identified with wander

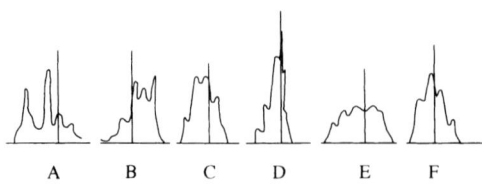

Fig. 3. Random sequence of short-exposure images of a point object.

of the image center in a short exposure, and a remaining segment of the spectrum is identified with blurring of the image in a short exposure. Fried [1966] has shown how to remove the angular-tilt segment to obtain an optical transfer function that contains only the remaining segment of the angular spectrum (Manning [1993] gives a more comprehensive discussion using Zernike polynomials). It is then possible to modify the resulting long-exposure image so that the angular tilt is removed.

If we consider the object to be made up of a continuous collection of point sources, we would expect in general that each point source would have a different short-exposure image-intensity pattern. However, as we discussed in § 1.2, if the angular diameter of the object is small enough as viewed from the lens plane, then each short-exposure pattern will be approximately the same for all points on the object. In this case, $S_S(x, x') = S_S(x - x')$. This condition of isoplanicity is essential if most of the currently used short-exposure image-processing techniques are to be useful. If the condition of isoplanicity is not applicable – and in many cases it is not – image processing becomes exceptionally difficult for any but the most simple objects. To give the reader a feeling for the type of image processing that may be done with short exposures, we describe below a few of the methods currently in use. As examples, we mention (i) adaptive optics, (ii) speckle interferometry and speckle masking, and (iii) the shift-and-add method. After each method is described briefly, we state why it is necessary to understand the statistical properties of the impinging optical field to assess the utility of the image-processing technique. A more complete description will be given in § 3.3.

1.3.1. *Adaptive optics*

When isoplanicity is satisfied it is possible to make use of the image of a point-source intensity pattern, in the same isoplanatic patch as the object, to remove the distortion due to turbulence in each short exposure. This is the method used in adaptive optics (Hardy, Lefebvre and Koliopoulos [1977], Hardy [1978], Welsh and Gardner [1989]). This method is one of pre-detection image processing rather than post-detection image processing. The basic idea is to use data from a wave-front sensor to control the deformation of a deformable mirror to compensate for the random phase fluctuations introduced by turbulence in the atmosphere. The sensing and compensation must take place in times that are short compared to t_T (see fig. 4; taken from Hardy, Lefebvre and Koliopoulos [1977]). More elaborate systems can be designed to compensate for random intensity fluctuations as well, but most systems are designed only to correct for phase changes.

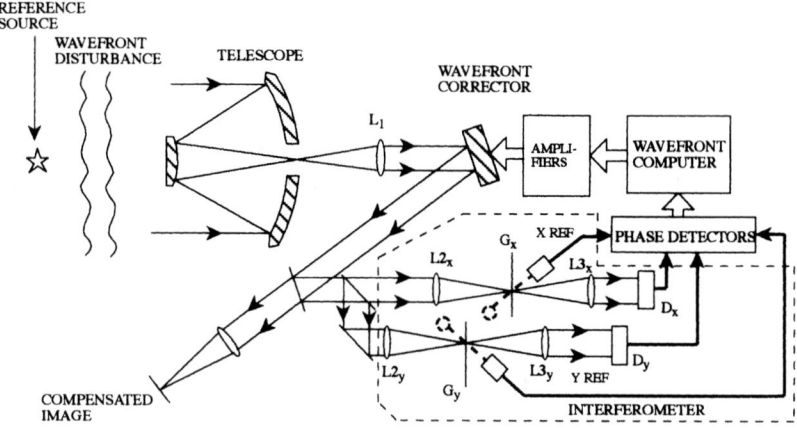

Fig. 4. Typical adaptive optics system (from Hardy, Lefebvre and Koliopoulos [1977]).

The area of the deformable mirror which is used to compensate for the phase fluctuations is divided into a finite number of sub-areas, and a phase correction is applied to each sub-area. To determine the number of sub-areas that are required, a calculation must be made to determine the characteristic spatial coherence length of the optical field impinging on the telescope. In addition, since most active optical systems neglect intensity fluctuations in the impinging optical field, a calculation must be made to find the scintillation index to determine when this is a good assumption. Both these calculations are now possible using the theory reviewed in § 2.

In principle, it is possible to use adaptive optics to correct for objects that do not satisfy the assumption of isoplanicity. The method would be to break the object up into many subsections, each of which lies in its isoplanatic patch. A different reference source would be required for each subsection. However, we are not aware of any system which currently corrects for non-isoplanatic regions of an object.

1.3.2. Speckle interferometry and speckle masking

An alternative method of making use of short-term exposures, termed speckle interferometry, was developed by Labeyrie [1970, 1976]. It is relatively insensitive to phase variations of the optical field impinging on the lens. The method requires the assumption of isoplanicity, an object with Gaussian statistics for the optical field, and a reference point source. Further, the final result gives only the autocorrelation of the object intensity pattern;

a phase-retrieval algorithm (Dainty and Fienup [1987]) is needed to obtain the intensity distribution itself. The general idea is to find the spatial power spectrum of the itensity pattern of each short-exposure image, to average over many short exposures and then to make use of the point-source spatial power spectrum to find the spatial power spectrum of the undistorted object. The reason that this method is superior to analyzing the long exposure of the image-intensity pattern is that the average of the short exposures of the point-source spatial power spectra is shown to be relatively insensitive to phase fluctuations in the impinging optical field.

To determine the validity of the above approach, it is necessary to calculate the four-point spatial coherence of the optical field impinging on the lens. This was done by Korff [1973] using a rather strong assumption about the statistics of the impinging field. However, it is possible today to do the calculation again using an improved theory. A more complete theory will also take into account the error introduced into the phase-retrieval algorithm by the correlation of the intensities at two points of the optical field.

The need for a phase-retrieval algorithm is a distinct disadvantage of speckle interferometry. Lohmann, Weigelt and Wirnitzer [1983], Lohmann and Wirnitzer [1984], and Bartelt, Lohmann and Wirnitzer [1984] showed how to make use of the three-point intensity correlation of the imaged optical field to obtain the object intensity. As in the case of speckle interferometry, it is necessary to know the properties of a higher-order coherence function of the impinging optical field to determine the validity of the method.

1.3.3. *Shift-and-add method*

It is sometimes possible to separate the overall movement of the image from the blurring of the details of the image. For a long exposure, this separation is of little advantage since the long-exposure averaging performs averages over all movements. However, in a sequence of short exposures it may be possible to detect and separate out the overall movements. In addition, once the overall image movement is detected, it is possible to choose the least perturbed images. The essence of the shift-and-add method (Bates and McDonnel [1986]) is to take this sequence of short exposures, superimpose them about a common center (usually determined as the brightest point in the image), and average the results. If this is done for the image of a point source, we obtain what may be called a short-exposure image-transfer function.

Fried [1966] showed how to predict the short-exposure transfer function, and has subsequently discussed the probability of having a relatively distortion-free short-exposure image (Fried [1978]). As a practical matter, the shift-and-add method works for relatively uncomplicated object patterns, since it is very difficult to locate the same point in a sequence of short exposures. Nevertheless, the method is often effective for a pattern of a few point objects.

1.4. GENERAL APPROACH

In § 2 we present a discussion of the theory that has been developed to study the propagation of light in a turbulent atmosphere. We emphasize those aspects of the work that are necessary to properly develop and analyze image-processing techniques. In the course of this discussion, we shall mention the limitations of various aspects of the theory to enable the reader to assess the validity of many of the analyses given in the literature. As in all fields of science, there is a tendency to use simple formulations beyond the region in which they are valid.

In § 3, a number of image-processing techniques are treated. The techniques are separated into long-exposure (§ 3.2) and short-exposure (§ 3.3) procedures. The methods are analyzed using the results obtained in § 2, and we point out their limitations. In particular, we stress the importance of the assumption of isoplanicity required in many of the short-exposure methods. The condition of isoplanicity using earth-based systems is most often applicable to the imaging of objects that lie well outside the atmosphere. Unfortunately, it is usually not applicable for imaging over horizontal paths within the atmosphere or for imaging objects on the earth from space-based systems.

§ 2. Propagation of Light Through the Atmosphere

In this section we summarize theoretical and experimental work that has been performed on the propagation of light through the atmosphere. In § 2.1, we discuss the statistical properties of the index-of-refraction field. In § 2.2, we present the propagation problem from the point of view of coherence theory, and begin by discussing the equations governing the two- and four-point coherence functions. We next give solutions of the equations that have been obtained, and compare them with relevant experimental data. The two-point phase and amplitude coherence functions and the Rytov approximation are discussed in § 2.3. We also consider the relation between these coherence functions and the two-point coherence function obtained using the assumption of joint Gaussian statistics for the phase- and log-amplitude. The condition of isoplanicity is treated in § 2.4 to supplement the descriptive discussion given in § 1.2. A summary of the results of § 2.1 to § 2.4 is given in § 2.5.

2.1. STATISTICAL PROPERTIES OF THE INDEX-OF-REFRACTION FIELD IN THE ATMOSPHERE

As light propagates through the turbulent atmosphere, it encounters random fluctuations of the index-of-refraction field. These fluctuations are related principally to the fluctuations in the temperature field, which in turn are determined by the turbulent air flows. There is a direct relation between the statistical properties of the temperature field and the index-of-refraction field. This relation is used to find the statistical properties of the index-of-refraction field, because measurements of the temperature field are simpler to perform. The index-of-refraction, neglecting water vapor, is given by

$$n = 1 + (77.6 \times 10^{-6})(p/T)(1 + 7.52 \times 10^{-3} \lambda^{-2}), \tag{2.1}$$

where T is the absolute temperature, p is the pressure in millibars and λ is the wavelength in micrometers. For constant pressure and wavelength, we have, by differentiation,

$$n' = A_T(T) T', \tag{2.2}$$

where n' and T' are the index-of-refraction and temperature fluctuations, respectively. In imaging problems, we let p, T and λ take on representative values, so that A_T is a constant. For example, for $p = 1000$ mb, $T = 300°$K and $\lambda = 0.5$ μm, we have $n' = 8.6 \times 10^{-7} T'$.

The statistical information we require for imaging problems is the two-point correlation function of the index-of-refraction fluctuations. This function is defined as

$$\sigma(\mathbf{x}_1, \mathbf{x}_2, \tau) = \{n'(\mathbf{x}_1, t) n'(\mathbf{x}_2, t + \tau)\}, \tag{2.3}$$

where the curly brackets indicate a time average over a time that is long compared to t_T. For a stationary and statistically homogeneous and isotropic medium, the correlation function depends only on τ and the magnitude of the separation vector $\mathbf{s} = \mathbf{x}_1 - \mathbf{x}_2$. We shall consider here only the case $\tau = 0$ and will set $\{n\} = 1$.

We define the structure function $D(x_1, x_2)$ as

$$D(x_1, x_2) = \{[n'(x_1, t) - n'(x_2, t)]^2\}. \tag{2.4}$$

In § 2.2 we use ensemble averages rather than time averages to formulate the statistical problem. In this paper we shall always assume an ergodic hypothesis and equate time and ensemble averages. As stated, if the turbulence is statistically homogeneous, the correlation function $\sigma(x_1, x_2)$ depends only upon s. In this case we define the spectrum $\Phi(\kappa)$ associated with the correlation function as

$$\Phi(\kappa) = (1/2\pi)^3 \int_{-\infty}^{\infty} \sigma(s) \exp[-i\kappa \cdot s] \, ds. \tag{2.5}$$

If the turbulence is also statistically isotropic, σ and Φ depend upon the magnitudes of s and κ respectively.

The turbulent field may be thought of as being composed of a superposition of eddies whose characteristic sizes range from a minimum length l_m (the inner scale) to a maximum length l_M (the outer scale). The smallest-size eddies represent the scale at which dissipation occurs, and the largest-size eddies are the scales at which energy is introduced into the turbulent field. As a result of the nonlinearity of the fluid-mechanical process, energy is transferred from the largest eddies to the smallest eddies, where the energy is dissipated. In the atmosphere there is an intermediate-scale range, termed the Kolmogorov region, which is dependent only on the rate of energy flow and is independent of the fluid viscosity.

The forms of the largest eddy sizes are dependent on the mechanism by which the eddies are created (e.g., wind blowing over a rough surface), and generally the eddy field is neither statistically homogeneous nor isotropic. Until recently, with the increased use of large telescopes, the big eddies did not play an important role in turbulent distortion of an image. However, with the use of very large telescopes, increased attention is being paid to their effect (McKechnie [1992]). Unfortunately, since the eddy structure is so dependent on the mechanism of its creation, very little is known about the spectrum of the large eddies. Most research that includes the effect of large eddies makes use only of a characteristic size l_M which typically may be between 1 and 10 meters near the earth's surface. The reader should be very cautious about using any models which give a form to the large-eddy spectrum. Usually they are just arbitrarily assumed forms that are used to join the large-eddy region to the Kolmogorov region and made to decay sufficiently fast so that the total energy in the eddy field is finite.

The eddy field representing the Kolmogorov and dissipative regions is generally statistically homogeneous and isotropic. The Kolmogorov region is the most important range of eddies for determining image distortion in most applications. The characteristic sizes of eddies in this range usually lie between approximately 1 cm and 1 m near the earth's surface. Fortunately, in this region there is a simple form for both the correlation function and the wavenumber spectrum. The velocity eddy spectrum in this region was first found by Kolmogorov [1941] using simple dimensional arguments. It was then extended to the temperature spectrum by Obukhov [1949] and Corrsin [1951]. A very complete discussion of the derivation of the Kolmogorov spectrum is given by Tatarski [1971].

The forms of the spectrum and the structure function in the Kolmogorov range are

$$D(s) = C_n^2 s^{2/3}, \tag{2.6}$$

and

$$\Phi(\kappa) = 0.033 C_n^2 \kappa^{-11/3}. \tag{2.7}$$

The constant C_n^2 is termed the structure constant. The structure constant varies slowly in both space and time. Near the ground surface, it may vary between the extreme values of 10^{-13} and 10^{-17} m$^{-2/3}$, depending upon the weather conditions, time of day and season of the year. The structure constant also varies dramatically with altitude. A typical variation is given by Hufnagel [1974] and reproduced in table 1.

The form of the spectrum, as the eddy size is reduced from the Kolmogorov range to the dissipation range, is given by Hill and Clifford [1978]. The inner scale varies between approximately 1 mm and 1 cm. The spectrum is not presented by Hill and Clifford as a simple analytic form, but requires the numerical solution of an ordinary differential equation. For this reason it has not been widely used in the literature, and ad hoc forms have been used instead. A form widely used (Tatarski [1971]) is

$$\Phi(\kappa) = 0.033 C_n^2 \exp[-\kappa^2 l_m^2/4]/\kappa^{-11/3}, \tag{2.8}$$

but this form should only be used if the result is not strongly dependent upon the exact form of the spectrum. In addition, care should be taken to make sure that the l_m used corresponds to the inner scale measured experimentally. To our knowledge, very little is known about how l_m varies with altitude. Recent experimental work on the determination of l_m near the earth's surface is given by Consortini, Cochetti, Churnside and Hill [1993].

TABLE 1
C_n^2 as a function of altitude (data from Hufnagel [1974]).

Altitude	$C_n^2 \times 10^{16}$ (m$^{-2/3}$)
1	1.41
2	0.714
3	0.474
4	0.230
5	0.141
6	0.139
7	0.176
8	0.235
9	0.272
10	0.290
11	0.264
12	0.235
13	0.204
14	0.154
15	0.118
16	0.0824
17	0.0466
18	0.0344
19	0.0205
20	0.010

2.2. PROPAGATION OF THE COHERENCE FUNCTIONS IN THE ATMOSPHERE

The basic equations governing the propagation of the optical field in the turbulent atmosphere are Maxwell's equations. The turbulent medium is represented by a dielectric constant that varies slowly in space on the scale of the mean light wavelength, and slowly in time on a scale of the mean light period. As a result of these conditions, the light propagation is principally about a mean propagation direction, and we may use a scalar theory. The governing equation is then

$$\nabla^2 u(\mathbf{x}, t) = [1/c^2(\mathbf{x}, t)] \partial^2 u(\mathbf{x}, t)/\partial t^2. \tag{2.9}$$

We assume polarized radiation, and here $u(\mathbf{x}, t)$ is a component of the electric field perpendicular to the mean propagation direction. The random function $c(\mathbf{x}, t) = c_0/n(\mathbf{x}, t)$ is the local velocity of light, c_0 is the velocity of light in a vacuum, and $n(\mathbf{x}, t)$ is the local index of refraction.

To simplify the statistical formulation, it is usually assumed that $n(\mathbf{x}, t)$ is a random function only of space, and the time variation is replaced by an ensemble of realizations of the random index-of-refraction field. That is, instead of light propagating through a time-varying random medium, we assume that we have light propagating through M time-invariant realizations of the random medium where $M \gg 1$. The ith realization of the index of refraction of the random medium has the form $n_i(x)$. A realization of the turbulent atmosphere occurs in a time short compared to t_T (defined in § 1.3). The ensemble formulation is assumed to be valid if the index-of-refraction field and the light source are stationary. In this case, the process is assumed to be ergodic and time and ensemble averages are taken to be equal.

In addition, the light source is usually taken to be quasi-monochromatic so that $\Delta v/\bar{v} \ll 1$. Assuming that the difference of all important path lengths in the problem is much less than $c_0/\Delta v$, we can approximate the optical field $u(x, t)$ by

$$u(x, t) = \hat{u}(x) \exp[-i\omega_0 t], \tag{2.10}$$

where $\omega_0 = 2\pi\bar{v}$. The governing equation becomes the Helmholtz equation:

$$\nabla^2 \hat{u}(x) + k^2(x)\hat{u}(x) = 0, \tag{2.11}$$

where $k^2(x) = \omega_0^2/c^2(x)$. Next we write $k^2(x)$ in the form

$$k^2(x) = \bar{k}^2[1 + \varepsilon'(x)], \tag{2.12}$$

and assume that here

$$\varepsilon' = 2n', \qquad |\varepsilon'(x)| \ll 1. \tag{2.13}$$

Finally, we make use of the fact that the light propagation is principally about a mean propagation direction (the mean propagation direction is taken to be along the z-axis). This assumption is valid, since $\bar{k}l_m \gg 1$ and $|\varepsilon'(x)| \ll 1$. When the propagation direction is confined to a small angle about the z-axis, the Helmholz equation may be replaced by a parabolic equation. This is accomplished by writing

$$\hat{u}(x) = \bar{u}(x) \exp[-i\bar{k}z]. \tag{2.14}$$

Substituting the above form for $\hat{u}$ into eq. (2.11) yields the parabolic equation

$$2i\bar{k}\partial\bar{u}/\partial z + \nabla_T^2 \bar{u} + \bar{k}^2 \varepsilon' \bar{u} = 0. \tag{2.15}$$

Here ∇_T^2 is the transverse Laplacian taken with respect to the coordinates x_T and y_T. The term $\partial^2 \bar{u}/\partial z^2$ has been neglected in comparison to the term $2i\bar{k}\partial\bar{u}/\partial z$ to obtain the parabolic equation.

2.2.1. Definition of the coherence functions

The two-point and four-point coherence functions are defined as follows:

$$\Gamma_2(x_{T1}, x_{T2}, z) = \langle \bar{u}(x_{T1}, z)\bar{u}^*(x_{T2}, z) \rangle,$$

and

$$\Gamma_4(x_{T1}, x_{T2}, x_{T3}, x_{T4}, z) = \langle \bar{u}(x_{T1}, z)\bar{u}^*(x_{T2}, z)\bar{u}^*(x_{T3}, z)\bar{u}(x_{T4}, z) \rangle. \quad (2.16)$$

Here the brackets $\langle \, \rangle$ represent an average over an ensemble of realizations of the turbulent atmosphere. The complex conjugate is represented by *.

The two-point coherence function Γ_2 gives the intensity of the light when $x_{T1} = x_{T2}$. When the transverse coordinates are not equal, it is convenient to transform to sum and difference coordinates defined as

$$p_T = (1/2)[x_{T1} + x_{T2}],$$

and

$$s_T = [x_{T1} - x_{T2}]. \quad (2.17)$$

Next, we define the spatial Fourier transform of $\Gamma_2(p_T, s_T, z)$ with respect to s_T:

$$\tilde{\Gamma}_2(p_T, \kappa_T, z) = \int_{-\infty}^{\infty} \exp[-i\kappa_T \cdot s_T] \, \Gamma_2(p_T, s_T, z) \, ds_T. \quad (2.18)$$

The function $\tilde{\Gamma}_2(p_T, \kappa_T, z)$ is interpreted roughly as the energy in the light field propagating in the direction θ where $\theta_T = (\kappa_T/\bar{k})$ at position p_T. At position p_T, the characteristic angular spread of the light field θ_c may also be found from Γ_2 as $1/(\bar{k}s_c)$, where s_c is the characteristic decay distance of Γ_2.

In the four-point coherence function, we find the mean square of the intensity fluctuations when $x_{T1} = x_{T2} = x_{T3} = x_{T4}$. The correlation between the intensities at two separate transverse points is found from Γ_4 by setting $x_{T1} = x_{T2}$ and $x_{T3} = x_{T4}$.

If the light propagating through the turbulent atmosphere is from a laser, then it is a simple matter to determine the coherence functions of the light in the exit plane of the laser. For example, if the field represents a single-mode laser beam located in the plane $z = 0$, then we have on this plane:

$$\bar{u}(x_T, 0) = I^{1/2} \exp[-(x_T^2 + y_T^2)/a^2], \quad (2.19)$$

where a is the beam radius. The two-point coherence function is then

$$\Gamma_2(x_{T1}, x_{T2}, 0) = I \exp[-(x_{T1}^2 + y_{T1}^2 + x_{T2}^2 + y_{T2}^2)/a^2]. \quad (2.20)$$

If the light comes from a thermal source like a star or a heated object, then we must take into account the incoherence of the light at the source before it interacts with the atmosphere. The incoherence or partial coherence of the light from a thermal source is defined in terms of a suitable time average (Beran and Parrent [1974]). Therefore, to formulate the problem properly for thermal sources we should introduce a time average over a time interval that is short compared to t_S, but long compared to $1/\Delta v$. This is generally not done explicitly. Rather, a thermal source located on the plane $z = 0$ is represented in the form:

$$\Gamma_2(\boldsymbol{p}_T, \boldsymbol{s}_T, 0) = \beta I(\boldsymbol{p}_T, 0)\delta(\boldsymbol{s}_T), \tag{2.21}$$

where β is a constant that depends on the exact nature of the source. The delta function gives an infinite intensity, but since it always appears under an integral, it is a suitable representation. The function $I(\boldsymbol{p}_T, 0)$ is the finite intensity that would be measured on the source surface. In an actual thermal source, the coherence spreads spatially over at least a characteristic wavelength.

To determine Γ_4 on the plane $z = 0$ for an incoherent source, it is usually assumed that the statistics of the light are Gaussian and that Γ_4 is found in terms of Γ_2. The form used is

$$\Gamma_4(\boldsymbol{x}_{T1}, \boldsymbol{x}_{T2}, \boldsymbol{x}_{T3}, \boldsymbol{x}_{T4}, 0) = \Gamma_2(\boldsymbol{x}_{T1}, \boldsymbol{x}_{T2}, 0)\Gamma_2(\boldsymbol{x}_{T3}, \boldsymbol{x}_{T4}, 0)$$
$$+ \Gamma_2(\boldsymbol{x}_{T1}, \boldsymbol{x}_{T3}, 0)\Gamma_2(\boldsymbol{x}_{T2}, \boldsymbol{x}_{T4}, 0). \tag{2.22}$$

2.2.2. Equations governing the coherence functions Γ_2 and Γ_4

Equations governing the coherence functions Γ_2 and Γ_4 may be found in the following references: Beran [1970], Beran and Ho [1969], Shishov [1968], Tatarski [1971], Uscinski [1977], Ishimaru [1978] and Kravtsov [1992]. The reader will find different methods of derivation in the works cited, but the final results are the same. Molyneux [1971a, b] also treats the governing equations of higher-order coherence functions. In most derivations, the turbulent medium is assumed to be homogeneous and isotropic, but these restrictions may be relaxed to take into account local anisotropy and the variation of the atmospheric statistics with altitude.

The equations governing Γ_2 and Γ_4 are

$$\partial \Gamma_2(\boldsymbol{x}_{T1}, \boldsymbol{x}_{T2}, z)/\partial z - (i/2\bar{k})[\nabla_{T1}^2 - \nabla_{T2}^2]\Gamma_2(\boldsymbol{x}_{T1}, \boldsymbol{x}_{T2}, z)$$
$$+ \bar{k}^2[\bar{\sigma}(0) - \bar{\sigma}(\boldsymbol{s}_T)]\Gamma_2(\boldsymbol{x}_{T1}, \boldsymbol{x}_{T2}, z) = 0, \tag{2.23}$$

and

$$\partial \Gamma_4(\mathbf{x}_{T1}, \mathbf{x}_{T2}, \mathbf{x}_{T3}, \mathbf{x}_{T4}, z)/\partial z - (i/2\bar{k})[\nabla^2_{T1} - \nabla^2_{T2} - \nabla^2_{T3} + \nabla^2_{T4}]$$
$$\times \Gamma_4(\mathbf{x}_{T1}, \mathbf{x}_{T2}, \mathbf{x}_{T3}, \mathbf{x}_{T4}, z) + \bar{k}^2 \bar{F}(x_{ij}) \Gamma_4(\mathbf{x}_{T1}, \mathbf{x}_{T2}, \mathbf{x}_{T3}, \mathbf{x}_{T4}, z) = 0.$$
(2.24)

The function $\bar{\sigma}(x_{ij})$ is given by

$$\bar{\sigma}(x_{ij}) = (1/4) \int_{-\infty}^{\infty} \sigma_\varepsilon(x_{ij}, s_z) \, ds_z.$$

Here,

$$x_{ij} = |\mathbf{x}_{Ti} - \mathbf{x}_{Tj}|, \qquad \mathbf{x}_{ij} = \mathbf{x}_{Ti} - \mathbf{x}_{Tj},$$
$$\bar{F}(x_{ij}) = \bar{D}(x_{12}) + \bar{D}(x_{13}) + \bar{D}(x_{24}) + \bar{D}(x_{34}) - \bar{D}(x_{14}) - \bar{D}(x_{23}),$$
$$\bar{D}(x_{ij}) = \bar{\sigma}(0) - \bar{\sigma}(x_{ij}),$$

and

$$\sigma_\varepsilon(x_{ij}, z) = 4\sigma(x_{ij}, z).$$

In eq. (2.24) statistical homogeneity and isotropy is assumed for the turbulent medium but in eq. (2.23) we have only assumed statistical homogeneity.

The boundary conditions for the solution of the coherence equations are found from a knowledge of the field on the plane $z = 0$. In terms of the sum and difference coordinates given in eq. (2.17), the equation for Γ_2 becomes

$$\partial \Gamma_2/\partial z - (i/\bar{k}) \nabla_p \cdot \nabla_s \Gamma_2 + \bar{k}^2 [\bar{\sigma}(\mathbf{0}) - \bar{\sigma}(\mathbf{s}_T)] \Gamma_2 = 0. \tag{2.25}$$

It is convenient to use the following set of independent coordinates for the equation governing Γ_4:

$$\mathbf{R} = (1/4)[\mathbf{x}_{T1} + \mathbf{x}_{T2} + \mathbf{x}_{T3} + \mathbf{x}_{T4}], \qquad \mathbf{s} = (1/2)[\mathbf{x}_{T1} - \mathbf{x}_{T2} + \mathbf{x}_{T3} - \mathbf{x}_{T4}],$$
$$\mathbf{p} = (1/2)[\mathbf{x}_{T1} + \mathbf{x}_{T2} - \mathbf{x}_{T3} - \mathbf{x}_{T4}],$$

and

$$\mathbf{q} = [\mathbf{x}_{T1} - \mathbf{x}_{T2} - \mathbf{x}_{T3} + \mathbf{x}_{T4}]. \tag{2.26}$$

The equation governing Γ_4 is then

$$\partial \Gamma_4/\partial z - (i/\bar{k})[\nabla_R \cdot \nabla_q + \nabla_s \cdot \nabla_p] \Gamma_4 + \bar{k}^2 \bar{F}(\mathbf{q}, \mathbf{s}, \mathbf{p}) \Gamma_4 = 0, \tag{2.27}$$

where

$$\bar{F}(q, s, p) = \bar{D}(|s + q/2|) + \bar{D}(|s - q/2|) + \bar{D}(|p + q/2|)$$
$$+ \bar{D}(|p - q/2|) - \bar{D}(|s + p|) - \bar{D}(|s - p|).$$

2.2.3. Solution of the equation governing the two-point coherence function

It is not difficult to obtain a solution of eq. (2.25) for Γ_2. One method is to take the Fourier transform of the equation with respect to p_T (Whitman and Beran [1970], Beran and Whitman [1971]). The resulting equation is a first-order partial differential equation which can be solved by the method of characteristics. If the resulting integrals cannot be solved analytically or asymptotically, they can easily be solved numerically for any light source of interest. Here, we shall present some simple solutions for a plane wave, a point source, and a finite incoherent object. These solutions are all of use in imaging problems.

Plane-wave initial condition. Light propagating from a distant star may be approximated as a plane wave when it impinges on the upper atmosphere. If the statistics of the index-of-refraction field were independent of altitude, then eq. (2.23) would be the governing equation for the light propagation from the upper atmosphere $z = 0$ to the earth's surface. After discussing this solution, we shall show how the variation of the index-of-refraction field with altitude may be introduced.

For simplicity, we suppose that the plane wave with intensity I propagates initially in the z direction. Thus, the initial condition for $\Gamma_2(p_T, s_T, 0)$ is independent of p_T and s_T. Since the statistics of the index-of-refraction field are independent of absolute position p_T the solution $\Gamma_2(p_T, s_T, z)$ is dependent only upon s_T and z. Equation (2.23) then becomes

$$\partial \Gamma_2/\partial z = -\bar{k}^2[\bar{\sigma}(0) - \bar{\sigma}(s_T)]\Gamma_2. \tag{2.28}$$

The solution of eq. (2.28) is

$$\Gamma_2(s_T, z) = I \exp\{-\bar{k}^2 z[\sigma(0) - \bar{\sigma}(s_T)]\}. \tag{2.29}$$

When the index-of-refraction statistics are isotropic, the transverse vector s_T is replaced by the scalar distance s_T.

For $z = 0$, the coherence function Γ_2 is equal to a constant for all s_T; i.e., it represents a single plane wave. As z increases, however, Γ_2 becomes a function of s_T, and this means that the optical field is now composed of an angular spectrum of plane waves. The energy remaining in the original plane wave is found by setting $s_T = \infty$ in eq. (2.29). We find:

$$\Gamma_2(\infty, z) = I \exp[-\bar{k}^2 z \bar{\sigma}(\mathbf{0})]. \tag{2.30}$$

Thus, when $z = z_c \sim 1/[\bar{k}^2 \bar{\sigma}(\mathbf{0})]$, a significant fraction of the energy has been scattered into the angular spectrum of plane waves. In most imaging problems, we are interested in a propagation distance that is much greater than z_c. Since $\bar{\sigma}(l_M) = O[\bar{\sigma}(0)]$, this means that values of s_T for which $s_T \ll l_M$ determine the nature of decay of Γ_2 as a function of s_T. When $s_T = O(l_M)$, the function Γ_2 is essentially zero.

For propagation in the turbulent atmosphere, we use the Kolmogorov form which is valid when $s_T \ll l_M$. We find:

$$\bar{D}(s_T) = \bar{\sigma}(0) - \bar{\sigma}(s_T) = 1.45 C_n^2 s_T^{5/3}, \tag{2.31}$$

and therefore

$$\Gamma_2(s_T, z) = I \exp[-1.45 C_n^2 \bar{k}^2 z s_T^{5/3}]. \tag{2.32}$$

The characteristic value of s_T, s_c, which determines the decay of Γ_2, is $s_c = 1/[1.45 C_n^2 \bar{k}^2 z]^{3/5}$. The constant s_c is also the characteristic size of the eddies that contribute most to the form of the angular spectrum of plane waves which make up the optical field. Lastly, $\theta_c = 1/(\bar{k} s_c)$ is the characteristic angular spread of the angular spectrum. We find that $s_c = 0.48 r_0$, where r_0 is the parameter defined by Fried [1966].

For the imaging of a distant star, the star lies well outside the atmosphere and C_n^2 is a function of altitude. However, if C_n^2 varies slowly on a scale of l_M it may be shown (Tatarski [1971], Ishimaru [1978]) that $C_n^2 z$ may be replaced in eq. (2.32) by

$$\int_{-\infty}^{z} C_n^2(z') \, dz'.$$

In this case, the Fourier transform of the function $S(s)$ defined in eq. (1.1) is equal to $\Gamma_2(s_T, z)$.

The value of s_c obtained for vertical propagation in the atmosphere lies typically between about 2 and 20 cm. Thus, for most telescopes, $s_c/d_L \ll 1$, and the resolution of a telescope is affected severely by the atmosphere if no correction procedure is used.

If the turbulent effect is exceptionally strong (e.g., for off-zenith viewing or for a long horizontal path), the very small values of s_T begin to play an important role. If the value of s_c is calculated to be less than a few centimeters, then the structure function given in eq. (2.31), using the Kolmogorov range only, must be modified to include the dissipation range. To obtain accurate results, the Hill form (Hill and Clifford [1978]) should be used. To obtain an estimate of the effect of the dissipation-range eddies, eq. (2.8) may be used. We note that for an exceptionally strong turbulent effect, we may expand $\bar{D}(s_T)$ in a Taylor series in s_T, so that

$$\bar{D}(s_T) \approx -(1/2)\, d^2 \bar{\sigma}(s_T)/ds_T^2|_{s_T=0}\, s_T^2.$$

In this case, the argument in the exponent of $\Gamma_2(s_T, z)$ is proportional to s_T^2 rather than $s_T^{5/3}$. In some literature on the subject, this Gaussian form Γ_2 has been used. However, we emphasize that it is only valid when the turbulent effect is very strong.

In § 3.1.1, a series of references are given in which eq. (2.32) is verified experimentally when s_T is not too large.

Point-source initial condition. The point-source result is very similar to the plane-wave case (see Beran [1967], Tatarski [1971], Ishimaru [1978]). The solution for isotropic statistics is

$$\Gamma_2(s_T, z) = (\text{const}/z^2) \exp\left[-\bar{k}^2 \int_0^z [\bar{\sigma}(0) - \bar{\sigma}(ws_T/z)]\, dw\right]. \tag{2.33}$$

Here it is assumed that $\bar{k}|s_T \cdot p_T|/z \ll 2\pi$.

In the Kolmogorov range, we have

$$\Gamma_2(s_T, z) = (\text{const}/z^2) \exp[-(3/8)1.45 C_n^2 \bar{k}^2 z s_T^{5/3}]. \tag{2.34}$$

The difference between the plane-wave solution eq. (2.32) and eq. (2.34) is the spherical decay factor ($1/z^2$) and the factor of $\tfrac{3}{8}$ in the argument of the exponent.

The function Γ_2 may also be modified easily to include the dissipative range of eddies, if this is necessary. As in the plane-wave case, the coherence function $\Gamma_2(s_T, z)$ may be identified with the Fourier transform of the function $S(s)$ in eq. (1.1).

Finite incoherent source. Beran and Whitman [1972] give the solution for $\Gamma_2(p_T, s_T, z)$ for light from a circular incoherent object propagating in a random medium. The solution is of the form:

$$\Gamma_2(p_T, s_T, z) = \Gamma_{2o}(p_T, s_T, z)\Gamma_{2p}(p_T, s_T, z), \tag{2.35}$$

where $\Gamma_{2o}(p_T, s_T, z)$ is the coherence function of the light in the absence of a random medium, and $\Gamma_{2p}(p_T, s_T, z)$ is the coherence function of the light from a point source propagating in the random medium. The form of the solution is valid for any finite incoherent object.

The spatial Fourier transform of $\Gamma_2(\boldsymbol{p}_T, \boldsymbol{s}_T, z)$ yields eq. (1.1), since $\Gamma_{20}(\boldsymbol{p}_T, \boldsymbol{s}_T, z)$ is the Fourier transform of $I_0(\boldsymbol{x})$ by the van Cittert–Zernike theorem (see Beran and Parrent [1974]).

2.2.4. *Solution of the equation governing the four-point coherence function*

Unlike the two-point coherence function which is related to the Fourier transform of the function $S(\boldsymbol{x})$, the four-point coherence function enters into the image problem in a more indirect manner. As we noted previously, when all four points of the coherence function are equal, we obtain the mean-square intensity. When $x_1 = x_2$ and $x_3 = x_4$, we find the correlation of the light intensity at two separated points. This information allows us to determine if intensity fluctuations must be taken into account in image-processing techniques. The full fourth-order coherence function is necessary to properly evaluate the validity of the speckle interferometry method.

The equation governing the fourth-order coherence function [eq. (2.27)] is very difficult to solve except in the perturbation region. In this section, we shall first present the perturbation solution and a geometric-optics-type solution for an initial plane wave. We then outline the two-scale method which gives results in the multiple-scattering region. Finally, we mention recent numerical simulation results.

Perturbation solution – plane-wave initial condition. The perturbation solution holds when the function $[\langle I^2 \rangle - \langle I \rangle^2]^{1/2}$ is very small compared to the mean intensity $\langle I \rangle$. In this case, the scattering term $\bar{k}^2 \bar{F}(\boldsymbol{q}, \boldsymbol{s}, \boldsymbol{p}) \Gamma_4$ may be treated as a perturbation, and the solution may be obtained by using the Green's function for the diffractive terms. The details are given by Ho and Beran [1968].

The result is:

$$\Gamma_4(s_{Tij}, z) = I^2 \Bigg[1 + (\bar{k}^2 z)[\bar{\sigma}(s_{T12}) + \bar{\sigma}(s_{T13}) + \bar{\sigma}(s_{T24}) + \bar{\sigma}(s_{T34}) - 2\bar{\sigma}(0)]$$

$$- 16\pi^2 \int_0^\infty \Phi(\eta) J_0(\eta s_{T14}) \eta(\bar{k}/i\eta^2 z)(1 - \exp[-i\eta^2 z/\bar{k}]) \, d\eta$$

$$+ 16\pi^2 \int_0^\infty \Phi(\eta) J_0(\eta s_{T23}) \eta(\bar{k}/i\eta^2 z)(1 - \exp[i\eta^2 z/\bar{k}]) \, d\eta \Bigg],$$

(2.36)

where

$$s_{Tij} = |x_{Ti} - x_{Tj}|, \quad \text{and} \quad \bar{\sigma}(s_T) = 16\pi^2 \int_0^\infty J_0(\eta s_T)\Phi(\eta)\eta \, d\eta.$$

When $x_{T1} = x_{T2}$ and $x_{T3} = x_{T4}$, the function $\Gamma_4(s_{Tij}, z)$ reduces to $R_I(s_T, z) = \langle I(x_{T1}, z)I(x_{T2}, z)\rangle$, which is the correlation of the intensities between the points x_{T1} and x_{T2}.

When $z \ll \bar{k}l_m^2$,

$$\Gamma_4(s_{Tij}, z) = I^2 \left[1 + (\bar{k}^2 z)[\bar{\sigma}(s_{T12}) + \bar{\sigma}(s_{T13}) + \bar{\sigma}(s_{T24}) + \bar{\sigma}(s_{T34}) \right.$$

$$- \bar{\sigma}(s_{T14}) - \bar{\sigma}(s_{T23}) - 2\bar{\sigma}(0)]$$

$$+ (i8\pi^2 \bar{k}z^2) \int_0^\infty \Phi(\eta)\eta^3 [J_0(\eta s_{T14}) - J_0(\eta s_{T23})] \, d\eta$$

$$\left. + (8\pi^2 z^3/3) \int_0^\infty \Phi(\eta)\eta^5 [J_0(\eta s_{T14}) + J_0(\eta s_{T23})] \, d\eta \right]. \quad (2.37)$$

The terms proportional to z and z^2 in eq. (2.37) represent the effect of phase variations, and vanish when $s_{T1} = s_{T2}$ and $s_{T3} = s_{T4}$. The mean-square fluctuations in intensity and the correlation of intensities vary as z^3. For example,

$$\langle I^2(z)\rangle = I^2 \left[1 + (16\pi^2/3)z^3 \int_0^\infty \Phi(\eta)\eta^5 \, d\eta \right], \quad (2.38)$$

which is a classical result obtained using geometric optics. We note that the z^3 term in eq. (2.38) is independent of $\bar{k}$. In addition, the term depends much more on the smaller turbulent eddies than the z-dependent terms do. This latter effect occurs because the integral term contains a factor of η^5 compared to a factor of η when $\bar{\sigma}(s_{Tij})$ is expressed as an integral over $\Phi(\eta)$. If the quantity $\sigma_I^2 = [\langle I^2\rangle/\langle I\rangle^2 - 1] \ll 1$ and $z \gg \bar{k}l_m^2$, then we find from eq. (2.36), using a Kolmogorov spectrum, the result:

$$\sigma_I^2 = 1.24\bar{k}^{7/6} C_n^2 z^{11/6}.$$

Geometric-optics-type solution. It is often true that the terms of the form $\bar{k}^2 z \bar{\sigma}(s_{Tij})$ may be of order unity, while the z^2 and z^3 terms remain small compared to one. In this case, the diffraction terms in eq. (2.27) may be neglected, and we find the geometric-optics-type solution:

$$\Gamma_4(s_{Tij}, z) = I^2 \exp(\bar{k}^2 z [\bar{\sigma}(s_{T12}) + \bar{\sigma}(s_{T13}) + \bar{\sigma}(s_{T24}) + \bar{\sigma}(s_{T34}) - \bar{\sigma}(s_{T14})$$

$$- \bar{\sigma}(s_{T23}) - 2\bar{\sigma}(0)]). \quad (2.39)$$

The solution in eq. (2.39) is valid beyond the perturbation region for phase fluctuations, and is sometimes used as an approximation for Γ_4. The approximation is valid when the phase fluctuations are strong and the intensity fluctuations are weak in the region $z \ll \bar{k}l_m^2$. [If the perturbation solution given in eq. (2.36) is valid when $z \gg \bar{k}l_m^2$, then the solution given in eq. (2.39) is clearly incorrect since, in this case, the last two terms in eq. (2.36) may be neglected.] The phase approximation may be used because of the relation between phase and intensity fluctuations. If phase fluctuations are generated within the region $z \ll \bar{k}l_m^2$, these phase fluctuations will not result in intensity fluctuations until the optical field has propagated a further characteristic distance $\bar{k}l_m^2$.

The two-scale method. The two-scale method is an embedding procedure by which the moment equations are transformed into a hierarchy of first-order partial differential equations which can be solved iteratively. The results converge to known solutions in the perturbation and saturation limits ($z \to \infty$), and give results for all propagation distances. The iterative procedure may be constructed so that the zero-order approximation is the first term of a convergent series. In this section, we shall summarize the method of solution for the zero-order approximation for an initial plane wave. A review article discussing the method has been written by Beran, Frankenthal, Mazar and Whitman [1993]; the reader is referred to this paper for a complete list of references.

For an initial plane wave, the $\nabla_R \cdot \nabla_q \Gamma_4$ term in eq. (2.27) may be set equal to zero. We may also set $q = 0$, since it appears as a parameter in the equation. The resulting equation is

$$\partial \Gamma_4 / \partial z - (1/\bar{k}) \nabla_s \cdot \nabla_p \Gamma_4 + \bar{k}^2 \bar{F}(s, p) \Gamma_4 = 0. \qquad (2.40)$$

The above equation is very similar to eq. (2.25) governing Γ_2. The important difference is that the scattering term in eq. (2.25) contains only the s_T coordinate, while in eq. (2.40) $\bar{F}(s, p)$ contains both the coordinates s and p. In eq. (2.25) it is possible to take a Fourier transform with respect to the sum coordinate and obtain a first-order partial differential equation that may be solved by the method of characteristics. On the other hand, taking a Fourier transform with respect to s or p in eq. (2.40) yields an integro-partial differential equation. The fact that $\bar{F}$ depends on s and p in a rather complex manner is the source of our difficulty in solving eq. (2.40). The two-scale method that has been developed is a way of overcoming the joint dependence.

The method is illustrated using only one transverse coordinate. However, it is valid for three dimensions. Using non-dimensional variables, eq. (2.40) can be made to depend upon a single parameter when the turbulence contains an inner scale (but no outer scale, or a fixed outer scale). When the turbulence is described by a pure Kolmogorov structure function, the parameter may be absorbed into the non-dimensional length scales and does not appear. The following non-dimensional variables are used:

$$z_1 = [1/(\varepsilon^2 \bar{k} l_m^2)]z, \qquad s_1 = [1/\varepsilon l_m]s,$$

and

$$p_1 = [1/\varepsilon l_m]p, \tag{2.41}$$

where l_m is the inner scale of turbulence, and

$$\varepsilon = 1/[\bar{k}^3 \bar{\sigma}(0) l_m^2]^{1/2}.$$

Taking the Fourier transform of eq. (2.40) with respect to p_1, we find for the equation governing

$$M(s_1, \eta, z_1) = \int_{-\infty}^{\infty} \Gamma_4(s_1, p_1, z_1) \exp[ip_1 \cdot \eta] \, dp_1, \tag{2.42}$$

the result:

$$\partial M/\partial z_1 - \eta \partial M/\partial s_1 = (1/2\pi) \int_{-\infty}^{\infty} H(\varepsilon s_1, \omega) M(s_1, \eta - \varepsilon \omega, z_1) \, d\omega. \tag{2.43}$$

Here H is the Fourier transform of $h = \bar{F}/\bar{\sigma}(0)$, and we write $h(\varepsilon s_1, \varepsilon p_1)$.

Equation (2.43) is next embedded in a higher-dimensional space $[s_1, s_2, \eta, \eta_2, z_1]$, where $s_2 = \varepsilon s_1$ and $\eta_2 = \varepsilon \eta$. This is a formal procedure and introduces no approximations. After embedding, we find the equation

$$\partial M/\partial z_1 - \eta \partial M/\partial s_1 - \eta_2 \partial M/\partial s_2$$

$$= (1/2\pi) \int_{-\infty}^{\infty} H(s_2, \omega) M(s_1, s_2, \eta - \varepsilon \omega, \eta_2 - \varepsilon^2 \omega, z_1) \, d\omega. \tag{2.44}$$

The ε argument can be eliminated from the η argument of M, and still retain the same form of the equation, by taking the Fourier transform of M with respect to s_1 and η in eq. (2.44). The equation governing the transform

$$N(w, s_2, v, \eta_2, z_1)$$

$$= (1/2\pi) \int_{-\infty}^{\infty} \int_{-\infty}^{\infty} M(s_1, s_2, \eta, \eta_2, z_1) \exp[i(s_1 w - \eta v)] \, ds_1 \, d\eta \tag{2.45}$$

is

$$\partial N/\partial z_1 - w\partial N/\partial v - \eta_2 \partial N/\partial s_2$$

$$= (1/2\pi) \int_{-\infty}^{\infty} \exp[-i\varepsilon v\omega] H(s_2, \omega) N(w, s_2, v, \eta_2 - \varepsilon^2 \omega, z_1) \, d\omega. \quad (2.46)$$

We notice that, if the ε^2 argument were not present in the function N appearing under the integral, the integral could be evaluated easily. If ε is set equal to zero, we find:

$$\partial N/\partial z_1 - w\partial N/\partial v - \eta_2 \partial N/\partial s_2 - h(s_2, \varepsilon v) N = 0. \quad (2.47)$$

The solution of eq. (2.47), which is termed N_0, is the lowest-order approximation to eq. (2.46). It was first obtained by Uscinski [1982] by using a multiple-phase screen technique. It can be shown that eq. (2.47) is the lowest-order term of a convergent iterative procedure.

An inverse transform of the function $N_0(w, s_2, v, \eta_2, z_1)$ is taken with respect to w and v to find $M(s_1, s_2, \eta, \eta_2, z_1)$. Next, we set $s_2 = \varepsilon s_1$ and $\eta_2 = \varepsilon \eta$. The result is $M(s_1, \eta, z_1)$. The fourth-order coherence function Γ_4 is then found by taking a Fourier transform with respect to η. The final expression for Γ_4 is

$$\Gamma_4(s, p, z)$$

$$= (1/2\pi) \int_{-\infty}^{\infty} \int_{-\infty}^{\infty} \exp[i(y - p/(z/\bar{k})^{1/2})x]$$

$$\times \exp\left(\bar{k}^2 \bar{\sigma}(0) z \int_0^1 h[s/l_m - (z/\bar{k}l_m^2)^{1/2} xt, y(z/\bar{k}l_m^2)^{1/2}] \, dt\right) dx \, dy.$$

(2.48)

The expression for $\Gamma_4(s, p, z)$ in three dimensions is identical in form to eq. (2.48). Numerical solutions are available in both two and three dimensions for a Gaussian correlation function, the Kolmogorov structure function without an inner scale and the Kolmogorov structure function with an inner scale (see Beran, Frankenthal, Mazar and Whitman [1993]). It is not too difficult to evaluate eq. (2.48) or its three-dimensional counterpart. Most solutions are for the mean-square intensity ($s = \mathbf{0}, p = \mathbf{0}$), but some solutions are available when s or p is not equal to zero*.

*Note added in proof: The reader is referred to Furutsu [1992] for an alternative approach to deriving the results obtained in this section.

The zero-order solution for the scintillation index gives the correct perturbation solution and approaches unity as the propagation distance approaches infinity. In between these limits, it has the correct qualitative behavior, as has been observed in experiments (Whitman and Beran [1985]). Further, it predicts an increase in the peak value of the curve as the inner scale increases. A benchmark solution is necessary to determine the accuracy of the result.

Experimental data have too much scatter to permit a good evaluation of the accuracy of the zero-order solution. Hence, most effort has focused on comparing the results with other theoretical and numerical calculations. Since the two-scale method is an approximation to the solution of the moment equation, it is desirable to compare the zero-order approximation with numerical solutions of eq. (2.40). Unfortunately, at present there are no numerical solutions available for three dimensions, but only for two dimensions. There are, however, numerical simulations in three dimensions that have been obtained by replacing the continuous random medium by a series of phase screens (Martin and Flatte [1988, 1990]).

Comparison of the zero-order solution to a numerical solution of eq. (2.47) in two dimensions was performed by Gozani [1985] for the case of a zero inner scale. The agreement between the solutions is very good for the scintillation index and differs by only a few percentage points. The comparison is shown in fig. 3 of Whitman and Beran [1985]. Tur [1982] provided a numerical solution for a Gaussian correlation. The comparison is shown in fig. 2a of Frankenthal, Whitman and Beran [1984]. Here the agreement for the scintillation index is not very good, and the two solutions differ by almost 20% in the peak region.

For three dimensions, the comparison between the zero-order solution and the Martin–Flatte numerical simulation is shown in figs. 1 and 2 of Beran, Frankenthal, Mazar and Whitman [1993]. We see that while the zero-order solution is qualitatively in agreement with the numerical-simulation result, there is significant disagreement in the peak region.

Solutions are also available for the correlation of intensities where r is the distance between two points in the transverse plane. Some results are shown in Beran, Frankenthal, Mazar and Whitman [1993]. Comparisons for the correlation of intensities with results of Tur [1982] in two dimensions are given in Beran and Whitman [1987]. The comparison of the zero-order approximation and Tur's numerical evaluation of eq. (2.40) is satisfactory for most separation distances at all ranges.

It is important to point out that although $\Gamma_4(r, \mathbf{0}, z) = \Gamma_4(\mathbf{0}, r, z)$, this relation is not true for the zero-order approximation. In all calculations that have been performed it has been found that the two calculations give very similar results in the zero-order approximation. Taking $p = r$ does, however, give slightly better results than choosing $s = r$. Since the two-scale procedure is convergent, the final result after many iterations should give identical results regardless of the choice made.

To obtain results that are in better agreement with experiment using the two-scale method, a next-order approximation must be used. Beran, Frankenthal, Mazar and Whitman [1993] discuss the first-order calculations that have been made. They also consider the results that have been obtained when a point source, rather than a plane wave, is used as an initial condition. Mazar and Bronstein [1992] consider finite aperture effects in the zero-order approximation.

Finally, we mention that calculations have been made for the scintillation index for a plane wave (light from a star) impinging on the atmosphere (Beran and Whitman [1988b]). The structure constant C_n^2 is taken to vary with altitude and results are given as a function of the angle from the zenith.

Numerical-simulation results. Since analytical solutions of eq. (2.40) are so difficult to obtain, it would be very desirable to have numerical solutions of this equation. As we mentioned previously (see Tur [1982], Gozani [1985]), there are solutions available for the two-dimensional problem (one transverse dimension), but as of this writing we are unaware of any numerical solutions in three dimensions. However, Martin and Flatte [1988, 1990] have performed numerical simulations using the field equation (2.15).

The basic idea of the numerical simulation used by Martin and Flatte [1988, 1990] is to replace the continuous turbulent atmosphere by a finite number of phase screens. In each simulated realization of the turbulent medium, the field is given a random phase distribution as it propagates through each phase screen. The random phases introduced by each phase screen are determined by the statistics of the turbulence in the atmosphere. The method for doing this is well established. Between the phase screens, the field propagates in free space; i.e., it is governed by eq. (2.15) with ε' set equal to zero.

The validity of the method depends on the number of phase screens used and on the number of realizations. The method is very intensive computationally, but as computer power increases, more phase screens and realizations may be considered. The results obtained by Martin and Flatte are summarized in their articles (Martin and Flatte [1988, 1990]) and give results that are compatible with the analytic solutions that are available. It is difficult to obtain very good agreement with experiment since there is so much scatter in the experimental results, but the simulations are compatible with experiment (see Consortini, Cochetti, Churnside and Hill [1993]).

The Martin–Flatte method may be used to obtain the scintillation index for both an initial plane wave and a point source in three dimensions. In coming years, if researchers are willing to spend the time and effort to implement the method, it should be possible to obtain accurate results for the full fourth-order coherence function. In addition, since the field is simulated in each realization, the probability density function associated with the intensity can be determined. However, here the number of realizations must be exceptionally large in order to find the correct distribution for very large intensities (Ben Yosef and Goldner [1988]).

2.3. TWO-POINT PHASE AND LOG-AMPLITUDE COHERENCE FUNCTIONS

It is convenient to consider phase and log-amplitude variations by writing the spatial part of the field in the form (see Tatarski [1971], Ishimaru [1978]):

$$\hat{u}(x) = \exp[\psi(x)], \tag{2.49}$$

where

$$\psi(x) = i\phi(x) + \chi(x).$$

Here $\phi(x)$ represents the phase and $\chi(x)$ is called the log-amplitude. The amplitude $A(x)$ and $\chi(x)$ are related by the relation

$$\chi(x) = \ln[A(x)].$$

From the Helmholtz equation, we find the following equation for ψ:

$$\nabla^2 \psi + \nabla\psi \cdot \nabla\psi + \bar{k}^2(1 + \varepsilon')\psi = 0. \tag{2.50}$$

Equation (2.50) is a nonlinear partial differential equation, and has generally been solved in the literature by perturbation theory. We write (see Ishimaru [1978]):

$$\psi = \psi_0 + \psi_1,$$

where ψ_0 satisfies eq. (2.50) when $\varepsilon' = 0$. To lowest order in ε', one finds:

$$\psi_{10}(x) = [1/\hat{u}_0(x)] \int_{v'} G(x - x')\varepsilon'(x')\hat{u}_0(x') \, dx'. \tag{2.51}$$

Here G is the free-space Green's function of the Helmholtz equation.

The solution in eq. (2.50) for ψ_{10} is called the first Rytov solution. In the 1960s, there were claims that this solution was valid beyond the perturbation region, but it is now generally accepted that this is not true. The solution may, however, be superior to the Born approximation of the Helmholtz equation in certain cases.

Following Ishimaru [1978], we write:

$$\psi_{10} = \chi + iS_1, \tag{2.52}$$

and define the following coherence and cross-coherence functions:

$$B_\chi(\mathbf{x}_{T1}, \mathbf{x}_{T2}, z) = \langle \chi(\mathbf{x}_{T1}, z)\chi(\mathbf{x}_{T2}, z) \rangle,$$

$$B_S(\mathbf{x}_{T1}, \mathbf{x}_{T2}, z) = \langle S_1(\mathbf{x}_{T1}, z)S_1(\mathbf{x}_{T2}, z) \rangle,$$

and

$$B_{\chi S}(\mathbf{x}_{T1}, \mathbf{x}_{T2}, z) = \langle \chi(\mathbf{x}_{T1}, z)S_1(\mathbf{x}_{T2}, z) \rangle. \tag{2.53}$$

Solutions for the coherence functions are found by assuming the forward-scattering approximation and the fact that $z \gg l_M$. For an initial plane wave, it is found that

$$B_{\chi,S}(s_T, z) = (2\pi)^2 \int_0^z d\eta \int_0^\infty \mu \, d\mu \, J_0(\mu s_T) H_{r,i}^2(z-\eta, \mu) \Phi(\mu), \tag{2.54}$$

and

$$B_{\chi S}(s_T, z) = (2\pi)^2 \int_0^z d\eta \int_0^\infty \mu \, d\mu \, J_0(\mu s_T) H_r(z-\eta, \mu) H_i(z-\eta, \mu) \Phi(\mu). \tag{2.55}$$

Here,

$$H_r = \bar{k} \sin[(z-\eta)\mu^2/2\bar{k}], \qquad H_i = \bar{k} \cos[(z-\eta)\mu^2/2\bar{k}].$$

The above expression may be evaluated easily for the Kolmogorov spectrum. The function $B\chi(s_T, z)$ is proportional to $R_I(s_T, z) - I^2$ in the Rytov approximation.

For most imaging problems, the perturbation or Rytov approximation is not adequate. It is usually necessary to know the phase coherence function when the phase fluctuations are large. In general this is not possible, but if the propagation distance is confined to the geometric-optics region [see eq. (2.39)], large phase fluctuations may be treated. In the geometric-optics region, $z \ll \bar{k}l_m^2$, we find:

$$B_S(s_T, z) = (4\pi^2 \bar{k}^2 z) \int_0^\infty \mu J_0(\mu s) \Phi(\mu) \, d\mu = (2\bar{k}^2 z) \int_0^\infty \sigma[(s^2 + x^2)^{1/2}] \, dx. \tag{2.56}$$

The above result is the same as that obtained by using geometric optics. In the forward-scatter approximation, geometric optics yields

$$S_1(x_T, z) = (\bar{k}^2/2) \int_0^z \varepsilon'(x_T, z') \, dz'. \tag{2.57}$$

Using eq. (2.57) in the definition for B_S, we find eq. (2.56). However, the geometric-optics analysis is not restricted to small fluctuations in the phase.

2.3.1. Use of Gaussian statistics

The reader should be cautioned about an approach that is often used in the literature. To find the two-point coherence function, eq. (2.49) is sometimes used. For example, one may write

$$\Gamma_2(x_{T1}, x_{T2}, z) = \langle \exp[\psi(x_{T1}, z) + \psi^*(x_{T2}, z)] \rangle. \tag{2.58}$$

In its present form, eq. (2.58) is not very useful. However, if it is assumed that $\psi(x)$ is distributed Gaussianly, then we find

$$\Gamma_2(x_{T1}, x_{T2}, z) = \exp[-\langle (\psi(x_{T1}, z) - \psi(x_{T2}, z))^2 \rangle]. \tag{2.59}$$

The expression for Γ_2 may then be found in terms of phase and log-amplitude coherence functions in addition to the cross-coherence function between the phase and log-amplitude.

The problem with the above formulation is two-fold:

(i) There is no experimental or other evidence that the joint probability density function of phase and log-amplitude at two separate points is distributed Gaussianly except in the perturbation region. In the geometric-optics region, the phase may be considered to be distributed Gaussianly because of the form of eq. (2.57). There is some evidence that the single point log-amplitude is distributed Gaussianly until somewhat beyond the Rytov region of applicability. However, over the entire propagation range, the two-point joint probability density function of phase and log-amplitude is not distributed Gaussianly. The reader is referred to articles by Phillips and Andrews [1981], Churnside and Clifford [1987], Churnside and Hill [1988], and Churnside and Frehlich [1988] for a discussion of the probability density of the log-amplitude.

(ii) Even if the two-point joint probability density function of the phase and log-amplitude was distributed Gaussianly, there are no analytic formulas for the phase and log-amplitude coherence functions beyond the perturbation region. The argument that the perturbation or Rytov solutions are valid for long ranges because $|\varepsilon'(x)| \ll 1$ is not correct. The relevant condition is $\langle \varepsilon'^2 \rangle \bar{k}^2 z l_c \ll 1$, where l_c is the characteristic eddy size of the eddies that contribute most to the scattering in the range interval z.

It should be noted that the above approach does give the correct result for Γ_2 for a plane-wave initial condition. This occurs because of the specific form of the Rytov solution. It should not be inferred from this result that satisfactory results for other initial conditions will be obtained. For the fourth-order case the Gaussian assumption generally gives erroneous results outside the region of weak turbulence (see, e.g., Tur and Beran [1983]).

2.4. ISOPLANICITY

We shall present here a quantitative treatment of isoplanicity to supplement our brief discussion in § 2.1. The approach will be based on Fried [1982]. We assume that there is a reference point source at position x_r (see fig. 5). The light from this reference source impinging on the lens is used in a short exposure to correct for the phase fluctuations caused by turbulence in the atmosphere for all points on the object. If the transverse separation

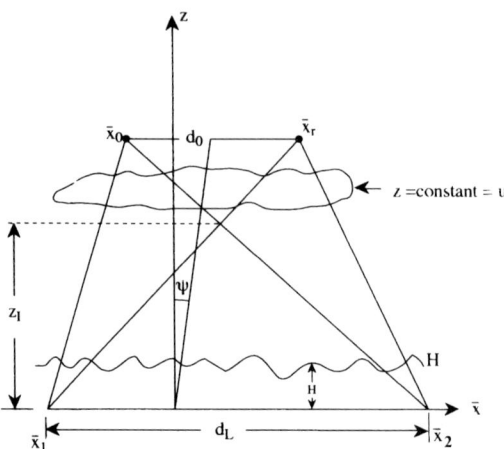

Fig. 5. Geometry for analyzing the conditions of isoplanicity.

between the reference point x_r and any object point x_o is sufficiently close (i.e., within an isoplanatic patch), then the phase corrections introduced from the reference source will cancel the random phase fluctuations in the light from the object point. If the separation is too large the cancellation will be imperfect.

The field of the light impinging on the lens point x from the object point x_o is

$$\bar{u}(x, x_o) = A \exp[i\phi(x, x_o)], \tag{2.60}$$

where A is a constant. The correction introduced at the point x by the reference source at x_r is

$$\bar{u}(x, x_r) = A \exp[-i\phi(x, x_r)]. \tag{2.61}$$

The coherence function in a single realization of the turbulence is denoted Γ_{2S}. The function Γ_{2S} of the light from the object point at x_o, corrected for phase errors using the reference source at x_r, is

$$\Gamma_2(x_1, x_2) = |A|^2 \exp[i(\phi(x_1, x_o) - \phi(x_1, x_r))]$$
$$\times \exp[-i(\phi(x_2, x_o) - \phi(x_2, x_r))]. \tag{2.62}$$

The mean-square value of the phase term in Γ_{2S}, $(\Delta\phi)^2$, is given by the expression

$$(\Delta\phi)^2 = \langle(\phi_1 - \phi_2 - \phi_3 + \phi_4)^2\rangle, \tag{2.63}$$

where

$$\phi_1 = \phi(x_1, x_o), \quad \phi_2 = \phi(x_1, x_r), \quad \phi_3 = \phi(x_2, x_o),$$

and

$$\phi_4 = \phi(x_2, x_r). \tag{2.64}$$

When $(\Delta\phi)^2 \ll \pi^2$, the isoplanatic condition is satisfied. In order to find $(\Delta\phi)^2$ there are ten terms to be evaluated (four autocorrelation terms of the form $\langle\phi_1\phi_1\rangle$, and six cross-correlation terms of the form $\langle\phi_1\phi_2\rangle$). For simplicity, we consider here only the case where the object and the reference source are at equal altitudes; i.e., $z_o = z_r = L$. With no loss of generality, we may take the image points to lie along the x-axis. In order to proceed, the following four assumptions are made:

(i) It is assumed that the light propagation is calculated using geometric optics. This implies that the ray from the source point to the measurement point is a straight line and the phase accumulation along the path is given by the expression

$$\bar{k} \int_{\text{path}} n \, ds.$$

(ii) All propagation paths are approximately at zenith angle ψ relative to the z-axis. As we see from table 1, the index-of-refraction fluctuations cease to have a significant effect at an altitude H which lies between 20 and 30 km. This means that all lateral distances are small in comparison to $H \sec \psi$.

(iii) The structure function for the index-of-refraction fluctuations has a Kolmogorov form, and

(iv) $H < z_0$.

Without the assumptions of geometric optics, we are aware of no calculations in the literature which perform this type of analysis. However, this assumption is non-trivial, and we should not expect the solution to give more than qualitative results beyond the region where geometric optics is valid. The final results given below agree with our physical intuition, but the numerical constants are very approximate at best. The reader will find extensive numerical calculations in Fried [1982].

We denote $d_0 = x_r - x_o$ and $d_L = x_2 - x_1$. Using the assumptions stated above, the expression for $(\Delta\phi)^2$ is found to be

$$(\Delta\phi)^2 = 2 \times 2.9 \bar{k}^2 \sec\psi \int_0^\infty C_n^2(u)$$
$$\times [d_L^{5/3} + (u\theta \sec\psi)^{5/3} - (1/2)|d_L^2 + 2d_L u\theta c \sec\psi$$
$$+ (u\theta \sec\psi)^2|^{5/6} - (1/2)|d_L^2 - 2d_L u\theta c \sec\psi$$
$$+ (u\theta \sec\psi)^2|^{5/6}] \, du, \qquad (2.65)$$

where c is the cosine of the angle between d_0 and the x-axis.

The minimum values of $(\Delta\phi)^2$ occur at $\theta = 0$ for $d_L > \theta u \sec\psi$, and at $d_L = 0$ for $d_L < \theta u \sec\psi$. To minimize the phase error, we shall expand the integrand in a Taylor series and demand that the remainder be much smaller than unity. In the first case (a), we expand around $\theta = 0$, and in the second case (b), we expand around $d_L = 0$. We find the following results:

(a) For $z_1 = d_L/\theta \gg H \sec\psi$, $\theta \approx 0$, $(\Delta\phi)^2$ is found to be approximately $2(\theta/\theta_0)^{5/3}$, where θ_0 is the isoplanatic angle defined as (Fried [1982]):

$$\theta_0 = \left[2.9(\sec\psi)^{8/3} \int_0^\infty C_n^2(u) u^{5/3} \, du \right]^{-5/3}. \qquad (2.66)$$

We may obtain an estimate of θ_0 by setting $u^{5/3}$ equal to $H^{5/3}$ in eq. (2.66) and then using the definition of Fried's coherence length r_0. We then find

$$\theta_0 \sim r_0/H(\sec\psi), \tag{2.67}$$

and the condition $(\Delta\phi)^2 \ll \pi^2$ becomes

$$\theta \ll r_0/H(\sec\psi). \tag{2.68}$$

We next note that if we choose the telescope size d_L to be just large enough to resolve the length d_0, so that $\bar{k}d_0 d_L/z = \pi$, then the condition on z_1 becomes

$$\theta \ll [\pi/\bar{k}H \sec\psi]^{1/2}. \tag{2.69}$$

As we stated in § 1.2, the inequalities given in eqs. (2.68) and (2.69) are the two conditions that must be satisfied if isoplanicity is to be a valid assumption.

(b) For $z_1 = d_L/\theta \ll H \sec\psi$, we find $(\Delta\phi)^2$ is approximately $2 \times 6.88 \times (d_L/r_0)^{5/3}$. In this case we have isoplanicity if $\bar{k}d_L^2 \ll \pi H$ and $d_L \ll r_0$. Here the random phase introduced into the light field from any point on the object is approximately a constant across the lens. In addition, the light rays from the different object points are uncorrelated on the lens. This case is not of practical importance when viewing astronomical objects.

2.4.1. Horizontal propagation paths

In the case of a horizontal propagation path, C_n^2 is a constant and H is equal to the propagation distance z. Again we expand the integral in eq. (2.65) around $\theta = 0$ and $d_L = 0$. We find, as special cases of the previous calculation:

(a) $\theta \ll r_0/z$, and $\theta \ll (\pi/\bar{k}z)^{1/2}$, where

$$r_0 = [2.9/(6.88\bar{k}^2 C_n^2 z)]^{-3/5}.$$

In terms of d_0, these conditions become $d_0 \ll r_0$ and $d_0 \ll [\pi z/\bar{k}]^{1/2}$. These conditions are generally very restrictive in practical imaging situations.

(b) For $d_L/\theta \ll z$, which is the near-field zone using the resolution condition, we find $d_L \ll r_0$ and $d_0 \gg [\pi z/\bar{k}]^{1/2}$.

2.5. SUMMARY OF RESULTS OF § 2 AND THEIR IMPORTANCE TO THE IMAGING PROBLEM

In § 2.1, we discussed the statistical properties of the index-of-refraction field. Definitions were given for the two-point correlation [eq. (2.3)], the structure function [eq. (2.4)] and spectrum [eq. (2.5)]. The Kolmogorov forms were given in eqs. (2.6) and (2.7). An altitude-dependent form for C_n^2 was given in table 1. The inner and outer scales were considered at the conclusion of the section.

The definition of the coherence functions Γ_2 and Γ_4 were given in § 2.2.1 and their relevance to the imaging problem discussed. In § 2.2.2 the equations governing the coherence functions in the turbulent atmosphere were given. Some solutions of the equation governing Γ_2 were presented in § 2.2.3. Solutions of the four-point coherence function were discussed in § 2.2.3.

The solution for Γ_2 shown in eq. (2.34) is very important, since the Fourier transform of Γ_2 is proportional to $S(x)$ given in eq. (1.1). The characteristic decay length associated with Γ_2, denoted s_c, allows one to determine if the distortion introduced by turbulence in a long-exposure image is significant compared to the diffraction of the lens of diameter d_L. When $s_c/d_L \ll 1$, turbulence dominates the distortion of the image, and when $s_c/d_L \gg 1$, turbulent distortion is of no importance. Even for short exposures the ratio s_c/d_L determines the importance of turbulent distortion since the short-exposure response has, on the average, the same angular spread as the long-exposure response. We note that if C_n^2 varies slowly with altitude, C_n^2 may be replaced by its mean value.

It is also important to realize that isoplanicity is not important for long exposures. In the discussion of a finite incoherent source, it was pointed out that Γ_2 is the product of the coherence function of the object in the absence of the atmosphere and the coherence function resulting from a point source of light that has passed through the turbulent atmosphere. The result does not depend on the angular size of the object. This is not true for short exposures unless the condition of isoplanicity is satisfied.

To assess the importance of intensity fluctuations (as opposed to phase fluctuations) in image distortion, the four-point coherence function must be considered. When all the points are at the same location, the four-point coherence function reduces to the mean square of intensity fluctuations. When the intensity fluctuations are small, an expression for the four-point moment is given in eq. (2.36), and in eq. (2.38) for the intensity fluctuations in the geometric-optics regime. Conversely, eq. (2.36) can be used to determine if the intensity fluctuations are indeed small, since we may assume that the result is correct if, for a given C_n^2, $\bar{k}$ and propagation distance z, we find that $[\langle I^2 \rangle / \langle I \rangle^2 - 1]^{1/2} \ll 1$.

When the intensity fluctuations are small, we may assume that they have little effect on image distortion. When the fluctuations are large, it is unclear just how important a role they play in image distortion since it is known that phase fluctuations play the dominant role. Much more work is necessary to determine how important large-intensity fluctuations are in such techniques as speckle interferometry. We note, however, that while in vertical viewing of an astronomical object the intensity fluctuations are generally small, there are some days when the intensity fluctuations are large. In addition, the intensity fluctuations are generally large for far-off-zenith viewing and for horizontal paths longer than many kilometers. When the intensity fluctuations are large, they may be found using the two-scale method or the numerical-simulation technique described in § 2.2.4.

In § 2.3, expressions were given for the two-point phase and log-amplitude coherence functions, using the lowest-order Rytov approximation when the fluctuations are small. Generally, the expressions for the phase-coherence functions are of limited utility in imaging problems and the geometric-optics approximation is of more use.

To relate the two-point coherence function to the phase and log-amplitude coherence functions, it is sometimes assumed that the two-point joint probability of phase and log amplitude is distributed Gaussianly beyond the perturbation region. There is no evidence for this, and any analyses based on this assumption should be used with great caution. There is, however, justification in assuming that the phase fluctuations may be distributed Gaussianly. This is certainly correct in the region where geometric optics is valid. A knowledge of the statistical properties of phase fluctuations is important in adaptive optics. It is necessary to know over what distance in the lens plane the phase fluctuations remain correlated after the light passes through the turbulent atmosphere. Within this distance, no adaptive corrections are required.

Manning [1993; see § 3.4.2] shows how one may correct the function $S(x)$ for phase distortion by expanding the random phase field in terms of Zernike polynomials. The correction procedure depends upon a knowledge of the two-point coherence function of the phases. However, in order to obtain usable relations, the assumption of joint Gaussian statistics is invoked in his analysis. Unless the log-amplitude fluctuations may be neglected, the method is subject to the criticism given above.

In § 2.4, we summarized a quantitative method for determining the conditions necessary for the assumption of isoplanicity to be valid. The assumption of isoplanicity is necessary for most of the short-exposure image-processing methods currently in use. It is not required for long-exposure procedures. Finally, we noted that for horizontal paths the assumption of isoplanicity is very restrictive.

§ 3. Imaging of Objects in the Turbulent Atmosphere

In this section, we formulate the imaging problem and summarize a number of processing techniques. An analytic formulation of the basic imaging equations is given in § 3.1. Using the equations presented in § 3.1 and the material given in § 2, in § 3.2 and § 3.3 we discuss a number of imaging methods currently in use. Long-exposure techniques are discussed in § 3.2, and short-exposure methods in § 3.3. We present some concluding remarks in § 3.4.

The principal purpose of this review is to discuss the use of the knowledge gained from studies of the propagation of light through the atmosphere in designing and evaluating image-processing methods. Although it is obviously important to consider all noise sources in a full analysis of imaging systems, it is generally beyond the scope of this article to consider in detail more than the effects of the atmospheric turbulence. For example, photon noise (see Dyson [1975], Goodman [1985]) can often determine whether one system is better than another for imaging a certain class of objects, although both systems may be equally effective in overcoming the effects of atmospheric turbulence.

Section 3 cites many articles in which the systems treated here are analyzed in great detail, and experimental results or computer simulations are presented. Our discussion is focused on the way in which the systems overcome the effects of atmospheric turbulence and on the limitations inherent in the methodologies. We do not present experimental results for most of these methods since the reader will be unable to properly evaluate the experiment in the absence of a greatly detailed presentation of the experimental conditions. For every method we consider, however, the reader will find in the references experimental evidence that the method is useful to image some class of objects. In most cases, the reader will be referred to a specific published experimental result for the method discussed. It should be pointed out, however, that despite the many references to the techniques discussed, the amount of experimental data (as opposed to computer-simulated results) is limited.

3.1. BASIC IMAGING EQUATIONS

The basic imaging equations are well known. The reader is referred to the following references: Born and Wolf [1964], Beran and Parrent [1974], Marathay [1982], Goodman [1985], and Manning [1993]. For simplicity, here we shall consider only the imaging of incoherent objects. The light is assumed to be filtered, so that $\Delta v/\bar{v} \ll 1$. We shall use a thin circular lens of diameter d_L, focal length f and no aberrations (see fig. 1). We shall distinguish between long and short exposures.

The thin lens introduces a phase factor of the form:

$$\exp[-i\bar{k}(x^2 + y^2)/2f].$$

Thus, the relation between the coherence function $\Gamma_2(x_L, z)$ impinging on the lens and the intensity in the focal plane $I(x_i)$ has the following form:

$$I(x_i) = \text{const} \cdot \int_{-\infty}^{\infty} \int_{-\infty}^{\infty} \Gamma_{2S}(x_{L1}, x_{L2}, z) P(x_{L1}) P^*(x_{L2})$$
$$\times \exp[(i\bar{k}/f)[x_i \cdot (x_{L2} - x_{L1})]] \, dx_{L1} \, dx_{L2}, \qquad (3.1)$$

where $P(x_L)$ is unity over the lens and zero outside the lens. Here, we use the notation Γ_{2S} to indicate that this is a short exposure and there is only one realization of the turbulent field. When the exposure is long, Γ_2 replaces Γ_{2S} in eq. (3.1) and the expression then represents an average over many realizations of the turbulent field. In § 3.1.1, we shall consider long exposures and in § 3.1.2 short exposures. We note that if $\Gamma_{2S}(x_{L1} - x_{L2}, z)$ and the characteristic coherence length of Γ_2 is small compared to d_L, then the intensity in the focal plane is the Fourier transform of Γ_2.

3.1.1. *Long exposures*

The relation between the coherence function at the lens plane and the intensity of the object was discussed in § 2.2.3 in the subsection on a finite incoherent object. For long exposures, the coherence function at the lens plane can be written as [see eq. (2.35)]

$$\Gamma_2(x_{L1}, x_{L2}, z) = \Gamma_{2o}(x_{L1}, x_{L2}, z) \Gamma_{2p}(x_{L1}, x_{L2}, z), \qquad (3.2)$$

where Γ_{2o} is the coherence of the incoherent object if the turbulent atmosphere was not present, and Γ_{2p} is the coherence function resulting from light from a point source propagating in the turbulent atmosphere. The coherence function Γ_{2o} is related to the intensity distribution of the incoherent object $I(x_o)$ by the relation

$$\Gamma_{2o}(x_{L1}, x_{L2}, z) = C_{VZ} \int_{-\infty}^{\infty} I(x_o) \exp[-i\bar{k}(x_{L1} - x_{L2}) \cdot x_o/z] \, dx_o, \qquad (3.3)$$

where (see Goodman [1985], p. 209):

$$C_{VZ} = (C_V/z^2)\exp[i\bar{k}(|x_{L1}|^2 - |x_{L2}|^2)/2z]. \tag{3.4}$$

Here, C_V is a constant.

If $\bar{k}d_L^2/2z \ll 1$, then the phase term in C_{VZ} may be set equal to zero. In this case Γ_{2o}, and hence Γ_2, is dependent only on $x_{L1} - x_{L2}$. We then introduce sum and difference coordinates in the integrals in the expression for $I(x_i)$ and integrate over the sum coordinate, since $P(x_L)$ is a known function. The resultant expression,

$$\Gamma_{2p}(x_{L1} - x_{L2}, z) \int_{-\infty}^{\infty} P(x_{L1})P^*(x_{L1} + (x_{L1} - x_{L2}))\,dx_{L1}, \tag{3.5}$$

is proportional to the Fourier transform of $S(x)$, $\tilde{S}(x_{L1} - x_{L2})$, in eq. (1.1).

Note that Γ_{2p} depends only upon $x_{L1} - x_{L2}$ because the turbulent atmosphere is assumed to be statistically homogeneous in all planes perpendicular to the mean propagation direction. These conditions are not related to the concept of isoplanicity discussed in § 2.4.

The function Γ_{2p} is given in eq. (2.34) for a point source located in the turbulent atmosphere. If we consider light from a distant star, then the light from a point source behaves like a plane wave when it impinges on the upper atmosphere. In this case, $C_n^2 z$ in eq. (2.34) is replaced by $\int_{-\infty}^{z} C_n^2(z')\,dz'$. Whichever expression is used, we note that Γ_{2p} has the form

$$\Gamma_{2p}(s_T, z) = \text{const} \cdot \exp[-C_s s_T^{5/3}]. \tag{3.6}$$

Thus the decay of Γ_{2p} as a function of s_T is very rapid. Under typical atmospheric conditions, $C_s s_T^{5/3}$ is of order unity when $s_T = 10$ cm. This means that to obtain diffraction-limited behavior for a lens of 1 m (needed to resolve an object with angular diameter of $\pi \times 10^{-7}$ radians), we need to know accurately the form of Γ_{2p} for values of s_T that are a factor $\exp[-(10)^{5/3}]$ smaller than the magnitude of $\Gamma_{2p}(0, z)$.

Measurements for finding Γ_{2p} have been made using interferometric techniques. The earliest experiments were performed for horizontal laser-beam propagation (see Wessely and Bolstad [1970], Bertolotti, Muzii and Sette [1970]), giving information on the atmospheric conditions close to the ground. For these experiments, the propagation paths were assumed to be short enough that only phase variations had to be considered. Later, Γ_{2p} for stellar sources was investigated (Roddier and Roddier [1973], Kelsall [1973], Dainty and Scaddan [1974, 1975], Roddier [1976], Brown and Scaddan [1979]). In these experiments, the random intensity variations were taken into account. The behavior of Γ_{2p} was found to vary according to the $\frac{5}{3}$ law [eq. (3.6)] for smaller separation distances. For increased separation distances, in most cases, the exponent was measured as less than $\frac{5}{3}$. In fig. 6 we show results from Bertolotti, Muzii and Sette [1970] for horizontal propagation. The vertical coordinates of the B and C curves have been multiplied by 10 relative to the other curves. The theoretical $\frac{5}{3}$ law is shown as a dotted line. In fig. 7, results of Roddier and Roddier [1973] are given for two different stars, Aldebaran and Altair. Curve A is for Aldebaran ($\psi =$

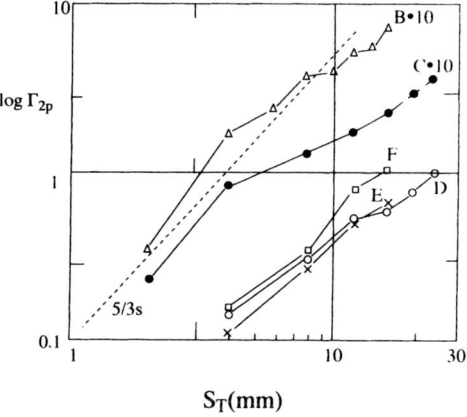

Fig. 6. Five recordings of Γ_{2p} for different turbulence conditions.

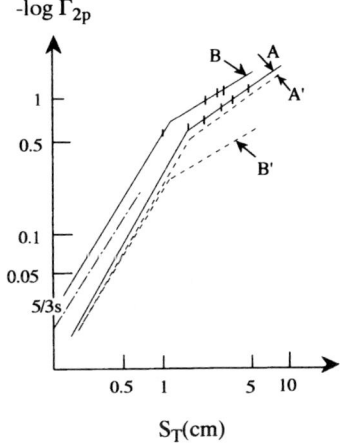

Fig. 7. Experimental results for Γ_{2p} for two different stars.

30°) and curve B is for Altair ($\psi = 70°$). A' and B' are the same curves after a zenith-angle correction of $\cos \psi$. The theoretical $\frac{5}{3}$ law is also shown. Large-scale inhomogeneities, spatial anisotropy in the case of viewing far from the zenith (fig. 7), non-stationarity of the atmosphere, and low light levels were reported as causes that affected the experimental results.

Instead of finding the function Γ_{2p}, its Fourier transform can be measured by an analysis of a star-trail image spread recorded on a photographic film (King [1971], Walters, Flavier and Hines [1979]). In general, the shape of Γ_{2p} that was obtained agreed with the $\frac{5}{3}$ law for smaller s_T. In figs. 8 and 9, we show the results of Walters, Flavier and Hines [1979] for night-time and day-time conditions. The theoretical curve $\exp[-C_S s_T^{5/3}]$ is adjusted to fit at the e^{-1} point. The system instrument function has been removed and is

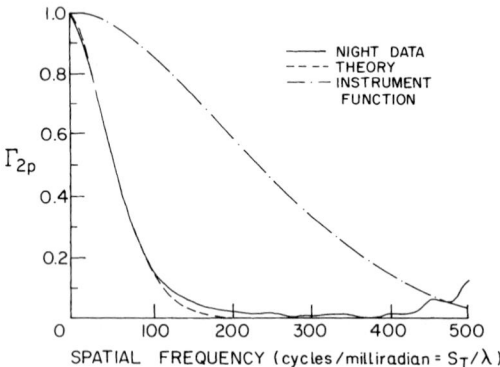

Fig. 8. The function Γ_{2p} obtained for night-time data.

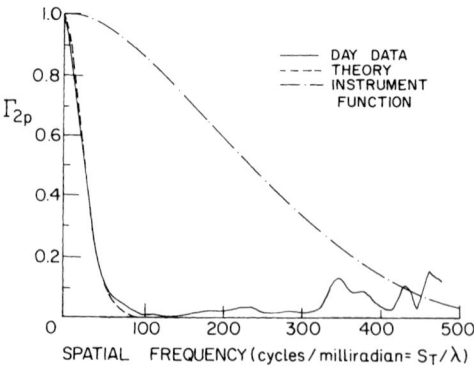

Fig. 9. The function Γ_{2p} obtained for day-time data.

shown for comparison. During the experiments, it appeared necessary to take into account factors such as nonlinearity of the photographic emulsion (which requires calibration of the film exposure characteristics), Rayleigh-scattered red light during day-time measurements, and aerosols.

Apart from the experimental difficulties that arose in the measurement of Γ_{2p}, there existed a strong variability in the coherence length r_0. Experiments showed that there was a factor of 2–3 in the diurnal variation of r_0 and large standard deviations (see Wessely and Bolstad [1970], Fried and Mevers [1974], Roddier [1976], Walters, Flavier and Hines [1979]). The coherence length r_0 is very dependent on location, time of day and season.

We have presented the experimental results in some detail to emphasize the practical difficulties encounterd in determining Γ_{2p} for large separation distances s_T. Without an accurate determination of Γ_{2p} for large s_T, one cannot perform high-resolution imaging with large telescopes (i.e., $s_c/d_L \ll 1$) using long-exposure methods. The short-exposure techniques provide a solution to this difficulty when the condition of isoplanicity is satisfied.

3.1.2. *Short exposures*

The difference between long and short exposures for incoherent imaging is in the form of Γ_2 that appears in eq. (3.1). In general, no separation like that given in eq. (3.2) is possible. The equation for $\Gamma_{2S}(x_{L1}, x_{L2}, z)$ in terms of $I(x_o)$ may be written in terms of a stochastic Green's function, $G_S(x_o, x_{L1}, x_{L2}, z)$, in the form:

$$\Gamma_{2S}(x_{L1}, x_{L2}, z) = \int_{-\infty}^{\infty} G_S(x_o, x_{L1}, x_{L2}, z) I(x_o) \, dx_o. \tag{3.7}$$

Here the Green's function contains within it the random effects of the turbulent atmosphere and the diffraction effects of propagating over a distance z. In general, G_S depends upon all three coordinates x_o, x_{L1} and x_{L2}.

Little can be done unless we restrict our attention to a small region Δx about some central point in the object, say x_{oo}, where the stochastic portion of the Green's function depends only on x_{oo}. Let the characteristic dimension of the region $x_{oo} + \Delta x$ be d_0. This region is termed an isoplanatic patch if d_0 satisfies the conditions given in § 2.4. In this case, we may write the Green's function in the form:

$$G_S(x_o, x_{L1}, x_{L2}, z) = G_{SI}(x_{L1}, x_{L2}, z) C_{VZ} \exp[-\bar{k}(x_{L1} - x_{L2}) \cdot x_o/z], \tag{3.8}$$

where G_{SI} does not depend upon x_o, and we find

$$\Gamma_{2S}(x_{L1}, x_{L2}, z) = \Gamma_{2o}(x_{L1}, x_{L2}, z) G_{SI}(x_{L1}, x_{L2}, z). \tag{3.9}$$

The Green's function $G_{SI}(x_{L1}, x_{L2}, z)$ is a very complicated function of x_{L1} and x_{L2} that depends on the particular realization of the turbulent atmosphere we are observing. We remember, however, that the average of $G_{SI}(x_{L1}, x_{L2}, z)$ over many realizations of the turbulent atmosphere is $\Gamma_{2p}(x_{L1} - x_{L2}, z)$.

The product form for Γ_{2S} given in eq. (3.9), which is based on the assumption of isoplanicity discussed in § 2.4, is the starting point for a number of the image-processing techniques we shall discuss in § 3.2.

3.2. LONG-EXPOSURE IMAGE-PROCESSING TECHNIQUES

The techniques presented in this section are designed to invert the long-exposure imaging equation [eq. (1.1)], where $S(x)$ is not known exactly and a noise term is added to the right-hand side of eq. (1.1). The equation often considered in the literature (Cannon, Trussel and Hunt [1978], Bates and McDonnel [1986], Meinel [1986]) is

$$\langle I(x) \rangle = \int S(x - x') I_o(x') \, dx' + N(x), \tag{3.10}$$

where $N(x)$ is a noise term that contains the effect of a variety of noise sources. It may also contain the effect of our imperfect knowledge about the function $S(x)$.

In general, the function $S(x)$ is assumed to be known from either measurement or theory. At the conclusion of § 3.1.1 we gave a brief survey of the accuracy with which Γ_{2p} and $S(x)$ may be determined from experiment and theory. The accuracy with which $S(x)$ can be determined is not sufficient to eliminate the effects of atmospheric turbulence when imaging astronomical objects using large ground-based telescopes. To overcome this problem, the short-exposure techniques discussed in § 3.3 have been developed. However, to image objects on the ground from space-based systems, or for imaging objects over horizontal paths within the atmosphere, the long-exposure techniques are often adequate. In the next two subsections we consider the demands on image-processing techniques for space-based images of ground objects and images taken over horizontal paths. In § 3.2.3 we shall discuss the inversion of eq. (3.10) with and without noise.

3.2.1. Space-based systems imaging ground objects

The problem of viewing objects on the earth from space is usually, as a practical matter, less difficult than viewing objects in space from an imaging system on the ground. Consider the configuration given in fig. 10. Beran and Whitman [1972] have shown that as light that has passed through a turbulent medium recedes from the medium, the characteristic coherence length of the light s_c increases. For $z \gg H$ and vertical viewing, we find that $s_{cz} = (z/H)s_{cH}$, where H is between 20 and 30 km.

If we assume that s_{cH} is about 10 cm, then s_{cz} of the light impinging on an imaging system orbiting at 250 km is about 1 m. A space lens of 1 m (the minimum lens size where the turbulent effect would begin to be a problem) can resolve a distance d_0 of about 8 cm at 250 km. Therefore, unless exceptionally stringent resolution requirements are placed on orbital systems viewing ground objects, atmospheric turbulence effects should not be a major problem. (It should be noted, however, that the lens size needed to resolve a given distance d_0 on earth is also linearly dependent on z. Thus, while a 2 m lens at 500 km will not be significantly affected by the turbulence in the atmosphere, it can still only resolve a distance d_0 of about 8 cm.)

The reader can, of course, choose other conditions where the effect of turbulence may be much more significant. For example, special care should be taken when considering far-off-vertical viewing, since the additional path length in the turbulent atmosphere may significantly reduce the value of s_c.

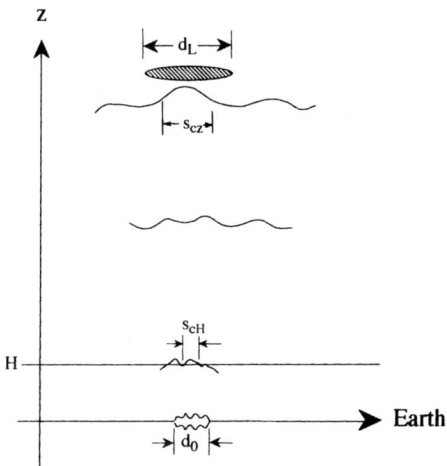

Fig. 10. Imaging by a space-based system.

It is important to calculate Γ_2 for each geometric configuration, since the coherence properties of the light change with propagation path. When z is not significantly greater than H the formulas given by Beran and Whitman [1972] can be used to find Γ_2 for the light impinging on the space-based system.

The assumption of isoplanicity cannot be satisfied easily for space-based imaging systems viewing objects on the ground. Isoplanicity depends upon the turbulent atmosphere being close to the imaging system and far from the object to be imaged. On the other hand, as we stated above, unless very small detail of the ground object is required, or far-off-vertical viewing is needed, atmospheric turbulence does not play the dominant role in image distortion that it does in large ground-based viewing systems.

3.2.2. Horizontal paths within the atmosphere

In this subsection we consider imaging along horizontal paths near ground level. To estimate the effects of atmospheric turbulence, we take the Fourier transform of eq. (3.10). We find:

$$\langle \tilde{I}(\kappa) \rangle = \tilde{S}(\kappa) \tilde{I}_o(\kappa) + \tilde{N}(\kappa), \tag{3.11}$$

where the function $\tilde{S}(\kappa)$ is proportional to Γ_{2p} which is given in eq. (3.6.).

The importance of the effects of turbulence on the image is determined by the ratio $\alpha = s_c/d_L$, where s_c is the characteristic coherence length of the light impinging on the lens. When $\alpha \gg 1$, the effects of turbulence are unimportant, and when $\alpha \ll 1$, the effects of turbulence are very difficult to overcome. For $\alpha \sim 1$, the effects of turbulence may be removed by the long-exposure techniques discussed below. Setting the exponent equal to one in eq. (2.34), we find:

$$s_c = 1/[(3/8)1.45 C_n^2 \bar{k}^2 z]^{3/5}. \tag{3.12}$$

In table 2, we present a chart for s_c as a function of C_n^2 and z for $k = 10^7$ m^{-1}. A similar calculation was made by Levanon [1992].

It is seen that in many imaging configurations, α is of order unity or greater. This is the class of problems for which the techniques discussed in the next section should be effective. When the weather conditions yield very strong turbulence and very fine resolution is required, the methods will be ineffective. We note that in the case of viewing an astronomical object with a very large ground-based telescope (e.g., 7 meters), the ratio α is always $\ll 1$ and these long-exposure techniques are not effective.

TABLE 2

s_c versus C_n^2 and z. Note that $C_n^2 = 10^{-14}$ (m$^{-2/3}$) corresponds to very strong turbulence and $C_n^2 = 10^{-17}$ (m$^{-2/3}$) corresponds to very weak turbulence. To resolve an object separation distance of 1 cm at 1 km requires a lens diameter of 3.1 cm. To resolve an object separation distance of 1 cm at 10 km requires a lens diameter of 31 cm.

s_c (cm)	C_n^2 (m$^{-2/3}$)	z (km)
1.2	10^{-14}	1
5	10^{-15}	1
20	10^{-16}	1
80	10^{-17}	1
0.5*	10^{-14}	5
1.9	10^{-15}	5
13	10^{-16}	5
30	10^{-17}	5
0.3*	10^{-14}	10
1.2	10^{-15}	10
5	10^{-16}	10
20	10^{-17}	10

*For $s_c < 1$ cm the dissipative range of eddies should be included for very accurate results.

3.2.3. Inversion procedures

There is a very extensive literature which discusses the inversion of eq. (3.10) to obtain $I_o(x)$ from the measured function $\langle I(x) \rangle$. Much of this literature deals with imaging problems that are unrelated to atmospheric turbulence, but the inversion techniques are applicable to any problem where the function $S(x)$ is assumed to be known and there is some statistical information about the noise $N(x)$.

We shall assume here that $S(x)$ can be determined with sufficient accuracy so that inversion is practical. Thus we shall be considering problems in which α is not too small. We expect that the function $\tilde{S}(\kappa)$ is approximately equal to the form of Γ_{2p} given in eq. (3.6). The constant multiplying $s_T^{5/3}$ in the exponent is assumed to be found from experiment. The variation of $\tilde{S}(\kappa)$ from the precise form given in eq. (2.36) may be included as part of the noise $N(x)$.

Beyond giving the form of $S(x)$ that is to be expected in eq. (3.10), there is little the development discussed in § 2 can provide. The noise represented in the function $N(x)$ derives from causes other than turbulence in the atmosphere. We shall, however, summarize some of the inversion methods that are used to find $I_o(x)$. We consider first neglecting the noise and then show how its effect may be included.

Neglect of noise ($N(x) = 0$). In the absence of noise, we can use eq. (3.11) to find $\tilde{I}_o(\kappa)$ from the relation

$$\tilde{I}_o(\kappa) = \langle \tilde{I}(\kappa) \rangle / \tilde{S}(\kappa). \tag{3.13}$$

The object intensity $I_o(x)$ can then be obtained by taking the inverse Fourier transform of $\tilde{I}_o(\kappa)$. However, this method is very sensitive to small errors in measurement, and iteration schemes have been developed which are more effective. Meinel [1986] surveys a number of these algorithms; we shall very briefly review his approach.

First we note that since measurements are generally digitalized, eq. (3.10) is usually written in the discrete form

$$i_m = \sum_k s_{mk} o_k + n_m, \tag{3.14}$$

Here, o_k represents the object intensity at k points on the object. The index k covers a full two-dimensional array. The variable i_m represents the image intensity at m points. The matrix s_{mk} is the point-spread function which we consider to be invariant spatially with its sum normalized to unity. The noise is given by the variable n_m.

As in the continuous case, the object intensity may be found directly by matrix inversion. However, in many problems the matrix s_{mk} is nearly singular, and therefore other inversion methods are preferred. To develop a typical algorithm in the absence of noise, Meinel [1986] writes the identity

$$0 = (i_m - a_m)^p, \tag{3.15}$$

where

$$a_m = \sum_k s_{mk} o_k, \tag{3.16}$$

and p is a real number.

In eq. (3.15), o_m is added to both sides of the equation, and the iterative relation

$$o_m^{(n+1)} = K[o_m^{(n)} + (i_m - a_m^{(n)})^p] \tag{3.17}$$

is obtained. Here $o_m^{(0)}$ is some reasonable guess for the object-intensity distribution and K is a scaling factor.

In a similar manner, Meinel [1986] develops a number of other iterative algorithms that have been used previously by other investigators. The algorithm given in eq. (3.15) does not demand that o_m be positive. An algorithm where o_m is positive can be obtained by exponentiating both sides of eq. (3.15) and multiplying both sides of the resultant equation by o_m. The convergence of the algorithms is very algorithm-dependent, and the reader is referred to the articles cited by Meinel [1986].

Inclusion of noise. In this subsection we present a few methods used to eliminate the effects of the noise $N(x)$.

Wiener filter. To remove the effects of the noise function $N(x)$ in an optimum way, a Wiener filter is sometimes used to invert eq. (3.10) (see Cannon, Trussel and Hunt [1978], Papoulis [1965]). We assume a solution of the form

$$\hat{I}_o(x) = \int Q(x - x') \langle I(x') \rangle \, dx', \tag{3.18}$$

where the function $Q(x)$ is determined by minimizing the error

$$\delta = \int |\hat{I}_o(x) - I_o(x)|^2 \, dx. \tag{3.19}$$

After the minimization, we find that

$$\tilde{Q}(\kappa) = \tilde{S}^*(\kappa) |\tilde{I}_o(\kappa)|^2 / [|\tilde{I}_o(\kappa)|^2 |\tilde{S}(\kappa)|^2 + \Phi_N(\kappa)], \tag{3.20}$$

where $\Phi_N(\kappa)$ is the spectrum of the noise function $N(x)$.

The success of the method depends on an a priori knowledge of the power spectrum of the object $I_o(x)$ and the noise $N(x)$. When this information is available, the function $\hat{I}_o(x)$ is often a useful approximation to $I_o(x)$.

Maximum likelihood or maximum entropy. Meinel [1986] summarizes a number of works that develop imaging algorithms using maximum likelihood or maximum entropy. The reader is referred to basic papers by Jaynes [1968], Frieden [1972], Frieden and Wells [1978], Meinel [1988] and several references in Meinel [1986]. The basic idea of maximum likelihood is to maximize the probability function

$P(i_m | o_k),$

defined as the probability of measuring an image i_m (at all m points) given o_k (at all k points), with respect to o_k. The probability function is often assumed to be Poisson distributed or Gaussian and all the points have independent probability functions.

For Poisson statistics, we have

$$P(i_m | o_k) = \prod_m [a_m^{i_m}][\exp(-a_m)]/i_m! \qquad (3.21)$$

The maximization is easier to perform using ln P. We find:

$$\partial \ln P / \partial o_m = \sum_k s_{mk}(i_k/a_k) - 1 = 0. \qquad (3.22)$$

A recursive algorithm is found from eq. (3.22) by adding unity to both sides of the equation, raising to a power p and multiplying both sides of the equation by o_m. This yields

$$o_m^{(j+1)} = K o_m^{(j)} \left[\sum_k s_{mk} \left(i_k \bigg/ \sum_n s_{kn} o_n^{(j)} \right) \right]^p. \qquad (3.23)$$

A number of other algorithms can also be derived using eq. (3.22) as a basis.

In place of using the probability function alone, one may maximize the sum of the probability function and an entropy function of the form

$$-\sum_m o_m \ln(o_m). \qquad (3.24)$$

In addition, one may add any constraint on the image that is available. Meinel [1986] summarizes the various algorithms that may be obtained. (See Frieden and Wells [1978] for experimental results. The enhanced results are far from being diffraction-limited.)

In all these formulations, we wish to emphasize again that the only information gained from studying the propagation function is the form of $S(x)$. The noise is assumed to arise from causes other than atmospheric turbulence. In some recent work, however, Bakut, Poll'skikh, Ryakhin, Sviridov and Ustinov [1984] (see also the discussion in Manning [1993]) assume that the joint probability density of the m image intensities is jointly Gaussian, but with correlation between locations in the image plane. This requires that, in addition to the function $S(x)$, it is necessary to know the correlation of intensities between any two points in the image plane. Bakut, Poll'skikh, Ryakhin, Sviridov and Ustinov [1984] assume that intensity fluctuations may be neglected in the lens plane in order to obtain this correlation function.

3.3. SHORT-EXPOSURE IMAGE-PROCESSING TECHNIQUES

In this section we shall review a number of methods that have been developed to overcome the effects of atmospheric turbulence on ground-based imaging systems. As we have stated repeatedly, the effectiveness of most of the methods is restricted to astronomical objects or satellites that satisfy the condition of isoplanicity.

3.3.1. Adaptive optics

In this subsection, we elaborate on the introduction we gave in § 1.3.1. The basic idea of adaptive optics is to change the function $P(x_L)$ in eq. (3.1) from a constant function to a new function which compensates for the phase and amplitude fluctuations introduced by the turbulence in the atmosphere. This compensation must be done in a time which is short compared to t_T, and a new compensation must be performed for each new realization of the index-of-refraction field. Most present systems compensate only for the phase change, which is the dominant effect, and we shall restrict our attention here to phase compensation only.

We can detect phase fluctuations introduced by atmospheric turbulence by measuring the phase change that results when light from a point source near the object impinges on the lens. By "near the object", we mean that the point source lies within the same isoplanatic patch (see § 2.4) as the object. This point source may be a natural object like a star, an artificially created source (see Foy and Laberyie [1985], Thompson and Gardner [1990], Gardner, Welsh and Thompson [1987], Primmerman, Murphy, Page, Zollars and Barclay [1991], Fugate, Fried, Ameer, Broeke, Browne, Roberts, Ruane and Taylor [1991], Fugate [1993]), or even a bright glint located on the object itself. Without the assumption of isoplanicity, no compensation can be made.

Substituting eqs. (3.8) and (3.9) into eq. (3.1), we find:

$$I(x_i) = \text{const} \cdot \int_{-\infty}^{\infty} \int_{-\infty}^{\infty} \Gamma_{2o}(x_{L1}, x_{L2}, z) G_{SI}(x_{L1}, x_{L2}, z) P(x_{L1}) P^*(x_{L2})$$

$$\times \exp[(i\bar{k}/f)[x_i \cdot (x_{L1} - x_{L2})]] \, dx_{L1} \, dx_{L2}. \qquad (3.25)$$

From eq. (3.3), we see that Γ_{2o} is a function only of the difference coordinate $\Delta x_L = x_{L1} - x_{L2}$, so that eq. (3.25) becomes

$$I(x_i) = \text{const} \cdot \int_{-\infty}^{\infty} \Gamma_{2o}(\Delta x_L) Q_R(\Delta x_L) \exp[(i\bar{k}/f)\{x_i \cdot \Delta x_L\}] \, d\Delta x_L,$$

$$\qquad (3.26)$$

where

$$Q_R(\Delta x_L) = \int_{-\infty}^{\infty} G_{SI}(x_{L1}, \Delta x_L, z) P(x_{L1}) P^*(x_{L1} + \Delta x_L) \, dx_{L1}.$$

The function $Q_R(\Delta x_L)$ contains the effect of the random phase fluctuations. To eliminate the random phase fluctuations, $P(x_{L1})P^*(x_{L1} + \Delta x_L)$ must be changed to compensate for the phase fluctuations in $G_{SI}(x_{L1}, \Delta x_L)$. Hardy, Lefebvre and Koliopoulos [1977] present a workable system that accomplishes this effect. The entire March 1977 issue of the *Journal of the Optical Society of America* (Vol. 67) is devoted to adaptive optics and provides an excellent background to the subject, see also two special sessions on Atmospheric Compensation Technology in the January 1994 and February 1994 issues of the *Journal of the Optical Society of America* (Vol. 11, Nos. 1 and 2).

The key to a workable system is the real-time measurement of the wave-front slope at a large number of locations on the telescope aperture. That is, the change in phase between all the locations must be determined. As we stated previously, the number of locations required is determined by the characteristic coherence length of the light impinging on the telescope (see § 2.2.3). In Hardy's system, the wave-front slope was measured using a shearing interferometer. Welsh and Gardner [1989] present a performance analysis of adaptive-optics systems using laser-guide stars and slope sensors. They review the previous literature and point out the many sources of error that must be studied in order to properly evaluate a system. The system they analyzed consists of an aperture, a wave-front-slope sensor, a deformable mirror and a linear control law. The mean-square residual phase error across the aperture, the point spread function, the optical transfer function and the Strehl ratio are used as measures of performance.

Recent results measuring the efficacy of adaptive optics are presented in Fugate, Fried, Ameer, Broeke, Browne, Roberts, Ruane and Taylor [1991], Fugate [1993], Primmerman, Murphy, Page, Zollars and Barclay [1991], Kane, Gardner and Thompson [1991], Roddier, Graves, McKenna and Northcott [1991], Roddier, Northcott, Graves, McKenna and Roddier [1993], and Lloyd-Hart, Dehany, McLeod, Wittman, Colucci, McCarthy and Angel [1993]. We note that since multiple telescopes are used in modern astronomical applications, the compensation problem must be extended to include inter-telescope effects. Angel, Wizinovich, Lloyd-Hart and Sandler [1990] discuss this problem and treat it with the use of neural networks.

Experimental results are given in fig. 2 of Primmerman, Murphy, Page,

Zollars and Barclay [1991] and in fig. 1 of Fugate, Fried, Ameer, Broeke, Browne, Roberts, Ruane and Taylor [1991].

3.3.2. Image sharpening

Muller and Buffington [1974] proposed a method of image processing that is based on maximizing a function S_I of the form

$$S_I = \int I^2(x) \, dx, \tag{3.27}$$

where $I(x)$ is the short-exposure image in the image plane. They show that the maximum value of S_I is obtained when the random-phase fluctuations introduced by atmospheric turbulence are set equal to zero. Thus, if one introduces a real-time adjustable phase-shifter in the lens plane and can also calculate S_I in real time, it should be possible to sharpen the image by adjusting appropriately the phase-shifter to compensate for atmospheric turbulence. The number of locations at which compensation is necessary is determined by the characteristic coherence length of the impinging light, just as in adaptive optics. Successful performance of the procedure requires an algorithm to adjust the phases to quickly and efficiently maximize S_I. In a number of papers (Buffington, Crawford, Muller, Schwemin and Smits [1977a], Buffington, Crawford, Muller and Orth [1977b], Pollaine, Buffington and Crawford [1979]), it was demonstrated that the concept can be implemented. The method presents an alternative to the adaptive-optics pre-detection procedure described in § 3.2.1. Here it is not necessary to have a reference source.

As we shall show, the proof that maximizing S_I yields a sharpened image requires the assumption of isoplanicity and the neglect of intensity fluctuations (see Appendix A of Muller and Buffington [1974]). Muller and Buffington present more general functions than S_I that may be maximized, but we shall restrict our attention here to S_I. The expression for $I(x)$ is given by eq. (3.1), where Γ_{2S} is given by eq. (3.7). If we invoke the assumption of isoplanicity, then Γ_{2S} is given by eq. (3.9). Finally, assuming that the random intensity fluctuations caused by atmospheric turbulence may be neglected, we may write G_{SI} in the form

$$G_{SI}(x_{L1}, x_{L2}, z) = \text{const} \cdot \exp[i\phi(x_{L1}, x_{L2}, z)]. \tag{3.28}$$

Substituting the expression for $I(x)$ in eq. (3.27) yields an expression that may be simplified considerably by performing a number of integrations and recognizing that they are delta functions. The details are in Appendix A of

Muller and Buffington [1974]. The final expression (with a simple change of variables) is reduced to a six-fold integral of the form

$$S_I = \text{const} \cdot \int \ldots \int |\tilde{I}_o(w_1, w_2)|^2$$

$$\times \exp[\phi(u_1, v_1) - \phi(u_1 - w_1, v_1 - w_2)) + \phi(u_2, v_2)$$

$$- \phi(u_2 + w_1, v_2 + w_2)] \, du_1 \, du_2 \, dv_1 \, dv_2 \, dw_1 \, dw_2. \quad (3.29)$$

Here the tilde indicates a Fourier transform (the z coordinate has been suppressed in the ϕ expressions). The integrand is positive except for the exponential term. If the ϕ terms are constant or linear functions of its arguments, the exponent is zero and S_I is maximized. A linear variation of the phase only shifts the origin of the image, but does not distort it. No proof is given by Muller and Buffington [1974] that no other phase functions will yield a zero exponent. However, assuming that the phase functions may be expanded in a sum of polynomials leads to that conclusion. Experimental results are given in figs. 2 and 3 of Buffington, Crawford, Muller and Orth [1977b].

3.3.3. *Speckle interferometry*

Speckle interferometry (Laberyie [1970, 1976]) is an imaging procedure that is insensitive to the random phase changes introduced by the turbulence in the atmosphere. For the method to work, the light from the source has to obey Gaussian statistics and the object has to be small enough to satisfy the isoplanicity condition. The general method of approach is to take a long series of short-exposures and find the spatial power spectrum of the image. This is done by Fourier-transforming the image, squaring the result and averaging over all the short exposures. This function is shown to be relatively insensitive to the random phase fluctuations introduced by turbulence. Making use of the spatial spectrum and a phase-retrieval algorithm, the object image may be restored.

Basic approach. The starting point of the speckle-interferometry method is the imaging equation with the assumption of isoplanicity for the short-exposure point-spread function $S_S(x - x')$ [see eq. (1.2)]. The function S_S changes with each realization of atmospheric turbulence. The imaging equation we use is

$$I(x) = \int_A S_S(x - x') I_o(x') \, dx'. \quad (3.30)$$

We next take the Fourier transform of eq. (3.30). Since eq. (3.30) is a convolution, this yields the algebraic equation

$$\tilde{I}(\kappa) = \tilde{S}_S(\kappa)\tilde{I}_o(\kappa), \tag{3.31}$$

where the tilde indicates a Fourier transform. Here, κ is the transform variable of x_i multiplied by $f/\bar{k}$. It is important to note that setting $\kappa = \Delta x_L$, the function $\tilde{I}_o(\kappa)$ is proportional to the coherence function $\Gamma_{2o}(\Delta x_L)$ that is obtained by an ideal measurement in the lens plane. The function $\tilde{I}(\kappa)$ is proportional to the coherence function in the lens plane. The measurement is, in this case, performed using a lens and a short exposure in the image plane. The function $\tilde{S}_S(\kappa)$ represents the combined transfer function of the lens and the atmosphere.

To obtain the spatial energy spectrum, we multiply eq. (3.31) by its complex conjugate. We find:

$$|\tilde{I}(\kappa)|^2 = |\tilde{S}_S(\kappa)|^2 |\tilde{I}_o(\kappa)|^2. \tag{3.32}$$

Finally, we average both sides of eq. (3.32) over a large number of short exposures. We then have

$$\langle |\tilde{I}(\kappa)|^2 \rangle = \langle |\tilde{S}_S(\kappa)|^2 \rangle |\tilde{I}_o(\kappa)|^2. \tag{3.33}$$

One way to obtain $|\tilde{I}_o(\kappa)|$ from eq. (3.33) is to divide both sides of the equation by $\langle |\tilde{S}_S(\kappa)|^2 \rangle$ and take the square root of both sides of the equation. We then use a phase-retrieval technique like that described in Dainty and Fienup [1987] to find $\tilde{I}_o(\kappa)$. Alternatively, we can take the Fourier transform of both sides of eq. (3.33) and process the autoconvolution function of $I_o(x)$. However we proceed, the success of the method depends upon the fact that the function $\langle |\tilde{S}_S(\kappa)|^2 \rangle$ may be shown to be relatively insensitive to the random phase fluctuations introduced by atmospheric turbulence. We caution the reader, however, that $\langle |\tilde{S}_S(\kappa)|^2 \rangle$ may be sensitive to the random intensity fluctuations when these fluctuations are strong.

To show that the function $\langle |\tilde{S}_S(\kappa)|^2 \rangle$ is relatively insensitive to the random phase fluctuations, we consider its form. By direct calculation, we find the following expression:

$$\langle |\tilde{S}_S(\kappa)|^2 \rangle = \text{const} \cdot \int_{-\infty}^{\infty} \Gamma_4(\kappa, \kappa') A_L(\kappa, \kappa') \, d\kappa', \tag{3.34}$$

where

$$\Gamma_4(\kappa, \kappa') = \langle \bar{u}(\kappa'')\bar{u}^*(\kappa + \kappa'')\bar{u}^*(\kappa' + \kappa'')\bar{u}(\kappa + \kappa' + \kappa'') \rangle, \tag{3.35}$$

and

$$A_L(\kappa, \kappa') = \int_{-\infty}^{\infty} P(\kappa'')P(\kappa + \kappa'')P^*(\kappa' + \kappa'')P(\kappa + \kappa' + \kappa'')\,d\kappa''. \quad (3.36)$$

The function Γ_4 is the fourth-order coherence function of the light field in the lens plane. If the object is incoherent and the turbulent fluctuations are statistically homogeneous in the horizontal direction, this function is independent of absolute position in the lens plane. That is, in eq. (3.35), Γ_4 is independent of κ''. To determine the effect of the random phase fluctuations on the right-hand side of eq. (3.34), we must know something about the length scales of A_L and Γ_4. For simplicity, we choose P to be unity within the circle of the lens (diameter d_L) and zero outside the circle. We find that the magnitude of κ and κ' are at most of order d_L.

Finding the characteristic length scales associated with Γ_4 is not a simple task, since the Γ_4 contains length scales associated with both the random phase fluctuations and the random intensity fluctuations. However, if we neglect the intensity fluctuations, we find that Γ_4 has the form:

$$\Gamma_4(\kappa, \kappa') = \langle \exp\{i[\phi(0) + \phi(\kappa) + \phi(\kappa') + \phi(\kappa + \kappa')]\} \rangle, \quad (3.37)$$

where ϕ represents the random phase fluctuations (since Γ_4 is independent of κ'', we have set $\kappa'' = 0$). As stated previously, the random phase fluctuations may often be considered to be distributed Gaussianly. In this case, eq. (3.37) assumes the form:

$$\Gamma_4(\kappa, \kappa') = \exp\{-2[D_\phi(\kappa) + D_\phi(\kappa') - D_\phi(\kappa + \kappa') - D_\phi(\kappa - \kappa')]\}, \quad (3.38)$$

where

$$D_\phi(\kappa) = \langle \phi(0)^2 \rangle - \langle \phi(0)\phi(\kappa) \rangle, \quad (3.39)$$

and we denote the characteristic scale associated with $\exp[-4D_\phi(\kappa)]$ to be l_ϕ. We shall first consider the case when $|\kappa| \gg d_L$. The behavior of $\langle |\tilde{S}_S(\kappa)|^2 \rangle$ in this limit determines how well we are able to image the small details of the object. As a result of this behavior, as we shall point out below, the processing in speckle interferometry is superior to that of traditional imaging using a long exposure. When $|\kappa| \gg d_L$, the function Γ_4 behaves approximately as

$$\Gamma_4(\kappa, \kappa') = \exp[-4D_\phi(\kappa')]. \quad (3.40)$$

In addition, since $l_\phi \ll d_L$, the variable κ' may be set equal to zero in eq. (3.36). We obtain finally:

$$\langle |\tilde{S}_S(\kappa)|^2 \rangle = \text{const} \cdot T_L(\kappa), \quad (3.41)$$

where

$$T_L(\kappa) = \int_{-\infty}^{\infty} |P(\kappa'')|^2 P(\kappa + \kappa'')|^2 \, d\kappa''. \tag{3.42}$$

The constant in eq. (3.41) depends upon the constant in eq. (3.34) and the integral over the right-hand side of eq. (3.40) with respect to κ'.

The importance of eq. (3.41) lies in the fact that for large values of $|\kappa|$, the transfer function $\langle |\tilde{S}_S(\kappa)|^2 \rangle$ behaves like the transfer function of the lens alone, $T_L(\kappa)$. This is in contrast to the long-exposure ordinary imaging case [see eq. (3.6)] in which turbulence dominates for large separation distances [to compare eq. (3.42) with eq. (3.6), set, $s_T = |\kappa|$ in eq. (3.6)]. The penalty that is paid in speckle interferometry for this greatly improved resolution is that the multiplicative constant in eq. (3.41) is very small. The constant is of order $(s_c/d_L)^2$, where s_c is the characteristic coherence length of the light field impinging on the lens.

Intensity fluctuations. The form for Γ_4 used in eq. (3.38) assumes that intensity fluctuations may be neglected. Korff [1973] did a more thorough analysis using the Rytov approximation and the assumption that the log-amplitude and phase were jointly Gaussianly distributed but uncorrelated. However, as we pointed out in § 2.3, these assumptions are not valid unless the intensity fluctuations are small.

Normally, in astronomical viewing in the vicinity of the zenith, the scintillation index is much less than unity. Sometimes, however, under bad seeing conditions, the scintillation index may approach unity. For far-off-zenith viewing conditions, however, the scintillation index is always near unity (Beran and Whitman [1988b]). Thus, unless viewing is restricted to the near zenith, on good seeing days, the effect of intensity fluctuations must be considered. To the authors' knowledge, there has been no analysis done in which the effects of large-intensity fluctuations are properly considered in speckle interferometry. Such an analysis could be done using the two-scale method outlined in § 2.2.

Phase retrieval. In practice, speckle interferometry has proved to be a viable procedure for obtaining near-diffraction-limited resolution in many applications. However, as stated above, the procedure does not give the intensity of the object directly, but requires a phase-retrieval algorithm to obtain $I_o(x)$ from $|\tilde{I}(\kappa)|$. There is an extensive literature on the subject (Fienup [1982], Dainty and Fienup [1987], Nakajima [1988]), and the methods appear to yield good results in two dimensions. A phase-retrieval algorithm is required because speckle interferometry treats the power spectrum of $\tilde{I}(\kappa)$. If a three-fold product of $\tilde{I}(\kappa)$ were considered instead, the phase of the function $\tilde{I}_o(\kappa)$ could be obtained, and $I_o(x)$ could be found directly. Lohmann, Weigelt and Wirnitzer [1983] considered such a technique; we discuss this method in the next section. We also refer the reader to Schulz and Snyder [1992] for image recovery using higher-order correlation functions. Experimental results are given in figs. 8 and 9 of Laberyie [1976].

3.3.4. Speckle masking (triple correlation theory)

To consider speckle masking, we return to eqs. (3.30) and (3.31) (see Lohmann, Weigelt and Wirnitzer [1983], Bartelt, Lohmann and Wirnitzer [1984], Wirnitzer [1985]). The quantity considered is either the function

$$\langle I^{(3)}(x, x') \rangle = \left\langle \int_{-\infty}^{\infty} I(x'') I(x + x'') I(x' + x'') \, dx'' \right\rangle, \tag{3.43}$$

or its Fourier transform, termed the bispectrum,

$$\langle \tilde{I}^{(3)}(\kappa, \kappa') \rangle = \langle \tilde{I}(\kappa) \tilde{I}(\kappa') \tilde{I}(-\kappa - \kappa') \rangle. \tag{3.44}$$

In terms of the Fourier transform of the object intensity $\tilde{I}_o(\kappa)$, we find from eqs. (3.31) and (3.44) the relation:

$$\langle \tilde{I}^{(3)}(\kappa, \kappa') \rangle = \langle \tilde{S}_S^{(3)}(\kappa, \kappa') \rangle \tilde{I}_o^{(3)}(\kappa, \kappa'). \tag{3.45}$$

The function $\tilde{I}_o^{(3)}(\kappa, \kappa')$ is the bispectrum of the object in the absence of atmospheric turbulence and assuming an infinite lens. The function $\langle \tilde{S}_S^{(3)}(\kappa, \kappa') \rangle$ is the bispectrum associated with the function $\tilde{S}_S(\kappa)$.

In the absence of turbulence, it can be shown that the function $\tilde{S}_S^{(3)}(\kappa, \kappa')$ is real. In this case, the phase of the function $\tilde{I}_o^{(3)}$ is the same as the measured function $\tilde{I}^{(3)}$. If we write

$$\tilde{I}_o^{(3)}(\kappa, \kappa') = |\tilde{I}_o^{(3)}(\kappa, \kappa')| \exp[\phi_o^{(3)}(\kappa, \kappa')] \tag{3.46}$$

with a similar expression for $\tilde{I}^{(3)}(\kappa, \kappa')$, we have

$$\phi^{(3)}(\kappa, \kappa') = \phi_o^{(3)}(\kappa, \kappa'). \tag{3.47}$$

Next, writing

$$\tilde{I}_o(\kappa) = |\tilde{I}_o(\kappa)| \exp[i\phi_o(\kappa)], \tag{3.48}$$

we have the phase relation:

$$\exp[\phi^{(3)}(\kappa, \kappa')] = \exp[i\{\phi_o(\kappa) + \phi_o(\kappa') + \phi_o(-\kappa - \kappa')\}]. \tag{3.49}$$

From eq. (3.49), a recursion relation may be developed to find the phase $\phi_o(\kappa)$ for all values of κ from a knowledge of $\phi^{(3)}(\kappa, \kappa')$ for all values of κ and κ'. With the phase information obtained from $\phi^{(3)}(\kappa)$, we have sufficient information to find completely $\tilde{I}_o(\kappa)$ and hence by Fourier transformation, $I_o(x)$. No phase-retrieval algorithm is needed. When random phase fluctuations resulting from atmospheric turbulence are present, Lohmann, Weigelt and Wirnitzer [1983] claim that eq. (3.49) still holds approximately and may be used to find the phase of $\tilde{I}_o(\kappa)$. They conclude that eq. (3.49) is correct by showing that the function $\langle \tilde{S}^{(3)}(\kappa, \kappa') \rangle$ is real. In Appendix B of their paper, they analyze an equation that is a generalization of eq. (3.35). The effect of turbulence on the light field enters the analysis through a sixth-order coherence function. To proceed, the authors assume that the light field has Gaussian statistics. However, except for very long propagation paths in the atmosphere, the light field does not have Gaussian statistics and hence the authors' analysis is not appropriate for most imaging problems.

Lohmann, Weigelt and Wirnitzer [1983] present experimental data showing the utility of their method. It is quite possible that the function $\langle \tilde{S}^{(3)}(\kappa, \kappa') \rangle$ is approximately real for most imaging systems despite the non-Gaussianity of the light field. Since the technique is often used, it seems that further analysis is appropriate to find the limitations of the procedure. We stress again that the validity of the method depends upon the assumption of isoplanicity and the neglect of intensity fluctuations.

Finally, we note that speckle-masking (or speckle-interferometry) techniques can be combined with the pre-detection compensation of adaptive optics discussed in § 3.2.3. A recent paper by Roggeman and Matson [1992] considers this combined method.

See fig. 5 of Lohmann, Weigelt and Wirnitzer [1983] for experimental results.

The Knox–Thompson method. From eq. (3.31) we may form the product $\tilde{I}(\kappa)\tilde{I}(\kappa')$ instead of the magnitude squared presented in eq. (3.32). This is essentially the function that Knox and Thompson [1974] used in their image-processing technique. The method takes into account phase variations, and lies somewhere between speckle interferometry and speckle masking. Ayers, Northcott and Dainty [1988] present a comparison between the Knox–Thompson method and speckle masking, and point out the differences in the phase-determination procedure. We note again that in our review of the speckle methods we have not discussed the photon limitations that occur because of low light levels. Ayers, Northcott and Dainty [1988] pay especial attention to this problem.

3.3.5. Intensity correlations and the three-point correlation function

Intensity correlations (intensity interferometry, imaging, correlography). The advantage of using a lens to view an object is the obvious fact that on the film in the focal plane of the lens we have an image of the object. However, Hanbury Brown and Twiss [1954, 1974] showed many years ago that it is possible to obtain an image of an object by measuring the correlation of the intensities in the lens plane. To show the essence of their method, it is necessary to introduce a very short time average to define the correlation of intensities. The formulation in § 2 did not explicitly make use of time averages, since in all the previous work we could use the approximation given in eq. (2.10). Strictly speaking, we should have introduced a very short time average when we discussed an incoherent source.

In the absence of a turbulent atmosphere, the correlation of intensities, R_{IT}, is defined as

$$R_{\text{IT}}(x_{\text{L}1}, x_{\text{L}2}) = \overline{I(x_{\text{L}1}, t)I(x_{\text{L}2}, t)}, \tag{3.50}$$

where the overbar indicates a time average over a time which is long compared to $1/\Delta\nu$. In order to obtain an image from a measurement of R_{IT}, we assume that the statistics of the light field from the object are Gaussian. From this assumption it follows that the statistics of the light field at the lens are also Gaussian, since the propagation equations are linear. (This is also true if the light passes through atmospheric turbulence in a single short exposure.) When the light statistics are Gaussian, the relationship between R_{IT} and Γ_{2T}, the two-point coherence function averaged over a very short time, is

$$R_{\text{IT}}(x_{\text{L}1}, x_{\text{L}2}) = I_{\text{T}}(x_{\text{L}1})I_{\text{T}}(x_{\text{L}2}) + |\Gamma_{\text{2T}}(x_{\text{L}1}, x_{\text{L}2})|^2. \tag{3.51}$$

We also assume that the light field at the object is incoherent and Γ_{2T} depends only on the difference $x_{\text{L}1} - x_{\text{L}2}$. From a measurement of $R_{\text{IT}} - I_{\text{T}}^2$ we can determine $|\Gamma_{\text{2T}}|$. Since Γ_{2T} is the Fourier transform of the image, we can determine the image by using a phase-retrieval algorithm. If the function Γ_{2T} is real, the image can be found directly. The method has been demonstrated to work (see Hanbury Brown and Twiss [1974]).

When a turbulent atmosphere is present, the formulas for a short exposure are the same as those given above. (Here Γ_{2T} is identified with Γ_{2S}.) If, in addition, the assumption of isoplanicity is satisfied and random intensity fluctuations are neglected, then the coherence function at the lens plane is of the form given in eq. (3.9), where

$$G_{\text{SI}}(p_{\text{T}}, s_{\text{T}}, z) = \exp[i\phi(p_{\text{T}}, s_{\text{T}}, z)]. \tag{3.52}$$

Since only the magnitude of Γ_{2T} is measured, the method is independent of random phase fluctuations introduced by the turbulence in the atmosphere.

The above technique may be used if laser light impinges upon an object whose surface is rough on the scale of a light wavelength and the object is slightly translated or rotated to create many realizations of reflected light. In this case, the statistics of the light are Gaussian when averaged over many realizations, and the object light has been termed pseudo-Gaussian (Goodman [1985], Idell and Fienup [1986]). The method was shown by Idell and Fienup [1986] to be effective in restoring images under laboratory conditions with simulated atmospheric turbulence.

It is not difficult to show how to include the effect of intensity fluctuations due to turbulence. In eq. (3.9), G_{SI} depends on the intensity fluctuations, and the difference $R_{IT} - I_T^2$ is proportional to $|\Gamma_{2oT}|^2 |G_{SI}|^2$. Some preliminary work on the effects of the intensity fluctuations and non-isoplanicity was reported by Beran and Whitman [1988a].

Three-point intensity correlation functions. To obtain phase information for the function Γ_2, it is necessary to consider the three-point correlation of intensities. The short-time average of the three-point correlation function in the lens plane is defined as

$$R_{I3T}(x_{L1}, x_{L2}, x_{L3}) = \overline{I(x_{L1}, t) I(x_{L2}, t) I(x_{L3}, t)}. \tag{3.53}$$

If we assume isoplanicity and neglect intensity fluctuations caused by atmospheric turbulence, then when the light statistics are Gaussian, we have

$$\begin{aligned} R_{I3T}(x_{L1}, x_{L2}, x_{L3}) = 3I^3 &+ I[|\Gamma_{2T}(x_{L1}, x_{L2})|^2 \\ &+ |\Gamma_{2T}(x_{L1}, x_{L3})|^2 + |\Gamma_{2T}(x_{L2}, x_{L3})|^2] \\ &+ |\Gamma_{2T}(x_{L1}, x_{L2})| |\Gamma_{2T}(x_{L1}, x_{L3})| |\Gamma_{2T}(x_{L2}, x_{L3})| \\ &\times \cos[\phi(x_{L1}, x_{L2}) + \phi(x_{L1}, x_{L3}) + \phi(x_{L2}, x_{L3})]. \end{aligned}$$
(3.54)

From eq. (3.54) we see that, in contrast to a measurement of R_{IT}, a measurement of R_{I3T} yields phase information about the coherence function Γ_{2T}. There is some ambiguity in the determination of the phase information since only the cosine term and not the cosine and sine terms are measured, but by-and-large, a significant amount of phase information can be obtained (see Marathay [1986]).

To the authors' knowledge, there has not been much work done using the three-point intensity correlation function measured in the lens plane. The problem with the method is mainly the low signal-to-noise ratio as compared to the speckle-masking procedure discussed in § 3.3.4.

3.3.6. Shift-and-add method

The shift-and-add method is discussed in the book by Bates and McDonnell [1986] and in articles by Hunt, Fright and Bates [1983] and Sinton, Davey and Bates [1986]. The attraction of the method lies in its simplicity, but, as Bates and McDonnell [1986] state, the underlying theory is little understood.

In this method, to obtain an improved image intensity in the focal plane, we consider a large number of short-exposure images $I_m(x)$, each given by the expression

$$I_m(x) = \int_{-\infty}^{\infty} S_m(x - x') I_o(x') \, dx' + N_m(x). \tag{3.55}$$

Here $S_m(x)$ is the point-spread function in the mth realization of the turbulent atmosphere and $N_m(x)$ is a noise term. We next choose the maximum value of $I_m(x)$ in each realization. We assume that the maximum position occurs at x_M. Next we translate, but do not rotate, the image by a distance x_M to the origin of coordinates and add all the images. This yields, for the corrected image intensity $I(x)$,

$$I(x) = (1/M) \sum_{m=1}^{M} I_m(x + x_M). \tag{3.56}$$

The method works well when the object is a discrete number of point objects, especially if one of the point objects has a significantly greater light intensity than the others. When the object does not have a distinct maximum, the method is less successful. The authors referenced above discuss improvements to the basic equation given above, and the reader is referred to their articles for a thorough treatment of the technique.

3.4. CONCLUDING REMARKS

In § 3 we have presented a number of image-processing techniques that depend upon both long-exposure and short-exposure methods. The long-exposure methods (§ 3.2) were developed many decades ago and are designed to invert eq. (3.10) to obtain the object intensity $I_o(x)$ from measurement of the image $\langle I(x) \rangle$. The methods depend upon a knowledge of the point-spread function $S(x)$ [or its Fourier transform $\tilde{S}(\kappa)$], which is usually obtained from measurement of the light from a reference point source located near the object. In most approaches, the noise includes all measurement errors except those caused by turbulence in the atmosphere; the effect of turbulence is contained in the function $S(x)$.

The main advantage of using eq. (3.10) is that the reference point source used to determine $S(x)$ does not have to be very close to the object that is being imaged. It is only necessary that the light from the reference source pass through atmospheric turbulence that has the same statistical properties as the light from the object to be imaged. $S(x)$ is the average over many short exposures $S_S(x)$ and the assumption of isoplanicity in each short exposure is NOT required. The principal disadvantage of using eq. (3.10) is that $S(x)$ [or $\tilde{S}(\kappa)$] cannot be measured with sufficient accuracy to allow for high-resolution imaging (see § 3.1.1).

To overcome the difficulty of determining $S(x)$ accurately, short-exposure techniques (§ 3.3) have been developed and implemented in the past few decades. The main disadvantage of the short-exposure methods is that they require the assumption of isoplanicity in each short exposure. This restricts the angular diameter of the astronomical objects that may be viewed. The isoplanatic requirement also limits very severely the utility of short-exposure procedures for imaging along horizontal paths in the atmosphere, or for imaging ground objects from space-based systems. Perhaps the major advance required in short-exposure imaging in the future would be the development of a practical method of imaging large-angle objects.

The theory presented in § 2 is developed sufficiently to allow most of the image-processing procedures presently in use to be evaluated properly. However, considerable care must be taken to make sure that the theory used is appropriate for the imaging configuration that is analyzed. For example, the Kolmogorov spectrum is not always adequate, joint statistics are often not Gaussian, and intensity fluctuations cannot always be neglected.

References

Angel, J.R.P., P. Wizinowich, M. Lloyd-Hart and D. Sandler, 1990, Adaptive optics for array telescopes using neural-network techniques, Nature **348**, 221–224.

Ayers, G.R., M.J. Northcott and J.C. Dainty, 1988, Knox-Thompson and triple-correlation imaging through atmospheric turbulence, J. Opt. Soc. Am. A **5**, 963–985.

Bakut, P.A., S.D. Poll'skikh, A.D. Ryakhin, K.N. Sviridov and N.D. Ustinov, 1984, Statistical synthesis of algorithms for the optimum processing of the image of an astronomical object, Radioeng. & Electron Phys. **29**(9), 104–110.

Bartelt, H., A.W. Lohmann and B. Wirnitzer, 1984, Phase and amplitude recovery from bispectra, Appl. Opt. **23**, 3121–3129.

Bates, R.H.T., and M.J. McDonnell, 1986, Image Restoration and Reconstruction (Clarendon Press, Oxford).

Ben Yosef, N., and E. Goldner, 1988, Sample size influence on optical scintillation analysis. 1: Analytic treatment of the higher order irradiance moments, Appl. Opt. **27**, 2167–2171.

Beran, M.J., 1967, Propagation of a spherically symmetric mutual coherence function through a random medium, IEEE Trans. Antennas & Propag. **AP-15**, 66–69.

Beran, M.J., 1970, Propagation of a finite beam in a random medium, J. Opt. Soc. Am. **60**, 518–521.

Beran, M.J., and T.L. Ho, 1969, Propagation of the fourth-order coherence function in a random medium (a nonperturbative formulation), J. Opt. Soc. Am. **59**, 1134–1138.

Beran, M.J., and G.B. Parrent Jr, 1974, The Theory of Partial Coherence (SPIE, Palos Verdes Estates, CA). First published 1964 (Prentice-Hall, Englewood Cliffs, NJ).

Beran, M.J., and A.M. Whitman, 1971, Asymptotic theory for beam propagation in a random medium, J. Opt. Soc. Am. **61**, 1044–1050.

Beran, M.J., and A.M. Whitman, 1972, Propagation of radiation from a finite incoherent source in a random medium, Opt. Acta **19**, 701–707.

Beran, M.J., and A.M. Whitman, 1987, Effect of the atmosphere on measurements of correlations of intensity, in: Digital Image Recovery and Synthesis, Proc. SPIE **828**, 122–126.

Beran, M.J., and A.M. Whitman, 1988a, Non-isoplanatic effects in imaging through turbulent media, in: Optical, Infrared, and Millimeter Wave Propagation Engineering, Proc. SPIE **926**, 306–310.

Beran, M.J., and A.M. Whitman, 1988b, Scintillation index calculations using an altitude-dependent structure constant, Appl. Opt. **27**, 2178–2182.

Beran, M.J., S. Frankenthal, R. Mazar and A.M. Whitman, 1993, Applications of the two-scale embedding technique, in: Wave Propagation in Random Media, eds V.I. Tatarski, A. Ishimaru and V.U. Zavorotny (SPIE/Institute of Physics, Bellingham, WA/Bristol, UK).

Bertolotti, M., L. Muzii and D. Sette, 1970, Correlation measurements on partially coherent beams by means of an integration technique, J. Opt. Soc. Am. **60**, 1603–1607.

Born, M., and E. Wolf, 1964, Principles of Optics (MacMillan, New York).

Brown, D.S., and R.J. Scaddan, 1979, An interferometer for efficient measurement of the atmospheric MFT, Observatory **99**, 125–129.

Buffington, A., F.S. Crawford, R.A. Muller and C.D. Orth, 1977b, First observatory results with an image sharpening telescope, J. Opt. Soc. Am. **67**, 304–305.

Buffington, A., F.S. Crawford, R.A. Muller, A.J. Schwemin and R.G. Smits, 1977a, Correction of atmospheric distortion with an image-sharpening telescope, J. Opt. Soc. Am. **67**, 298–303.

Buffington, A., F.S. Crawford, R.A. Muller, A.J. Schwemin and R.G. Smits, 1977c, Correction of atmospheric distortion with an image-sharpening telescope, J. Opt. Soc. Am. **67**, 298–305.

Cannon, T.M., H.J. Trussel and B.R. Hunt, 1978, Comparison of image restoration methods, Appl. Opt. **17**, 3384–3390.

Churnside, J.H., and S.F. Clifford, 1987, Log-normal Rician probability-density function of optical scintillations in the turbulent atmosphere, J. Opt. Soc. Am. A **4**, 1923–1930.

Churnside, J.H., and R.G. Frehlich, 1988, Probability density function measurements of optical scintillations in the atmosphere, in: Optical, Infrared, and Millimeter Wave Propagation Engineering, Proc. SPIE **926**, 172–178.

Churnside, J.H., and R.J. Hill, 1988, Probability density of irradiance scintillations for strong path-integrated refractive turbulence, J. Opt. Soc. Am. A **4**, 727–733.

Consortini, A., F. Cochetti, J.H. Churnside and R.J. Hill, 1993, Inner-scale effect on irradiance variance measured for weak to strong atmospheric scintillation, J. Opt. Soc. Am. A **2**, 2133–2143.

Corrsin, S., 1951, On the spectrum of isotropic temperature fluctuations in isotropic turbulence, J. Appl. Phys. **22**, 469–473.

Dainty, J.C., and J.R. Fienup, 1987, Phase retrieval and image reconstruction for astronomy, in: Image Recovery, ed. H. Stark (Academic Press, New York) 231–275.

Dainty, J.C., and R.J. Scaddan, 1974, A coherence interferometer for direct measurement of the atmospheric transfer function, Mon. Not. R. Astron. Soc. **167**, 69P–73P.

Dainty, J.C., and R.J. Scaddan, 1975, Measurements of the atmospheric transfer function at Mauna Kea, Hawaii, Mon. Not. R. Astron. Soc. **170**, 519–532.

Dyson, F.J., 1975, Photon noise and atmospheric noise in active optical systems, J. Opt. Soc. Am. **65**, 551–558.

Fienup, J.R., 1982, Phase retrieval algorithms: A comparison, Appl. Opt. **21**, 2758–2769.

Foy, R., and A. Laberyie, 1985, Feasibility of adaptive telescope with laser probe, Astron. & Astrophys. **152**, 129–131.

Frankenthal, S., A.M. Whitman and M.J. Beran, 1984, Two-scale solutions for intensity fluctuations in strong scattering, J. Opt. Soc. Am. A **1**, 585–597.

Fried, D.L., 1966, Optical resolution through a randomly inhomogeneous medium for very long and very short exposures, J. Opt. Soc. Am. **56**, 1372–1379.

Fried, D.L., 1978, Probability of getting a lucky short-exposure image through turbulence, J. Opt. Soc. Am. **68**, 1651–1658.

Fried, D.L., 1982, Anisoplanatism in adaptive optics, J. Opt. Soc. Am. **72**, 52–61.

Fried, D.L., and G.E. Mevers, 1974, Evaluation of ro for propagation down through the atmosphere, Appl. Opt. **13**, 2620–2622.

Frieden, B.R., 1972, Restoring from maximum likelihood and maximum entropy, J. Opt. Soc. Am. **62**, 511–518.

Frieden, B.R., and D.C. Wells, 1978, Restoring with maximum entropy. III. Poisson sources and backgrounds, J. Opt. Soc. Am. **68**, 93–103.

Fugate, R.Q., 1993, Laser beacon, Opt. & Photonic News, June, pp. 14–19.

Fugate, R.Q., D.L. Fried, G.A. Ameer, B.R. Broeke, S.L. Browne, P.H. Roberts, R.E. Ruane and G.A. Taylor, 1991, Measurement of atmospheric wavefront distortion using scattered light from a laser guide star, Nature **353**, 144–146.

Furutsu, K., 1992, Random Media and Boundaries (Springer, Berlin).

Gardner, C.S., B.M. Welsh and L.A. Thompson, 1987, Experiments on laser guide stars at Mauna Kea Observatory for Adaptive Imaging in Astronomy, Nature **328**, 229–231.

Goodman, J.W., 1985, Statistical Optics (Wiley, New York).

Gozani, J., 1985, Numerical solution of the fourth-order coherence function of a plane wave propagating in a two-dimensional Kolmogorovian medium, J. Opt. Soc. Am. A **2**, 2144–2151.

Hanbury Brown, R., and R.Q. Twiss, 1954, A new type of interferometer for use in radio astronomy, Philos. Mag. **45**, 663–682.

Hanbury Brown, R., and R.Q. Twiss, 1974, The Intensity Interferometer (Taylor and Francis, London).

Hardy, J.W., 1978, Active optics: A new technology for the control of light, Proc. IEEE **66**, 651–697.

Hardy, J.W., J.E. Lefebvre and C.L. Koliopoulos, 1977, Real-time atmospheric compensation, J. Opt. Soc. Am. **67**, 360–369.

Hill, R.J., and S.F. Clifford, 1978, Modified spectrum of atmospheric temperature fluctuations, J. Opt. Soc. Am. **68**, 892–897.

Ho, T.L., and M.J. Beran, 1968, Propagation of the fourth-order coherence function in a random medium, J. Opt. Soc. Am. **58**, 1135–1341.

Hufnagel, R.E., 1974, Variations in atmospheric turbulence, Techn. Digest, Optical Propagation through Turbulence (Optical Society of America, Washington, DC) paper WA1.

Hunt, B.R., W.R. Fright and R.H.T. Bates, 1983, Analysis of shift-and-add method for imaging through turbulent media, J. Opt. Soc. Am. **73**, 456–465.

Idell, P.S., and J.R. Fienup, 1986, Imaging correlography, a new approach to active imaging, in: Proc. 12th DARPA Strategic Space Symp., Naval Post Graduate School, Monterey, CA, Oct. 28–30 (Riverside Research Inst.).

Ishimaru, A., 1978, Wave Propagation and Scattering in Random Media, 2 Volumes (Academic Press, New York).

Jaynes, E.T., 1968, Prior probabilities, IEEE Trans. Syst. Sci. Cybern. **SSSC-4**, 227–241.

Kane, T.J., C.S. Gardner and L.A. Thompson, 1991, Effects of wavefront sampling speed on the performance of adaptive astronomical telescopes, Appl. Opt. **30**, 214–221.

Kelsall, D., 1973, Optical seeing' through the atmosphere by an interferometric technique, J. Opt. Soc. Am. **63**, 1472–1484.

King, I.R., 1971, The profile of a star, Pub. Astron. Soc. Pac. **83**, 199–201.

Knox, K.T., and B.J. Thompson, 1974, Recovery of images from atmospherically degraded short-exposure photographs, Astron. J. **193**, L45–L48.

Kolmogorov, A., 1941, The local structure of turbulence in an incompressible viscous fluid for very large Reynolds numbers, C.R. Acad. Sci. URSS **30**, 301.

Korff, D., 1973, Analysis of a method for obtaining near-diffraction-limited information in the presence of atmospheric turbulence, J. Opt. Soc. Am. **63**, 971–980.

Kravtsov, Y.A., 1992, Propagation of electromagnetic waves through a turbulent atmosphere, Rep. Prog. Phys. **55**, 39–112.

Labeyrie, A., 1970, Attainment of diffraction limited resolution in large telescopes by Fourier analysing speckle patterns in star images, Astron. & Astrophys. **6**, 85–87.

Labeyrie, A., 1976, High resolution techniques in optical astronomy, in: Progress in Optics, Vol. XIV, ed. E. Wolf (North-Holland, Amsterdam) pp. 49–87.

Levaron, A., 1992, private communication.

Lloyd-Hart, M., R. Dehany, B. McLeod, D. Wittman, D. Colucci, D. McCarthy and R. Angel, 1993, Direct 75 milliarcsecond images from multiple mirror telescope with adaptive optics, Astron. J. **402**, L81–L84.

Lohmann, A.W., and B. Wirnitzer, 1984, Triple correlations, Proc. IEEE **72**, 889–901.

Lohmann, A.W., G. Weigelt and B. Wirnitzer, 1983, Speckle masking in astronomy: Triple correlation theory and applications, Appl. Opt. **22**, 4028–4037.

Manning, R.M., 1993, Stochastic Electromagnetic Image Propagation and Adaptive Compensation (McGraw-Hill, New York).

Marathay, A.S., 1982, Elements of Optical Coherence Theory (Wiley, New York).

Marathay, A.S., 1986, Phase function of spatial coherence from multiple intensity correlations, in: Advanced Technology Optical Telescopes III, Proc. SPIE **628**, 273–276.

Martin, J.M., and S.M. Flatte, 1988, Intensity images and statistics from numerical simulation of wave propagation in 3-D random media, Appl. Opt. **27**, 2111–2126.

Martin, J.M., and S.M. Flatte, 1990, Simulation of point-source scintillation through three-dimensional random media, J. Opt. Soc. Am. A **7**, 838–847.

Mazar, R., and A. Bronstein, 1992, Finite aperture effects on intensity fluctuations in random media, Opt. Commun. **15**, 365–369.

McKechnie, T.S., 1992, Atmospheric turbulence and the resolution limits of large ground-based telescopes, J. Opt. Soc. Am. A **9**, 1937–1954.

Meinel, E.S., 1986, Origins of linear and non-linear recursive restoration algorithms, J. Opt. Soc. Am. A **3**, 787–799.

Meinel, E.S., 1988, Maximum-entropy image restoration: Lagrange and recursive techniques, J. Opt. Soc. Am. A **5**, 25–29.

Molyneux, J.E., 1971a, Propagation of the N-order coherence function in a random medium. The governing equations, J. Opt. Soc. Am. **61**, 248–255.
Molyneux, J.E., 1971b, Propagation of the Nth-order coherence function in a random medium, II. General solutions and asymptotic behavior, J. Opt. Soc. Am. **61**, 369–377.
Muller, R.A., and A. Buffington, 1974, Real-time correction of atmospherically degraded telescope images through image sharpening, J. Opt. Soc. Am. **64**, 1200–1210.
Nakajima, N., 1988, Phase retrieval using the logarithmic Hilbert transform and the Fourier-series expansion, J. Opt. Soc. Am. A **5**, 257–261.
Obukhov, A.M., 1949, Structure of the temperature field in a turbulent flow, Izv. Akad. Nauk SSSR, Ser. Geogr. Geofiz. **13**, 58–69.
Papoulis, A., 1965, Probability, Random Variables, and Stochastic Variables (McGraw-Hill, New York).
Phillips, R.L., and L.C. Andrews, 1981, Measured statistics of laser-light scattering in atmospheric turbulence, J. Opt. Soc. Am. **71**, 1440–1445.
Pollaine, S., A. Buffington and F.S. Crawford, 1979, Measurement of the size of the isoplanatic patch using a phase-correcting telescope, J. Opt. Soc. Am. **69**, 84–89.
Primmerman, C.A., D.V. Murphy, D.A. Page, B.G. Zollars and H.T. Barclay, 1991, Compensation of atmospheric optical distortion using a synthetic beacon, Nature **353**, 141–143.
Roddier, C., 1976, Measurements of the atmospheric attenuation of spectral components of astronomical images, J. Opt. Soc. Am. **66**, 478–482.
Roddier, F., 1981, The effect of atmospheric turbulence in optical astronomy, in: Progress in Optics, Vol. XIX, ed. E. Wolf (North-Holland, Amsterdam) pp. 281–376.
Roddier, C., and F. Roddier, 1973, Correlation measurements of complex amplitude of stellar plane waves perturbed by atmospheric turbulence, J. Opt. Soc. Am. **63**, 661–663.
Roddier, F., J.E. Graves, D. McKenna and M.J. Northcott, 1991, The University of Hawaii adaptive optics system. I. General approach, in: Active and Adaptive Optical Systems, ed. M.A. Ealley, Proc. SPIE **1542**, 248–253.
Roddier, F., M.J. Northcott, J.E. Graves, D.L. McKenna and D. Roddier, 1993, One-dimensional spectra of turbulence-induced Zernicke aberrations: time-delay and isoplanicity error in partial adaptive compensation, J. Opt. Soc. Am. A **10**, 957–965.
Roggemann, M.C., and C.L. Matson, 1992, Power spectrum and Fourier phase spectrum estimation by using fully and partially compensating adaptive optics and bispectrum postprocessing, J. Opt. Soc. Am. A **9**, 1525–1535.
Schulz, T.J., and P.L. Snyder, 1992, Image recovery from correlations, J. Opt. Soc. Am. A **9**, 1266–1272.
Shishov, V.I., 1968, Theory of wave propagation in random media, IVUZ-Radiophysics **11**, 866–874.
Sinton, A.M., B.L.K. Davey and R.H.T. Bates, 1986, Augmenting shift-and-add with zero-and-add, J. Opt. Soc. Am. A **3**, 1010–1017.
Tatarski, V.I., 1971, The effects of the turbulent atmosphere on wave propagation, Publ. TT-68–50464 (National Technical Information Service, Springfield, VA).
Thompson, L.A., and C.S. Gardner, 1990, Design and performance analysis of adaptive optical telescopes using laser guide stars, Proc. IEEE **78**, 1721–1743.
Tur, M., 1982, Numerical solutions of the fourth moment equation of a plane wave propagating in a random medium, J. Opt. Soc. Am. **72**, 1683–1691.
Tur, M., and M.J. Beran, 1983, Wave propagation in random media: A comparison of two theories, J. Opt. Soc. Am. **73**, 1343–1349.
Uscinski, B.J., 1977, The Elements of Wave Propagation in Random Media (McGraw-Hill, New York).
Uscinski, B.J., 1982, Intensity fluctuations in a multiple scattering medium. Solution of the fourth moment equation, Proc. R. Soc. (London) Ser. A **380**, 137–169.

Walters, D.L., D.L. Flavier and J.R. Hines, 1979, Vertical atmospheric MTF measurements, J. Opt. Soc. Am. **69**, 828–837.

Welsh, B.M., and C.S. Gardner, 1989, Performance analysis of adaptive-optics systems using laser guide stars and slope sensors, J. Opt. Soc. Am. **6**, 1913–1923.

Wessely, H.W., and J.O. Bolstad, 1970, Interferometric technique for measuring the spatial-correlation function of optical radiation fields, J. Opt. Soc. Am. **60**, 678–682.

Whitman, A.M., and M.J. Beran, 1970, Beam spread of laser light propagating in a random medium, J. Opt. Soc. Am. **60**, 1595–1602.

Whitman, A.M., and M.J. Beran, 1985, Two-scale solution for atmospheric scintillation, J. Opt. Soc. Am. A **2**, 2133–2143.

Wirnitzer, R., 1985, Bispectral analysis at low light levels and astronomical speckle masking, J. Opt. Soc. Am. A **2**, 14–21.

VI

DIGITAL HALFTONING: SYNTHESIS OF BINARY IMAGES

BY

OLOF BRYNGDAHL, THOMAS SCHEERMESSER and FRANK WYROWSKI

*Physics Department,
University of Essen,
45117 Essen,
Germany*

CONTENTS

		PAGE
§ 1.	INTRODUCTION AND TRENDS	391
§ 2.	THE QUANTIZATION PART OF AN IMAGE-PROCESSING SYSTEM	394
§ 3.	SPECTRAL CHARACTERISTICS OF BINARIZATION TECHNIQUES	400
§ 4	ANALYSIS OF HALFTONE PROCEDURES WITH EXCESSIVE RESOLUTION	423
§ 5.	HALFTONE PROCEDURES WITH RESTRICTED RESOLUTION BELOW SAMPLING RASTER	442
§ 6.	HALFTONE RESOLUTION IN EXCESS OF SAMPLING RASTER	456
§ 7.	SYSTEM CONSIDERATIONS OF IMAGE QUANTIZATION GAIN IN IMPORTANCE	460
ACKNOWLEDGEMENTS		461
REFERENCES		461

§ 1. Introduction and Trends

1.1. DIGITAL TECHNIQUES ON ADVANCE

Handling of information in audible as well as visual form has become increasingly digital in nature over the last couple of decades. This trend has been accompanied by an ever more dominating role of digital concepts and techniques in recording, storage, transmission, and presentation of sound and images. The information is contained in the particular variations of physical quantities (electronic, magnetic, optical, etc.). To accommodate transducer problems between the different modes in which the audio-visual information is imposed; e.g., distribution of energy and material properties, appropriate coding schemes have been introduced and developed.

In this overview, we concentrate on some aspects of pictorial information in connection with prints and displays of images. In particular, we will describe tone and detail reproduction of binarized two-dimensional graytone scenes. Most of the techniques described may be extended to quantization to more than two levels.

1.2. NARROWING THE SUBJECT TO IMAGE CHARACTERIZATION

Our visualization of the physical world influences in a flexible and adjustable way the mental image impression. It depends on, and is influenced by, a number of processes and factors. The visual system is an ingenious and intriguing combination of information gathering, processing, sorting, and interpretation procedures. The image-forming and -receiving parts are understood rather well, but details of the further handling and processing of the formed signals are still unresolved in their proceeding to higher cortical centra.

Direct physical measurements cannot be performed on pictorial sensations. Thus, an absolute judgement on perceived image characteristics like quality seems inappropriate and inadequate. Our apprehension and appreciation of visual information is subjective and influenced by psychological factors. Factors dependent on these phenomena will not be considered here.

We aim to concentrate on a couple of physical characteristics of printed and displayed images, and on those kinds of images where electronic/computational means are employed in their formation. These physical parameters are connected mainly with the spatial sampling and intensity quantization of the pictorial information. Factors and phenomena related to color and time variation will not be discussed.

1.3. THE HALFTONE TECHNIQUE

The halftone technique, which converts a continuous tone image into a binary one, was patented by Talbot [1852]. Over the years, this technique has been refined and implemented in the graphic arts as a two-step photographic process. A screen function (carrier) is introduced, and the binarization (thresholding) is performed by using a high-contrast photographic material. The halftone technique, in which a screen and the photographic process are used to convert a graytone picture into a quantized and in particular a binary one, has been developed to perfection. The procedure is elaborate, and relies on profound professional skills.

The halftone technique is a two-dimensional (2D) information-encoding scheme which introduces a spatially structured texture. Configurational changes of this binary structure will cause its average surface coverage (irradiance, intensity, density, reflectivity, etc.) over local areas to vary. Representation of the pictorial information in the form of gray-level variation is influenced by the choice of parameters of the halftone technique. Part of the strength and success of the halftone technique seems to result from its match to the low- and bandpass character of the visual system. As long as the introduced texture of the image has a uniform nature, we are to a large degree able to accept and disregard it.

In the established photographic halftoning, screen functions are used that result in approximately round or elliptical equidistant dots in a regular rectangular or hexagonal arrangement. As long as the spatial details of the screen function are well above the resolution of the following detector (e.g., the visual system), the halftone technique functions satisfactorily.

1.4. DIGITAL HALFTONING – A NEW BRANCH OF IMAGE SYNTHESIS

In the last two decades, electronic devices and systems have had a significant impact on image information processing and display technologies. The introduction and development of ever newer electronically addressed printer systems and quantized/binary display media has required the incorporation of advancements and refinements of the quantization/binarization schemes for pictorial information. The halftone technique was converted to accommodate and utilize these developments. Mere application and simulation of the existing techniques was not possible. They had to be adapted and further developed. Moreover, novel versions were required to meet the characteristics of the new media; e.g., their restricted space bandwidth products. Gradually, more sophisticated computational abilities were added, together with ideas of how to convert 1D modulated signals into their 2D counterparts. Additional degrees of freedom were gained and used both to optimize the process and to conform it advantageously to available hardware. This trend has led to the introduction of new concepts.

The result became a new breed: an electronic version of the halftoning technique. A new area was established; many new concepts and developments were introduced which influenced the field in many respects (Schroeder [1969], Klensch, Meyerhofer and Walsh [1970], Knowlton and Harmon [1972], Judice, Jarvis and Ninke [1974], Stucki [1975], Gard [1976], Jarvis, Judice and Ninke [1976], Roetling [1977a], Stoffel and Moreland [1981], Ulichney [1987], Sahoo, Soltani, Wong and Chen [1988]). Following advancements in computer systems were sophistications and improvements in electronic halftoning. The availability of sizeable and inexpensive processors and rapidly accessible and addressable mass storage, together with advancements in peripheral computer equipment (e.g., scanners, printers, and electronic camera and display systems) have made their impact. We now face matured techniques. Many of these technologies function in a binary fashion.

The field of electronic halftoning has profited from the introduction and development of electronic and optoelectronic means to binarize graytone images. Incorporation of ever more sophisticated software offers possibilities to utilize existing hardware efficiently, and is able to introduce flexibility to allow optimization in different situations and for particular purposes.

The inclusion of discreteness in the form of spatial sampling and quantization produced a need for refined software to optimize utilization of existing hardware. The flexibility to include processing of the image information locally and/or globally has opened new extensions of, and possibilities in, the field of electronic halftoning. A cost-effective trend toward an extended and efficient use of the hardware seems to be to put more emphasis on the software – the computer as well as physical software; i.e., on a variety of concepts which are specific to (and only possible to implement with) synthetic means. With the trend to simplify and generalize digital image-producing processes, a fruitful marriage – digital halftoning – was established.

1.5. THREE GENERATIONS OF ELECTRONIC HALFTONING

In this development are incorporated some distinct classes of procedures characterized by the degree of computational efforts: i.e., complexity of algorithms. The signs of the onset of the field of electronic halftoning appeared in the 1960s. The first-generation techniques grew out of simulations of the photographic halftoning realization, and soon incorporated typical electronic ingredients (Bayer [1973]).

A decade later, a second generation of binarization techniques was initiated by the presentation of the idea of error correction and diffusion (Floyd and Steinberg [1975]). This marked the transition from passive to active procedures in which the pictorial information is able to influence the halftoning method and the final binary result. Local processing which can be executed sequentially is the typical ingredient for this generation of techniques.

In the 1980s, a third generation of techniques took shape gradually. Here, the real strength of massive computing power began to reveal enormous possibilities. Global processing of an iterative character became the trend (Broja, Wyrowski and Bryngdahl [1989]).

The characteristics of the electronic devices penetrating into the field of image documentation and presentation require additional and new developments in image quantization/binarization procedures. To conform to these demands, the creation and introduction of ever more sophisticated software is considered, and opens new dimensions in image synthesis. In the following, we present some of our views of this field, which is in continuous progress. We are now only at the beginning of a powerful development which offers adaptability and conformity to desired results and available means.

§ 2. The Quantization Part of an Image-Processing System

2.1. THE MODEL CONSIDERED

For our purposes, it is preferable to consider image quantization as a part of an overall system which can be presented in the form of a model. In the following, we concentrate on image binarization. Our arguments may be generalized to multi-level quantization.

The integration of digital halftoning into the total system must be considered in analysis as well as in development of image-binarization methods. A model of the overall imaging system is illustrated in fig. 1. We proceed with a list of comments on the different steps of this model.

1st step. $f(x)$ indicates the intensity of a graytone image; i.e., a real positive quantity. Among the parameters excluded from the considerations which follow are color and time. x is a vector $x = (x, y)$, and it is assumed that the distribution $f(x)$ is restricted to the real, positive interval $[0, 1]$ without loss of generality.

With the restriction to electronic procedures, $f(x)$ has to be converted into a data set stored in computer memory. This step is indicated by the operator $\mathscr{S}$. From this operation there results a 2D data set $f(m)$ with

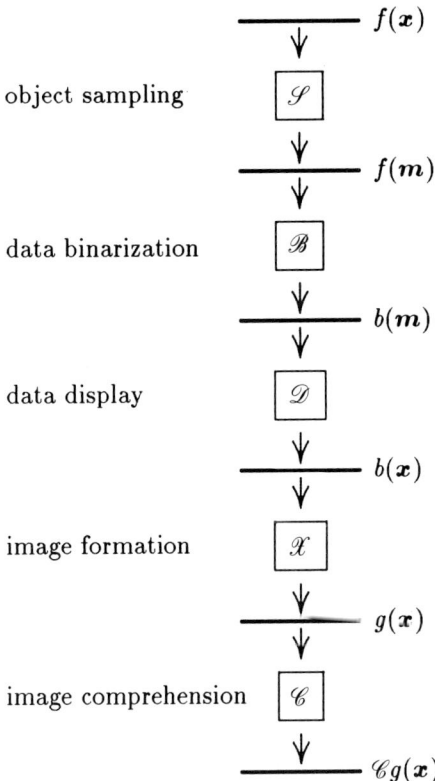

Fig. 1. Model of the system, of which image quantization is a part.

$m = (m, n)$, where m and n are integers. We make the specific choice:

$$\mathscr{S}f(x) = f(m) = f(x) \operatorname{comb}[x, \delta x]; \tag{2.1}$$

i.e., equidistant sampling of $f(x)$, with

$$\operatorname{comb}[x, \delta x] = \sum_m \delta(x - m\delta x)\delta(y - n\delta y) \tag{2.2}$$

and the sampling grid $\delta x = (\delta x, \delta y)$. In this sense, eq. (2.1) represents a sampling which results in the data set $f(m)$ of the sampling values of $f(x)$.

The conditions are, of course, slightly different in reality. For example, mean values are recorded by scanning systems, the outputs from imaging systems are modified by their transfer functions, etc. These effects can be considered, but they are unimportant for our purposes in the following.

The first step in fig. 1; i.e., the operator $\mathscr{S}$ may be entirely superfluous in some situations. This is the case when $f(m)$ is directly at hand in the form of a data set (e.g., by processes confined within the computer, in synthetically produced images, and the like). However, in the analysis of the total system, a continuous function $f(x)$ is helpful even in this case (see § 3). We then define $f(x)$ as

$$f(x) = \sum_m f(m) \operatorname{rect}[x - m\delta x, \delta x] \operatorname{rect}[y - n\delta y, \delta y], \tag{2.3}$$

with $\operatorname{rect}[a, b] = 1$ for $|a| \leq \frac{1}{2}b$, and 0 otherwise. This means that each $f(m)$ is coordinated with a rectangular pixel with the area $\delta x \times \delta y$. With eq. (2.3), a continuous counterpart to $f(m)$ is formed, and $f(m)$ can be interpreted as a sampled version of $f(x)$.

2nd step. $f(m)$ is binarized by a digital halftone technique. The operator $\mathscr{B}$ describes this step. The result is:

$$b(m) = \mathscr{B}f(m), \tag{2.4}$$

with $b(m) \in \{0, 1\}$. The operator $\mathscr{B}$ is the key issue presented in this overview. Several types of binarization procedures $\mathscr{B}$ exist. Their description is the subject of § 3.

3rd step. In the following we will focus on printing and display systems. In these situations, the data set $b(m)$ is converted into an image-intensity distribution. This transformation is described by the display operator $\mathscr{D}$. The result is:

$$b(x) = \mathscr{D}b(m), \tag{2.5}$$

where $b(x)$ indicates the intensity of a binary image. We assume that for

$b(m) = 1$, $b(m)$ is changed into a binary pulse:

$$p(x) = \begin{cases} 1 & \text{for } x \in \mathbb{P}, \\ 0 & \text{otherwise,} \end{cases} \tag{2.6}$$

where $\mathbb{P}$ represents the area covered by each pulse; i.e., the pulse shape due to $p(x) \in \{0, 1\}$. Further, we assume that (i) the pulse shape is independent of x, and (ii) the sizes of the pulses do not extend beyond the sampling grid unit $(\delta x, \delta y)$, which ensures that (i) $\mathbb{P} \neq \mathbb{P}(x)$ and (ii) there is no overlap between the pulses. Under these restrictions, we have

$$b(x) = b(m) * p(x), \tag{2.7}$$

where $b(x) \in \{0, 1\}$ and $*$ indicates a convolution.

In the following, we choose $\mathbb{P}$ to be rectangular and of size $\delta x \times \delta y$; i.e., these rectangular cells correspond to the sampling grid units. More general pulse shapes are considered in § 6.

4th step. In this step, the binary image $b(x)$ is fed to the end stage of the system by an optical imaging system described by the operator $\mathscr{X}$. This process results in:

$$g(x) = \mathscr{X} b(x), \tag{2.8}$$

where $g(x)$ is the light distribution obtained in this step, and $b(x)$ can be considered to be an incoherent light distribution (a coherent light distribution can be treated in an analogous way). This specifies $\mathscr{X}$, which, for example, may constitute the image-forming portion of the visual system. We obtain:

$$g(x) = \mathscr{X} b(x) = b(x) * h(x), \tag{2.9}$$

with

$$h(x) = \mathscr{F}^{-1} H(u), \tag{2.10}$$

where

$$H(u) = \begin{cases} H(u) & \text{for } |u| \leq u_d, \\ 0 & \text{otherwise.} \end{cases} \tag{2.11}$$

The operator $\mathscr{F}^{-1}$ indicates an inverse Fourier transform. $H(u)$ is the power spectrum of the intensity-impulse response of the system $\mathscr{X}$ (Goodman [1968]). The spatial frequency is represented by $u = (u, v)$.

The output $g(x)$ of this system is not binary. u_d in eq. (2.11) is the cutoff spatial frequency of the imaging part of the detector system. In the following,

u_d is an important parameter which we use to systematize the properties and potentialities of the different binarization methods.

5th step. The final part of the system in the model of fig. 1 is represented by the operator $\mathscr{C}$, which incorporates additional factors and characteristics of the detector system which may not be specified exactly. Major problems of the analysis and synthesis of image-binarization methods are manifested in this step. For visual observation of the distribution $b(x)$, we describe the image $g(x)$ formed on the retina by $\mathscr{X}$. The preprocessing at the retinal level and the further extensive processing, comprehension, and judgement in the cortex are included in $\mathscr{C}$.

The output $\mathscr{C}g(x)$ is rather uncertain in physical terms. The prevailing conception is that $\mathscr{C}$ enhances or reduces different features of perceived images; e.g., edges, shapes, textures, etc. This last step has until now been too speculative to be modelled satisfactorily for the present purpose. However, this dubious formulation is sufficient for the following consideration.

With the model presented,

$$\mathscr{C}\mathscr{X}\mathscr{D}\mathscr{B}\mathscr{S}f(x) \tag{2.12}$$

is the output of the total system. The combined operator $\mathscr{D}\mathscr{B}\mathscr{S}$ constitutes the binarization step, whereas $\mathscr{C}\mathscr{X}$ comprises the detection and comprehension steps of the formed binary image.

For the total system [i.e., eq. (2.12)], we observe that $\mathscr{S}$, $\mathscr{D}$ and $\mathscr{X}$ are defined operators with the exception of the free parameters: δx is the grid used in $\mathscr{S}$ and $\mathscr{D}$; $\mathbb{P}$ is the pulse shape for the output in $\mathscr{D}$, which is assumed to be fixed; u_d is the cutoff frequency in $\mathscr{X}$.

On the other hand, $\mathscr{C}$ is unknown apart from aspects which are only diffusely describable. Further, $\mathscr{B}$ is to be defined and its parameters are to be determined in such a way that a desired output will result from the total system. A key decision is to formulate a desired output result. This is in no way a trivial task to perform with purely physical arguments.

2.2. ANALYSIS OF THE MODEL

The function of the system will result in a desired output. The ideal output is defined as follows. A graytone picture $f(x)$ is presented on a medium which only allows quantized/binary values. In most situations, we would like the appearance of the observed quantized/binary image to look the same as the observed original graytone picture; i.e., as if no quantization/

binarization had been performed. With the given operators, we can define the ideal output as

$$\mathscr{C}\mathscr{X}f(x) = \mathscr{C}\mathscr{X}b(x), \tag{2.13}$$

where $\mathscr{C}\mathscr{X}f(x)$ is the output without, and $\mathscr{C}\mathscr{X}b(x)$ the output with, digital quantization/binarization performed by the procedure $\mathscr{D}\mathscr{B}\mathscr{S}$. We describe $b(x)$ by

$$b(x) = \mathscr{D}\mathscr{B}\mathscr{S}f(x). \tag{2.14}$$

Equation (2.13) is based on the fact that the quantization/binarization serves solely for presentation and not for processing of the input picture $f(x)$. This is in contrast to an image-processing situation where an inequality in eq. (2.13) is hoped for. This means that the ideal quantization/binarization is here defined in such a way that the operator $\mathscr{D}\mathscr{B}\mathscr{S}$ does not influence the output of the total system considered. To accomplish this, at least approximately, the free parameters of $\mathscr{D}\mathscr{B}\mathscr{S}$ must be adjusted to match the features of $\mathscr{C}\mathscr{X}$. The operators $\mathscr{D}$ and $\mathscr{S}$ are defined according to eqs. (2.5) and (2.1). We have specified the cells to be rectangular (see end of 3rd step in § 2.1), and thus the resolution δx of the display remains as a free parameter of operators $\mathscr{D}$ and $\mathscr{S}$. Further freedoms of $\mathscr{D}\mathscr{B}\mathscr{S}$ are those related to the binarization method $\mathscr{B}$.

The characteristics of $\mathscr{C}\mathscr{X}$ are (i) the cutoff frequency u_d of the detector, which is the essential feature for our purpose, (ii) the particular shape of the transfer function $H(u)$ for $|u| \leqslant u_d$, and (iii) certain peculiarities of the image-comprehension operator $\mathscr{C}$.

The significance of u_d is evident from the consideration that to satisfy eq. (2.13) [i.e., $\mathscr{C}\mathscr{X}f(x) = \mathscr{C}\mathscr{X}b(x)$] a sufficient condition is the validity of the equality

$$F(u) = B(u) \quad \text{for } |u| \leqslant u_d, \tag{2.15}$$

where $F(u)$ and $B(u)$ are the Fourier transforms of $f(x)$ and $b(x)$. In consequence of eq. (2.15), there follows [see eqs. (2.9)–(2.11)]:

$$\mathscr{X}f(x) - \mathscr{X}b(x), \tag{2.16}$$

and accordingly, eq. (2.13). From eq. (2.15) we conclude that the spectral effects of $\mathscr{B}$ are decisive. These are examined further in § 3.

Another important parameter, u_g, occurs in halftoning when using a specific algorithm. If all the spectral quantization noise below u_g is removed, then for a decoding with $u_d \leqslant u_g$ no distortions are transferred. In most cases, u_g can be chosen within a specific range by adjustment of the algorithmic parameters, but not every u_g can be realized in a satisfactory way. There is a method-dependent upper limit for this parameter.

In § 4, specific investigations of the relationship between δx, u_d, u_g, and $\mathscr{B}$ are performed in order to fulfill eqs. (2.15) and (2.16) for the procedures presented in § 3. The intention is to be able to indicate the maximum possible u_g for the different methods $\mathscr{B}$.

Equations (2.15) and (2.16) are no more than sufficient requirements for satisfying eq. (2.13). Thus, it is also conceivable to apply relaxed conditions when including the image-comprehension operator $\mathscr{C}$. This is necessary when $u_d \leq u_g$ cannot be achieved, as will be shown in § 5 and § 6.

§ 3. Spectral Characteristics of Binarization Techniques

As indicated in § 2.2, the effect of $\mathscr{B}$ has a profound influence on the spatial frequency spectrum of the binary image. It is therefore of fundamental concern to consider the validity of eq. (2.15). In this section, we present a description of different binarization methods together with their spectral characteristics. As indicated in the considerations in connection with eq. (2.15) the deviation of $B(u)$ from $F(u)$ is of general interest. In particular this is the case in the lowpass region. To investigate this departure, we introduce

$$Q(u) = B(u) - F(u), \qquad (3.1)$$

which is the spectrum of the quantization noise. The corresponding quantization noise introduced by the procedure $\mathscr{B}$ is

$$q(x) = b(x) - f(x). \qquad (3.2)$$

In eq. (3.1), $Q(u)$ constitutes a convenient and suitable quantity to characterize the effect of different binarization techniques $\mathscr{B}$ on the pictorial distribution $f(x)$ (e.g., Kermisch and Roetling [1975], Allebach [1981], Algie [1983], Ulichney [1987, 1988], Broja, Wyrowski and Bryngdahl [1993]). In the following, we will consider continuous or sampled distributions, depending on the effect and method under discussion.

The procedures $\mathscr{B}$ described in the following are classified according to the relationship between $\mathscr{B}$ and $f(x)$. In § 3.1, passive techniques are presented; i.e., methods in which $\mathscr{B}$ is independent of $f(x)$. Active techniques, where $\mathscr{B}$ depends on $f(x)$, are then considered. In § 3.2 we discuss the active direct techniques, and in § 3.3, the active iterative ones.

3.1. PASSIVE HALFTONE TECHNIQUES

The description of passive techniques is divided into two parts. An analysis of the relationships between $f(x)$, $b(x)$, and $q(x)$ in the spatial domain and $F(u)$, $B(u)$, and $Q(u)$ in the frequency domain, all based on continuous distributions, is given in § 3.1.1 and § 3.1.2. A presentation of the effect caused by the sampling is given in § 3.1.3.

3.1.1. *Binarization by constant threshold*

A simple binarization method $\mathscr{B}_t$, the hardclip, constitutes application of a constant threshold level. For this case, we have

$$\mathscr{B}_t f(x) = b(x) = \text{step}[f(x) - t] = \tfrac{1}{2}\{1 + \text{sgn}[f(x) - t]\}, \tag{3.3}$$

where

$$\text{sgn}[f(x) - t] = \begin{cases} 1, & f(x) - t \geq 0, \\ -1, & f(x) - t < 0, \end{cases} \tag{3.4}$$

and the threshold is $t \in [0, 1]$. Usually, $t = \tfrac{1}{2}$ is chosen. For this case the operator $\mathscr{B}_t$ is illustrated in fig. 2.

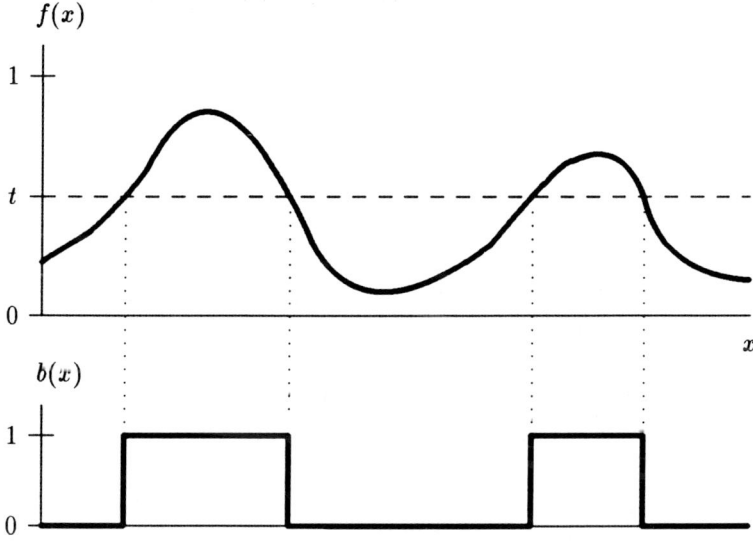

Fig. 2. Illustration of how a graytone function $f(x)$ is converted into a binary function $b(x)$ using a constant threshold level t.

In order to analyze the quantization noise $q(x)$ for application of the operator $\mathscr{B}_t$, the transform method can be used (see Goodman [1968], Broja and Bryngdahl [1993]).

With the assumption that we can conceive $b(x)$ as

$$b(x) = \hat{b}(f(x)), \tag{3.5}$$

where $\hat{b}$ is the step-function according to eq. (3.3),

$$b(x) = \int_{-\infty}^{+\infty} \hat{B}(\alpha) \exp[2\pi i \alpha f(x)] \, d\alpha \tag{3.6}$$

is obtained. $\hat{B}(\alpha)$ is the Fourier transform of the step-function. In this case, $\hat{B}(\alpha) \approx 0$ for $|\alpha| \geqslant \alpha_0$, and it is possible to replace the infinite limits of the integration of eq. (3.6) with finite ones:

$$b(x) = \int_{-\alpha_0}^{\alpha_0} \hat{B}(\alpha) \exp[2\pi i \alpha f(x)] \, d\alpha. \tag{3.7}$$

To calculate $B(u)$, we form the Fourier transform of the expression in eq. (3.7) and exchange the order of integration; i.e.,

$$B(u) = \int_{-\alpha_0}^{\alpha_0} \hat{B}(\alpha) \, d\alpha \int_{-\infty}^{\infty} \int_{-\infty}^{\infty} \exp[2\pi i \alpha f(x)] \exp[-2\pi i u x] \, dx \, dy. \tag{3.8}$$

To evaluate the second integral, its first exponential factor can be expanded in a series. The result is

$$B(u) = \sum_{j=0}^{\infty} c_j [\overset{j}{*} F(u)], \tag{3.9}$$

where $\overset{j}{*}$ denotes a j-fold convolution and

$$c_j = \frac{(2\pi i)^j}{j!} \int_{-\alpha_0}^{\alpha_0} \hat{B}(\alpha) \alpha^j \, d\alpha.$$

With use of eq. (3.9), the quantization noise spectrum can be expressed in the form

$$Q(u) = B(u) - F(u) = \sum_{j=0}^{\infty} c'_j [\overset{j}{*} F(u)], \tag{3.10}$$

with

$$c'_j = \begin{cases} c_j - 1, & j = 1, \\ c_j, & \text{otherwise.} \end{cases}$$

From eq. (3.10) we conclude that the quantization noise spectrum is conformed to the spectrum of the graytone image. It is thus concentrated in the lowpass region for a typical graytone image which itself has mainly low-frequency components. Moreover, it is smeared by multiple autoconvolutions.

3.1.2. Binarization by carrier techniques

A concept used frequently in binarization of pictures is the introduction of a carrier. The constant threshold level of the method in § 3.1.1 is here modified to be periodic, but independent of the pictorial distribution $f(x)$; i.e., the threshold t is exchanged by $t(x)$, with $t(x) \in [0, 1]$. We denote this method by $\mathscr{B}_c$, and obtain

$$\mathscr{B}_c f(x) = b(x) = \text{step}[f(x) - t(x)]$$
$$= \tfrac{1}{2}\{1 + \text{sgn}[\tilde{f}(x) - \tilde{t}]\} = \hat{b}(\tilde{f}(x)), \qquad (3.11)$$

with

$$\tilde{f}(x) = f(x) - t(x) + \tilde{t}. \qquad (3.12)$$

Binarization by a carrier technique is illustrated in fig. 3.

The objective of the substitution [eq. (3.12)] is to obtain eq. (3.11) in the same form as eq. (3.3) for $\mathscr{B}_t$. It is essential that $\tilde{t} = \text{const.}$ is independent of the location. Thus, the operator $\mathscr{B}_c$ can be described by

$$\mathscr{B}_c f(x) = \mathscr{B}_{\tilde{t}} \tilde{f}(x), \qquad (3.13)$$

with $\mathscr{B}_{\tilde{t}}$ defined according to eq. (3.3).

The quantization noise spectrum can be obtained by the substitutions $t \to \tilde{t}$ and $f \to \tilde{f}$ in eq. (3.3):

$$Q(u) = \sum_{j=0}^{\infty} c_j' [\overset{j}{*} \tilde{F}(u)]. \qquad (3.14)$$

The composition of $\tilde{F}(u)$ is

$$\tilde{F}(u) = F(u) - T(u) + \tilde{t}\delta(u). \qquad (3.15)$$

The principal properties of $Q(u)$ are now: (i) it is dependent on $F(u)$ as well as $T(u)$, and (ii) for a periodic carrier $T(u)$ consists of δ-peaks, and the noise spectrum is centered at the locations of the frequencies of the periodic carrier. An example is the cosine-carrier:

$$t(x) = \tfrac{1}{2}\{1 + \cos[2\pi u_0 x]\}, \qquad (3.16)$$

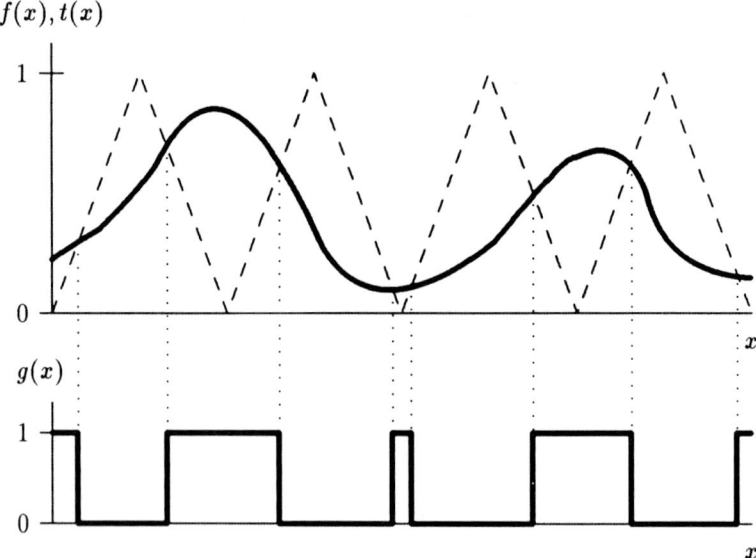

Fig. 3. Illustration of how a graytone function $f(x)$ is converted into a binary function $b(x)$ using a periodically varying threshold function $t(x)$.

with the spectrum

$$T(u) = \tfrac{1}{2}\delta(u) + \tfrac{1}{4}\delta(u - u_0) + \tfrac{1}{4}\delta(u + u_0). \tag{3.17}$$

This results in:

$$Q(u) = c'_0 \delta(u) + c'_1 \{F(u) - \tfrac{1}{4}\delta(u - u_0) - \tfrac{1}{4}\delta(u + u_0)\}$$

$$+ c'_2 \{F(u) * F(u) - \tfrac{1}{2} F(u - u_0) - \tfrac{1}{2} F(u + u_0) + \tfrac{1}{16}\delta(u - 2u_0) + \ldots\}$$

$$+ \ldots, \tag{3.18}$$

with the specific choice $\tilde{t} = \tfrac{1}{2}$.

It is evident that higher frequencies are introduced into the quantization noise spectrum by the carrier, and these may be even higher than those frequencies of the carrier itself. For an accurate analysis, it is necessary to determine the c_j's. This has been performed for a triangular-shaped carrier (Kermisch and Roetling [1975]).

We may conclude that in addition to the introduction of higher frequencies, $|Q(u)|$ is decreased in the lowpass region compared with its value in the case of the operator $\mathcal{B}_t$. With the introduction of higher frequencies it is logical that the contribution to the quantization noise spectrum will decrease in the lowpass region, because a portion of the noise energy is transferred to higher orders.

When using a carrier, it is important to know the relation between the sampling grid and the carrier period. The sampling can cause aliasing (Allebach and Liu [1977], Allebach [1979]); i.e., higher orders of the quantization noise distribution are introduced not only by the carrier but also by the sampling. Thus, the sampling must also be considered; we will turn to this problem next.

3.1.3. *Influence of sampling on passive techniques*

In § 3.1.1 and § 3.1.2, passive halftone techniques applied to continuous functions were described, and the corresponding expressions of the quantization noise spectrum $Q(u)$ were given. However, the actual distributions are discrete, and this will modify the quantization noise spectra.

Under the first step in § 2.1, a sampling operator $\mathscr{S}$ was introduced in eq. (2.1). Our concern here is to transfer the previous analysis of $Q(u)$ to its sampled version

$$Q(k) = \mathscr{F}\mathscr{S}q(x), \tag{3.19}$$

where $\mathscr{F}$ is the Fourier transform operator. Although a Fourier transform of $\mathscr{S}q(x)$ leads to a continuous distribution in general, it is convenient to consider the discrete version $Q(k)$ obtained by the fast Fourier transform (FFT) algorithm. No loss of information occurs, and the FFT is carried out easily by a computer. In the following discussion, $\mathscr{F}$, when applied to a discrete distribution, denotes an FFT. Application of eq. (2.1) results in:

$$Q(k) = \mathscr{F}[q(x)\,\text{comb}(x, \delta x)] = Q(u) * \text{comb}(u, \delta x^{-1}), \tag{3.20}$$

with $\delta x^{-1} = (\delta x^{-1}, \delta y^{-1})$.

The convolution in eq. (3.20) indicates that the quantization noise spectrum is now repeated with the period δx^{-1}, which will introduce aliasing. The consequences of the aliasing on $Q(u)$, and accordingly on $Q(k)$, are discussed for the different procedures in the following subsections.

Constant-threshold procedure. The application of the binarization operator $\mathscr{B}_t$ as defined in eq. (3.3) introduces a quantization noise spectrum which is located about the center of the lowpass region. With increasing distance from this center, $Q(u)$ decreases comparatively fast. From this, it follows that the repetition caused by the sampling will introduce only insignificant qualitative changes in the lowpass portion.

Periodic-carrier (dither) procedure. Periodic carriers are defined on a raster by a matrix. This matrix corresponds to one period, and the matrix elements indicate the values of the carrier at the raster points within the period. Figure 4 shows an example of such a matrix representation of a carrier.

There are different possibilities for determining the matrix elements:
(i) The straightforward, simply sampled data of an analytic carrier function; e.g., triangular-, sawtooth-, square-shaped period, etc. This also comprises the digital version of raster photography (Bryngdahl [1978], Holladay [1980], Karim and Liu [1980]).
(ii) Ordered dither procedures: the matrix elements are chosen dependent on specific objectives; e.g., generation of a particular texture within a period (Limb [1969], Lippel and Kurland [1971], Bayer [1973], Lippel [1976], Allebach and Stradling [1979], Anastassiou and Pennington [1982], Carlsohn and Besslich [1984], Mrusek, Just and Bryngdahl [1988], Rao and Arce [1988], Sullivan, Ray and Miller [1991]).

To describe the sampled quantization noise spectrum $Q(k)$ of the sampled noise distribution $q(m)$, the operator $\mathscr{B}_c$ as given in eq. (3.11) can be used.

According to eqs. (3.14) to (3.18), the application of the binarization operator $\mathscr{B}_c$ introduces a quantization noise spectrum that is not only located about a center as in the case of $\mathscr{B}_t$, but several centers localized at the frequencies of the carrier and at integer multiples of these frequencies which correspond to the higher spectral orders of the carrier frequency.

According to the repetition described by eq. (3.20), we have a situation where the different spectral orders due to the carrier will influence one another.

To describe the effect of the transition from $Q(u)$ to $Q(k)$, the essential quantity is the relation between the period p_c of the carrier and the sampling

0	8	2	10
12	4	14	6
3	11	1	9
15	7	13	5

Fig. 4. Matrix representation of one period of a periodic carrier. The values of the ordered dither are given in parts of sixteenths.

interval δx (Just, Hauck and Bryngdahl [1986]). We consider here the 1D case. The extension to 2D is straightforward. When the ratio

$$\frac{p_c}{\delta x} = n \in \mathbb{N} \tag{3.21}$$

is an integer, we face the following situation. The central points of the repeated $Q(u)$ terms introduced by the convolution with the comb-function [see eq. (3.20)] superpose exactly on the central points of the original spectrum $Q(u)$. From this, it follows that the central points of the distribution $Q(k)$ are also located at integer multiples of the carrier frequency. However, a modification arises due to aliasing. The characteristic higher spectral orders remain, but the distributions at their locations are altered by the superposed orders. In particular, one region has changed drastically; viz., the lowpass region. The nth orders of the neighbour repetitions superpose onto the origin.

The lowpass portion of $Q(u)$ is reduced distinctly by the introduction of a carrier compared to the constant-threshold case. The sampling gives rise to circumstances in which the repetition introduces further noise contributions in the lowpass domain of $Q(k)$ compared to $Q(u)$. These particulars are illustrated in fig. 5.

We proceed to the case

$$\frac{p_c}{\delta x} = \frac{n_1}{n_2} \notin \mathbb{N}, \tag{3.22}$$

with $n_1, n_2 \in \mathbb{N}$ and $n_1 > 2n_2$; i.e., there exists a rational ratio between the period of the carrier and the sampling distance (Just and Bryngdahl [1987]). The situation is now modified in a way such that the central points of the orders of the repeated $Q(u)$ are no longer allocated straight to the original $Q(u)$. As indicated in fig. 21, only the higher orders of the $\pm jn_2$th repetition ($j \in \mathbb{N}$) of the repeated $Q(u)$ overlap. All of the remaining orders are located in between the higher orders of the original $Q(u)$.

We conclude from the preceding discussion that the higher orders of $Q(k)$ are here not only centered at multiples of the carrier frequency, but at multiples of $\delta x(p_c n_2)^{-1}$. This results in many more orders which are packed more densely. Thus, $Q(k)$ appears more homogeneous than $Q(u)$, as is evident from fig. 21. Furthermore, the first order that affects the center of $Q(u)$ is the n_1th order of the n_2th repetition. The n_1th order is rather weak due to its distant location. Thus, the introduction of a rational carrier will reduce the influence of the sampling on the narrow lowpass region. This behavior is demonstrated in fig. 21. A simple way to introduce the rational carrier in

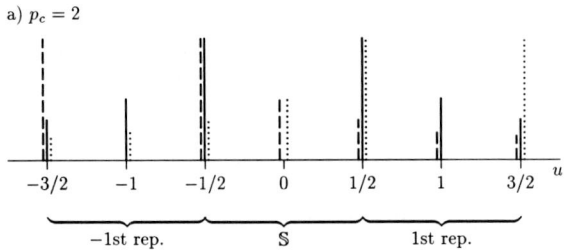

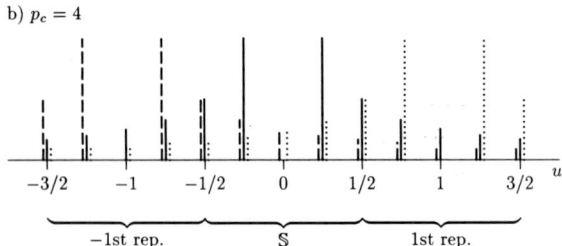

Fig. 5. Illustration of the orders of the quantization noise spectrum introduced by $\mathscr{B}_c$. This spectrum, with all its orders, is repeated periodically. The central spectrum is represented by solid drawn bars; the one shifted to the right by dotted drawn bars, and the one shifted to the left by dashed bars. Higher repetitions are omitted for clearness. The period of the carrier is in (a) $p_c = 2$ and in (b) $p_c = 4$.

2D is by a mere rotation of a carrier relative to the sampling raster (Just, Hauck and Bryngdahl [1986]). To a good approximation, an irrational carrier can be approached. Their exact realizations are not possible using a computer.

Random-carrier procedure. The carrier threshold $t(x)$, which was assumed above to be periodic, can be introduced as a random distribution (Roberts [1962], Bryngdahl [1973], Allebach and Liu [1976], Allebach [1978]). When $t(x)$ in eq. (3.11) is a homogeneous random distribution, $T(u)$ will also be a homogeneous distribution. From this, it follows that $Q(u)$ and its repetition $Q(k)$ [see eq. (3.20)] will be homogeneous random distributions as well. Thus, with a random carrier, sampling does not introduce any additional significant effect.

3.2. ACTIVE DIRECT HALFTONE TECHNIQUES

Procedures $\mathscr{B}_f$ are defined as active when the binarization operator $\mathscr{B}_f$ is dependent on the image distribution $f(x)$. In comparison with the passive techniques, the operators $\mathscr{B}_f$ are more complicated. They must be executed in the computer; i.e., they are genuine electronic halftone methods which act on a sampled distribution of $f(x)$. Passive techniques, on the other hand, may be interpreted as sampled versions of continuously formulated optical processes.

Formally, it is conceivable to consider all active techniques as carrier methods. In doing so, the threshold is dependent on the location x and the image $f(x)$; i.e., active techniques can be explicitly formulated as:

$$\mathscr{B}_f f(x) = \text{step}\{f(x) - t[x; f(x)]\}, \qquad (3.23)$$

with $0 \leqslant t[x; f(x)] \leqslant 1$. The dependence of the carrier on $f(x)$ results in an increased flexibility to influence the quantization noise spectrum $Q(u)$. How this freedom can be exploited constructively is open to the user to decide.

The specification of the carrier $t[x; f(x)]$ defines the operator $\mathscr{B}_f$ and can be ensued either directly or indirectly. The carriers introduced directly result in direct techniques, while those introduced indirectly result in iterative techniques.

We now turn to the case of carriers introduced directly. In § 3.2.1 and § 3.2.2, two procedures are mentioned specifically.

3.2.1. *Phase-modulated carrier procedure*

There exist possibilities of modifying the periodic-carrier procedure (Eschbach and Hauck [1984, 1985, 1987b]). A phase-modulated version was suggested by Just and Bryngdahl [1988]. The expression for the sampled version of the carrier can be rearranged to the form:

$$t[m; f(m)] = \cos[\pi f(m)] - \cos[2\pi(m/p_1 + n/p_2) - f(m)/2] + f(m), \qquad (3.24)$$

where p_1/m and p_2/n are the periods of the carrier and the term $f(m)/2$ indicates the entered phase.

Equation (3.24) represents the mathematical description of a modified version of a triangular-shaped carrier. For a description of the quantization noise in the image, we refer to Just and Bryngdahl [1988].

3.2.2. *Error-diffusion procedure*

The error-diffusion technique was introduced by Floyd and Steinberg [1975, 1976]. The fundamental idea is based on a sequential algorithm which corrects for binarization errors made locally by diffusing them, in a weighted fashion, to not-yet processed neighbor pixels before proceeding. Due to the digital character of this algorithm, only sampled distributions occur; e.g., $f(m)$; $t[m; f(m)]$, etc. In addition, its sequential execution is essential; i.e., the carrier $t[m; f(m)]$ corresponds only to those values of $f(m)$ which have already been determined and which are included in the carrier formula. The carrier can be expressed in the form:

$$t[m; f(m)] = t_0 - \sum_{m' \in \Omega} d(m')e(m-m'), \tag{3.25}$$

whereby the error is

$$e(m) = \{f(m) - t[m; f(m)] + t_0\} - b(m), \tag{3.26}$$

and the binarization step is

$$b(m) = \text{step}\{f(m) - t[m; f(m)]\}. \tag{3.27}$$

In general, $t_0 = 0.5$ is introduced. The weights with which the errors are diffused are represented by $d(m)$, and Ω is a set of indices m' for which $d(m') \neq 0$.

The error-diffusion procedure is illustrated in fig. 6. An example of the set Ω and $d(m')$ as suggested by Floyd and Steinberg [1976] is shown in fig. 7.

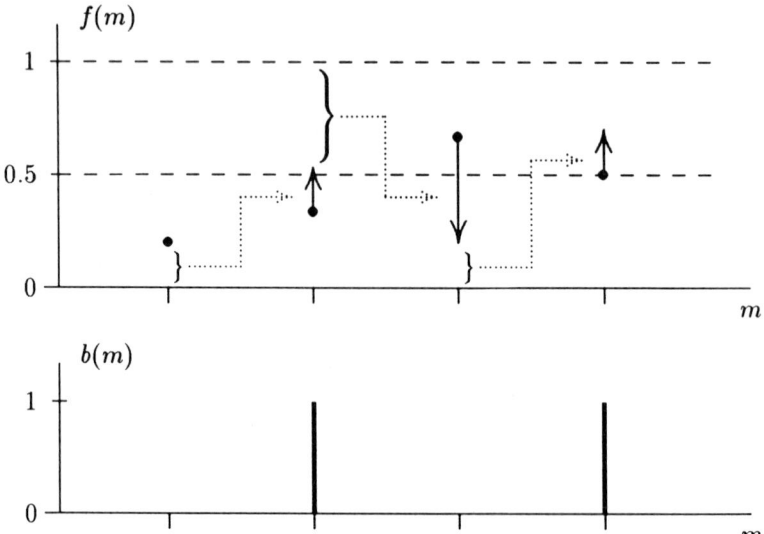

Fig. 6. Illustration of how a graytone distribution $f(m)$ is converted into a binary one $b(m)$ using the error-diffusion algorithm. The error made by binarization of a pixel is diffused to the next pixels before they are binarized sequentially.

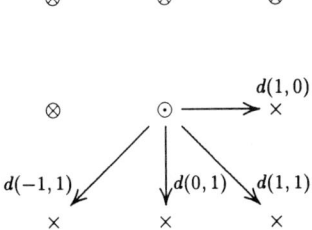

Fig. 7. Illustration of weights $d(m)$ which are applied to distribute the error at the binarized pixel indicated by ⊙ to not-yet processed pixels indicated by ×. The weights proposed by Floyd and Steinberg [1976] are: $d(1,0) = \frac{7}{16}$, $d(1,1) = \frac{1}{16}$, $d(0,1) = \frac{5}{16}$ and $d(-1,1) = \frac{3}{16}$.

The effect of the error-diffusion procedure on the quantization noise spectrum can be described analytically in terms of filter theory (Weissbach and Wyrowski [1992a]); i.e.,

$$Q(k) = \bar{Q}(k)H(k), \tag{3.28}$$

where $\bar{Q}(k)$ depends on the picture $f(m)$ and the diffusion weights $d(m)$. The filter $H(k)$ is given by

$$H(k) = \sum_{m' \in \Omega} d(m') \exp[2\pi i m' k] - 1, \tag{3.29}$$

which is dependent only on the diffusion weights $d(m')$. Thus, the quantization noise spectrum $Q(k)$ can be shaped by the filter function $H(k)$ from the choice of the diffusion weights.

In particular, it is possible to introduce a k_0 to obtain a zero point of the filter; i.e.,

$$H(k_0) = 0. \tag{3.30}$$

According to eqs. (3.28) and (3.30), the quantization noise spectrum fulfills $Q(k_0) = 0$. In general, the contributions to $Q(k)$ in the surroundings of u_0 are low, and it is possible in practice to form a region of low values of the quantization noise spectrum by choice of the diffusion weights.

Analytic and iterative methods have been suggested to determine $d(m)$ and obtain a desired filter function $H(k)$ (Van den Bulck [1992], Fetthauer, Weissbach and Bryngdahl [1992], Weissbach and Wyrowski [1992a]). There exists a high degree of flexibility in the choice of filter function $H(k)$ and the manipulation of the quantization noise spectrum.

The numerical stability of any $H(k)$ is not secured. Thus, in the synthesis of a certain filter (i.e., a certain set of diffusion weights), the analysis of the stability must be included (Broja, Eschbach and Bryngdahl [1986], Weissbach and Wyrowski [1992b]).

Several modifications of the error-diffusion algorithm have been suggested over the years (Dalton, Arce and Allebach [1982], Witten and Neal [1982], Billotet-Hoffmann and Bryngdahl [1983], Stevenson and Arce [1985], Knuth [1987], Eschbach [1990], Eschbach and Knox [1991], Velho and Miranda-Gomes [1991], Fan [1993], Knox and Eschbach [1993]). Two particular types are: (i) The combination of the error diffusion with a carrier procedure (Billotet-Hoffmann and Bryngdahl [1983], Kurosawa, Tsuchiya, Maruyama, Ohtsuka and Nakazato [1986]) and (ii) the consideration of overlap among the pixels in the output image (Stucki [1981]); i.e., observation of aspects of the operator $\mathscr{D}$ [see eq. (2.5)].

So far, procedures with carriers introduced directly, $t[m; f(m)]$, have been described. We now turn to those cases where the carrier is introduced indirectly; i.e., iterative techniques.

3.3. ACTIVE ITERATIVE HALFTONE TECHNIQUES

In § 3.2, active direct techniques were defined. Equation (3.23) serves that purpose. The active nature is a result of the dependence of the carrier on x and the picture $f(x)$. This formalism was introduced mainly for historical reasons. In the past, the carrier methods were focused on. Then the carrier notion made sense, as it did for the error-diffusion technique. However, for specification of iterative procedures, a formal introduction of a carrier is not appropriate. This is because in these procedures the carrier is introduced indirectly during the iteration. Here it is only possible to specify the carrier after the execution of the iterative halftone technique, and then only for a particular process and a particular graytone picture.

Useful as the carrier formalism was for classifying the halftone procedures of the past, it is unsuitable for systematizing iterative methods. An adequate representation of the iterative techniques is the adaptation of an operator formalism described below. In doing so, we must keep in mind that iterative procedures are executed in a computer and thus performed on sampled pictures. In most of these procedures, the spectra of the pictures are implicated and controlled, and this is also carried out in a sampled mode. However, it is to be noted that the spectra of the printed and displayed images (i.e., for those spectral effects of interest) are continuous.

Although only a sampled spectrum can be influenced in a digitally implemented iterative procedure, it is possible to affect the characteristics of the continuous spectrum. This is accomplished by oversampling with at least twice the Nyquist rate. In this way an iterative control of the continuous spectrum is obtained to an ample approximation.

3.3.1. Definition of iterative techniques

In an iterative halftone technique, an operator $\tilde{\mathscr{B}}$ is applied successively to a start distribution $b^{(0)}(m)$ until a binary distribution is attained after N cycles:

$$b(m) = \tilde{\mathscr{B}}^N b^{(0)}(m); \tag{3.31}$$

$b^{(0)}(m)$ and the distributions obtained by employing the operator $\tilde{\mathscr{B}}$ are not necessarily binary. The end result alone has to be binary.

The start distribution is provided by applying another operator $\hat{\mathscr{B}}$ on the graytone image $f(m)$:

$$b^{(0)}(m) = \hat{\mathscr{B}} f(m). \tag{3.32}$$

The operator of the entire binarization process $\mathscr{B}_f$, which defines an active method, is composed of these two operators; i.e.,

$$b(m) = \mathscr{B}_f f(m) = \tilde{\mathscr{B}}^N \hat{\mathscr{B}} f(m). \tag{3.33}$$

It ought to be observed that application of $\tilde{\mathscr{B}}$ or $\hat{\mathscr{B}}$ alone can, but does not have to, produce a binarization. Only their joint employment $\tilde{\mathscr{B}}^N \hat{\mathscr{B}}$ must ensure a binarization/quantization.

The specific spectral characteristics of the procedure and the formed images cannot be indicated until the operator $\mathscr{B}_f$ has been specified. To simplify the notation, $\mathscr{B}$ replaces $\mathscr{B}_f$ in the following.

Several kinds of iterative techniques have been suggested over the last few years (e.g., Eschbach and Hauck [1987a], Mitsa and Parker [1992], Peli [1991], Rolleston and Cohen [1992]). Among the widely recognized ones are the iterative Fourier transform algorithm [IFTA] (Broja, Wyrowski and Bryngdahl [1989]), the simulated annealing algorithm [SAA] (Carnevali, Coletti and Patarnello [1985]), the direct binary search algorithm [DBS] (Seldowitz, Allebach and Sweeney [1987]), fuzzy logic (Hsueh, Chern and Chu [1991]), and those based on neural-network concepts (Anastassiou and Kollias [1988], Tuttass and Bryngdahl [1993]).

3.3.2. Iterative Fourier transform algorithm

The iterative Fourier transform algorithm (IFTA) allows the introduction of constraints in the image as well as in the spectrum during execution of the halftone process. The IFTA is based on successive Fourier and inverse-

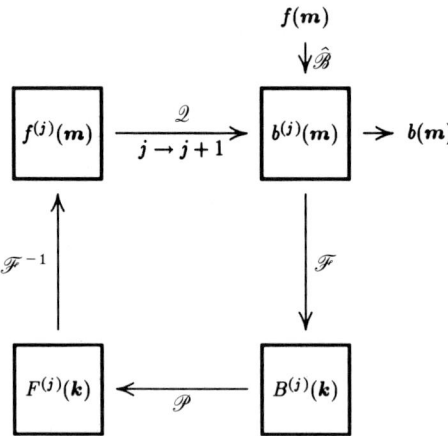

Fig. 8. Illustration of the iterative Fourier transform algorithm. Constraints are introduced in the image by the operator $\mathcal{Q}$, and in the spectrum by the operator $\mathcal{P}$ during the iterations.

Fourier transforms ($\mathcal{F}$ and $\mathcal{F}^{-1}$), and modifications of the spectrum by the operator $\mathcal{P}$ and of the image by the operator $\mathcal{Q}$; i.e., $\tilde{\mathcal{B}} = \mathcal{Q}\mathcal{F}^{-1}\mathcal{P}\mathcal{F}$. Figure 8 illustrates the different steps of the IFTA. With this algorithm, it is possible to generate binary images having spectra which satisfy certain constraints.

The separate operators which constitute a cycle of the IFTA are:

$\mathcal{Q}$: The restrictions in the image domain are accomplished by the operator $\mathcal{Q}$, which provides for the binary distribution after at least N iteration cycles. $\mathcal{Q}$ must be chosen in such a way that stagnation of the algorithm is avoided (see below). Further, it is noted that we frequently have $\hat{\mathcal{B}} = \mathcal{Q}$; i.e., even the start distribution is generated by $\mathcal{Q}$.

$\mathcal{F}$: The transformation from the image to the spectrum domain is achieved by the Fourier transform operator $\mathcal{F}$; viz., $B^{(j)}(k) = \mathcal{F}b^{(j)}(m)$.

$\mathcal{P}$: The restrictions in the spectrum domain are accomplished by the operator $\mathcal{P}$. Here $B^{(j)}(k)$ is modified; i.e., $F^{(j)}(k) = \mathcal{P}B^{(j)}(k) = \mathcal{P}\mathcal{F}b^{(j)}(m)$, in such a way that $F^{(j)}(k)$ fulfills the constraints in the spectrum domain. The spectral effects of $\tilde{\mathcal{B}}$ can be influenced directly by the operator $\mathcal{P}$. It is not possible to state a general prediction about the spectral effects. These effects are dependent on $\mathcal{P}$.

$\mathcal{F}^{-1}$: The transformation from the spectrum to the image domain is achieved by the inverse Fourier transform operator $\mathcal{F}^{-1}$; viz., $f^{(j)}(m) = \mathcal{F}^{-1}F^{(j)}(k)$. Note that $f^{(j)}(m)$ is not necessarily a binary distribution.

The desired goal is that the applied algorithm converges to an image which satisfies the restrictions in both domains; i.e., to a binary image which possesses the directly manipulated spectral features.

Two eventual impediments may occur:
(i) Does a solution exist at all? When no binary image exists with the desired properties, then as good an approximate solution as possible has to be propounded. To what extent an acceptable solution can be found is determined by the method and the defined criteria.
(ii) A stagnation problem of the iterative algorithm may arise. Even where a solution exists, it may not be possible to find it. It may happen that from a certain j onward

$$\tilde{\mathcal{B}}b^{(j)}(\mathbf{m}) = b^{(j)}(\mathbf{m}), \tag{3.34}$$

whereby the spectral constraints are not fulfilled; i.e., the algorithm stagnates.

Let us turn to an actual situation. An apparent choice of the quantization operator $\mathcal{Q}$ is the hardclip operator $\mathcal{B}_t$ (defined in § 3.1); i.e., $\mathcal{Q} = \mathcal{B}_t$. The natural expectation is that due to the iteration, the spectral effects of $\mathcal{B}_t$ can be manipulated in a desired way. However, the result is a stagnation, and this approach is not to be pursued. Another choice of $\mathcal{Q}$ has to be made. In particular, two examples of appropriate operators have been suggested:

(i) The quantization may be introduced stepwise in the iteration algorithm (Broja, Wyrowski and Bryngdahl [1989]). During each step,

$$b^{(j+1)}(\mathbf{m}) = \mathcal{Q}_c f^{(j)}(\mathbf{m})$$

$$= \begin{cases} 1, & f^{(j)}(\mathbf{m}) \geqslant 1 - \Delta_c \\ f^{(j)}(\mathbf{m}), & \Delta_c < f^{(j)}(\mathbf{m}) < 1 - \Delta_c \\ 0, & f^{(j)}(\mathbf{m}) \leqslant \Delta_c \end{cases} \tag{3.35}$$

is performed in the image plane. As shown in fig. 9, a range of graylevels is quantized in each iteration cycle. The range Δ_c is increased from 0 to 0.5 by steps every few cycles. In the last cycle, the hardclip $\mathcal{B}_t$ is applied.

(ii) Another way to overcome the stagnation is to introduce a local pseudo-random value $z^{(j)}(\mathbf{m}) \in [0, 1]$ in the iteration cycle j (Mrusek, Broja and Bryngdahl [1990], Scheermesser, Broja and Bryngdahl [1993a]):

$$b^{(j+1)}(\mathbf{m}) = \mathcal{Q}_r f^{(j)}(\mathbf{m})$$

$$= \begin{cases} 1, & f^{(j)}(\mathbf{m}) \geqslant 1 - \Delta_r \\ \text{step}[f^{(j)}(\mathbf{m}) - z^{(j)}(\mathbf{m})], & \Delta_r < f^{(j)}(\mathbf{m}) < 1 - \Delta_r \\ 0, & f^{(j)}(\mathbf{m}) \leqslant \Delta_r. \end{cases} \tag{3.36}$$

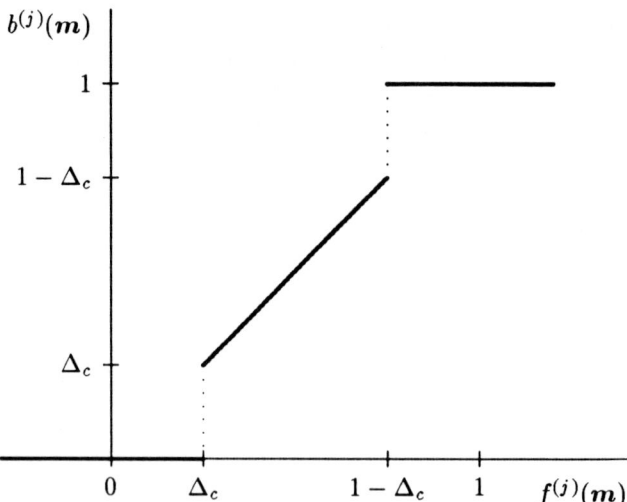

Fig. 9. Illustration of a stepwise introduction of the quantization to overcome stagnation in the iterative process. The range Δ_c is increased sequentially during the quantization process.

The effect of the operator $\mathcal{Q}_r$ is illustrated in fig. 10. The free parameter $\Delta_r \in [0, 0.5]$ can be increased from 0 to 0.5 stepwise, or can be held constant during the iteration. In the latter case, its specific choice (which depends on $\mathcal{P}$) is crucial if we are to optimize the result.

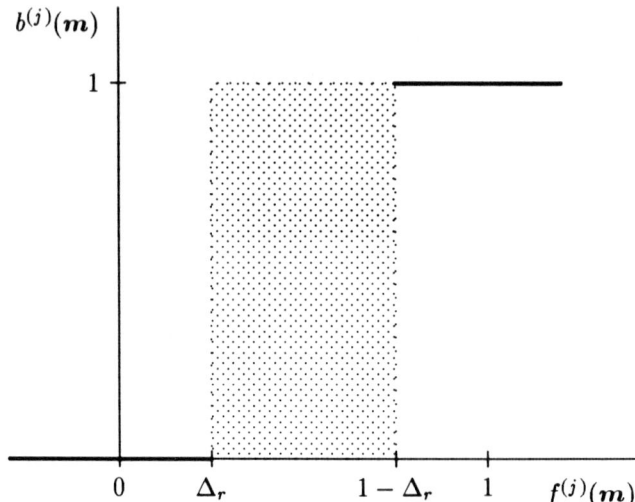

Fig. 10. Illustration of method to overcome stagnation in the iterative process by using a local random value within the range $[\Delta_r, 1 - \Delta_r]$.

In addition to the two possibilities described above, other approaches to choosing $\mathscr{Q}$ are also reasonable (see, e.g., Mrusek, Broja and Bryngdahl [1990]). Moreover, it is also possible to avoid a stagnation by choosing an appropriate operator $\mathscr{P}$ different from the one described above, whereby $\mathscr{Q} = \mathscr{B}_t$ is conceivable.

3.3.3. Simulated annealing algorithm

The simulated annealing algorithm (Kirkpatrick, Gelatt and Vecchi [1983]) is another choice of an iterative technique which has been applied in electronic halftoning.

The operator $\hat{\mathscr{B}}$ is used to form a start distribution $b^{(0)}(m)$ from $f(m)$. It can, for example, be furnished with a random dither, a hardclip, or something similar. Proceeding with $\hat{\mathscr{B}}$ from this binary distribution, a new binary distribution $b^{(j)\prime}(m)$ is calculated by a random reversal of one or more pixels in $b^{(j)}(m)$. It is tested whether $b^{(j)\prime}(m)$ fulfills some predetermined quality criterion (cost function) better than $b^{(j)}(m)$. If this is the case, $b^{(j+1)}(m) = b^{(j)\prime}(m)$ is accepted unconditionally; if not, $b^{(j+1)}(m) = b^{(j)\prime}(m)$ is accepted with a certain probability which depends on the quality criterion and adapted control parameters. Otherwise, the iteration is continued with $b^{(j+1)}(m) = b^{(j)}(m)$. The process of finding new distributions and accepting or rejecting them according to the given criterion is performed as indicated in the flowchart of fig. 11.

Theoretically, a $b(m)$ can be obtained by this method which satisfies optimally the selected quality criterion. However, the calculation efforts required are comparatively high and reaching the theoretical optimum is guaranteed only after an infinite number of calculations.

3.3.4. Neural-network algorithms

Ideas and concepts borrowed from neural-network structures have been adapted and applied in situations of electronic halftoning (Anastassiou [1988], Ling and Just [1991], Tuttass and Bryngdahl [1993]). The graytone picture $f(m)$ is converted into a binary image $b(m)$ via weight matrices combined customarily with a nonlinearity. A feature presented generally is that all of the elements of the input image are connected to all of the pixels of the output image.

There exist different possibilities of modifying the procedure. An obvious extension is the multi-layer network to broaden the number of feasible operations. In this way, effects of increased complexity can be optimized. A

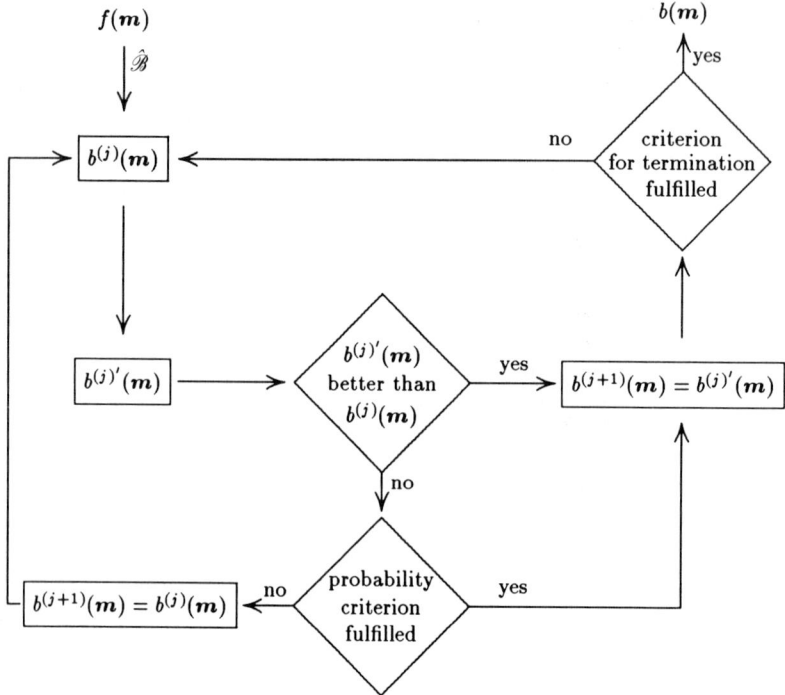

Fig. 11. Flowchart of the simulated annealing algorithm.

situation for N layers with the successive conversion

$$f(m) \to b^{(0)}(m) \to \ldots \to b^{(N-1)}(m) \to b(m), \tag{3.37}$$

is illustrated in fig. 12.

The problem to be solved is the determination of weight matrices, by means of which it is possible to attain a binary version $b(m)$ with desired

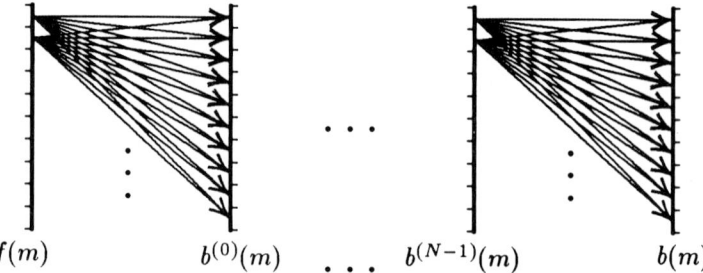

Fig. 12. Illustration of a neural-network algorithm in which $b(m)$ is obtained after N level conversions.

features from a graytone picture $f(\boldsymbol{m})$. With respect to the introduction of the weight matrices, there are in particular two situations:
  (i) Conventional neural networks: the weights are determined in an iterative fashion. The selected single- or multi-level network is then put to work as conceptualized.
  (ii) Hopfield networks (Hopfield [1982]): the ouptut is coupled to the input of the network by a feedback loop which makes it a true iterative procedure. The system is able to arrive at a stable end-result; i.e., $b^{(j+1)}(\boldsymbol{m}) = b^{(j)}(\boldsymbol{m})$. The distribution obtained in this way minimizes a certain energy function which is related to a quality criterion similar to the cost function for simulated annealing.

Thus, this procedure can also be used to control the spectral effects of the binarization operator $\mathscr{B}$.

As indicated above, it is possible to specify the spectral effects of the iterative techniques. These effects are dependent on the specific choice of operators, and in particular on the operator $\mathscr{P}$ for IFTA and the cost and energy functions for simulated annealing and neural network methods. One of the properties that distinguishes the iterative procedures from the direct ones is the possibility of explicitly influencing and manipulating factors with spectral consequences.

3.4. FOUNDATIONS FOR ANALYSIS OF HALFTONE TECHNIQUES

The spectral features of different types of halftone techniques were specified in §§ 3.1–3.3. The spectral characteristics of these methods constitute the basis on which to discuss the fulfillment of the relation [cf. eq. (2.13)]:

$$\mathscr{C}\mathscr{X}f(\boldsymbol{m}) = \mathscr{C}\mathscr{X}b(\boldsymbol{m}). \tag{3.38}$$

So far, some qualitative considerations of eq. (3.38) have been dealt with. To be able to perform a quantitative argument, we turn to some necessary definitions.

In electronic halftoning the binarization $\mathscr{B}$ is accomplished by a computer, and the result is the sampled distribution $b(\boldsymbol{m})$. This in turn has the effect that the spectrum $B(\boldsymbol{k})$ of $b(\boldsymbol{m})$ is periodic. The prevailing circumstances are illustrated in fig. 13.

The partition $\mathbb{S}$, which is one period of the periodic spectrum, entirely determines $B(\boldsymbol{k})$ and accordingly $B(\boldsymbol{u})$. Consequently, to discuss the effect of the binarization operator $\mathscr{B}$ on the distribution over the total spectral plane, it is sufficient to consider the effect on the partition $\mathbb{S}$.

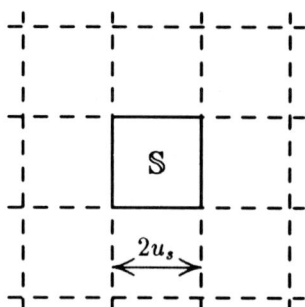

Fig. 13. The spectrum $B(k)$ of the sampled image $b(m)$ is repeated in both dimensions. The extent of the innermost period $\mathbb{S}$ is $2u_s$.

The size of the area $\mathbb{S}$ is defined by the sampling distance δx, which determines the resolution of the display medium. We introduce a figure of resolution, $2u_s$, which is equal to the spatial frequency extent of $\mathbb{S}$ (see fig. 13). $\mathbb{S}$ is assumed to be quadratic. The resolution of the display is the same in the x and y directions; i.e., $\delta x = \delta y$.

The requirement which must be met on u_s depends on the resolution of the detector, which is indicated quantitatively by u_d. It is important to consider the relation between the cutoff frequency u_d of the detector and the spectral resolution u_s of the display.

Dependent on the relative values of u_d and u_s, it is logical to look at some different situations. To begin with, we consider the case where the resolution of the detector is inferior to the resolution of the image presented; i.e.,

$$u_d < u_s. \tag{3.39}$$

This case is illustrated in fig. 14. The area $|u| \leq u_d$ is indicated by $\mathbb{D}$. The

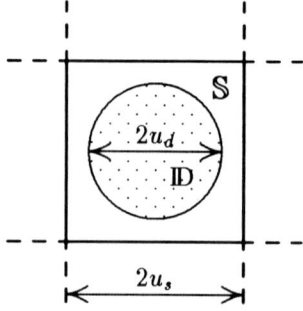

Fig. 14. Illustration of the case where the resolution of the detector u_d is less than the resolution of the image u_s.

system resolves the frequencies $|u| \leq u_d$ within the circular area $\mathbb{D}$, which is located inside the period $2u_s$.

A sufficient condition to fulfill eq. (3.38) is that

$$F(u) = B(u), \quad u \in \mathbb{D}. \tag{3.40}$$

The question arises: under what conditions is it possible to satisfy eq. (3.40)? From the description of the methods given above, it is evident that the quantization noise $q(m)$ produced will modify the spectrum; viz., the quantization noise spectrum $Q(u)$ is introduced. Thus, it is necessary to detach spatially $Q(u)$ from $\mathbb{D}$ to fulfill eq. (3.40). To achieve this separation, space is needed, and the question is how much space outside of $\mathbb{D}$ has to be available. Dependent on the method, it is possible to formulate this quantitatively. Equation (3.40) can be realized approximately for

$$u_d \leq u_g, \tag{3.41}$$

with

$$u_g \leq \alpha_{\max}(\mathscr{B}) u_s, \tag{3.42}$$

and $\alpha_{\max}(\mathscr{B}) \leq 1$. Some methods need more space to accommodate the noise spectrum than others; i.e., the constant $\alpha_{\max}$ is dependent on $\mathscr{B}$. Furthermore, $\alpha_{\max}(\mathscr{B})$ is an estimation. It depends on the method as well as on $f(x)$. For special cases, even $\alpha_{\max}(\mathscr{B}) > 1$ can be achieved, due to the circular symmetry of the transfer function. The determination of the values of $\alpha_{\max}(\mathscr{B})$ for the selection of $\mathscr{B}$'s presented in this section is the essential subject of § 4.

In § 4.1, the procedure to determine $\alpha_{\max}(\mathscr{B})$ experimentally is presented. The value of $\alpha_{\max}(\mathscr{B})$ indicates what portion of the available resolution of the display can be used to present the image $f(m)$ with two levels, and what portion the quantization noise spectrum will occupy. Thus, $\alpha_{\max}(\mathscr{B})$ is an appropriate parameter to imply to what degree a display can be utilized in practice to present pictorial information.

In practice, we face the situation that u_d is given by the resolution of the visual system or other detector used. In order to fulfill eq. (3.38) making use of the sufficient condition (2.16) [i.e., fulfillment of condition (3.40)], in general u_0 must satisfy eqs. (3.41) and (3.42). Consequently, a display with a sufficiently high resolution and/or a binarization method $\mathscr{B}$ with a high enough value of $\alpha_{\max}(\mathscr{B})$ to utilize optimally the attainable resolution of the display are vital.

Extreme cases can also occur where conditions (3.41) and (3.42) cannot be fulfilled because only $u_d > \alpha_{\max}(\mathscr{B})u_s$ is attainable (for example, due to hardware limitations), then

$$\alpha_{\max}(\mathscr{B})u_s < u_d \leq u_s. \tag{3.43}$$

This case is illustrated in fig. 15 and will be treated in § 5.

Because $Q(u)$ cannot be separated completely from $\mathbb{D}$, it is evident that compromises are necessary. In particular for frequencies represented by the hatched area in fig. 15, it is not possible to fulfill $F(u) = B(u)$. Obviously, in this situation eqs. (2.16) and (3.40) are not fulfilled to a good approximation. Nevertheless, to fulfill eq. (3.38) approximately, it is possible to incorporate the image-comprehension operator $\mathscr{C}$ by using the freedom to manipulate $Q(u)$ inside and outside of $\mathbb{D}$. This is an important factor which needs to be examined systematically in the future. Several parameters which cannot be expressed totally in physical terms influence the judgement of the perceived image. Artistic and pleasing values lead us to diffuse descriptions and evaluations. However, different valuable and important suggestions have been proposed in the past where diverse aspects of $\mathscr{C}$ were considered. How these factors can be adapted to the operator $\mathscr{B}$ will be the subject of § 5.

So far, the case $u_d \leq u_s$ has been considered. Of course, the case

$$u_d > u_s \tag{3.44}$$

is also possible; i.e., the resolution of the display is lower than the resolution of the detector system (see fig. 16). Here, more than one period is detected, and

(i) the pixels of the display are resolved, and
(ii) for $u_d \geq \sqrt{2}u_s$, all of the quantization noise spectrum $Q(u)$ is located within $\mathbb{D}$; i.e., the quantization noise is fully perceived. Consequently, eq. (3.38) is not fulfilled.

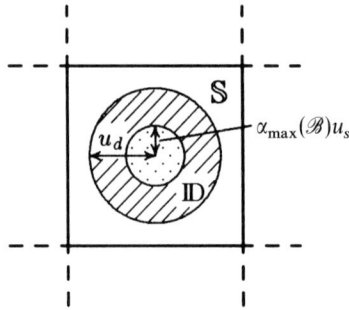

Fig. 15. Illustration of the case where the resolution of the detector u_d is larger than the maximum radius of a practically noisefree lowpass region $\alpha_{\max}(\mathscr{B})u_s$, but smaller than the extent of $\mathbb{S}$.

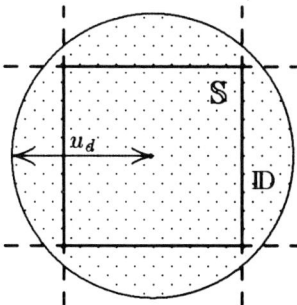

Fig. 16. Illustration of the case where the resolution of the detector system u_d exceeds the frequency extent of the image.

Features of this case, where the shapes of the pixels are also relevant, are considered in § 6.

§ 4. Analysis of Halftone Procedures with Excessive Resolution

In § 3, different types of quantization methods were presented together with the typical characteristics of their quantization noise spectra. We now turn to the consideration of how those noise spectra can be conformed to the features of the detector system; i.e., to reduce $Q(u)$ within the area $\mathbb{D}$.

In this section, it is assumed that the desired noisefree lowpass region is small enough to allow realization by a specific method; viz., $u_d < \alpha_{\max}(\mathscr{B})u_s$. In other words, if the extent of the transfer function of the detector system is less than the maximum usable spectral region of the electronic halftoning technique, the image can be perceived almost free of distortions.

The problem is what percentage of the spectrum of the quantized image is available for an optimization procedure. Among the specific questions to be considered are the actual values of $\alpha_{\max}$ of the different procedures, how well $\mathscr{B}$ can utilize the available resolution, and how effectively it is possible to suppress the noise contributions in $\mathbb{D}$.

All images formed by electronic halftoning techniques are inherently discrete, which causes an aliasing effect. However, due to the periodicity of the spectrum, we may confine the analysis to the innermost period $\mathbb{S}$ of the spectrum.

4.1. TRANSFER FUNCTION AND QUALITY CRITERIA

To evaluate quantitatively the quantized images from a physical point of view, we consider the transfer function of the detector system and numerical quality specifications. The transfer model presumed is the special, idealized case of a diffraction-limited, incoherent optical system with a circular exit pupil. The optical transfer function which is circularly symmetric is (Goodman [1968]):

$$H(\boldsymbol{u}) = \begin{cases} \dfrac{2}{\pi} \left\{ \cos^{-1}\left(\dfrac{|\boldsymbol{u}|}{u_d}\right) - \dfrac{|\boldsymbol{u}|}{u_d} \sqrt{1 - \left(\dfrac{|\boldsymbol{u}|}{u_d}\right)^2} \right\}, & \text{for } |\boldsymbol{u}| \leq u_d, \\ 0, & \text{otherwise,} \end{cases} \quad (4.1)$$

where u_d is the cutoff frequency (see § 3.4). The shape of this transfer function is shown in fig. 17. Some of its properties are:

$$H(\boldsymbol{u}) \neq 0 \quad \text{for } \boldsymbol{u} \in \mathbb{D}; \quad H(0, 0) = 1; \\ H(\boldsymbol{u}) \leq H(0, 0); \quad \text{and} \quad H(\boldsymbol{u}) = 0 \quad \text{for } |\boldsymbol{u}| > u_d. \quad (4.2)$$

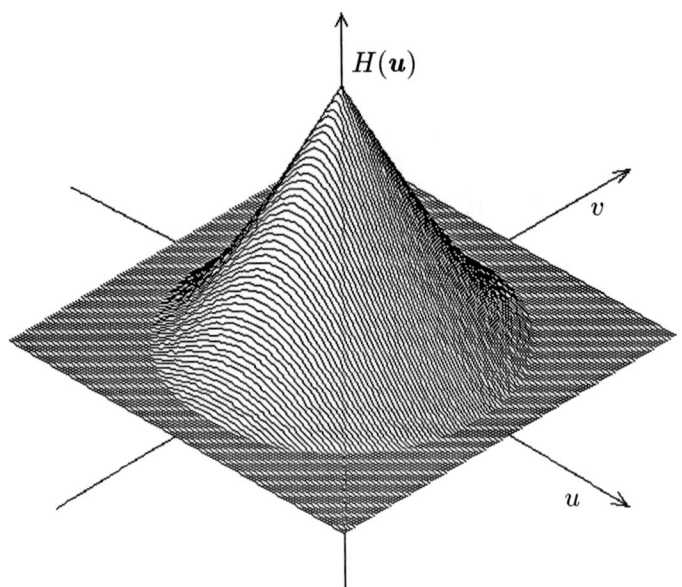

Fig. 17. Plot of the optical transfer function $H(\boldsymbol{u})$ of the detector system.

The desired intent is to fulfill eq. (3.40), and therefore:

$$Q(\mathbf{u}) = 0, \quad \text{for } |\mathbf{u}| \leq u_d. \tag{4.3}$$

In general, this is impossible to accomplish. A residual noise contribution will ordinarily remain in the region $\mathbb{D}$. As a measure for the residual spectral-noise contributions in $\mathbb{D}$, the standard deviation σ can be used. The normalized value of this mean squared deviation is given by

$$\sigma^2 = \sum_{\mathbf{k} \in \mathbb{D}_k} |Q(\mathbf{k})|^2 \Big/ \sum_{\mathbf{k} \in \mathbb{D}_k} 1 \tag{4.4}$$

for sampled spectra. For continuous spectra, the summation is to be replaced by an integral. The set $\mathbb{D}_k$ is defined as $\mathbb{D}_k = \{\mathbf{k} = (k, l) | (k\delta u, l\delta v) \in \mathbb{D}\}$, with the sampling rate $\delta\mathbf{u} = (\delta u, \delta v)$ in the Fourier domain.

Further, the dependence of the transfer function $H(\mathbf{u})$ on the frequency $\mathbf{u}$ implies that the variation of $Q(\mathbf{u})$ in $\mathbb{D}$ has an influence. To take into account that the significance of the spectral noise contributions decreases with increasing frequency, a weighted measure is used:

$$\sigma_H^2 = \sum_{\mathbf{k} \in \mathbb{D}_k} |H(\mathbf{k})Q(\mathbf{k})|^2 \Big/ \sum_{\mathbf{k} \in \mathbb{D}_k} H(\mathbf{k}). \tag{4.5}$$

These measures quantify the presence of spectral noise components: σ^2, the total noise contributions in $\mathbb{D}$, and σ_H^2, the weighted noise contributions in $\mathbb{D}$.

4.2. CARRIER METHODS

4.2.1. *Carrier with integer period ratio between carrier and raster*

For simplicity, the situation considered is a quadratic, periodic sampling raster with the periods $\delta x = \delta y = 1$. A periodic image carrier with the period p_c is applied. In this section, we suppose the ratio between the two periods to be an integer; viz.,

$$p_c/\delta x = p_c \in \mathbb{N}. \tag{4.6}$$

A factor to consider is the proper choice of p_c of the quantization procedure to optimize the lowpass spectral region to suit a desired transfer function. In this respect, it is favorable to choose p_c to be as small as possible to shift the peaks in the noise spectrum (see § 3.1.2) away from the d.c. peak and achieve a large lowpass region. The sampling allows carrier periods $p_c \geq 2$. However, the periodic repetition of the spectrum will cause aliasing effects in the form of unwanted overlap of higher orders from adjacent spectra.

The smaller p_c is chosen, the lower are the orders of the contiguous repeated noise spectra that will intrude into $\mathbb{D}$.

It is necessary to make a compromise for selection of an appropriate p_c. The actual superposition of orders from repeated spectra is illustrated in the next few examples. Three centrally repeated spectra with the relative dislocation of the period p_c are shown. Figure 5a represents schematically the composite noise spectrum for $p_c = 2$. This is the smallest useful carrier period which is an integer multiple of the raster period. The fundamental frequency component of the quantization noise is located at the border of the innermost spectral period; i.e., $u = p_c^{-1} = \pm 0.5$, and the higher harmonics (solid bars) are all superposing neighbor periods in the periodic spectrum. However, the second harmonics of the adjacent spectra (dashed and dotted bars) superpose and disturb the central d.c. peak and surroundings in the lowpass region.

So far, the prospects do not look too encouraging. However, some image aspects improve with increasing carrier periods. The situation with $p_c = 4$ is illustrated in fig. 5b. Noise of the fundamental frequency is now brought closer to the d.c. peak; viz., $u = p_c^{-1} = \pm 0.25$. Superposing onto the center are now the 4th harmonics of the adjacent orders. This restricts the useful lowpass region. On the other hand, the disturbances by centrally located higher harmonics are reduced compared with the case where $p_c = 2$. Assuming a circular $\mathbb{D}$, the available lowpass area is reduced to about 20% of the spectrum.

The tendency persists with further increase of the carrier period. For $p_c = 8$, the useful region is reduced to 5% with the 8th harmonics in centrum, and for $p_c = 16$ to 1.2% and 16th harmonics. Scans across the noise spectra of actual binarizations are shown in fig. 18. The similarities with the schematic diagrams of fig. 5 are obvious.

The different examples are here illustrated by binarizations of the graytone picture shown in fig. 19 using a pyramid-shaped carrier (two crossed 1D triangular carriers). The carrier is renormalized to secure a linear tone reproduction. In fig. 20, the conditions for $p_c = 8$ are depicted.

These considerations indicate that it is possible to predict an $\alpha_{\max}(\mathscr{B})$ for this halftoning procedure. Its value is about 0.25 or somewhat smaller, with the consequence that only about 5% of the spectrum area can be utilized assuming a circular symmetric transfer function. This is a compromise between the size of the lowpass region and the strength of contributions of the noise components around the d.c. peak. We found this compromise acceptable for $p_c \geqslant 4$. Of course, this is a rough estimate. It depends strongly on the quality requirements.

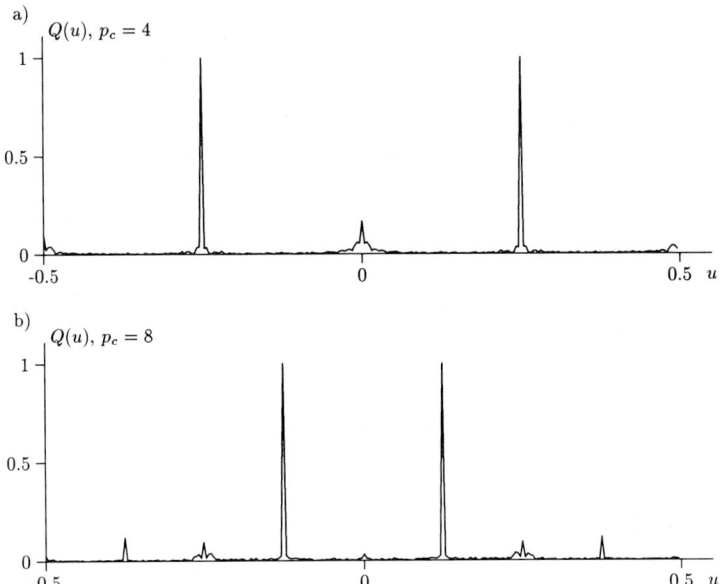

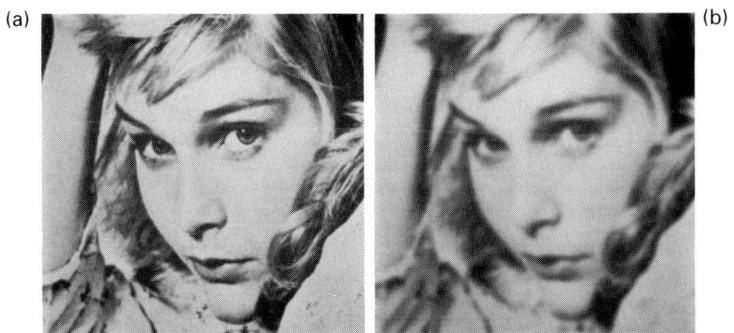

Fig. 18. Cross-section of noise spectrum of binarized image: (a) for $p_c = 4$, and (b) for $p_c = 8$.

Fig. 19. Graytone picture used in the examples. The original is shown in (a), and after being filtered with $H(u)$ in (b).

4.2.2. Carrier with noninteger period ratio between carrier and raster

The carrier period can also be chosen in such a way that it is not an integer multiple of the raster period (Just, Hauck and Bryngdahl [1986]). Then it is possible to attain a more uniform noise spectrum. This noninteger period ratio can be a rational or an irrational number. When the noninteger ratio is a rational fraction $p_c = n_1/n_2 \in \mathbb{Q}$ with $n_1, n_2 \in \mathbb{N}$, the peaks of the

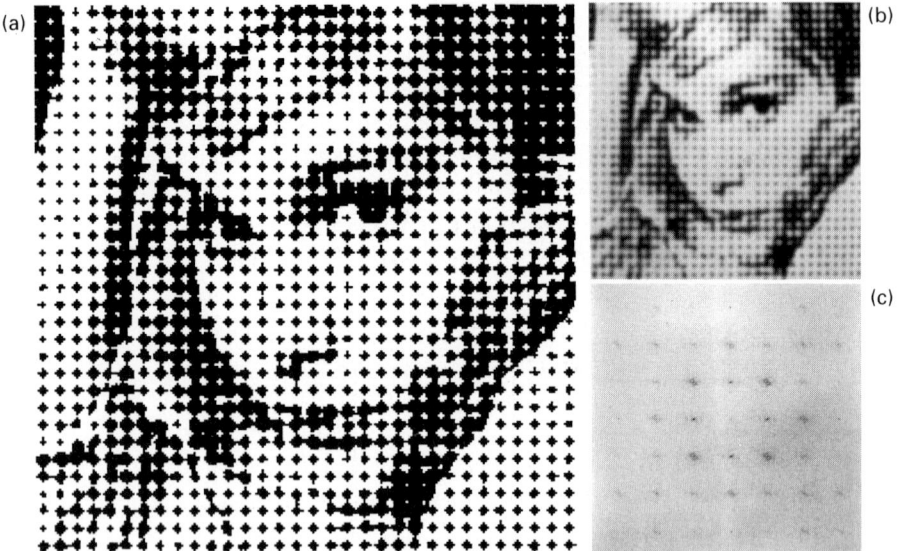

Fig. 20. Binarization using a pyramid carrier with $p_c = 8$: (a) binary image with 256×256 pixels; (b) version of (a) after filtering with $H(u)$; (c) quantization noise spectrum of (a) in reversed polarity.

jn_2th ($j \in \mathbb{Z}$) repeated noise spectrum will superpose the peaks of the central spectrum.

A conceivable and favorable ratio would be an irrational $p_c \in \mathbb{R}$, $p_c \notin \mathbb{Q}$; then the peaks of the diverse repeated noise spectra would not coincide. However, this case is unattainable with a computer due to its limited precision.

To optimize the innermost portion of the lowpass region with respect to size and suppression of its share of quantization noise, the following strategy is reasonable (Scheermesser, Wyrowski and Bryngdahl [1993c]):

(i) The frequency of the carrier is chosen as high as possible to allocate prominent peaks of the central noise spectrum as far as possible from the d.c. peak. This would imply a p_c value close to 2.

(ii) The peaks of the individual repeated spectra ought to be distributed as uniformly as possible, which would result in less adding up of noise components of the different repetitions.

Following these tactics, it is possible to proceed as in § 4.1.1. Under the presumption that no harmonics below the 8th of the repeated spectra are tolerated in the closest proximity to the d.c. peak, $p_c = 8/3$ is a plausible choice. Then the 8th harmonic of the third repeated noise spectrum and the

16th harmonic of the sixth repetition, etc., will overlap onto the d.c. peak. As mentioned above, contributions from the $(3j)$th repeated spectrum are detrimental in this particular case.

The quantization noise spectrum from the chosen example is illustrated in fig. 21. Only the three innermost adjacent repetitions are considered; their peaks contribute the most. The tendency obtained with this choice of p_c is a uniform spread among the peaks of the noise spectrum. The stronger peaks appear in the outer, and the weaker in the inner portion of the spectrum, and a lowpass region the width of which equals $\frac{1}{4}$ is essentially free from noise components. Examples of how this procedure functions when put into practice are shown in figs. 22a and 23.

In the chosen case, n_1 and n_2 are rather small. This will cause the spectral noise contributions to build up in relatively few locations, especially onto the d.c. peak. To improve the situation, an irrational p_c close to $\frac{8}{3}$ may be chosen; e.g., $p_c = e \approx 2.718$. Then the spectral noise peaks will ideally no longer superpose. The appearance of the noise spectrum will occur more uniformly, and its peaks will be broadened. However, the coarse structure of the noise spectrum will remain, as is evident in figs. 22b and 24. The tendency of homogenization of the noise spectrum is noticeable from a comparison between the curves for $\frac{8}{3}$ and e in fig. 22.

We may conclude, in the case where the spectral noise peaks up to the 3rd harmonics are considered, that for $p_c = e$ a considerably more uniform noise spectrum is obtained within $\mathbb{D}$, as is possible with $p_c = 8$ (see § 4.2.1). The usable area of the spectrum is about 5%, and $\alpha_{\max}(\mathscr{B}) \approx 0.25$. The peaks corresponding to the 3rd harmonics appear isolated and relatively far from the d.c. peak, and may be acceptable in some cases. Then $\alpha_{\max}(\mathscr{B}) \approx 0.5$ is

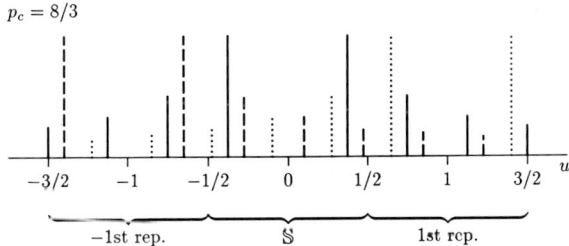

Fig. 21. Illustration of the orders of the quantization noise spectrum introduced by $\mathscr{B}_c$. This spectrum, with all its orders, is repeated periodically. The central spectrum is represented by solid bars, the one shifted to the right by dotted bars, and the one shifted to the left by dashed bars. The period of the carrier is $p_c = \frac{8}{3}$.

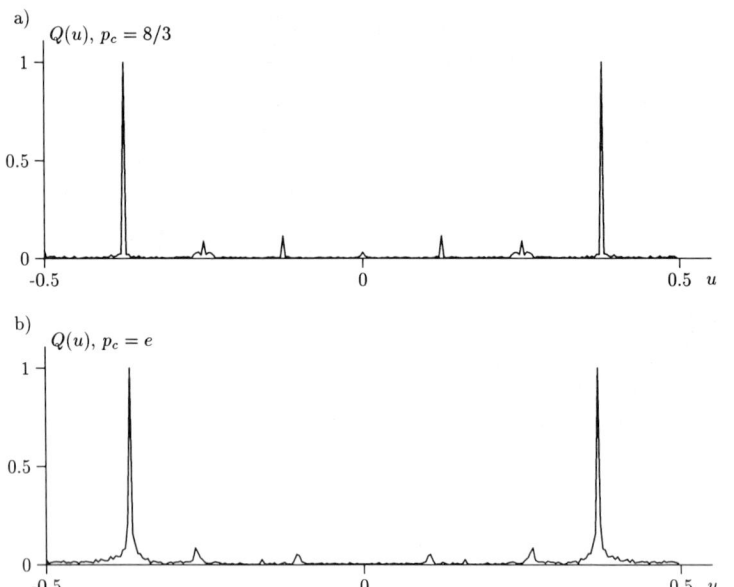

Fig. 22. Cross-section of noise spectrum of binarized image: (a) for $p_c = \frac{8}{3}$, and (b) for $p_c = e$.

Fig. 23. Binarization using a pyramid carrier with $p_c = \frac{8}{3}$: (a) binary image with 256 × 256 pixels; (b) version of (a) after filtering with $H(u)$; (c) quantization noise spectrum of (a) in reversed polarity.

Fig. 24. Binarization using a pyramid carrier with $p_c = e$: (a) binary image with 256×256 pixels; (b) version of (a) after filtering with $H(u)$; (c) quantization noise spectrum of (a) in reversed polarity.

obtained for a fairly homogeneous noise spectrum in $\mathbb{D}$, and a usable area of the innermost spectrum would be about 20% of the total.

In general, we may state that when carrier procedures are used for image quantization, it is possible to influence the lowpass region of the noise spectrum via the choice of the carrier frequency.

Filter-encoded carrier-treated images are shown in figs. 20, 23 and 24. They have been processed with the idealized system characterized by the transfer function of eq. (4.1); i.e., the carrier images were Fourier-transformed, multiplied with the selected $H(k)$'s for $u_d = 0.25$, and inversely Fourier-transformed. The tendency is evident that for $p_c = \frac{8}{3}$ (or even more pronounced for $p_c = e$), a larger lowpass region with higher image definition is permissible, while for $p_c = 8$ a lowpass larger than $u = \frac{1}{8}$ results in strong image distortions. This is in agreement with the findings in § 4.2.

4.3. ERROR-DIFFUSION METHOD

We continue the evaluation of binary images with regard to their frequency content, and focus upon images processed by the error-diffusion procedure. Here too, it is only possible to influence indirectly the quantization noise spectrum; viz., by the choice of the weights by which the corrected errors are accounted for, a modification of the error feedback (Sullivan, Miller and Pios [1993]), or a modulation of the threshold (Daels, Easton and Eschbach [1991], Knox and Eschbach [1993]). These weights define a filter function (see § 3.2.2) which influences the spectrum of the formed quantization noise. The weights by which the errors are diffused may be optimized in such a way that desired features are forced upon the quantization noise (Fetthauer, Weissbach and Bryngdahl [1992], Weissbach and Wyrowski [1992a]).

The optimization goal considered here is to shape the quantization noise spectrum to achieve a minimum mean-square deviation σ^2 in region $\mathbb{D}$. To determine the corresponding weights, (i) an optimization procedure like the steepest-descent method, simulating annealing, or a similar algorithm can be applied, or (ii) they can be determined analytically from the filter function.

The determined weights must be controlled to result in a stable method; i.e., that no error oscillations and devious images occur. If necessary, the weights must be readjusted to guarantee stability of the quantization procedure. This will then result in a modified filter function.

Some restraints of the procedure are imperative:

(i) It is preposterous to overfill or even fill the image extent with weight factors. For practical reasons, it makes sense to optimize only a limited number of weights. This introduces restrictions on the realization of conceivable filter functions. In consequence of this, when the optimization is performed for a confined $\mathbb{D}$, the quantization noise will be reduced automatically in its surroundings too. An example of this is illustrated in fig. 25. Mainly high-frequency noise is formed, in spite of the fact that a lowpass with a small $u_g = \frac{1}{16}$ was optimized. Here, and in the following examples, 31 weights were calculated analytically and modified to assure stability. $u_g = \frac{1}{16}$ can be regarded as the lower limit; i.e., lowpass regions with a very small extent cannot be realized without controlling higher frequencies as well.

(ii) The other limit is that arbitrarily extended lowpass regions cannot be controlled.

When increasing $\mathbb{D}$, the result shows a tendency to approach that of a hardclip process. This can be explained by considering $\sigma_{\mathbb{S}}^2$ over the entire innermost part of the spectrum. Expressed in continuous distributions, we obtain:

$$\sigma_{\mathbb{S}}^2 = c \int_{\mathbb{S}} |F(u) - B(u)|^2 \, du, \qquad (4.7)$$

with $c = \{\int_{\mathbb{S}} 1 \, du\}^{-1}$. The corresponding deviation in the spatial domain is

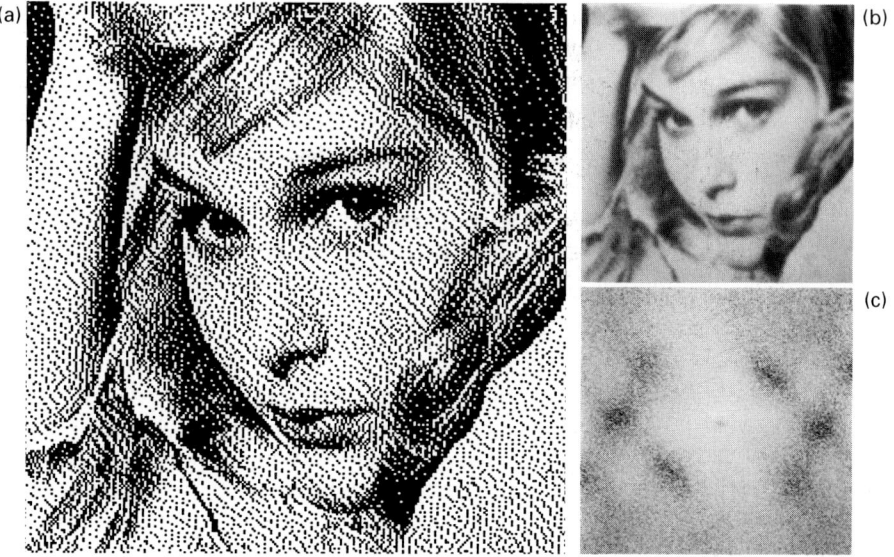

Fig. 25. Binarization using an optimized error-diffusion algorithm with 31 weights for $u_g = \frac{1}{16}$: (a) binary image with 256 × 256 pixels; (b) version of (a) after filtering with $H(u)$; (c) quantization noise spectrum of (a) in reversed polarity.

given by

$$\sigma_{\mathbb{S}}^2 = c \int |f(x) - b(x)|^2 \, dx, \tag{4.8}$$

using Parseval's theorem. The integration in eq. (4.8) is extended over the entire image. The sum of the integrand in discrete form or the integral as expressed by eq. (4.8) contains squares of moduli, so that expression (4.8) has an absolute minimum value when the quantization error which is introduced in each location is minimum. This is the characteristic of the hardclip process; i.e., a hardclip minimizes $\sigma_{\mathbb{S}}^2$.

For an optimization process operating properly, it is to be expected that the result will resemble a hardclip when the whole spectrum is controlled, although there are special cases where the absolute minimum value of σ^2 is ambiguous. Consequently low-frequency noise occurs in conventional images where the image spectrum is concentrated in and around its d.c. peak (cf. § 3). This effect is not restricted to the error-diffusion method but is valid for any optimization procedure; e.g., for the iterative Fourier transform algorithm which is illustrated below. The drawback of this behavior consists of the central concentration of the noise components which is unfavorable with regard to the considered transfer function $H(u)$. Indeed, when the size of the lowpass area is increased the noise contributions accumulate in the centrum of the spectrum, with the result that homogeneous white and black areas appear in the image. This behavior is illustrated in fig. 26 for $u_g = \frac{1}{2}$; i.e., the lowpass area is extended to the very edge of the spectrum.

From this consideration, we conclude that the error-diffusion method functions satisfactorily in a range between the extremes mentioned above. The range $\frac{1}{6} \leqslant u_g \leqslant \frac{1}{4}$ has been found empirically where barely any noise frequencies appear in the lowpass region and on its outside it is rather homogeneous. This is verified for $u_g = \frac{1}{4}$ in fig. 27. The quantization noise is even reduced within $\mathbb{D}$ in comparison with the well-known process with weights according to Floyd and Steinberg [1975]. Figure 28 presents an image binarized according to Floyd and Steinberg's algorithm. On the other hand, for $u_g \geqslant \frac{1}{4}$, continuous white or black surfaces turn up.

Optimization with predictable, acceptable results seem possible to obtain for $\frac{1}{6} \leqslant u_g \leqslant \frac{1}{4}$ by the error-diffusion method. In particular, the value of σ^2 is here noticeably lower than for optimization by carrier methods. We obtain

$$\alpha_{\max}(\mathscr{B}_{\mathrm{ED}}) \approx 1/2, \qquad (4.9)$$

Fig. 26. Binarization using an error-diffusion algorithm for $u_g = \frac{1}{2}$: (a) binary image with 256 × 256 pixels; (b) version of (a) after filtering with $H(u)$; (c) quantization noise spectrum of (a) in reversed polarity.

Fig. 27. Binarization using an error-diffusion algorithm for $u_g = \frac{1}{4}$: (a) binary image with 256 × 256 pixels; (b) version of (a) after filtering with $H(u)$; (c) quantization noise spectrum of (a) in reversed polarity.

Fig. 28. Binarization using the original error-diffusion algorithm: (a) binary image with 256 × 256 pixels; (b) version of (a) after filtering with $H(u)$; (c) quantization noise spectrum of (a) in reversed polarity.

and the area of the spectrum available for image information is about 20% with rotationally symmetric H.

4.4. ITERATIVE FOURIER TRANSFORM METHOD

Optimization procedures also exist where global transformations of the image information are considered. Among these, the iterative Fourier transform algorithm (IFTA) has proven to be powerful, flexible, and convenient. It is discussed here, and is used as a representative for a class of optimization methods (e.g., IFTA, SAA, DBS) which may render comparable results.

The IFTA was described in § 3.3.2. To apply this procedure, the operators $\mathcal{P}$ and $\mathcal{Q}$ must be defined in order to satisfy the boundary conditions. A proper choice of $\mathcal{P}$ forces the conditions in the spectrum to be fulfilled:

$$F^{(j)}(k) = \mathcal{P}B^{(j)}(k) = \begin{cases} F(k) & \text{for } k \in \mathbb{D}, \\ B^{(j)}(k) & \text{for } k \notin \mathbb{D}. \end{cases} \tag{4.10}$$

The operator in the spatial domain is the random clip function [see eq. (3.36)]. We have:

$$b^{(j+1)}(m) = \mathcal{Q}_r f^{(j)}(m), \tag{4.11}$$

where Δ_r is increased from 0 to $\frac{1}{2}\int_\mathbb{D} 1\, d\mathbf{k}/\int_\mathbb{S} 1\, d\mathbf{k}$ (or more) during the iteration process.

For the iterative procedure, the limits of operation are:
(i) In contrast to the conditions valid for the error-diffusion method, it is now possible to optimize without problems in small lowpass regions. Outside the controlled area a homogeneous noise spectrum (white noise) accumulates. An example with $u_g = \frac{1}{16}$ is shown in fig. 29.
(ii) For the same reason as in the case of the error-diffusion method, results are obtained with large controlled areas which resemble hard-clipped images. Figure 30 illustrates the case with $u_g = \frac{1}{2}$. However, here this effect sets in for a slightly larger controlled area than in the case of error diffusion; viz., for u_g values above 0.3. Thereafter, continuous white and black areas will also occur in this case.

Typical for this procedure is the sharp outline of the controlled area from the remains of the spectrum and the homogeneity of the noise spectrum outside the controlled area.

For the iterative Fourier transform method presented here, we obtain:

$$\alpha_{\max}(\mathcal{B}_{\text{IFTA}}) \approx 0.6, \tag{4.12}$$

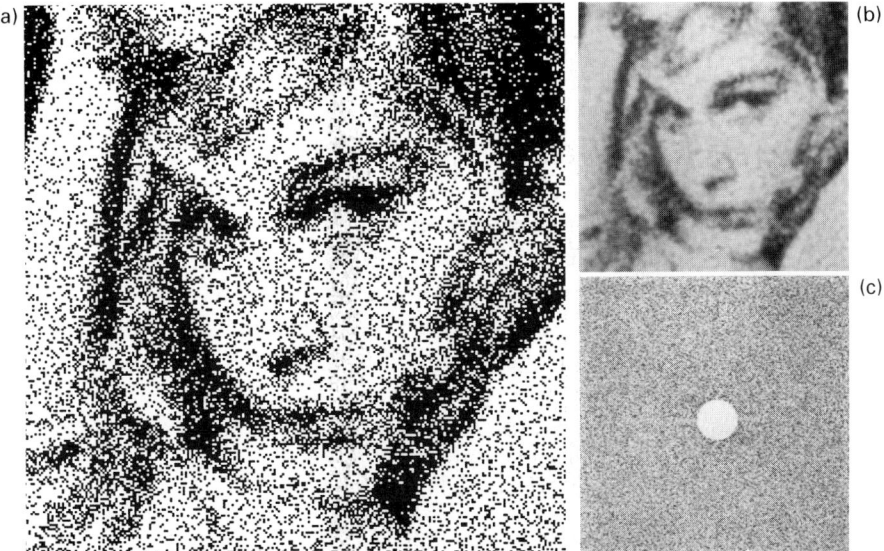

Fig. 29. Binarization using the iterative Fourier transform algorithm with $u_g = \frac{1}{16}$: (a) binary image with 256×256 pixels; (b) version of (a) after filtering with $H(u)$; (c) quantization noise spectrum of (a) in reversed polarity.

Fig. 30. Binarization using the iterative Fourier transform algorithm with $u_g = \frac{1}{2}$: (a) binary image with 256×256 pixels; (b) version of (a) after filtering with $H(u)$; (c) quantization noise spectrum of (a) in reversed polarity.

which corresponds to a usable area of the total spectrum of about 28%. An example is shown in fig. 31 for $u_g = 0.3$.

4.5. COMPARATIVE REMARKS ON QUANTIZATION METHODS

Some values of the mean squared deviation σ^2 in $\mathbb{D}$ are listed for the different types of methods and different sizes of the detector area in table 1. The entered procedures are the favorable carrier method with $p_c = e$, the standard error-diffusion algorithm by Floyd and Steinberg [1975], and the optimizations with ED (error diffusion) and IFTA (iterative Fourier

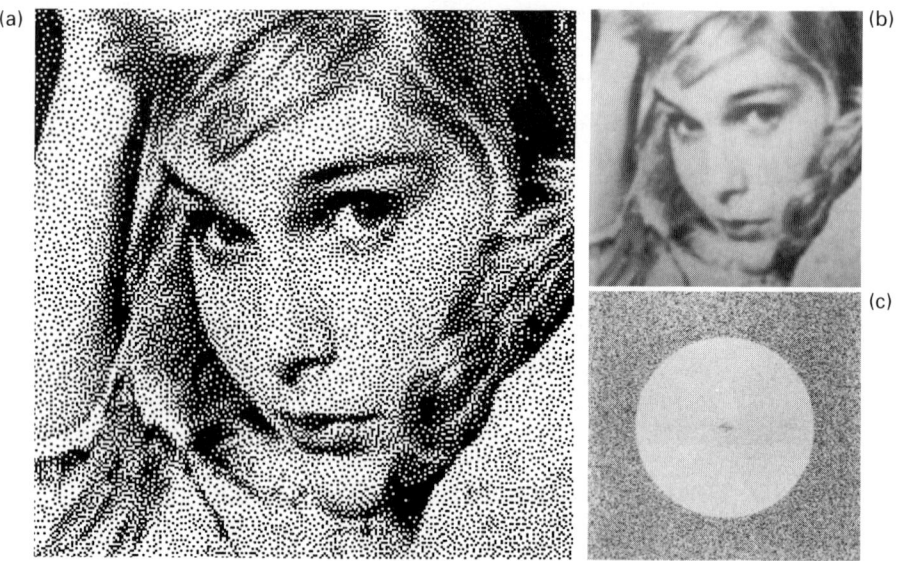

Fig. 31. Binarization using the iterative Fourier transform algorithm with $u_g = 0.3$: (a) binary image with 256 × 256 pixels; (b) version of (a) after filtering with $H(u)$; (c) quantization noise spectrum of (a) in reversed polarity.

TABLE 1
Values of the mean squared deviation σ^2 in $\mathbb{D}$ for different binarization algorithms and different values of u_d.

	$u_d = \frac{1}{16}$	$u_d = \frac{1}{8}$	$u_d = \frac{1}{4}$
Carrier ($p_c = e$)	49.9	63.7	66.4
Original ED	14.8	13.4	23.1
Optimized ED	7.9	2.6	8.1
IFTA	1.5	5.2	16.1

transform algorithm), where u_g was adjusted to match u_d. From table 1 we may conclude that it is possible to achieve great reductions in the σ^2 values by the optimization procedures compared with carrier methods. Furthermore, they eliminate the regular pixel configuration which is still discernible for the e-carrier, because the noise contributions become distributed fairly uniformly outside of $\mathbb{D}$ without any centers. It is noticeable that for large control areas comparable or better results are obtained by the optimized ED than by the IFTA.

For further comparison, another representation is displayed in fig. 32, in which σ^2 is plotted as a function of the size of the lowpass area. In contrast to table 1, this is carried out for one particular image which was binarized with $u_g = \frac{1}{8}$, for which the considered methods function satisfactorily. σ^2 was calculated within a circular lowpass region for increasing radius of the circle. The finer details of the curves in fig. 32 are dependent on the special choice of parameters. However, the tendencies are obvious and are in accord with the data of table 1.

Outside the lowpass area, distinct differences in the behavior of the different methods occur (see fig. 32). This is the subject of § 5.

4.6. MODIFICATION OF RESIDUAL DETECTED NOISE

Some additional considerations of the noise within the detector area are discussed in this section. So far, the intention has been to eliminate the noise

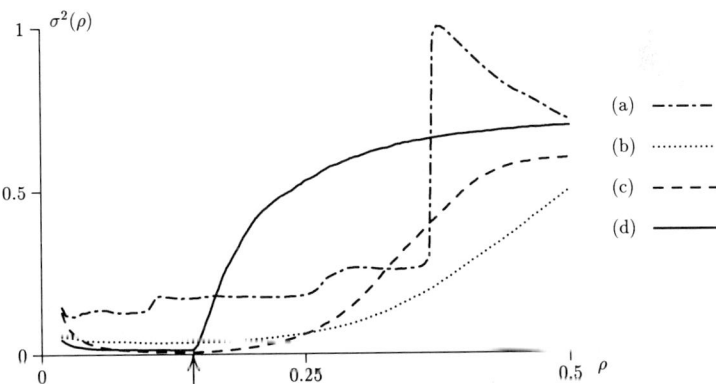

Fig. 32. Comparison of residual spectral noise in $\mathbb{D}$ in form of σ^2 versus lowpass region of radius ρ. Image was binarized using (a) carrier method ($p_c = e$), (b) original error-diffusion algorithm, (c) optimized error-diffusion algorithm ($u_g \approx \frac{1}{8}$ indicated by arrow), and (d) iterative Fourier transform algorithm ($u_g \approx \frac{1}{8}$).

contributions in the detector region; i.e., $Q(u) = 0$ for $u \in \mathbb{D}$. However, this is in general not possible, and some residual noise persists.

No direct influence has been exercised on the spatial distribution of the quantization noise within $\mathbb{D}$. In the presentation in § 4 up to here, it was as far as possible aimed at its reduction. Due to the particular shape of the transfer function $H(u)$, it is possible that the spatial noise distribution influences the result of the decoding process.

It has been noted in § 4 that for the iterative methods presented above, a noise characteristic similar to that of the hardclip is obtained when σ^2 is minimized in a large area. The noise will be concentrated at the d.c. peak with increase of the controlled area $\mathbb{D}$. This is undesirable when considering the shape of $H(u)$. It is preferable that the residual noise which must be tolerated is located at the boundary of $\mathbb{D}$ and not in its center. Although the total noise contribution estimated by σ^2 will then turn out comparatively higher, it may be possible to reduce the noise weighted by $H(u)$. The optimization problem is then to distribute the noise in such a way that σ_H^2 [cf. eq. (4.5)] is reduced as much as possible. An appropriate procedure to this end is the IFTA, which is flexible and allows modified operation in the spectrum with an operator $\mathscr{P}_\xi$ instead of $\mathscr{P}$ (cf. § 4.4).

An empirically found, favorable operator is

$$F^{(j)}(u) = \mathscr{P}_\xi B^{(j)}(u) = F(u) + \xi(u)[B^{(j)}(u) - F(u)], \tag{4.13}$$

with $\xi(u) = 1$ for $u \notin \mathbb{D}$. $\xi(u)$ is chosen to be a positive scalar function: $\xi(u) \in [0, 1]$. The noise is weighted with this function. Outside of $\mathbb{D}$, the noise is not influenced. The operator $\mathscr{P}$ is contained as the special case:

$$\xi(u) = \begin{cases} 0 & \text{for } u \in \mathbb{D}, \\ 1 & \text{for } u \notin \mathbb{D}. \end{cases} \tag{4.14}$$

For the choice of a $\xi(u)$ adapted to $H(u)$, the following factors are to be considered: $\xi(u)$ ought to (i) be rotationally symmetric, (ii) increase from the center towards the boundary, and (iii) be chosen to minimize σ_H^2.

To estimate the effect of noise modification, some tests were performed with

$$\xi(u) = (|u|/u_g)^\kappa, \quad \kappa \in \mathbb{R}^+. \tag{4.15}$$

Figure 33 illustrates a cross-section of a special $\xi(u)$. A minimum of σ_H^2 was reached for $\kappa \approx 1.2$. In fig. 34, images obtained in this manner are reproduced.

Values of σ^2 and σ_H^2 have been calculated for different methods with a fixed $u_d = 0.25$ and $u_g = 0.25$, where possible, which are listed in table 2.

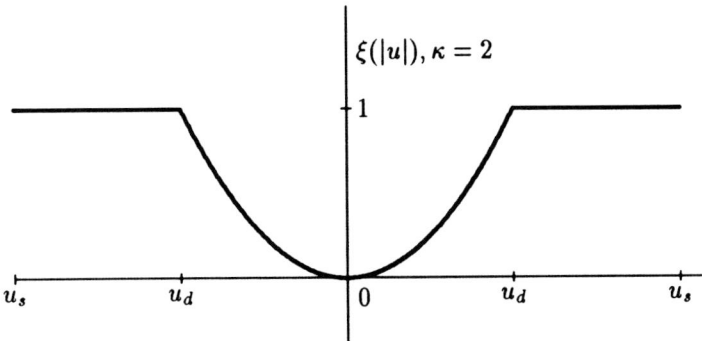

Fig. 33. The function $\xi(u)$ was used to control the allowed noise contribution within $\mathbb{D}$.

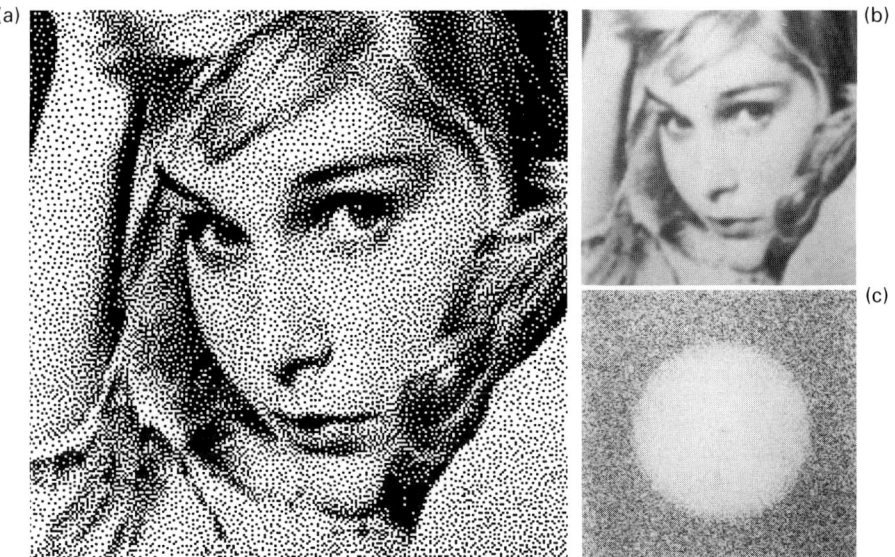

Fig. 34. Binarization using the iterative Fourier transform algorithm when weighting the noise within $\mathbb{D}$: (a) binary image with 256×256 pixels; (b) version of (a) after filtering with $H(u)$; (c) quantization noise spectrum of (a) in reversed polarity.

From the table, it is evident that for the IFTA, σ_H^2 is reduced to half its value, with a choice of $\xi(u)$ according to eq. (4.15). On the other hand, σ^2 is raised, as would be expected. This verifies the possibility of rearranging the residual noise within $\mathbb{D}$. Even functions other than an exponential can be applied. Especially suitable seems a $\xi(u)$ which equals 0 in a small lowpass area around the d.c. peak and which rises outwards in the rest of $\mathbb{D}$ contrary

TABLE 2

Values of σ^2 and σ_H^2 in $\mathbb{D}$ for different binarization algorithms ($u_d = 0.25$ and $u_g = 0.25$).

	σ^2	σ_H^2
Carrier ($p_c = e$)	66.4	27.31
Original ED	23.1	6.62
Optimized ED	8.1	4.05
IFTA	16.1	10.36
IFTA ($\xi \sim |u|^{1.2}$)	44.7	5.28

to the variation of H. The difficulty is to introduce the least possible noise with the desired characteristic.

With this, we close the analysis of the case where $u_g \leqslant \alpha_{\max}(\mathscr{B})u_s$, and continue next with $\alpha_{\max}(\mathscr{B})u_s < u_d \leqslant u_s$, which would require $u_g > \alpha_{\max}(\mathscr{B})u_s$.

§ 5. Halftone Procedures with Restricted Resolution below Sampling Raster

We now turn to the case where the detector system is able to resolve frequencies above those which are possible to control by the halftone algorithm, but still below the sampling frequency; i.e., $\alpha_{\max}(\mathscr{B})u_s < u_d < u_s$. Thus, higher frequencies are perceived than the algorithm can handle, and it is necessary to look for a compromise in order to reduce satisfactorily the detected quantization noise.

The effect of $\alpha_{\max}(\mathscr{B}) < u_d$ was described in the previous sections. In carrier methods, the structure of the carrier becomes visible. In the optimized error-diffusion and iterative procedures, it would be necessary to choose u_g according to $\alpha_{\max}(\mathscr{B}) < u_d \leqslant u_g$, which leads to images which resemble hardclip due to excessive control of the spectrum.

In the present situation, we must abandon the strict requirement that the spectra of the graytone and halftone images remain approximately identical within $\mathbb{D}$; i.e., $B(u) \approx F(u)$ for $u \in \mathbb{D}$. We can no longer expect the results of images decoded with $H(u)$ to be almost the same. In this case, a compromise must be reached. In certain parts of the spectrum, noise contributions must be accepted, and in those regions image information must be sacrificed. On the other hand, the information in other regions can be protected by avoiding the introduction of noise.

In general, the influence of the noise may vary according to its location in the spectrum. The amount and structure of the noise may even change locally.

The question to be raised is how to proceed. A universal answer does not exist, but the answer will depend on the particular purpose of the image coding. A few possible procedures are presented in the following. To master the complex boundary conditions, an algorithm of high flexibility is preferable. Further, a minimum burden of algorithmic difficulties is desired to grant freedom to concentrate on physical effects. The iterative Fourier transform algorithm is applied with appropriate modifications. It is chosen (although it may not be optimal in all situations) because it is rather easy to handle.

5.1. ADAPTATION OF NOISE SPECTRUM TO TRANSFER FUNCTION

In a way similar to that described in § 4.6, the distribution of the noise contribution in $\mathbb{D}$ may be modified to form a favorable arrangement with respect to $H(u)$. However, the goal is now not to obtain the least possible noise power in $\mathbb{D}$, but rather to reshape the remaining noise in $\mathbb{D}$. The noise distribution should be optimized with regard to $H(u)$ and the total amount of noise inside $\mathbb{D}$ is not of primary interest.

With this presumption, a criterion is needed where $H(u)$ is considered. A proper requirement appears to be to create a binary image with

$$\min_{B(k)} \sigma_H^2 = \min_{B(k)} \left(\frac{\sum_{k \in \mathbb{D}_k} |H(k)Q(k)|^2}{\sum_{k \in \mathbb{D}_k} H(k)} \right), \quad (5.1)$$

which is not trivial to evaluate. However, it is clear to interpret; the closer the noise component can be confined to the boundary of $\mathbb{D}$, the more propitious are the conditions due to the shape of $H(u)$.

When the IFTA is applied, the operator in the frequency domain $\mathscr{P}$ must be changed to $\mathscr{P}_p$ as compared to eqs. (4.10) and (4.13). The multiplication of $Q(u)$ with a function $\xi(u)$ as in eq. (4.13) to form the noise spectrum is on the whole of no particular advantage, because the necessary noise power is not known in advance. Too strong restrictions are likely, and they will lead to convergence problems. A preferred procedure is to leave the adjustment to the algorithm itself. During the iteration cycle, the original value is introduced with the probability $p(u) \in [0, 1]$ or its discrete equivalent $p(k)$. During each cycle there is determined, at each position k in the spectrum, how high the probability is that the original value is introduced. With this probability there is executed

$$F^{(j)}(\boldsymbol{k}) = \mathscr{P}_p B^{(j)}(\boldsymbol{k}) = \begin{cases} F(\boldsymbol{k}), & z^{(j)}(\boldsymbol{k}) \leq p(\boldsymbol{k}), \\ B^{(j)}(\boldsymbol{k}), & z^{(j)}(\boldsymbol{k}) > p(\boldsymbol{k}), \end{cases} \quad (5.2)$$

where $z^{(j)}(\boldsymbol{k}) \in [0, 1]$ represents pseudo-random numbers.

The spectral noise distribution can be influenced by the probability distribution $p(\boldsymbol{u})$. Problems do not arise due to introduction of values in isolated positions; in each cycle other values are introduced at different locations. Moreover, the values in the spectrum are coupled by oversampling and the binary character of the image. A relatively smooth spectral noise distribution is created, the form of which can be controlled by $p(\boldsymbol{u})$.

Even in this case the requirements may be too strong, and the results may resemble that of a hardclip; i.e., the original values are introduced too frequently. To avoid this, the value of the integral has to be sufficiently small; i.e.,

$$\int_S p(\boldsymbol{u}) \, d\boldsymbol{u} < p_{\text{ref}}, \quad (5.3)$$

with some reference value p_{ref}. Enough noise contributions must be allowed.

The problem may be stated by the questions:

(i) In what way is the form of $Q(\boldsymbol{u})$ dependent on the shape of $p(\boldsymbol{u})$?

(ii) Which $p(\boldsymbol{u})$ is optimum in order to fulfill condition (5.1)?

It is possible to give empirical indications as a remark to these questions. The consideration of eq. (5.3) [i.e., when the value of $\int p(\boldsymbol{u}) \, d\boldsymbol{u}$ is sufficiently small], results in the observation that the spectral noise distribution behaves on the average like

$$Q(\boldsymbol{u}) \sim \frac{1}{p(\boldsymbol{u}) + p_0}, \quad (5.4)$$

where p_0 is a constant. This is demonstrated in fig. 35 for different radially symmetric probability distributions $p(\boldsymbol{u})$. Radial sections from the centrum outwards of $p(\boldsymbol{u})$ and $1/Q(\boldsymbol{u})$ are shown with values averaged in the azimuthal direction. The example in fig. 35a resembles the transfer function of eq. (4.1). It is evident that the curves of fig. 35 are similar when the offset p_0 and the different scaling are disregarded.

From these examples, it is clear that the expected spectral noise distribution is known from the chosen shape of the probability distribution $p(\boldsymbol{u})$.

Two examples of binarized images are reproduced here, whereby the $p(\boldsymbol{u})$ functions were chosen to be the transfer function $H(\boldsymbol{u})$ raised to the power

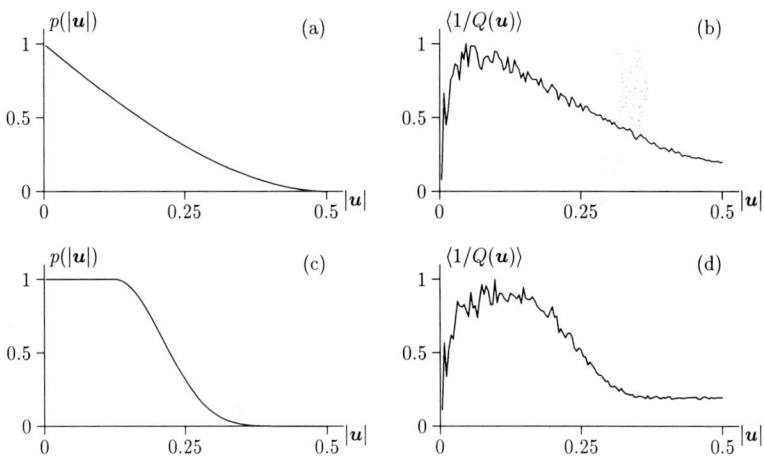

Fig. 35. Examples of the influence on the noise spectrum. A probability distribution $p(\mathbf{u})$ related to the transfer function $H(\mathbf{u})$ was introduced. With $p(\mathbf{u})$, it is possible to influence the spectral noise distribution $Q(\mathbf{u})$. With the $p(\mathbf{u})$ of (a), the distribution (averaged in the azimuthal direction) of (b) is obtained, and with (c) the curve (d).

κ. With the parameter κ it is possible to globally enhance or reduce the spectral noise distribution in order to satisfy the integral condition (5.3). The value of the integral must be small enough to allow the algorithm to converge, but not too small to ensure that only a low amount of noise is introduced. We look for the optimum value of κ for each given $p(\mathbf{u})$. Parameter κ is chosen to be small enough to just prevent black and white clusters of pixels forming in the image (i.e., that no appearances similar to hardclip occur), but large enough to minimize σ_H^2 as much as possible. This is performed empirically. An example is shown in fig. 36 for $\kappa = 1.25$.

The lowpass region is important, and it is possible to exclude spectral noise from it and allow the noise components to increase gradually towards the boundary of the spectrum. Probability functions for this case can be constructed with a constant central portion and a decreasing outer part. In the example given in fig. 35c, the following function was chosen:

$$p(\mathbf{u}) = \begin{cases} 1, & |\mathbf{u}| \leq u_s/4, \\ \left[\frac{1}{2} + \frac{1}{2}\cos\left\{\frac{4}{3}\pi\left(\frac{|\mathbf{u}|}{u_s} - \frac{1}{4}\right)\right\}\right]^\kappa, & u_s/4 < |\mathbf{u}| \leq u_s, \\ 0, & |\mathbf{u}| > u_s. \end{cases} \quad (5.5)$$

Fig. 36. Binarization using the iterative Fourier transform algorithm for the probability distribution $p(\boldsymbol{u})$ of fig. 35a: (a) binary image with 256×256 pixels; (b) version of (a) after filtering with $H(\boldsymbol{u})$; (c) quantization noise spectrum of (a) in reversed polarity.

An optimum was found for $\kappa = 4$ in this case. The corresponding binary image is shown in fig. 37.

To the question of which $p(\boldsymbol{u})$ is optimum for the σ_H^2, we may state the following. In the experiments we have described here as well as other experiments using the iterative Fourier transform algorithm, roughly the same minimum of σ_H^2 was obtained. Nevertheless, the visual appearances of the quantized images in the experiments performed show dissimilarities (cf. figs. 36 and 37). It seems that the condition (5.1) must be exercised with caution. The much simplified assumption that only the energy of the spectrally weighted noise is of significance seems apparently to be correct only in a limited sense.

At this point, the combined effects due to the image-comprehension operator $\mathscr{C}$ gain in influence (Roetling [1976b, 1977b], Naesaenen [1984]). Isolated image features become evident. Edges and details are perceived increasingly (Roetling [1976a]).

In the case where further, far-reaching adaptations and optimizations shall be performed, it is necessary to depart from the general consideration of the transfer function for modelling the quantization noise. Additional information of the image spectrum, the operator $\mathscr{C}$, and other factors must be integrated in the quantization process.

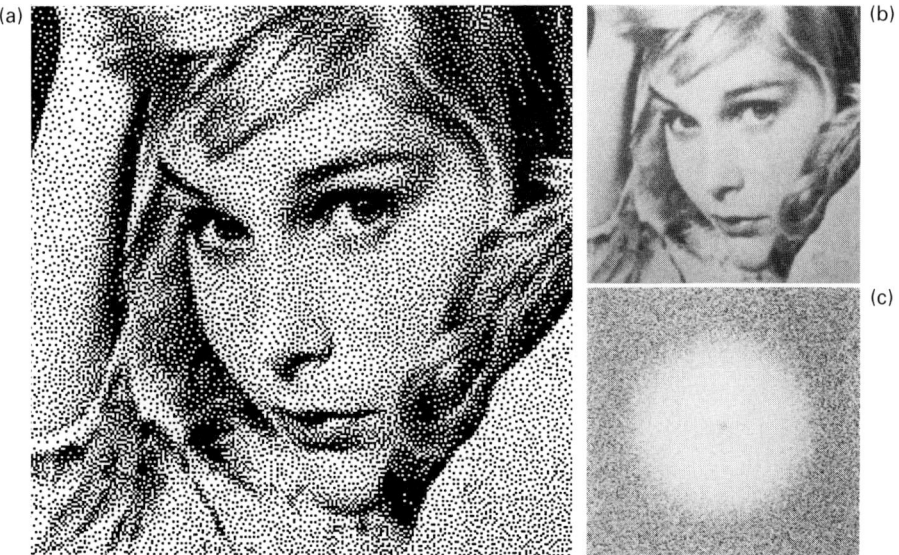

Fig. 37. Binarization using the iterative Fourier transform algorithm for the probability distribution $p(\boldsymbol{u})$ of fig. 35c: (a) binary image with 256×256 pixels; (b) version of (a) after filtering with $H(\boldsymbol{u})$; (c) quantization noise spectrum of (a) in reversed polarity.

5.2. ADJUSTMENT OF HALFTONE PROCEDURE TO IMAGE SPECTRUM

In the preceding section, it was shown that the transfer function $H(\boldsymbol{u})$ should not be the only criterion to be applied when the decoding region becomes larger than is allowed by $u_d \leqslant \alpha_{\max}(\mathscr{B})$. Images were found to be different despite the fact that they possessed comparable values of σ_H^2.

Another approach by which to come to a compromise between the desired information being retained and the noise being accepted is the following. The parts of the image that form important constituents of the image information and the noise spectrum are to be recognized. These characteristics are then to be preserved during the quantization procedure with or without consideration of the transfer function. Decisions regarding the importance of image constituents ought to be coupled to knowledge or assumptions about $\mathscr{C}$, because the image-forming part of the visual system alone is unable to decode the binary image in the desired fashion. With these precautions, the visual system may be able to extract the tendered information while $H(\boldsymbol{u})$ permits a major part of the spectrum to be transferred in a weighted form.

Some problems arise when trying to reach these goals. Identification and determination of essential parts of the image are difficult to define. Once defined, the quantization procedure must be carried out under fulfillment of the requirement of preserving the defined information.

The test object used throughout this review consists mainly of low-frequency components. This means that an adjustment of the quantization procedure to the image information in principle leads to the same goal as before; i.e., something like a lowpass control. Therefore, another test object is chosen here to demonstrate the apparent effects; viz., the synthetic cosine test object of fig. 38 with varying frequency in the horizontal direction and varying contrast in the vertical direction. This test object has a distinctly different character compared with the girl's face. In particular, high-frequency components form a significant part of the information. In the case where the object of fig. 38 is binarized irrespective of its content (e.g., an undisturbed lowpass is aimed at), then the higher frequencies of the image will be submerged in high-frequency noise.

For comparative judgement, we use the passive binarization procedure with a pyramidal carrier of period e. This example is shown in fig. 39. The convolution interpretation for the spectral noise distribution (cf. § 3) is manifested: viz., the spectrum possesses strong contributions along the x-axis. Due to the multiple convolutions with the spectral peaks of the carrier, stripes occur in the noise spectrum. The strong high-frequency noise disturbs the perception for the high-frequency portions of the image. This is clearly observed in the image and its decoded version in fig. 39.

We turn now to binarization techniques in which the object information is considered.

Fig. 38. Additional graytone test object used in § 5. The original is shown in (a) and after being filtered with $H(u)$ in (b).

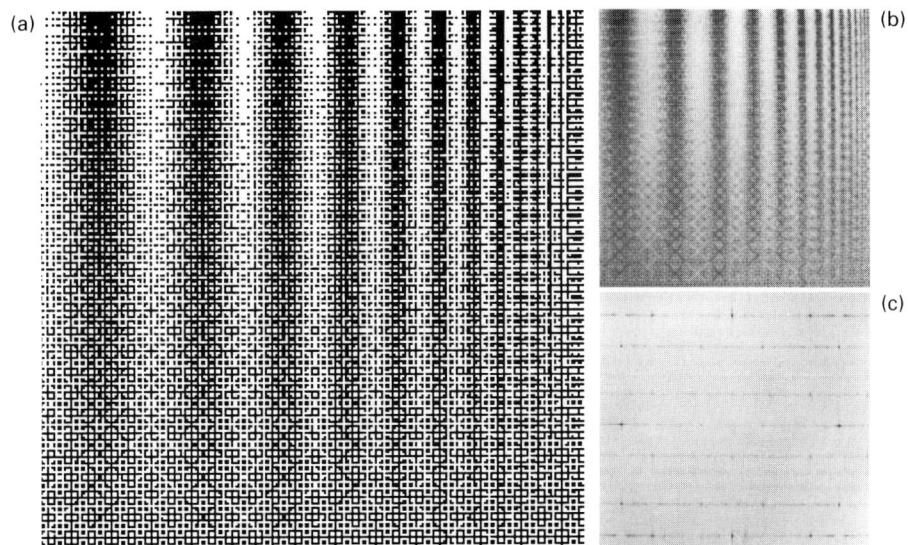

Fig. 39. Binarization using a pyramid carrier with $p_c = e$: (a) binary image with 256×256 pixels; (b) version of (a) after filtering with $H(u)$; (c) quantization noise spectrum of (a) in reversed polarity.

5.2.1. Error-diffusion procedure

A well-known binarization procedure where the object information is taken into account, even if only in an indirect way, is the error-diffusion method. This technique, extended by a filtering scheme as described in § 3, results in a strikingly good appearance of the quantized image.

The characteristics due to the Floyd and Steinberg [1975, 1976] error-diffusion algorithm include in particular two components: (i) an edge enhancement which is favorable in so far as the edges appear sharper (Knox [1989]), and (ii) a tendency to form regular textures in portions of weakly varying graylevels which result in a homogeneous appearance of gray areas. Both of these effects are noticeable in fig. 40. To the far right, black and white lines occur almost exclusively, due to the edge enhancement. To the very bottom, a chequered texture appears at the graylevel 0.5.

These effects may also lead to unwanted consequences:
  (i) high-contrast details which are not present in the original object may show up due to edge enhancement, and
  (ii) borders between regions of different textures may result in the formation of false contours.

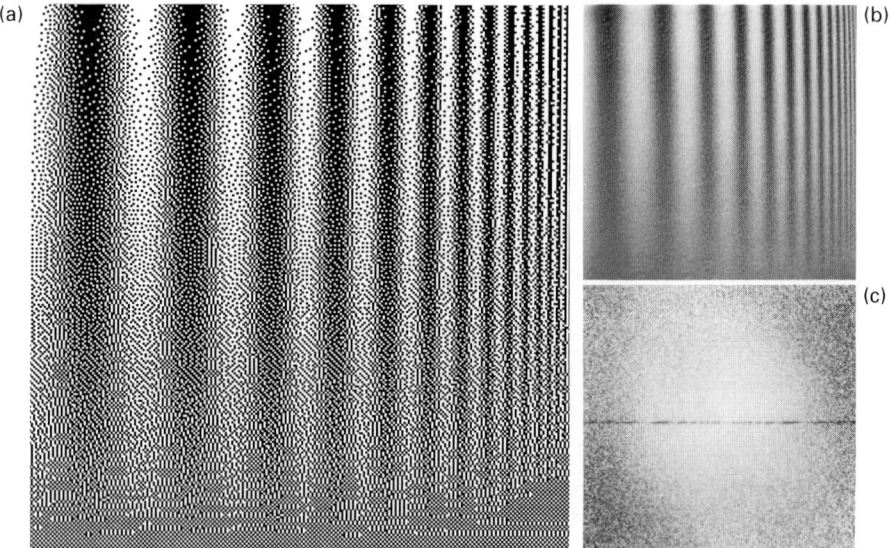

Fig. 40. Binarization using the original error-diffusion algorithm: (a) binary image with 256 × 256 pixels; (b) version of (a) after filtering with $H(u)$; (c) quantization noise spectrum of (a) in reversed polarity.

Although a chequered texture is not resolved, because its spectral-noise contributions are located in the corners of the spectrum and thus outside the support of H, its borders are easy to recognize in the decoded image, and have a disturbing appearance.

5.2.2. Iterative procedure

The iterative procedures discussed up to now are particularly well suited for imprinting special characteristics onto the binary image. This implies that the possibility exists of introducing suitable spectral boundary conditions to assure significant image features.

An observation of plausible, effective consequences is that frequency components of high amplitude are important for the appearance of the image, and may serve to identify portions of the spectrum which contain important information about the image (Scheermesser, Broja and Bryngdahl [1993a]). The procedure is now analogous to the control of the lowpass area, namely to check all spectral values which possess adequately high amplitudes.

By defining a threshold t_a, we obtain the following expressions of the control areas in which the amplitude is higher and lower than t_a (see fig. 41):

$$\Omega_a = \{k \mid |F(k)| \geq t_a\}, \qquad \Omega_n = \{k \mid |F(k)| < t_a\}. \tag{5.6}$$

The operator $\mathscr{P}$ [cf. eq. (4.10)] is here changed to $\mathscr{P}_a$:

$$F^{(j)}(k) = \mathscr{P}_a B^{(j)}(k) = \begin{cases} F(k), & k \in \Omega_a, \\ B^{(j)}(k), & k \in \Omega_n. \end{cases} \tag{5.7}$$

In the image space the same operator $\mathscr{Q}_r$ can be used as before. This is one possible suggestion. For example, among other choices one could regard very low or even zero amplitudes in case they were significant for the observed property of the image.

In the example presented, the size of Ω_a is determined by t_a. Here a restriction on a maximum is given as before. However, the restricted maximum is not a maximum frequency, but rather a maximum area in the spectrum. It was found empirically that the control area should not exceed 30–40% of the spectrum.

An example of the application of this procedure is shown in fig. 42. The cosine object was binarized with the operator $\mathscr{P}_a$ according to eq. (5.7). The control area is depicted in fig. 43 for a specific choice of $\mathscr{P}_a$. In this case, an area of 30% of the spectrum is controlled. Obviously, the horizontal axis and its close surroundings are controlled along its full length. Further, portions exist where only every second point (e.g., along the vertical axis), or every fourth point is controlled. Nevertheless, from the noise spectrum it is evident not only that the noise components in these controlled points decline, but also that homogeneous portions are formed in the noise spectrum.

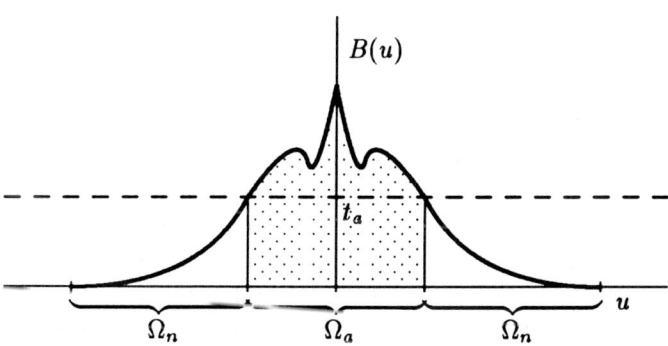

Fig. 41. Introduction of control areas dependent on the amplitude in object spectrum.

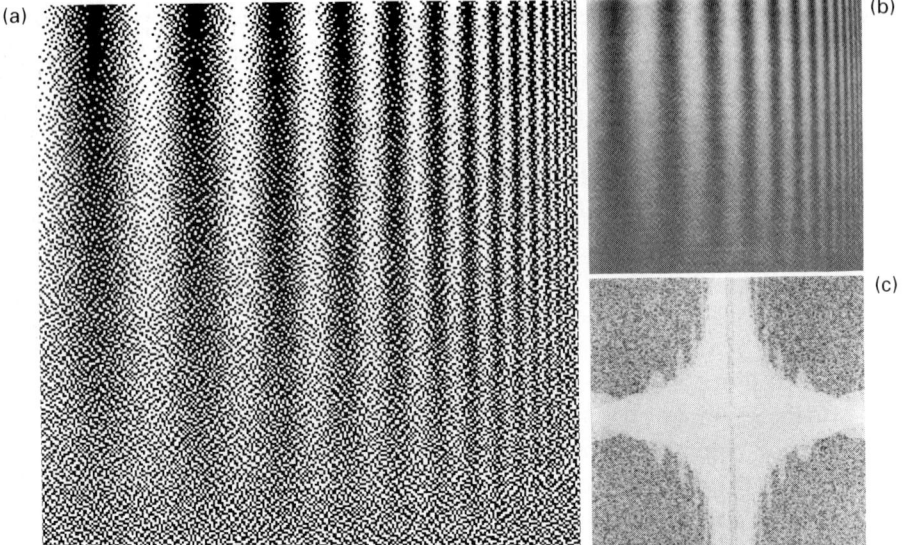

Fig. 42. Binarization using the iterative Fourier transform algorithm with control area according to fig. 43: (a) binary image with 256 × 256 pixels; (b) version of (a) after filtering with $H(u)$; (c) quantization noise spectrum of (a) in reversed polarity.

The reason for the homogeneous, subsided noise components is to be found in the systems of equations of $B(u)$ which are obtained from the oversampling and the binarization (Scheermesser, Broja and Bryngdahl [1993b]). A linear system is attained in the former, and a nonlinear system of the values of the points of the spectrum in the latter case. This implies that a coupling between these values exists which is of a local nature. From this, it follows that by controlling a few values the points in between are implicitly controlled too.

The effects in the image are shown in fig. 42. The cosine variation is reproduced relatively faithfully up to the highest frequencies. In contrast to a binarization with the error-diffusion technique, neither edge enhancement nor texture alterations occur. Instead, slightly coarser structures are the result of the noise components of low frequency in the diagonal directions. This is the consequence of the redistribution of the controlled area from being a centered lowpass to becoming one with an adapted form.

We conclude that it is possible to adapt the quantization procedure to the image and its spectrum to be able to exploit existing freedoms with the goal of conveying as much information as possible from the graytone object to the binary image.

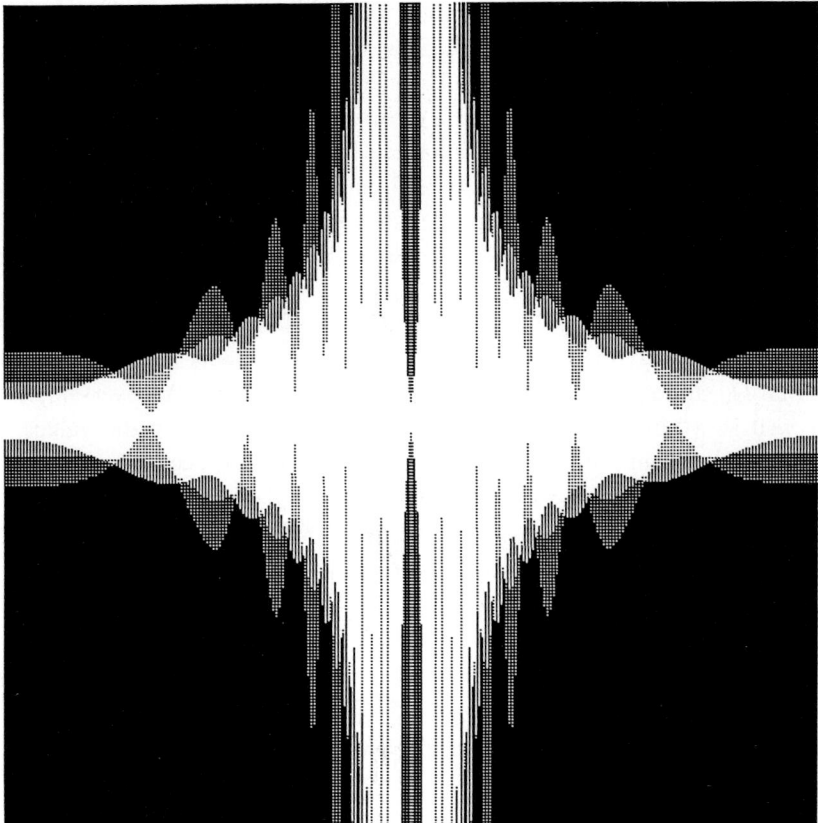

Fig. 43. Control area of spectrum for the example of fig. 42.

5.2.3. Control of phase in spectrum

It is known that the information contained in the phase of the image spectrum is of particular significance in specifying the edges and the details of the image (Oppenheim and Lim [1981]). This renders possible the application of another type of boundary condition in the spectrum; i.e., to adjust the phase of the spectrum of the binarized image to that of the original object (Broja, Wyrowski and Bryngdahl [1993]). The phase is then controlled in a certain region in addition to the region in which the amplitude and phase are checked. In this way the unavoidable quantization noise can be introduced in the spectral amplitude while the spectral phase is retained. This allows for an extensive adaptation to features of the image spectrum.

The result is an enhancement of edges and details of the image, because in the corresponding locations the phase is no longer random but adapted to the original object via the phase. The effect is similar to the edge-enhancement characteristic obtained by the error-diffusion procedure.

The strength of the edge enhancement can be influenced by a relative change of the controlled areas. In addition to Ω_a and Ω_n (see § 5.5.2), an area Ω_p is defined in which the phase of the spectrum of the binarized image will be equal to the phase of the spectrum of the original object. This can be accomplished with a threshold $t_p \leq t_a$ (Scheermesser, Broja and Bryngdahl [1993a]) according to § 5.2.2, or as a bandpass, a lowpass, or something else.

It is found that the introduction of noise components in the amplitude within Ω_p can only be performed restrictively (cf. § 5.2.2). An oversampling as well as a binarization leads to a local coupling between the values of neighboring points of the spectrum. In Ω_p a coupling occurs between the amplitude and the phase values. The control of the phase within Ω_p comprises implicitly the control of the amplitudes within Ω_p. The coupling may be weakened by giving up the oversampling. Then the noise components will increase within Ω_p for the same, or even a better, adaptation of the phase.

A possible realization is to modify $\mathscr{P}_a$ of eq. (5.7) to $\mathscr{P}_p$:

$$F^{(j)}(\mathbf{k}) = \mathscr{P}_p B^{(j)}(\mathbf{k}) = \begin{cases} F(\mathbf{k}), & \mathbf{k} \in \Omega_a, \\ |B^{(j)}(\mathbf{k})| e^{i\phi(k)}, & \mathbf{k} \in \Omega_p, \\ B^{(j)}(\mathbf{k}), & \mathbf{k} \in \Omega_n, \end{cases} \quad (5.8)$$

with

$$F(\mathbf{k}) = |F(\mathbf{k})| e^{i\phi(k)},$$

and

$$B^{(j)}(\mathbf{k}) = |B^{(j)}(\mathbf{k})| e^{i\psi(k)}.$$

Otherwise, the same iterative Fourier transform algorithm is used as described above.

An example is shown in fig. 44, which was aimed intentionally to turn out rather extreme to illustrate the effect. Ω_a consisted of a circular lowpass area equal to 5%, and Ω_p an annular bandpass equal to 20% of the spectrum (see fig. 45). The edge enhancement is seen clearly in the binary and decoded images. In comparison with images in § 4.4 (fig. 31), where a circular lowpass

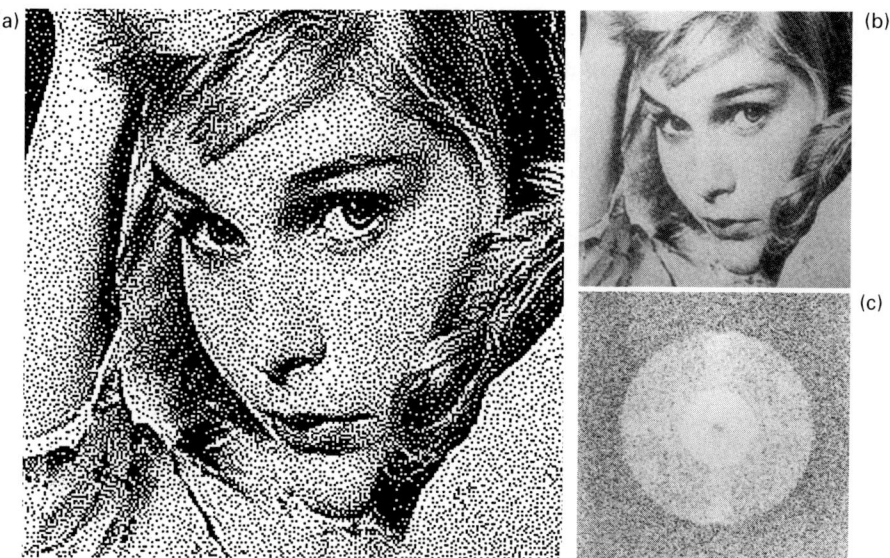

Fig. 44. Binarization using the iterative Fourier transform algorithm with control area Ω_a for amplitude and phase and Ω_p for phase (see fig. 45): (a) binary image with 256 × 256 pixels; (b) version of (a) after filtering with $H(u)$ for $u_d = u_s$; (c) quantization noise spectrum of (a) in reversed polarity.

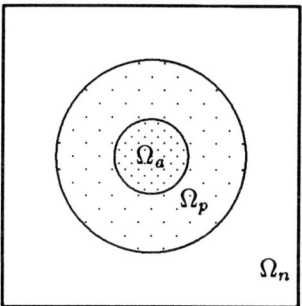

Fig. 45. Control areas of spectrum for example of fig. 44.

area of comparable size was applied, the difference is noticeable. However, the decoding performed in fig. 44 was different from the one used in the previous sections. Here the decoding was executed with a transfer function which included almost the total spectrum (viz., $u_d = u_s$), and not only the lowpass area, because the intention now was different. We are not interested in a lowpass reproduction, but rather in emphasizing an intended selection of information and its effect on the decoded image. Further, the amplitudes of the noise components are noticeably lower within Ω_p than within Ω_n, due to the coupling between amplitude and phase.

We have seen that it is possible to use spectral phase control in the binarization of images. The phase information due to edges and details can be incorporated in the quantization procedure to enhance edges. A sensation of improvement of image acuity can be procured. It is possible to influence this effect by the choice of the operator $\mathscr{P}_p$.

In this section, we have commented on the adaptation of the quantization process to the image spectrum to indicate the feasibility of digression from the lowpass criterion. Some examples were presented. Of course, many additional possibilities and combinations exist. They are all based on assumptions about $\mathscr{C}$; e.g., to favor details, capability to suppress noise optically present, etc.

§ 6. Halftone Resolution in Excess of Sampling Raster

Here we address the situation where the support of the transfer function of the detection or viewing system is larger than the innermost period of the spectrum of the binarized image; i.e., $u_d > u_s$ (see fig. 16). Portions of the repeated spectrum will then influence the decoded image. Some important aspects which are known to occur in this situation will only be mentioned. The questions and topics under consideration are rather complex. The characteristics of the applied equipment now play an important role for the formation of the resolved individual dots of the image. Specific aspects of the apparatus and their influence on the quantization process must be taken into account.

6.1. PERCEPTION OF HALFTONE DOTS

The sampling in the spatial domain results in a repetition of the spectrum. A support of the transfer function $H(u)$ larger than the innermost period of the repeated spectrum has the consequence that neighboring repetitions contribute to the perceived image. Individual pixels will now be observed, contrasting with the previous situation in which an average over several neighboring pixels was formed and detected.

In the ideal case, sampling implies that δ-distributions are considered in the sampling points in the spatial domain. A filtering with $H(u)$ causes the δ of each point to be convolved with the impulse response of the eye, $h(x)$;

i.e., a resemblance to a somb²-function (Gaskill [1978]) is obtained instead of a δ-distribution. Thus, the individual pixels are blurred.

Examples are shown in figs. 46. A 32×32 pixel section of the binary image of fig. 44 is depicted. The section reproduced is enlarged strongly to render apparent the shape of the individual somb-distributions which would be formed on the retina when regarding the shape of the impulse response $h(x)$. In fig. 46a, the cutoff frequency of the transfer function is three times larger than the sampling frequency; i.e., in the u- and v-directions, one extra period is contributing additionally. The somb-functions are here rather extended, and neighboring halftone dots start to overlap. In fig. 46b, the

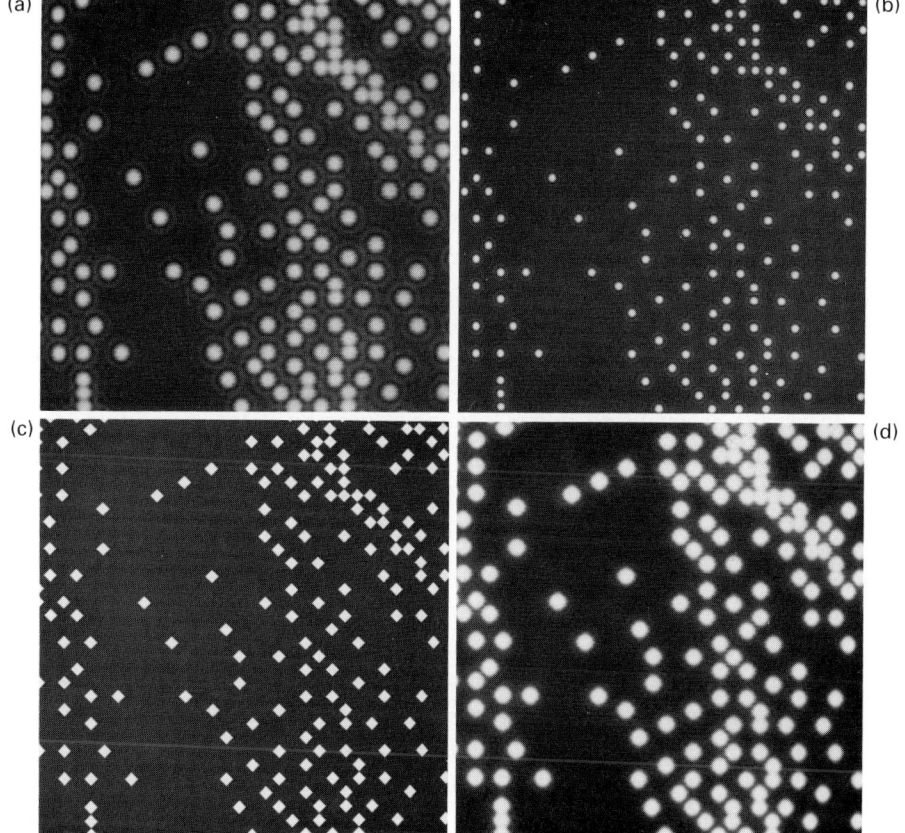

Fig. 46. Examples for the case when $u_d > u_s$. In (a), $u_d = 3u_s$, and in (b), $u_d = 6u_s$ for δ dots. In (c), $u_d \to \infty$, and in (d), $u_d = 4u_s$ for diamond-shaped dots.

cutoff frequency is six times the sampling frequency. The dots are now almost isolated; i.e., spatially separated.

6.2. FORMATION OF HALFTONE DOTS

In reality, no δ-distributions occur, but extended dots of certain shapes do. The actual dot-intensity profile depends on the process by which it is formed. For example, the dots may vary from being Gaussian-shaped on CRTs, in prints and in photographs, to square pixels on LCD displays.

In the spatial domain, the pixel is further convolved with the impulse response. The pixel form is more or less preserved. The higher the resolution of the image-forming system [i.e., the larger the support of the transfer function $H(u)$], the better the shape of the pixel is perceived. The effect is illustrated in figs. 46c and d, where the same image section as above is presented. Diamond-shaped dots are employed. In fig. 46c, an infinitely extended transfer function is applied. In fig. 46d, the cutoff frequency of the transfer function is four times the sampling frequency, and overlap between smeared neighbor pixels is evident.

The shape of the halftone dots implies, in the frequency domain, a modulation of the total spectrum with the Fourier transform of the dot profile. For example, a square shape results in a global $sinc^2$- and a circular dot in a global $somb^2$-variation. In principle, this modulation can be taken into account in the design of the quantization procedure (in particular for the IFTA).

Especially when using image-forming processes of low resolution, the shape of the halftone dots is important for the perceived impression, because the dot profile has a considerable influence on the information contained in the image.

6.3. OVERLAP OF HALFTONE DOTS

For several image-forming apparatuses and processes, the halftone dot size does not conform to the sampling raster (Roetling and Holladay [1979], Allebach [1980], Goertzel and Thompson [1987]). It may even be necessary to allow dot overlap when the dot shape does not permit formation of a mosaic. Otherwise, no complete solid areas can be produced in the image.

The consequence of the overlap of neighboring halftone dots is different depending on the dot arrangement. The relative positions of the dots influ-

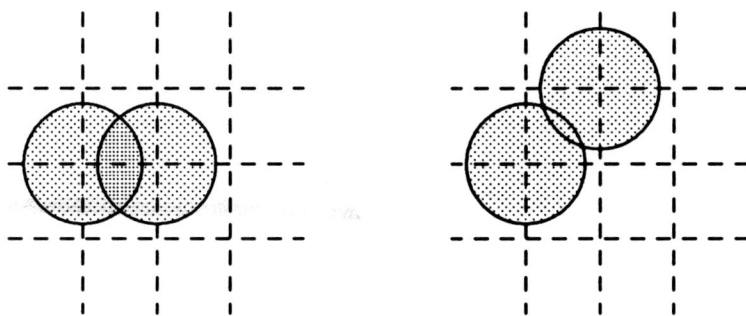

Fig. 47. Dot-overlap dependence on dot location.

ence the overlap; e.g., dots in points located diagonally in a square raster and neighboring dots in general overlap differently (see fig. 47). Furthermore, the intensity of overlapping halftone dot profiles is in general not addable, but nonlinear due to saturation and other effects. This leads to a dependence of the average intensity distribution on the texture configuration.

6.4. ERRORS IN POSITION AND SHAPE OF HALFTONE DOTS

The assumed defined position and shape of the halftone dots in the quantization procedure is a mathematical abstraction. In reality, any apparatus or process possesses a limited precision which will cause errors (Melnychuck and Shaw [1988]). Statistical fluctuations exist in the position, size and shape of the halftone dots which may also lead to fluctuations in the overlap of neighboring dots. Furthermore, the halftone dots are frequently made up of a statistical distribution of many microscopic elements. To improve the description of the halftone procedure in this respect, the corresponding parameters would have to be treated as stochastic variables.

In this section, some aspects have been referred to which become relevant when the resolution of the procedure is of the order of, or in excess of, the sampling period. In principle, these effects occur for $u_d < u_s$ as well. However, the lower the cutoff frequency u_d of the detector, the less the influence of the halftone dot shape will be. On the other hand, dot overlap is also an important factor when $u_d < u_s$, because of its effect on the resulting intensity distribution. The features listed are important to consider when specific characteristics of apparatus and processes influence the binarization procedure.

§ 7. System Considerations of Image Quantization Gain in Importance

The application of digital means to perform image quantization indicated a powerful and diversified development trend with no end in sight. The new additional ingredients and computational possibilities turned out to add immense flexibilities and extra aspects in almost all respects from the technological at one end of the spectrum to the artistic features at the other. The realization of a high degree of algorithm sophistication and complexity has been, or is, within reach.

The field of image synthetization and quantization expanded to an additional dimension when it became digital. Image processing can be included conveniently in the quantization algorithm which may consider different systems aspects. This all means that the field of image quantization has undergone an extensive expansion, which is mirrored in this overview.

The chosen presentation is closely related to factors and phenomena which form the specific model, where image quantization is an integrated link in an image-processing system. The description presented has ingredients and considerations which are connected and linked closely to the physical, and in particular, the optical characteristics of the system under examination, which in this overview was almost always the visual system.

In this overview, image binarization is regarded as a part of a defined system covering the process from an original input object to a detected and perceived image distribution. The major portion of the desired, undisturbed portion of the transmitted information is then made up of the lowpass part of the electronically halftoned image. Subjective consideration of image-quality criteria has been suppressed or disregarded intentionally. Factors related to specific applications and certain types of object information have not been examined particularly. Instead, we have tried to emphasize existing or introduced degrees of freedom and extra free parameters to allow control of, or influence on, various image properties and algorithm characteristics. The local and global influence of pixel distributions connected with sampling and quantization has been considered. As far as possible, the same object was used throughout in the examples. A compromise in the number of pixels used (256×256) was made in order to show algorithm peculiarities and the influence on different image properties; i.e., the individual pixels are still resolvable.

The spectrum of digital image-quantization algorithms which have been developed in the last three decades opens a large range from sequential, pixel-wise processing to sophisticated iterative methods based on extensive calculations. The variety of these methods is growing constantly by incorporation of additional powerful computations, additional ideas, new suggestions, and interesting inventions. There exists an almost unlimited number of possibilities to extend and modify the methods described in this review.

The situation treated here has been limited to quantization of graytone images. The considerations and evaluations in this particular processed feature also demonstrate the principles which can be applied to influence and modify other image parameters and properties as well. The competing effects which occur frequently in image processing are illustrated here by the opposing requirements and desire to optimize detail resolution as well as graytone rendition.

In this overview we have presented examples from major classes of algorithms which have been suggested in the field of digital electronic halftoning. In actual applications, practicality and compatibility aspects must be considered. The selected algorithm must be developed and optimized to suit the particular situation. The specific requirements, needs and tolerances will in general narrow the choice of a suitable algorithm.

In more general type of processing, further parameters, aspects and variables like multi-level quantization (Kurosawa and Kotera [1989], Broja, Michalowski and Bryngdahl [1990], Zeggel, Weissbach and Bryngdahl [1993]), color (Heckbert [1982], Gentile, Walowit and Allebach [1990], Miller and Sullivan [1990], Dixit [1991], Ling and Just [1991], Kolpatzik and Bouman [1992], Wu [1992]), and temporal variations must be included. This multi-dimensional extension is in no way trivial. However, similarities between the ingredients and algorithms exist, and resemble those of the digital binarization of graytone images presented here.

Acknowledgements

We thank Ingo Kummutat and Michael Tluk for their professional help with the illustrations.

References

Algie, S., 1983, Comp. Vision & Image Proc. **24**, 329.
Allebach, J.P., 1978, Photogr. Sci. Eng. **22**, 89.
Allebach, J.P., 1979, J. Opt. Soc. Am. **69**, 869.
Allebach, J.P., 1980, Appl. Opt. **119**, 2513.
Allebach, J.P., 1981, in: Image Quality, Proc. SPIE **310**, 151.

Allebach, J.P., and B. Liu, 1976, J. Opt. Soc. Am. **66**, 909.
Allebach, J.P., and B. Liu, 1977, J. Opt. Soc. Am. **67**, 1147.
Allebach, J.P., and R.N. Stradling, 1979, Appl. Opt. **18**, 2708.
Anastassiou, D., and S. Kollias, 1988, in: Visual Communication & Imaging, Proc. SPIE **1001**, 1062.
Anastassiou, D., and K.S. Pennington, 1982, IBM J. Res. Dev. **26**, 687.
Bayer, B.E., 1973, in: IEEE Int. Conf. on Communications, Vol. 1, pp. 26–11.
Billotet-Hoffmann, C., and O. Bryngdahl, 1983, Proc. SID **24**, 253.
Broja, M., and O. Bryngdahl, 1993, J. Opt. Soc. Am. **10**, 554.
Broja, M., R. Eschbach and O. Bryngdahl, 1986, Opt. Commun. **60**, 353.
Broja, M., K. Michalowski and O. Bryngdahl, 1990, Opt. Commun. **79**, 280.
Broja, M., F. Wyrowski and O. Bryngdahl, 1989, Opt. Commun. **69**, 205.
Broja, M., F. Wyrowski and O. Bryngdahl, 1993, Opt. Commun. **95**, 205.
Bryngdahl, O., 1973, J. Opt. Soc. Am. **63**, 1098.
Bryngdahl, O., 1978, J. Opt. Soc. Am. **68**, 416.
Carlsohn, M.F., and P.W. Besslich, 1984, in: Proc. VIIth Conf. on Applications of Digital Image Processing, Proc. SPIE **504**, 303.
Carnevali, P., L. Coletti and S. Patarnello, 1985, IBM J. Res. Dev. **29**, 569.
Daels, K., R.L. Easton and R. Eschbach, 1992, Opt. Commun. **87**, 93.
Dalton, J.C., G.R. Arce and J.P. Allebach, 1982, Proc. SPIE.
Dixit, S.S., 1991, Computer & Graphics **15**, 561.
Eschbach, R., 1990, J. Opt. Soc. Am. **7**, 708.
Eschbach, R., and R. Hauck, 1984, Opt. Commun. **52**, 165.
Eschbach, R., and R. Hauck, 1985, Opt. Commun. **54**, 71.
Eschbach, R., and R. Hauck, 1987a, Opt. Commun. **62**, 300.
Eschbach, R., and R. Hauck, 1987b, J. Opt. Soc. Am. **4**, 1873.
Eschbach, R., and K.T. Knox, 1991, J. Opt. Soc. Am. **8**, 1844.
Fan, Z., 1993, J. Electron. Imag. **2**, 62.
Fetthauer, F., S. Weissbach and O. Bryngdahl, 1992, Opt. Commun. **94**, 44.
Floyd, R., and L. Steinberg, 1975, SID Int. Sym. Digest of Tech. Papers 36.
Floyd, R., and L. Steinberg, 1976, Proc. SID **17**, 75.
Gard, R.L., 1976, Comp. Graphics & Image Proc. **5**, 151.
Gaskill, J.D., 1978, Linear Systems, Fourier Transforms, and Optics (Wiley, New York).
Gentile, R.S., E. Walowit and J.P. Allebach, 1990, J. Opt. Soc. Am. **7**, 1019.
Goertzel, G., and G.R. Thompson, 1987, IBM J. Res. Dev. **31**, 2.
Goodman, J.W., 1968, Introduction to Fourier Optics (McGraw-Hill, New York).
Heckbert, P.S., 1982, in: Proc. ACM Siggraph '82 Conf., Computer Graphics **16**, 297.
Holladay, T.M., 1980, Proc. SID **21**, 185.
Hopfield, J.J., 1982, Proc. Natl. Acad. Sci. USA **79**, 2554.
Hsueh, Y.C., M.G. Chern and C.H. Chu, 1991, Comp. & Graphics **15**, 397.
Jarvis, J.F., C.N. Judice and W.H. Ninke, 1976, Comp. Graph. & Im. Proc. **5**, 13.
Judice, C.N., J.F. Jarvis and W.H. Ninke, 1974, Proc. SID **14/15**, 161.
Just, D., and O. Bryngdahl, 1987, Opt. Commun. **64**, 23.
Just, D., and O. Bryngdahl, 1988, Opt. Commun. **68**, 179.
Just, D., R. Hauck and O. Bryngdahl, 1986, Opt. Commun. **60**, 359.
Karim, M.A., and H.-K. Liu, 1980, Opt. Lett. **5**, 132.
Kermisch, D., and P.G. Roetling, 1975, J. Opt. Soc. Am. **65**, 716.
Kirkpatrick, C., D. Gelatt Jr and M.P. Vecchi, 1983, Science **220**, 671.
Klensch, R.V., D. Meyerhofer and J.J. Walsch, 1970, RCA Rev. **31**, 517.
Knowlton, K., and L. Harmon, 1972, Comp. Graphics & Image Proc. **1**, 1.
Knox, K.T., 1989, Edge enhancement in error diffusion, SPSE Annual Meeting, Boston.

Knox, K.T., and R. Eschbach, 1993, J. Electron. Imag. **2**, 185.
Knuth, D.E., 1987, ACM Trans. Graphics **6**, 245.
Kolpatzik, B.W., and C.A. Bouman, 1992, J. Electron. Imag. **1**, 277.
Kurosawa, T., and H. Kotera, 1989, SID Japan Display, 616.
Kurosawa, T., H. Tsuchiya, Y. Maruyama, H. Ohtsuka and K. Nakazato, 1986, in: Proc IEEE 2nd Int. Conf. on Image Processing, p. 379.
Limb, J.O., 1969, Bell Sys. Tech. J. **48**, 2555.
Ling, D., and D. Just, 1991, in: Image Processing Algorithms & Techniques, Proc. SPIE **1452**, 10.
Lippel, B., 1976, Proc. SID **17**, 115.
Lippel, B., and M. Kurland, 1971, IEEE Trans. Commun. **COM-19**, 879.
Melnychuk, P., and R. Shaw, 1988, J. Opt. Soc. Am. **5**, 1328.
Miller, R., and J. Sullivan, 1990, in: SPSE's 3rd Ann. Conf., p. 149.
Mitsa, T., and K.J. Parker, 1992, J. Opt. Soc. Am. **9**, 1920.
Mrusek, R., M. Broja and O. Bryngdahl, 1990, Opt. Commun. **75**, 375.
Mrusek, R., D. Just and O. Bryngdahl, 1988, Opt. Commun. **67**, 16.
Naesaenen, R., 1984, IEEE Trans. Syst. Man & Cyber. **SMC-14**, 920.
Oppenheim, A.V., and J.S. Lim, 1981, Proc. IEEE **69**, 529.
Peli, E., 1991, J. Opt. Soc. Am. **8**, 625.
Rao, T.S., and G.R. Arce, 1988, J. Opt. Soc. Am. **5**, 1502.
Roberts, L.G., 1962, IRE Trans. Inf. Theory **IT-8**, 145.
Roetling, P.G., 1976a, J. Opt. Soc. Am. **66**, 985.
Roetling, P.G., 1976b, Proc. SID **17**, 111.
Roetling, P.G., 1977a, Phot. Sci. Eng. **21**, 60.
Roetling, P.G., 1977b, SID Int. Symp. Dig. Tech. Papers 8.
Roetling, P.G., and M. Holladay, 1979, J. Appl. Phot. Eng. **5**, 179.
Rolleston, R.J., and S.J. Cohen, 1992, J. Electron. Imag. **1**, 209.
Sahoo, P.K., S. Soltani, K.C. Wong and Y.C. Chen, 1988, Comp. Vis. Graph. and Image Proc. **41**, 233.
Scheermesser, T., M. Broja and O. Bryngdahl, 1993a, J. Opt. Soc. Am. **10**, 412.
Scheermesser, T., M. Broja and O. Bryngdahl, 1993b, Opt. Commun. **99**, 264.
Scheermesser, T., F. Wyrowski and O. Bryngdahl, 1993, Halftoning using two-dimensional carriers with noninteger period, J. Electron. Imag., submitted.
Schroeder, M.R., 1969, IEEE Spectrum **6**, 66.
Seldowitz, M.A., J.P. Allebach and D.W. Sweeney, 1987, Appl. Opt. **26**, 2788.
Stevenson, R.L., and G.R. Arce, 1985, J. Opt. Soc. Am. **2**, 1009.
Stoffel, P., and J.F. Moreland, 1981, IEEE Trans. Commun. **COM-29**, 1898.
Stucki, P., 1975, SID Symp., Digest of Papers, p. 34.
Stucki, P., 1981, A multiple-error correction computation algorithm for bi-level hard copy reproduction, Ph.D. Thesis (University of Zurich).
Sullivan, J., R. Miller and G. Pios, 1993, J. Opt. Soc. Am. **10**, 1714.
Sullivan, J., L. Ray and R. Miller, 1991, IEEE Trans. Syst. Man & Cyber. **SMC-21**, 33.
Talbot, W.F., 1852, Improvements in the art of engraving, British Patent No. 565.
Tuttass, T., and O. Bryngdahl, 1993, Opt. Commun. **99**, 25.
Ulichney, R.A., 1987, Digital Halftoning (MIT Press, Cambridge, MA).
Ulichney, R.A., 1988, Proc. IEEE **76**, 56.
Van den Bulck, P., 1992, IEEE Benelux & ProRISC Proc. of the Workshop on Cinc., Syst. and Signal Proc., 271.
Velho, L., and J. de Miranda Gomes, 1991, Comp. Graphics **25**, 81.
Weissbach, S., and F. Wyrowski, 1992a, Appl. Opt. **31**, 2518.
Weissbach, S., and F. Wyrowski, 1992b, Opt. Commun. **93**, 151.
Witten, I.H., and R.M. Neal, 1982, IEEE Trans. Comp. Graphics & Appl. **2**, 47.
Wu, X., 1992, ACM Trans. Graph. **11**, 348.
Zeggel, T., S. Weissbach and O. Bryngdahl, 1993, Opt. Commun. **100**, 67.

AUTHOR INDEX FOR VOLUME XXXIII

A

Abramovich, B.S. 40, *121*
Abramowitz, M. 218, 220, 224, *259*
Aceves, A.B. 205, 225, 227, 236, 238, 240, 257, *259*
Acklin, B. 258, *259*
Agarwal, G.S. 132, 162, 179, 180, 182, 183, 197, *198*
Agrawal, G.P. 205, 206, 210, 229, 238, *259*
Akhmanov, S.A. 131, *198*
Algie, S. 400, *461*
Allebach, J.P. 400, 405, 406, 408, 412, 413, 458, 461, *461–463*
Ambartsumian, V.A. 4, *121*
Ameer, G.A. 371–373, *385*
Anastassiou, D. 406, 413, 417, *462*
Anderson, P.W. 5, 47, *121*
Andrews, L.C. 351, *387*
Angel, J.R.P. 372, *383*
Angel, R. 372, *386*
Anzygina, T.N. 5, 63, *121*
Apostol, I. 264, 284, 286, *316*
Apresyan, L.A. 3, *121*
Aragone, C. 144, *198*
Arce, G.R. 406, 412, *462, 463*
Asch, M. 79, *121*
Ashcroft, N.W. 208, 211, 214, 231, 232, 236, *259*
Ayers, G.R. 379, *383*

B

Babkin, G.I. 3, 4, 9, 19, 30, 33, 46, 54, 71, *121, 122*
Bachor, H.A. 264, *313*
Bakut, P.A. 370, *384*
Ban, M. 135, *198*
Bandilla, A. 148, 156, 180, 182, *198, 201*
Banerjee, A. 171, *198*
Barabanenkov, Yu.N. 14, 15, *122*

Barclay, H.T. 371, 372, *387*
Barnett, S.M. 151, 152, 155, 176, *198, 200*
Barr, L.D. 279, 302, 303, *313*
Bartelt, H. 329, 378, *384*
Barthelemy, A. 132, *199*
Bates, R.H.T. 329, 364, 382, *384, 386, 387*
Bayer, B.E. 394, 406, *462*
Becker, F. 264, *313*
Bellman, R. 4, *122*
Ben Yosef, N. 349, *388*
Beran, M.J. 337, 339, 341, 342, 344, 346–348, 352, 358, 365, 366, 377, 381, *384, 385, 387, 388*
Bermejo, F.J. 175, *199*
Bertolotti, M. 178, 187, 190, *199–201*, 360, 361, *384*
Besslich, P.W. 406, *462*
Białynicka-Birula, Z. 164, *199*
Biedermann, K. 292, 293, *313*
Bilbault, J.M. 258, *259*
Billotet-Hoffmann, C. 412, *462*
Bishop, R.F. 164, *201*
Björk, G. 132, 194, 198, *202*
Blow, K.J. 160, 164, *199*
Bolstad, J.O. 360, 363, *388*
Bone, D.J. 264, *313*
Bone, P.M. 279, 302, *313*
Boone, P.M. 283, *313*
Born, M. 219, *259*, 358, *384*
Bouman, C.A. 461, *463*
Brandt, G.B. 283, *313*
Brangaccio, D.H. 264, *313*
Braunstein, S.L. 165, 170, *199*
Brisudová, M. 173, *199*
Broeke, B.R. 371–373, *385*
Broja, M. 394, 400, 402, 412, 413, 415, 417, 450, 452–454, 461, *462, 463*
Bronstein, A. 348, *386*
Brown, D.S. 360, *384*

Brown, G.M. 264, *313*
Brown, N. 264, *314*
Brown, T.G. 206, 253–255, 257, *260*
Browne, S.L. 371–373, *385*
Bruning, J.H. 264, *313*, *314*
Bryngdahl, O. 394, 400, 402, 406–409, 411–413, 415, 417, 427, 428, 432, 450, 452–454, 461, *462*, *463*
Buffington, A. 373, 374, *384*, *387*
Bugrov, A.G. 29, 69, 72, *122*
Burridge, R. 45, 77, 79, 113, *122*, *126*
Butters, J.N. 291, 293, 294, *313*
Butusov, N. 266, *315*
Bužek, V. 157, 162, 164, 165, 171, 176–178, 190, *199*, *202*
Byrd, P.F. 218, 220, *259*

C

Cada, M. 258, *259*
Calaba, R. 4, *122*
Callaway, J. 237, *259*
Cannon, T.M. 364, 369, *384*
Capinsky, W.S. 258, *259*
Carlsohn, M.F. 406, *462*
Carnevali, P. 413, *462*
Carpio-Valadez, J.M. 264, *315*
Carré, P. 264, *313*
Carruthers, P. 148, 151, 153, 154, *199*
Casti, J. 4, *122*
Chaba, A.N. 196, *199*
Chai, C.L. 190, *200*
Chalbaud, E. 144, *198*
Chandrasekhar, S. 18, *122*
Chaturvedi, S. 133, 135, 162, 193, *198*, *199*
Chen, W. 205, 211, 219, 224, 225, *259*
Chen, Y.C. 393, *463*
Chern, M.G. 413, *462*
Chernov, L.A. 16, *122*
Christodoulides, D.N. 205, 225, *259*
Chu, C.H. 413, *462*
Churnside, J.H. 333, 349, 351, *384*, *385*
Clifford, S.F. 333, 341, 351, *384*, *385*
Cochetti, F. 333, 349, *385*
Cohen, S.J. 413, *463*
Colet, P. 175, *199*
Coletti, L. 413, *462*
Collett, M.J. 132, 174, 196, 197, *199*, *201*
Colucci, D. 372, *386*
Consortini, A. 333, 349, *385*
Cooperman, G.D. 205, 216, 224, 239–241, 243, *260*

Corones, J.P. 13, 72, *122*
Corrsin, S. 333, *385*
Coste, J. 257, *259*, *260*
Coudé du Foresto, V. 279, 302, 303, *313*
Crawford, F.S. 373, 374, *384*, *387*
Creath, K. 264, 291, 294, 297, *313*, *316*
Cummings, F.W. 176, *200*

D

Da Costa, G. 291, *313*
Daels, K. 432, *462*
Dainty, J.C. 329, 360, 375, 377, 379, *383*, *385*
Dalton, J.C. 412, *462*
Danckaert, J. 257, *259*
Dandliker, R. 264, *316*, *317*
Daniel, D.J. 133, 136, 139, 140, 156, *199*
D'Ariano, G. 171, *199*
Davey, B.L.K. 382, *387*
David, D. 238, *259*
Davison, M.E. 72, *122*
De Angelis, C. 240, *259*
De Backer, L.C. 283, *313*
de Miranda Gomes, J. 412, *463*
De Muynck, W.M. 197, *200*
De Oliveira, F.A.M. 171, *200*
De Santo, J.A. 119, *126*
De Sterke, C.M. 205, 207, 216, 224, 230, 232, 235–241, 243, 246, 252, 253, 255, 258, *260*
De Vore, S.L. 281, *315*
Deb, B. 177, *199*
Dehany, R. 372, *386*
Desailly, R. 132, *200*
Ditkin, V.A. 34, *122*
Dixit, S.S. 461, *462*
Dodd, R.K. 205, 230, 232, 237, 238, *259*
Doerband, B. 265, 296, 302, *314*
Doucot, B. 33, *122*, *126*
Drazin, P.G. 205, 230, 237, 238, *259*
Drummond, P.D. 132, 133, 174, *199*
Dung, H.T. 177, *199*
Dupertuis, M.A. 258, *259*
Dutta Gupta, S. 257, *259*
Dyatlov, A.I. 40, *121*
Dyson, F.J. 358, *385*
Dzubur, A. 264, 303, *314*

E

Easton, R.L. 432, *462*
Eilbeck, J.C. 205, 230, 232, 237, 238, *259*

Ek, L. 292, 293, *313*
Eschbach, R. 409, 412, 413, 432, *462*, *463*

F
Fan, Z. 412, *462*
Fang, J.M. 132, *201*
Feld, N.S. 184, *201*
Feldman, S. 224, 239, 241, 243, *260*
Fetthauer, F. 411, 432, *462*
Fienup, J.R. 329, 375, 377, 381, *385*, *386*
Filipowicz, P. 176, *198*
Flatte, S.M. 347, 348, *386*
Flavier, D.L. 362, 363, *388*
Floyd, R. 394, 409–411, 434, 438, 449, *462*
Fobelets, K. 257, *259*
Fortus, V.M. 14, *122*
Fox, J. 279, 302, 303, *313*
Foy, R. 371, *385*
Frankenthal, S. 344, 346–348, *384*, *385*
Frankowski, G. 264, 272, 273, 302, *314*
Frehlich, R.G. 351, *384*
Freilikher, V.D. 5, 48, 79, 86, *122*, *123*
Freischland, K. 265, 296, 303, *314*
Freischland, K.R. 265, 296, 302, 303, *314*
Fried, D.L. 325, 327, 329, 330, 340, 352, 354, 363, 371–373, *385*
Frieden, B.R. 369, 370, *385*
Friedman, M.D. 218, 220, *259*
Fright, W.R. 382, *386*
Froehly, C. 132, *199*, *200*
Fugate, R.Q. 371–373, *385*
Fujii, Y. 206, *260*
Furutsu, K. 15, *122*, 346, *385*

G
Gabor, D. 263, 287, *314*
Gallagher, J.E. 264, *313*
Gantsog, Ts. 162, 165, 180, 182, 183, *199*, *201*
García Fernández, P. 175, *199*
Gard, R.L. 393, *462*
Gardiner, C.W. 90, 91, *122*
Gardner, C.S. 327, 371, 372, *385–388*
Garmire, E. 205, 206, 217–220, 224, 253, 257, *260*
Gaskill, J.D. 457, *462*
Gelatt Jr, D. 417, *462*
Gentile, R.S. 461, *462*
Gentile, T R 176, *199*
Gerry, C.C. 160, 162, 165, 175, 184, *199*
Gibbon, J.D. 205, 230, 232, 237, 238, *259*

Gibbs, H.M. 132, *199*, 205, 223, *259*
Glauber, R.J. 164, *201*
Glenn, W.H. 206, *260*
Glogower, J. 151, 153, *201*
Goertzel, G. 458, *462*
Goland, V.I. 27, 29, 37, 61, 66, 67, *122*, *123*
Goldner, E. 349, *388*
Goodman, J.W. 358, 360, 381, *385*, 397, 402, 424, *462*
Gozani, J. 347, 348, *385*
Graves, J.E. 372, *387*
Gredeskul, S.A. 5, 48, 63, 64, 79, 86, *122*, *123*, *126*
Green, R.J. 302, *314*
Greivenkamp, J.E. 264, 279, *314*
Grønbech-Jensen, N. 195, *199*
Gulin, O.E. 15, 27, 45, *123*
Guzev, M.A. 36, 38, 45, 51–54, 56, 73, 78, 79, *123*

H
Haake, F. 133, 134, *199*
Hall, J.L. 173, *202*
Hanbury Brown, R. 380, *384*
Hardy, J.W. 327, 328, 372, *385*
Hariharan, P. 264, 266, *314*, *317*
Harmon, L. 393, *462*
Harnad, J. 238, *259*
Hasegawa, A. 132, *199*
Hauck, R. 407–409, 413, 427, *462*
Haus, H.A. 132, 190, 195, *200*, 253, *259*
Hayslett, C.R. 264, *314*
He, J. 258, *259*
Heckbert, P.S. 461, *462*
Herbert, C.J. 206, 258, *259*
Herriott, D.R. 264, *313*
Hesselink, L. 264, 310, *316*
Hill, K.O. 206, *260*
Hill, R.J. 333, 341, 349, 351, *384*, *385*
Hillery, M. 132, 146, 197, *199*
Hines, J.R. 362, 363, *388*
Ho, S.T. 256, *259*
Ho, T.L. 337, 342, *384*, *385*
Hobson, W. 255, *260*
Holladay, M. 458, *463*
Holladay, T.M. 406, *462*
Hollberg, L.W. 173, *201*
Holmes, C.A. 133, 134, 139, 140, 156, 175, *200*
Honda, T. 264, 283, *315*
Hong, C.K. 146, *199*

Hopfield, J.J. 419, *462*
Horák, R. 158, 182, 187, 189, 190, *199*, *200*
Hradil, Z. 146, 149, 157, 187, 192, *200*
Hsueh, Y.C. 413, *462*
Hudson, J.A. 13, *123*
Hufnagel, R.E. 333, 334, *386*
Hughey, B.J. 176, *199*
Hunt, B.R. 364, 369, 382, *384*, *386*

I
Ichioka, Y. 264, 267, *314*
Idell, P.S. 381, *386*
Idesawa, M. 264, *317*
Imoto, N. 132, 149, 190, 194–196, 198, *200*, *202*
Ina, H. 264, 273, 278, 302, 303, 308, *316*
Infeld, E. 132, *200*
Inuiya, M. 264, 267, *314*
Ishimaru, A. 337, 340, 341, 349, 350, *386*
Islam, M.N. 255, 256, *259*, *260*
Iwaasa, Y. 264, 306, *316*

J
Jackiw, R. 147, *200*
Jackson, K.R. 240, *260*
Jarvis, J.F. 393, *462*
Javanainen, J. 176, *198*
Jaynes, E.T. 176, *200*, 369, *386*
Jex, I. 171, 176–178, 190, *199*
Joenathan, C. 264, *314*
John, S. 211, *260*
Johnson, D.C. 206, *260*
Johnson, R.S. 205, 230, 237, 238, *259*
Jonas, J.A. 264, *314*
Jones, R. 266, 291, 293, *313*, *314*
Jordan, K.E. 19, *123*
Joseph, R.I. 205, 225, *259*
Judice, C.N. 393, *462*
Just, D. 406–409, 417, 427, 461, *462*, *463*

K
Kafri, O. 284, *314*
Kagiwada, H.H. 4, *124*
Kalaba, R. 4, *124*
Kalal, M. 264, 309, *314*
Kane, T.J. 372, *386*
Karim, M.A. 406, *462*
Kárská, M. 141, 158, 167, 170, 189, 191, *200*, *201*
Kärtner, F.X. 138, *200*
Kaup, D.J. 225, 226, 228, *260*
Kawasaki, B.S. 206, *260*
Keiser, W. 265, 296, 303, *314*
Kelsall, D. 360, *386*
Kermisch, D. 400, 404, *462*
Kerr, D. 264, *315*
Khokhlov, R.V. 131, *198*
Kielich, S. 156, 157, 162, 164, 165, 180–183, 185, *200*, *201*
Kim, Fam Le 164, *201*
Kim, M.S. 171, *200*
Kimble, H.J. 173, *202*
King, I.R. 362, *386*
Kirkpatrick, C. 417, *462*
Kitagawa, M. 132, 149, 156, 160, 194–196, 198, *200*, *202*
Klein, N. 158, 176, *201*
Klensch, R.V. 393, *462*
Kleppner, D. 176, *199*
Klyatskin, V.I. 3, 4, 9, 10, 13–16, 19, 26, 27, 29, 30, 32–34, 36–38, 41, 45, 46, 48–54, 56, 58, 60–62, 66, 67, 71–73, 78, 79, 87, 92, 95, 97, 101, 108, 113, *121–125*
Knapp, R. 19, *125*
Knight, P.L. 162, 164, 171, 176, *198–200*, *202*
Knowlton, K. 393, *462*
Knox, K.T. 379, *386*, 412, 432, 449, *462*, *463*
Knuth, D.E. 412, *463*
Kobayashi, S. 264, 273, 278, 302, 303, 308, *316*
Kogelnik, H. 207, 209, 213, 214, 217, *260*
Kohler, W. 33, 40, 79, *121*, *125*
Kokal, J.V. 264, 271, *315*
Koliopoulos, C.L. 264, *314*, 327, 328, 372, *385*
Kollias, S. 413, 417, *462*
Kolmogorov, A. 333, *386*
Kolpatzik, B.W. 461, *463*
Korff, D. 329, 377, *386*
Koshel', K.V. 29, 45, 62, 69, *123–125*
Kotera, H. 461, *463*
Kozlov, V.F. 19, 54, *121*, *125*
Král, P. 171, 172, *200*
Kravtsov, Y.A. 337, *386*
Kravtsov, Yu.A. 3, 13, 15, 95, *121*, *122*, *125*, *126*
Kreider, K.L. 72, *125*
Kreis, T. 264, 278, 288, *314*
Křepelka, J. 148, 149, 151, 154, 156, 160, 167, 170, 172, 173, 178, 190, *200*, *201*

Kristensson, G. 72, *125*, *126*
Krueger, R.J. 72, *122*, *125*, *126*
Kryukov, D.I. 14, *122*
Kuechel, M. 265, 296, 302, 303, *314*
Kuechel, W. 265, 296, 302, *314*
Kujawinska, M. 264, 302, *314*, *316*
Kulkarny, V.A. 4, 90, 92, 97, *126*
Kumar, P. 179, *200*
Kurland, M. 406, *463*
Kurosawa, T. 412, 461, *463*
Kuznetsov, E.A. 225, 226, 228, *260*
Kwon, O.Y. 264, *314*

L

Laberyie, A. 371, *385*
Labeyrie, A. 328, 374, 378, *386*
Lederer, F. 257, *260*
Leendertz, J.A. 291, 294, *313*
Lefebvre, J.E. 327, 328, 372, *385*
Leith, E.N. 263, 272, *314*
LeMesurier, B. 20, *126*
Leoński, W. 195, *200*
Levaron, A. 366, *386*
Levi, A.F.J. 256, *259*
Lifshits, I.M. 5, 48, 63, 64, *126*
Lim, J.S. 453, *463*
Limb, J.O. 406, *463*
Ling, D. 417, 461, *463*
Lippel, B. 406, *463*
Liu, B. 405, 408, *462*
Liu, H.-K. 406, *462*
Livnat, A. 284, *314*
Lloyd-Hart, M. 372, *383*, *386*
Lohmann, A.W. 329, 378, 379, *384*, *386*
Løkberg, O.J. 291, *315*
Loomis, J.S. 264, *314*, *315*
Loudon, R. 151, 160, 164, *199*, *200*
Lukš, A. 133, 134, 136, 137, 139, 141, 142, 146–149, 151, 153–157, 160, 167, 170, 173, 178, 189, 190, 192, *200*, *201*
Luther-Davies, B. 264, 309, *314*
Lynch, R. 162, *200*
Lyubavin, L.Ya. 3, 9, 26, 27, 30, 71, *121*, *122*, *124*

M

Maas, A.M. 264, *316*
Machida, S. 132, 149, 194, 198, *202*
Machorro, R. 283, *315*
Macleod, H.A. 207, 209, 253, *260*

Macy Jr, W.W. 264, 268, 278, 302, *315*
Magome, N. 264, *315*
Maier, G.E.A. 264, *313*
Malacara, D. 263, 264, 281, 297, 302, 303, 306, *315*, *316*
Malacara, Z. 303, *316*
Malakhov, A.N. 13, *126*
Malcuit, M.S. 206, 258, *259*
Mandel, L. 146, *199*
Maneuf, S. 132, *199*, *200*
Manning, R.M. 15, *126*, 327, 357, 358, 370, *386*
Marathay, A.S. 358, 381, *386*
Marburger, J.H. 205, 206, 217–220, 224, 253, 257, *260*
Martens, H. 197, *200*
Martin, J.M. 347, 348, *386*
Maruyama, Y. 412, *463*
Massie, N.A. 264, *315*
Matson, C.L. 379, *387*
Mayer, M. 265, 296, 303, *314*
Mazar, R. 344, 346–348, *384*, *386*
McCarthy, D. 372, *386*
McDaniel, S.T. 13, *126*
McDonnell, M.J. 329, 364, 382, *384*
McKechnie, T.S. 332, *386*
McKenna, D. 372, *387*
McKenna, D.L. 372, *387*
McLaughlin, D. 19, 113, *126*
McLeod, B. 372, *386*
McNeil, K.J. 174, *200*
Mecozzi, A. 133, 134, 164, 170, 185, *200*, *201*
Meinel, E.S. 364, 368–370, *386*
Melnychuk, P. 459, *463*
Meltz, G. 206, *260*
Mendoza-Santoyo, F. 264, *315*
Mendoza, B. 307, *315*
Mendoza, F. 307, *315*
Meng, H.X. 190, *200*
Mermin, N.D. 208, 211, 214, 231, 232, 236, *259*
Mertz, J.C. 173, *201*
Mertz, L.N. 264, 268, *315*
Messer, H.I. 264, *315*
Messiah, A. 143, 144, *200*
Mevers, G.E. 363, *385*
Meyerhofer, D. 393, *462*
Meystre, P. 176, *198*
Michalowski, K. 461, *462*
Mikhailov, A.V. 225, 226, 228, *260*

Milburn, G.J. 132–134, 136, 139, 140, 156, 157, 170, 175, 196, 197, *199–202*
Miller, R. 406, 432, 461, *463*
Mills, D.L. 205, 211, 219, 224, 225, *259, 260*
Mir, M.A. 160, 170, *200*
Miranowicz, A. 156, 157, 162, 164, 165, *200, 201*
Mitsa, T. 413, *463*
Mizrahi, V. 206, *260*
Molyneux, J.E. 337, *387*
Moore, A. 307, *315*
Moore, J.S. 184, *201*
Moreland, J.F. 393, *463*
Morey, W.W. 206, *260*
Morris, H.C. 205, 230, 232, 237, 238, *259*
Mrusek, R. 406, 415, 417, *463*
Muller, R.A. 373, 374, *384, 387*
Murphy, D.V. 371, 372, *387*
Mutoh, K. 264, 300, 306, *316*
Muzii, L. 360, 361, *384*

N

Naesaenen, R. 446, *463*
Nakadate, S. 264, 291, 294, 297, 298, 308, *315, 317*
Nakajima, N. 377, *387*
Nakazato, K. 412, *463*
Nayfeh, A. 232, *260*
Neal, R.M. 412, *463*
Newell, A.C. 225, 226, 228, *260*
Newton, R.G. 148, *200*
Nieto, M.M. 148, 151, 153, 154, *199*
Ninke, W.H. 393, *462*
Northcott, M.J. 372, 379, *383, 387*
Nugent, K.A. 264, 302, 309, *314, 315*

O

Obukhov, A.M. 333, *387*
Ochs Jr, R.L. 72, *126*
Ohtsuka, H. 412, *463*
Ohyama, N. 264, 283, *315*
Oppenheim, A.V. 453, *463*
Oreb, B.F. 264, *314, 317*
Orth, C.D. 373, 374, *384*
Ostrovskaya, G.V. 266, *315*
Ostrovsky, Y.I. 266, *315*
Ozrin, V.D. 15, *122*

P

Paczulp, G.A. 279, 302, 303, *313*
Page, D.A. 371, 372, *387*

Papanicolaou, G. 19, 20, 33, 40, 45, 48, 50, 77, 79, 113, *121–123, 125–127*
Papoulis, A. 369, *387*
Paprzycka, M. 184, *200*
Parker, D.H. 265, *315*
Parker, K.J. 413, *463*
Parrent Jr, G.B. 337, 342, 358, *384*
Pastur, L.A. 5, 48, 63, 64, *121, 126*
Patarnello, S. 413, *462*
Paul, H. 148, *198*
Pearton, S.J. 255, *260*
Pegg, D.T. 151, 152, 155, *198, 200*
Peli, E. 413, *463*
Pennington, K.S. 406, *462*
Perelomov, A.M. 165, 171, *200*
Peřina, J. 134, 146, 156, 158, 167, 170, 173, 178, 187, 189, 190, *199–201*
Peřinová, V. 133, 134, 136, 137, 139, 141, 142, 146–149, 151, 153–158, 160, 167, 170, 172, 173, 178, 189–192, *200, 201*
Pernigo, M. 164, *201*
Petris, A. 302, *316*
Peyraud, J. 257, *259, 260*
Phillips, R.L. 351, *387*
Phoenix, S.J.D. 160, 164, *199*
Piekara, A.A. 184, *201*
Pios, G. 432, *463*
Pollaine, S. 373, *387*
Poll'skikh, S.D. 370, *384*
Polyanskii, E.A. 119, *126*
Popa, D. 264, 283, 284, 286, *316*
Popov, G.V. 36, 40, 51, 52, *123, 126*
Postel, M. 79, *121, 125, 126*
Potasek, M.J. 132, *201*
Pratt, W.K. 273, *315*
Preater, R.W. 265, 307, *315*
Prelewitz, D.F. 206, 253–255, 257, *260*
Primmerman, C.A. 371, 372, *387*
Proctor, M. 258, *259*
Prudnikov, A.P. 34, *122*
Pryputniewicz, R.J. 264, *313*
Puri, R.R. 179, 180, 182, 183, *198*

R

Ramanujam, P.S. 195, *199*
Rammal, R. 33, *122, 126*
Ransom, P.L. 264, 271, *315*
Rao, T.S. 406, *463*
Ray, D.S. 177, *199*
Ray, L. 406, *463*
Razmi, M.S.K. 160, 170, *200*

Reid, G.T.   264, *315*
Reid, M.D.   178, 179, *201*
Reinisch, R.   257, *259*
Remoissenet, M.   258, *259*
Rempe, G.   158, 176, *201*
Richardson, J.   279, 302, 303, *313*
Risken, H.   133, 134, 176, 177, *199*, *201*
Ritze, H.-H.   156, 180, 182, *201*
Rixon, R.C.   264, *315*
Robert, B.D.   240, *260*
Roberts, L.G.   408, *463*
Roberts, P.H.   371–373, *385*
Robinson, D.W.   264, 294, 302, *314*, *315*
Roddier, C.   264, 278, 279, 302, 303, *313*, *315*, 360, 361, 363, *387*
Roddier, D.   372, *387*
Roddier, F.   264, 278, 279, 302, 303, *313*, *315*, 360, 361, 372, *387*
Rodriguez-Vera, R.   264, *315*
Rodriguez, S.   160, 175, *199*
Roetling, P.G.   393, 400, 404, 446, 458, *462*, *463*
Roggemann, M.C.   379, *387*
Rolleston, R.J.   413, *463*
Rosenfeld, D.P.   264, *313*
Rosetti, M.   171, *199*
Rowlands, G.   132, *200*
Roychoudhuri, C.   283, *315*
Ru, Q.-S.   264, 283, *315*
Ruane, R.E.   371–373, *385*
Ryakhin, A.D.   370, *384*
Rytov, S.M.   3, 13, 15, 95, *126*

S

Sahoo, P.K.   393, *463*
Saichev, A.I.   4, 13, 15, 17, 27, 48–50, 87, 101, *122*, *124*–*127*
Saito, H.   264, 291, 294, 297, 298, 308, *315*, *317*
Saito, S.   132, 196, 198, *200*, *202*
Salamo, S.   144, *198*
Salbut, L.A.   264, *316*
Saleh, B.E.A.   132, 147, *201*
San Miguel, M.   175, *199*
Sánchez-Mondragón, J.J.   264, *315*
Sandeman, R.J.   264, *313*
Sanders, B.C.   155, 196, 197, *201*
Sandler, D.   372, *383*
Sankey, N.D.   206, 253–255, 257, *260*
Savage, C.   133, 134, *199*
Scaddan, R.J.   360, *384*, *385*

Scheermesser, T.   415, 428, 450, 452, 454, *463*
Schenzle, A.   138, *200*
Schillke, F.   264, 272, 273, 302, *314*
Schleich, W.   164, *201*
Schmidt, J.   264, *316*
Schroeder, M.R.   393, *463*
Schulz, T.J.   378, *387*
Schwemin, A.J.   373, *384*
Schwendimann, P.   132, 179, *201*
Scott, M.R.   27, *127*
Seldowitz, M.A.   413, *463*
Selloni, A.   132, 179, *201*
Servin, M.   303, *316*
Sette, D.   360, 361, *384*
Shabat, A.B.   132, *202*
Shampine, L.F.   27, *127*
Shapiro, J.H.   154, 179, *200*, *201*
Shaw, R.   459, *463*
Shen, Y.R.   131, *201*, 210, 216, *260*
Sheng, P.   48, 50, 77, 79, *121*, *122*, *126*, *127*
Shevtsov, B.M.   14, 75, *122*, *127*
Shevtzov, B.M.   29, *122*
Shishkarev, A.A.   29, *125*
Shishov, V.I.   337, *387*
Shnider, S.   238, *259*
Shumovsky, A.S.   177, *199*
Sibilia, C.   178, 187, 190, *199*–*201*
Simova, E.S.   306, *316*
Singer, F.   132, *201*
Sinton, A.M.   382, *387*
Sipe, J.E.   205–207, 214, 216, 224, 230, 232, 235–241, 243, 246, 252, 253, 255, 258, *260*
Slavinskij, M.M.   17, *126*, *127*
Slettemoen, G.Å.   291–293, *315*, *316*
Slusarev, V.A.   5, 63, *121*
Slusher, R.E.   173, *201*, 255, 256, *259*, *260*
Smits, R.G.   373, *384*
Snyder, P.L.   378, *387*
Snyder, R.   264, 310, *316*
Sobolev, V.V.   18, *127*
Soccolich, C.E.   255, 256, *259*, *260*
Solano, C.   306, *316*
Solomon, S.   284, *316*
Soltani, S.   393, *463*
Spigler, R.   19, *123*
Spigler, R.J.   19, *127*
Spik, A.   264, 302, *314*, *316*
Srinivasan, V.   133, 135, 162, 193, *198*, *199*
Steel, M.J.   258, *260*
Stegeman, G.I.   214, *260*

Stegun, I.A.   218, 220, 224, *259*
Steinberg, L.   394, 409–411, 434, 438, 449, *462*
Stetson, K.   291, *316*
Stevenson, R.L.   412, *463*
Stobbe, I.   264, 272, 273, 302, *314*
Stoev, K.N.   306, *316*
Stoffel, P.   393, *463*
Stoler, D.   164, *201, 202*
Stradling, R.N.   406, *462*
Stremler   300, *316*
Stucki, P.   393, 412, *463*
Suematsu, M.   264, *316*
Suganuma, M.   265, 306, *316*
Sukhorukov, A.P.   131, *198*
Sulem, C.   19, 20, *126*
Sulem, P.-L.   19, 20, *126*
Sullivan, J.   406, 432, 461, *463*
Susskind, L.   151, 153, *201*
Suzuki, M.   264, *317*
Svelto, O.   131, *201*
Sviridov, K.N.   370, *384*
Swain, R.C.   265, 307, *315*
Swanter, W.   264, *314*
Sweeney, D.W.   413, *463*
Szłachetka, P.   173, *201*

T

Tai, K.   255, *260*
Takeda, M.   264, 273, 278, 300–303, 306, 308, *314*, *316*
Talbot, W.F.   392, *463*
Tan, S.M.   197, *201*
Tanaś, R.   132–134, 156, 157, 160, 162, 164, 165, 180–185, 195, *199–201*
Tappert, F.D.   13, *127*
Tara, K.   162, *198*
Tatarski, V.I.   333, 337, 340, 341, 349, *387*
Tatarskii, V.I.   3, 13, 15, 95, *126*, *127*
Taylor, G.A.   371–373, *385*
Teich, M.C.   132, 147, *201*
Temchenko, V.V.   45, *123*
Thacker, H.B.   132, *201*
Thalmann, R.   264, *316*
Thompson, B.J.   379, *386*
Thompson, G.R.   458, *462*
Thompson, L.A.   371, 372, *385–387*
Tischer, W.   264, 272, 273, 302, *314*
Titulaer, U.M.   164, *201*
Tiziani, H.J.   302, *314*

Tombesi, P.   133, 134, 164, 170, 175, 185, *200, 201*
Tominaga, M.   264, 267, 268, 302, 306, *316*
Toral, R.   175, *199*
Torroba, R.   264, *314*
Toyooka, S.   264, 267, 268, 302, 306, *316*
Trullinger, S.E.   205, 225, *260*
Trussel, H.J.   364, 369, *384*
Trutschel, U.   257, *260*
Tsuchiya, H.   412, *463*
Tsujiuchi, J.   264, 283, *315*
Tung, Z.   264, 303, *316*
Tur, M.   347, 348, 352, *387*
Tuttass, T.   413, 417, *463*
Twiss, R.Q.   380, *384*

U

Ulichney, R.A.   393, 400, *463*
Underwood, K.L.   264, *314*
Upatnieks, J.   263, 272, *314*
Uscinski, B.J.   337, 346, *387*
Ustinov, N.D.   370, *384*

V

Vadacchino, M.   171, *199*
Valley, J.F.   173, *201*
Van den Bulck, P.   411, *462*
Vecchi, M.P.   417, *462*
Velho, L.   412, *463*
Veretennicoff, I.   257, *259*
Vest, C.M.   266, *316*
Vidiella Barranco, A.   164, *199*
Virovlyanskij, A.L.   17, *127*
Vitrant, G.   257, *259*
Vlad, V.I.   264, 283, 284, 286, 302, 303, 306, 307, *315, 316*
Voicu, L.   306, *316*
Vourdas, A.   164, *201*
Vrooman, H.A.   264, *316*
Vrscay, E.R.   160, 175, *199*
Vukicevic, D.   264, 303, *314*

W

Wabnitz, S.   205, 225, 227, 236, 238, 240, 257, *259, 260*
Walker, J.G.   302, *314*
Walls, D.F.   132–134, 145, 174, 178, 179, 189, 196, 197, *199, 201*
Walowit, E.   461, *462*
Walsch, J.J.   393, *462*
Walters, D.L.   362, 363, *388*

Walther, H.   158, 176, *201*
Waters, J.P.   292, *316*
Wegmann, U.   265, 296, 302, *314*
Wegner, H.   264, *313*
Weigelt, G.   329, 378, 379, *386*
Weissbach, S.   411, 412, 432, 461, *462, 463*
Wells, D.C.   369, 370, *385*
Welsh, B.M.   327, 371, 372, *385, 388*
Werner, M.J.   176, 177, *201*
Wessely, H.W.   360, 363, *388*
White, A.D.   264, *313*
White, B.   19, 45, 48, 50, 77, 79, *121, 122, 125–127*
White, B.S.   4, 90, 92, 97, *126, 127*
Whitman, A.M.   339, 341, 344, 346–348, 365, 366, 377, 381, *384, 385, 388*
Wiedman, W.   265, 296, 302, 303, *314*
Wielinga, B.   175, *202*
Willemin, J.-F.   264, *317*
Williams, D.C.   264, 294, *315*
Wilson-Gordon, A.D.   162, *202*
Winful, H.G.   205, 206, 216–220, 224, 230, 235, 236, 239–241, 243, 253, 257, *260*
Wing, G.M.   4, 27, *122, 127*
Wirnitzer, B.   329, 378, 379, *384, 386*
Wirnitzer, R.   378, *388*
Witten, I.H.   412, *463*
Wittman, D.   372, *386*
Wizinowich, P.   372, *383*
Wojciak, J.   264, 302, *314*
Wolf, E.   219, *259*, 358, *384*
Womack, K.H.   264, 273, 274, *314, 317*
Wong, K.C.   393, *463*
Wu, H.   173, *202*
Wu, L.-A.   173, *202*
Wu, X.   461, *463*
Wyant, J.C.   264, *314, 317*
Wykes, C.   266, 291, 293, *313, 314*
Wyrowski, F.   394, 400, 411–413, 415, 428, 432, 453, *462, 463*

Y

Yablonovitch, E.   211, *260*
Yamamoto, Y.   132, 149, 156, 160, 190, 194–196, 198, *200, 202*
Yamashi, Y.   264, *317*
Yanagawa, T.   198, *202*
Yang, X.   184, *202*
Yariv, A.   132, *202*
Yaroschuk, E.V.   19, 54, *121, 125*
Yaroschuk, I.O.   19, 37, 40, 42, 54, 60, 61, *123, 124, 126, 127*
Yatagai, T.   264, *317*
Yoshizawa, T.   265, 306, *316*
Yuen, H.P.   167, 172, 175, *201, 202*
Yurke, B.   164, 173, *201, 202*

Z

Zakharov, Y.E.   132, *202*
Zamir, R.   224, 239, 241, 243, *260*
Zeggel, T.   461, *463*
Zernike, F.   263, *317*
Zhang, Z.   48, 50, *127*
Zhang, Z.M.   190, *200*
Zheng, X.   184, *202*
Zollars, B.G.   371, 372, *387*
Zwillinger, D.   4, 92, *127*

SUBJECT INDEX FOR VOLUME XXXIII

A
adaptive optics   327, 328, 371–373
angular spectrum of plane waves   326
atmospheric optics   3

B
band structure, photonic   211
Bessel function   39
Bethe–Salpeter equation   3
Bloch function   231, 232, 235, 236, 238
Born approximation   350
Bourre approximation   3
Bragg condition   44, 210
– frequency   210

C
Carson relation   300
characteristic functional   98
coherence function   330, 336–339, 341, 345, 349, 353, 356, 359
– length   363
coherent state, generalized   147
– –, squeezed   164
– –, SU(1,1)   165, 167, 171, 175
– –, two-photon   155, 167
coupled-mode equations   213–215, 217, 221, 228, 232, 234, 236, 257
– theory   212

D
degenerate parametric process   173
diffusion coefficient   33
digital halftoning   392–394
Dyson equation   3

E
error-diffusion method   431, 433

F
Fabry–Pérot cavity   205
Fano factor   149, 162, 195
Fokker–Planck equation   5, 33, 35, 38, 48, 65, 87–92, 95, 99, 106, 134, 141, 187, 191, 192
four-wave mixing   179
Fresnel transformation   282, 283

G
geometric optics approximation   4, 93, 342, 354
Green function   6–10, 13, 32, 71, 79–81, 118–120, 349, 363, 364

H
halftone technique   392, 412, 419
Hamming window   273
Hankel function   117, 120
Heisenberg uncertainty relation   144
Heisenberg–Langevin equation   189
Helmholtz equation   5, 6, 335, 349
Hermite polynomial   170
Hilbert transform   276, 277, 290, 291
hologram   264
holographic interferogram   266
– interferometry   282, 284, 303, 304
holography   263, 272, 281, 293
homodyne detection process   145
Hopfield model   419

I
image-processing system   394
ionosphere   3
isoplanicity   324, 352, 371
isoprobable curve   107

J
Jackiw state   147, 148
Jacobi elliptic function   218, 220, 224, 225
Jaynes–Cummings model   176, 177

K

Kerr effect 132, 185, 196
– medium 131, 132, 158, 159, 164, 167, 172, 175, 177, 179, 180, 182, 187, 188, 192, 194–197
– nonlinear interferometer 161
– nonlinearity 134, 185, 186, 195, 231
Knox–Thompson method 379
Kolmogorov region 332, 333
– spectrum 350, 383
– structure function 345, 346
Kontorovich–Lebedev transformation 55

L

Langevin equation 175, 179
Laplace transformation 78
Legendre function 34
– polynomial 114
lens transmittance 282

M

Mach–Zehnder interferometer 162, 196, 197
Markov random process 44, 48, 99
Maxwell equations 212, 334
Meller–Fock transformation 34

N

Neumann function 39
neural-network algorithm 417
nonclassical states, generation of 132
number state, squeezed and displaced 155, 171, 172

O

optical bistability 132, 174, 205
– fiber 205, 206
– information processing 263

P

parabolic approximation 81
– equation 3, 12, 13, 80, 81
paraxial approximation 131
Pauli matrices 219
period-doubling 243
phase operator 151
phase-space interference 164
photon antibunching 182
– statistics 138, 178
Poisson distribution 148, 178
Poissonian light 139

Q

Q-function 139
quantum nondemolition measurement 132
– optics 131, 133, 148

R

Rabi oscillation 176
radiation condition 6
radiative transfer, linear theory of 3
– –, statistical theory of 33
radio astronomy 321
random medium, layered 17, 69
raster photography 406
remote sensing 321
Riccati equation 21–23, 33, 64
rotating-wave approximation 176
Rydberg atom 176
Rytov approximation 330, 350, 357

S

sampling theorem 278
Schrödinger equation 5, 163
– –, nonlinear 19, 131, 205, 207, 230, 236–238
Schwartz inequality 143
Seidel coefficient 302
self-focusing 131, 132
self-phase modulation 131
self-pulsations 224, 243
self-trapping 131
simulated annealing algorithm 417
sine-Gordon equation 225
small-angle approximation 13
soliton, gap 205, 206, 257, 258
–, optical 132
–, – fiber 238
speckle interferometry 328, 329, 374, 377, 379
– masking 378
speckle-pattern 263, 293
– interferometry 281, 282
squeezed state 142, 144, 157
squeezing 157, 162, 167, 171, 176, 179, 189, 192
–, amplitude-squared 147, 160, 170
–, higher-order 146, 160
Stokes parameters 219, 224

T

Talbot distance 286

Thirring model   225, 226, 238
transfer function   273, 329, 447
turbulent atmosphere   321, 356–358, 364
– medium   321, 348

V
van Cittert–Zernike theorem   342

W
Wiener filter   369
Wigner function   164, 174

Z
Zernike coefficient   302
– polynomial   279–281, 327, 357

CONTENTS OF PREVIOUS VOLUMES

VOLUME I (1961)

I.	The Modern Development of Hamiltonian Optics, R. J. PEGIS	1–29
II.	Wave Optics and Geometrical Optics in Optical Design, K. MIYAMOTO	31–66
III.	The Intensity Distribution and Total Illumination of Aberration-Free Diffraction Images, R. BARAKAT	67–108
IV.	Light and Information, D. GABOR	109–153
V.	On Basic Analogies and Principal Differences between Optical and Electronic Information, H. WOLTER	155–210
VI.	Interference Color, H. KUBOTA	211–251
VII.	Dynamic Characteristics of Visual Processes, A. FIORENTINI	253–288
VIII.	Modern Alignment Devices, A. C. S. VAN HEEL	289–329

VOLUME II (1963)

I.	Ruling, Testing and Use of Optical Gratings for High-Resolution Spectroscopy, G. W. STROKE	1–72
II.	The Metrological Applications of Diffraction Gratings, J. M. BURCH	73–108
III.	Diffusion Through Non-Uniform Media, R. G. GIOVANELLI	109–129
IV.	Correction of Optical Images by Compensation of Aberrations and by Spatial Frequency Filtering, J. TSUJIUCHI	131–180
V.	Fluctuations of Light Beams, L. MANDEL	181–248
VI.	Methods for Determining Optical Parameters of Thin Films, F. ABELÈS	249–288

VOLUME III (1964)

I.	The Elements of Radiative Transfer, F. KOTTLER	1–28
II.	Apodisation, P. JACQUINOT, B. ROIZEN-DOSSIER	29–186
III.	Matrix Treatment of Partial Coherence, H. GAMO	187–332

VOLUME IV (1965)

I.	Higher Order Aberration Theory, J. FOCKE	1–36
II.	Applications of Shearing Interferometry, O. BRYNGDAHL	37–83
III.	Surface Deterioration of Optical Glasses, K. KINOSITA	85–143
IV.	Optical Constants of Thin Films, P. ROUARD, P. BOUSQUET	145–197
V.	The Miyamoto–Wolf Diffraction Wave, A. RUBINOWICZ	199–240
VI.	Aberration Theory of Gratings and Grating Mountings, W. T. WELFORD	241–280
VII.	Diffraction at a Black Screen, Part I: Kirchhoff's Theory, F. KOTTLER	281–314

VOLUME V (1966)

I.	Optical Pumping, C. COHEN-TANNOUDJI, A. KASTLER	1– 81
II.	Non-Linear Optics, P. S. PERSHAN	83–144
III.	Two-Beam Interferometry, W. H. STEEL	145–197
IV.	Instruments for the Measuring of Optical Transfer Functions, K. MURATA	199–245
V.	Light Reflection from Films of Continuously Varying Refractive Index, R. JACOBSSON	247–286
VI.	X-Ray Crystal-Structure Determination as a Branch of Physical Optics, H. LIPSON, C. A. TAYLOR	287–350
VII.	The Wave of a Moving Classical Electron, J. PICHT	351–370

VOLUME VI (1967)

I.	Recent Advances in Holography, E. N. LEITH, J. UPATNIEKS	1– 52
II.	Scattering of Light by Rough Surfaces, P. BECKMANN	53– 69
III.	Measurement of the Second Order Degree of Coherence, M. FRANÇON, S. MALLICK	71–104
IV.	Design of Zoom Lenses, K. YAMAJI	105–170
V.	Some Applications of Lasers to Interferometry, D. R. HERRIOT	171–209
VI.	Experimental Studies of Intensity Fluctuations in Lasers, J. A. ARMSTRONG, A. W. SMITH	211–257
VII.	Fourier Spectroscopy, G. A. VANASSE, H. SAKAI	259–330
VIII.	Diffraction at a Black Screen, Part II: Electromagnetic Theory, F. KOTTLER	331–377

VOLUME VII (1969)

I.	Multiple-Beam Interference and Natural Modes in Open Resonators, G. KOPPELMAN	1– 66
II.	Methods of Synthesis for Dielectric Multilayer Filters, E. DELANO, R. J. PEGIS	67–137
III.	Echoes at Optical Frequencies, I. D. ABELLA	139–168
IV.	Image Formation with Partially Coherent Light, B. J. THOMPSON	169–230
V.	Quasi-Classical Theory of Laser Radiation, A. L. MIKAELIAN, M. L. TER-MIKAELIAN	231–297
VI.	The Photographic Image, S. OOUE	299–358
VII.	Interaction of Very Intense Light with Free Electrons, J. H. EBERLY	359–415

VOLUME VIII (1970)

I.	Synthetic-Aperture Optics, J. W. GOODMAN	1– 50
II.	The Optical Performance of the Human Eye, G. A. FRY	51–131
III.	Light Beating Spectroscopy, H. Z. CUMMINS, H. L. SWINNEY	133–200
IV.	Multilayer Antireflection Coatings, A. MUSSET, A. THELEN	201–237
V.	Statistical Properties of Laser Light, H. RISKEN	239–294
VI.	Coherence Theory of Source-Size Compensation in Interference Microscopy, T. YAMAMOTO	295–341
VII.	Vision in Communication, L. LEVI	343–372
VIII.	Theory of Photoelectron Counting, C. L. MEHTA	373–440

VOLUME IX (1971)

I.	Gas Lasers and their Application to Precise Length Measurements, A. L. BLOOM	1– 30
II.	Picosecond Laser Pulses, A. J. DEMARIA	31– 71
III.	Optical Propagation Through the Turbulent Atmosphere, J. W. STROHBEHN	73–122
IV.	Synthesis of Optical Birefringent Networks, E. O. AMMANN	123–177
V.	Mode Locking in Gas Lasers, L. ALLEN, D. G. C. JONES	179–234
VI.	Crystal Optics with Spatial Dispersion, V. M. AGRANOVICH, V. L. GINZBURG	235–280
VII.	Applications of Optical Methods in the Diffraction Theory of Elastic Waves, K. GNIADEK, J. PETYKIEWICZ	281–310
VIII.	Evaluation, Design and Extrapolation Methods for Optical Signals, Based on Use of the Prolate Functions, B. R. FRIEDEN	311–407

VOLUME X (1972)

I.	Bandwidth Compression of Optical Images, T. S. HUANG	1– 44
II.	The Use of Image Tubes as Shutters, R. W. SMITH	45– 87
III.	Tools of Theoretical Quantum Optics, M. O. SCULLY, K. G. WHITNEY	89–135
IV.	Field Correctors for Astronomical Telescopes, C. G. WYNNE	137–164
V.	Optical Absorption Strength of Defects in Insulators, D. Y. SMITH, D. L. DEXTER	165–228
VI.	Elastooptic Light Modulation and Deflection, E. K. SITTIG	229–288
VII.	Quantum Detection Theory, C. W. HELSTROM	289–369

VOLUME XI (1973)

I.	Master Equation Methods in Quantum Optics, G. S. AGARWAL	1– 76
II.	Recent Developments in Far Infrared Spectroscopic Techniques, H. YOSHINAGA	77–122
III.	Interaction of Light and Acoustic Surface Waves, E. G. LEAN	123–166
IV.	Evanescent Waves in Optical Imaging, O. BRYNGDAHL	167–221
V.	Production of Electron Probes Using a Field Emission Source, A. V. CREWE	223–246
VI.	Hamiltonian Theory of Beam Mode Propagation, J. A. ARNAUD	247–304
VII.	Gradient Index Lenses, E. W. MARCHAND	305–337

VOLUME XII (1974)

I.	Self-Focusing, Self-Trapping, and Self-Phase Modulation of Laser Beams, O. SVELTO	1– 51
II.	Self-Induced Transparency, R. E. SLUSHER	53–100
III.	Modulation Techniques in Spectrometry, M. HARWIT, J. A. DECKER JR	101–162
IV.	Interaction of Light with Monomolecular Dye Layers, K. H. DREXHAGE	163–232
V.	The Phase Transition Concept and Coherence in Atomic Emission, R. GRAHAM	233–286
VI.	Beam-Foil Spectroscopy, S. BASHKIN	287–344

VOLUME XIII (1976)

I.	On the Validity of Kirchhoff's Law of Heat Radiation for a Body in a Nonequilibrium Environment, H. P. BALTES	1– 25
II.	The Case For and Against Semiclassical Radiation Theory, L. MANDEL	27– 68
III.	Objective and Subjective Spherical Aberration Measurements of the Human Eye, W. M. ROSENBLUM, J. L. CHRISTENSEN	69– 91
IV.	Interferometric Testing of Smooth Surfaces, G. SCHULZ, J. SCHWIDER	93–167
V.	Self-Focusing of Laser Beams in Plasmas and Semiconductors, M. S. SODHA, A. K. GHATAK, V. K. TRIPATHI	169–265
VI.	Aplanatism and Isoplanatism, W. T. WELFORD	267–292

VOLUME XIV (1976)

I.	The Statistics of Speckle Patterns, J. C. DAINTY	1– 46
II.	High-Resolution Techniques in Optical Astronomy, A. LABEYRIE	47– 87
III.	Relaxation Phenomena in Rare-Earth Luminescence, L. A. RISEBERG, M. J. WEBER	89–159
IV.	The Ultrafast Optical Kerr Shutter, M. A. DUGUAY	161–193
V.	Holographic Diffraction Gratings, G. SCHMAHL, D. RUDOLPH	195–244
VI.	Photoemission, P. J. VERNIER	245–325
VII.	Optical Fibre Waveguides – A Review, P. J. B. CLARRICOATS	327–402

VOLUME XV (1977)

I.	Theory of Optical Parametric Amplification and Oscillation, W. BRUNNER, H. PAUL	1– 75
II.	Optical Properties of Thin Metal Films, P. ROUARD, A. MEESSEN	77–137
III.	Projection-Type Holography, T. OKOSHI	139–185
IV.	Quasi-Optical Techniques of Radio Astronomy, T. W. COLE	187–244
V.	Foundations of the Macroscopic Electromagnetic Theory of Dielectric Media, J. VAN KRANENDONK, J. E. SIPE	245–350

VOLUME XVI (1978)

I.	Laser Selective Photophysics and Photochemistry, V. S. LETOKHOV	1– 69
II.	Recent Advances in Phase Profiles Generation, J. J. CLAIR, C. I. ABITBOL	71–117
III.	Computer-Generated Holograms: Techniques and Applications, W.-H. LEE	119–232
IV.	Speckle Interferometry, A. E. ENNOS	233–288
V.	Deformation Invariant, Space-Variant Optical Pattern Recognition, D. CASASENT, D. PSALTIS	289–356
VI.	Light Emission From High-Current Surface-Spark Discharges, R. E. BEVERLY III	357–411
VII.	Semiclassical Radiation Theory Within a Quantum-Mechanical Framework, I. R. SENITZKY	413–448

VOLUME XVII (1980)

I.	Heterodyne Holographic Interferometry, R. Dändliker	1– 84
II.	Doppler-Free Multiphoton Spectroscopy, E. Giacobino, B. Cagnac	85–161
III.	The Mutual Dependence Between Coherence Properties of Light and Nonlinear Optical Processes, M. Schubert, B. Wilhelmi	163–238
IV.	Michelson Stellar Interferometry, W. J. Tango, R. Q. Twiss	239–277
V.	Self-Focusing Media with Variable Index of Refraction, A. L. Mikaelian	279–345

VOLUME XVIII (1980)

I.	Graded Index Optical Waveguides: A Review, A. Ghatak, K. Thyagarajan	1–126
II.	Photocount Statistics of Radiation Propagating Through Random and Nonlinear Media, J. Peřina	127–203
III.	Strong Fluctuations in Light Propagation in a Randomly Inhomogeneous Medium, V. I. Tatarskii, V. U. Zavorotnyi	204–256
IV.	Catastrophe Optics: Morphologies of Caustics and their Diffraction Patterns, M. V. Berry, C. Upstill	257–346

VOLUME XIX (1981)

I.	Theory of Intensity Dependent Resonance Light Scattering and Resonance Fluorescence, B.R. Mollow	1– 43
II.	Surface and Size Effects on the Light Scattering Spectra of Solids, D. L. Mills, K. R. Subbaswamy	45–137
III.	Light Scattering Spectroscopy of Surface Electromagnetic Waves in Solids, S. Ushioda	139–210
IV.	Principles of Optical Data-Processing, H. J. Butterweck	211–280
V.	The Effects of Atmospheric Turbulence in Optical Astronomy, F. Roddier	281–376

VOLUME XX (1983)

I.	Some New Optical Designs for Ultra-Violet Bidimensional Detection of Astronomical Objects, G. Courtès, P. Cruvellier, M. Detaille, M. Saïsse	1– 61
II.	Shaping and Analysis of Picosecond Light Pulses, C. Froehly, B. Colombeau, M. Vampouille	63–153
III.	Multi-Photon Scattering Molecular Spectroscopy, S. Kielich	155–261
IV.	Colour Holography, P. Hariharan	263–324
V.	Generation of Tunable Coherent Vacuum-Ultraviolet Radiation, W. Jamroz, B. P. Stoicheff	325–380

VOLUME XXI (1984)

I.	Rigorous Vector Theories of Diffraction Gratings, D. Maystre	1– 67
II.	Theory of Optical Bistability, L. A. Lugiato	69–216
III.	The Radon Transform and its Applications, H. H. Barrett	217–286
IV.	Zone Plate Coded Imaging: Theory and Applications, N. M. Ceglio, D. W Sweeney	287–354
V.	Fluctuations, Instabilities and Chaos in the Laser-Driven Nonlinear Ring Cavity, J. C. Englund, R. R. Snapp, W. C. Schieve	355–428

VOLUME XXII (1985)

I.	Optical and Electronic Processing of Medical Images, D. MALACARA . . .	1– 76
II.	Quantum Fluctuations in Vision, M. A. BOUMAN, W. A. VAN DE GRIND, P. ZUIDEMA	77–144
III.	Spectral and Temporal Fluctuations of Broad-Band Laser Radiation, A. V. MASALOV	145–196
IV.	Holographic Methods of Plasma Diagnostics, G. V. OSTROVSKAYA, YU.I. OSTROVSKY	197–270
V.	Fringe Formations in Deformation and Vibration Measurements using Laser Light, I. YAMAGUCHI	271–340
VI.	Wave Propagation in Random Media: A Systems Approach, R. L. FANTE .	341–398

VOLUME XXIII (1986)

I.	Analytical Techniques for Multiple Scattering from Rough Surfaces, J. A. DESANTO, G. S. BROWN	1– 62
II.	Paraxial Theory in Optical Design in Terms of Gaussian Brackets, K. TANAKA	63–111
III.	Optical Films Produced by Ion-Based Techniques, P. J. MARTIN, R. P. NETTERFIELD	113–182
IV.	Electron Holography, A. TONOMURA	183–220
V.	Principles of Optical Processing with Partially Coherent Light, F. T. S. YU	221–275

VOLUME XXIV (1987)

I.	Micro Fresnel Lenses, H. NISHIHARA, T. SUHARA	1– 37
II.	Dephasing-Induced Coherent Phenomena, L. ROTHBERG	39–101
III.	Interferometry with Lasers, P. HARIHARAN	103–164
IV.	Unstable Resonator Modes, K. E. OUGHSTUN	165–387
V.	Information Processing with Spatially Incoherent Light, I. GLASER	389–509

VOLUME XXV (1988)

I.	Dynamical Instabilities and Pulsations in Lasers, N. B. ABRAHAM, P. MANDEL, L. M. NARDUCCI	1–190
II.	Coherence in Semiconductor Lasers, M. OHTSU, T. TAKO	191–278
III.	Principles and Design of Optical Arrays, WANG SHAOMIN, L. RONCHI . . .	279–348
IV.	Aspheric Surfaces, G. SCHULZ	349–415

VOLUME XXVI (1988)

I.	Photon Bunching and Antibunching, M. C. TEICH, B. E. A. SALEH	1–104
II.	Nonlinear Optics of Liquid Crystals, I. C. KHOO	105–161
III.	Single-Longitudinal-Mode Semiconductor Lasers, G. P. AGRAWAL	163–225
IV.	Rays and Caustics as Physical Objects, YU.A. KRAVTSOV	227–348
V.	Phase-Measurement Interferometry Techniques, K. CREATH	349–393

VOLUME XXVII (1989)

I. The Self-Imaging Phenomenon and Its Applications, K. Patorski 1–108
II. Axicons and Meso-Optical Imaging Devices, L. M. Soroko 109–160
III. Nonimaging Optics for Flux Concentration, I. M. Bassett, W. T. Welford, R. Winston 161–226
IV. Nonlinear Wave Propagation in Planar Structures, D. Mihalache, M. Bertolotti, C. Sibilia 227–313
V. Generalized Holography with Application to Inverse Scattering and Inverse Source Problems, R. P. Porter 315–397

VOLUME XXVIII (1990)

I. Digital Holography – Computer-Generated Holograms, O. Bryngdahl. F. Wyrowski 1–86
II. Quantum Mechanical Limit in Optical Precision Measurement and Communication, Y. Yamamoto, S. Machida S. Saito, N. Imoto, T. Yanagawa, M. Kitagawa, G. Björk 87–179
III. The Quantum Coherence Properties of Stimulated Raman Scattering, M. G. Raymer, I. A. Walmsley 181–270
IV. Advanced Evaluation Techniques in Interferometry, J. Schwider 271–359
V. Quantum Jumps, R. J. Cook 361–416

VOLUME XXIX (1991)

I. Optical Waveguide Diffraction Gratings: Coupling between Guided Modes, D. G. Hall 1–63
II. Enhanced Backscattering in Optics, Yu.N. Barabanenkov, Yu.A. Kravtsov, V. D. Ozrin, A. I. Saichev 65–197
III. Generation and Propagation of Ultrashort Optical Pulses, I. P. Christov 199–291
IV. Triple-Correlation Imaging in Optical Astronomy, G. Weigelt 293–319
V. Nonlinear Optics in Composite Materials. 1. Semiconductor and Metal Crystallites in Dielectrics, C. Flytzanis, F. Hache, M. C. Klein, D. Ricard, Ph. Roussignol 321–411

VOLUME XXX (1992)

I. Quantum Fluctuations in Optical Systems, S. Reynaud, A. Heidmann, E. Giacobino, C. Fabre 1–85
II. Correlation Holographic and Speckle Interferometry, Yu.I. Ostrovsky, V. P. Shchepinov 87–135
III. Localization of Waves in Media with One-Dimensional Disorder, V. D. Freilikher, S. A. Gredeskul 137–203
IV. Theoretical Foundation of Optical-Soliton Concept in Fibers, Y. Kodama, A. Hasegawa 205–259
V. Cavity Quantum Optics and the Quantum Measurement Process., P. Meystre 261–355

VOLUME XXXI (1993)

I.	Atoms in Strong Fields: Photoionization and Chaos, P. W. MILONNI, B. SUNDARAM	1–137
II.	Light Diffraction by Relief Gratings: A Macroscopic and Microscopic View, E. POPOV	139–187
III.	Optical Amplifiers, N. K. DUTTA, J. R. SIMPSON	189–226
IV.	Adaptive Multilayer Optical Networks, D. PSALTIS, Y. QIAO	227–261
V.	Optical Atoms, R. J. C. SPREEUW, J. P. WOERDMAN	263–319
VI.	Theory of Compton Free Electron Lasers, G. DATTOLI, L. GIANNESSI, A. RENIERI, A. TORRE	321–412

VOLUME XXXII (1993)

I.	Guided-Wave Optics on Silicon: Physics, Technology and Status, B. P. PAL	1–59
II.	Optical Neural Networks: Architecture, Design and Models, F. T. S. YU	61–144
III.	The Theory of Optimal Methods for Localization of Objects in Pictures, L. P. YAROSLAVSKY	145–201
IV.	Wave Propagation Theories in Random Media Based on the Path-Integral Approach, M. I. CHARNOTSKII, J. GOZANI, V. I. TATARSKII, V. U. ZAVOROTNY	203–266
V.	Radiation by Uniformly Moving Sources (Vavilov–Cherenkov effect, Doppler effect in a medium, transition radiation and associated phenomena), V. L. GINZBURG	267–312
VI.	Nonlinear Optical Processes in Atoms and in Weakly Relativistic Plasmas, G. MAINFRAY, C. MANUS	313–361

CUMULATIVE INDEX – VOLUMES I–XXXIII

ABELÈS, F., Methods for Determining Optical Parameters of Thin Films	II, 249
ABELLA, I.D., Echoes at Optical Frequencies	VII, 139
ABITBOL, C.I., see Clair, J.J.	XVI, 71
ABRAHAM, N.B., P. MANDEL, L.M. NARDUCCI, Dynamical Instabilities and Pulsations in Lasers	XXV, 1
AGARWAL, G.S., Master Equation Methods in Quantum Optics	XI, 1
AGRANOVICH, V.M., V.L. GINZBURG, Crystal Optics with Spatial Dispersion	IX, 235
AGRAWAL, G.P., Single-Longitudinal-Mode Semiconductor Lasers	XXVI, 163
ALLEN, L., D.G.C. JONES, Mode Locking in Gas Lasers	IX, 179
AMMANN, E.O., Synthesis of Optical Birefringent Networks	IX, 123
ARMSTRONG, J.A., A.W. SMITH, Experimental Studies of Intensity Fluctuations in Lasers	VI, 211
ARNAUD, J.A., Hamiltonian Theory of Beam Mode Propagation	XI, 247
BALTES, H.P., On the Validity of Kirchhoff's Law of Heat Radiation for a Body in a Nonequilibrium Environment	XIII, 1
BARABANENKOV, YU.N., YU.A. KRAVTSOV, V.D. OZRIN, A.I. SAICHEV, Enhanced Backscattering in Optics	XXIX, 65
BARAKAT, R., The Intensity Distribution and Total Illumination of Aberration-Free Diffraction Images	I, 67
BARRETT, H.H., The Radon Transform and its Applications	XXI, 217
BASHKIN, S., Beam-Foil Spectroscopy	XII, 287
BASSETT, I.M., W.T. WELFORD, R. WINSTON, Nonimaging Optics for Flux Concentration	XXVII, 161
BECKMANN, P., Scattering of Light by Rough Surfaces	VI, 53
BERAN, M.J., J. OZ-VOGT, Imaging through Turbulence in the Atmosphere	XXXIII, 319
BERRY, M.V., C. UPSTILL, Catastrophe Optics: Morphologies of Caustics and their Diffraction Patterns	XVIII, 257
BERTOLOTTI, M., see Mihalache, D.	XXVII, 227
BEVERLY III, R.E., Light Emission From High-Current Surface-Spark Discharges	XVI, 357
BJÖRK, G., see Yamamoto, Y.	XXVIII, 87
BLOOM, A.L., Gas Lasers and their Application to Precise Length Measurements	IX, 1
BOUMAN, M.A., W.A. VAN DE GRIND, P. ZUIDEMA, Quantum Fluctuations in Vision	XXII, 77
BOUSQUET, P., see Rouard, P.	IV, 145
BROWN, G.S., see DeSanto, J.A.	XXIII, 1
BRUNNER, W., H. PAUL, Theory of Optical Parametric Amplification and Oscillation	XV, 1
BRYNGDAHL, O., Applications of Shearing Interferometry	IV, 37
BRYNGDAHL, O., Evanescent Waves in Optical Imaging	XI, 167
BRYNGDAHL, O., F. WYROWSKI, Digital Holography – Computer-Generated Holograms	XXVIII, 1

BRYNGDAHL, O., T. SCHEERMESSER, F. WYROWSKI, Digital Halftoning: Synthesis of
  Binary Images — XXXIII, 389
BURCH, J.M., The Metrological Applications of Diffraction Gratings — II, 73
BUTTERWECK, H.J., Principles of Optical Data-Processing — XIX, 211

CAGNAC, B., see Giacobino, E. — XVII, 85
CASASENT, D., D. PSALTIS, Deformation Invariant, Space-Variant Optical Pattern
  Recognition — XVI, 289
CEGLIO, N.M., D.W. SWEENEY, Zone Plate Coded Imaging: Theory and Applications — XXI, 287
CHARNOTSKII, M.I., J. GOZANI, V.I. TATARSKII, V.U. ZAVOROTNY, Wave Propagation
  Theories in Random Media Based on the Path-Integral Approach — XXXII, 203
CHRISTENSEN, J.L., see Rosenblum, W.M. — XIII, 69
CHRISTOV, I.P., Generation and Propagation of Ultrashort Optical Pulses — XXIX, 199
CLAIR, J.J., C.I. ABITBOL, Recent Advances in Phase Profiles Generation — XVI, 71
CLARRICOATS, P.J.B., Optical Fibre Waveguides – A Review — XIV, 327
COHEN-TANNOUDJI, C., A. KASTLER, Optical Pumping — V, 1
COLE, T.W., Quasi-Optical Techniques of Radio Astronomy — XV, 187
COLOMBEAU, B., see Froehly, C. — XX, 63
COOK, R.J., Quantum Jumps — XXVIII, 361
COURTÈS, G., P. CRUVELLIER, M. DETAILLE, M. SAÏSSE, Some New Optical Designs
  for Ultra-Violet Bidimensional Detection of Astronomical Objects — XX, 1
CREATH, K., Phase-Measurement Interferometry Techniques — XXVI, 349
CREWE, A.V., Production of Electron Probes Using a Field Emission Source — XI, 223
CRUVELLIER, P., see Courtès, G. — XX, 1
CUMMINS, H.Z., H.L. SWINNEY, Light Beating Spectroscopy — VIII, 133

DAINTY, J.C., The Statistics of Speckle Patterns — XIV, 1
DÄNDLIKER, R., Heterodyne Holographic Interferometry — XVII, 1
DATTOLI, G., L. GIANNESSI, A. RENIERI, A. TORRE, Theory of Compton Free Electron
  Lasers — XXXI, 321
DE STERKE, C.M., J.E. SIPE, Gap Solitons — XXXIII, 203
DECKER JR, J.A., see Harwit, M. — XII, 101
DELANO, E., R.J. PEGIS, Methods of Synthesis for Dielectric Multilayer Filters — VII, 67
DEMARIA, A.J., Picosecond Laser Pulses — IX, 31
DESANTO, J.A., G.S. BROWN, Analytical Techniques for Multiple Scattering from
  Rough Surfaces — XXIII, 1
DETAILLE, M., see Courtès, G. — XX, 1
DEXTER, D.L., see Smith, D.Y. — X, 165
DREXHAGE, K.H., Interaction of Light with Monomolecular Dye Layers — XII, 163
DUGUAY, M.A., The Ultrafast Optical Kerr Shutter — XIV, 161
DUTTA, N.K., J.R. SIMPSON, Optical Amplifiers — XXXI, 189

EBERLY, J.H., Interaction of Very Intense Light with Free Electrons — VII, 359
ENGLUND, J.C., R.R. SNAPP, W.C. SCHIEVE, Fluctuations, Instabilities and Chaos in the
  Laser-Driven Nonlinear Ring Cavity — XXI, 355
ENNOS, A.E., Speckle Interferometry — XVI, 233

FABRE, C., see Reynaud, S. — XXX, 1
FANTE, R.L., Wave Propagation in Random Media: A Systems Approach — XXII, 341
FIORENTINI, A., Dynamic Characteristics of Visual Processes — I, 253

FLYTZANIS, C., F. HACHE, M.C. KLEIN, D. RICARD, PH. ROUSSIGNOL, Nonlinear Optics in Composite Materials. 1. Semiconductor and Metal Crystallites in Dielectrics — XXIX, 321
FOCKE, J., Higher Order Aberration Theory — IV, 1
FRANÇON, M., S. MALLICK, Measurement of the Second Order Degree of Coherence — VI, 71
FREILIKHER, V.D., S.A. GREDESKUL, Localization of Waves in Media with One-Dimensional Disorder — XXX, 137
FRIEDEN, B.R., Evaluation, Design and Extrapolation Methods for Optical Signals, Based on Use of the Prolate Functions — IX, 311
FROEHLY, C., B. COLOMBEAU, M. VAMPOUILLE, Shaping and Analysis of Picosecond Light Pulses — XX, 63
FRY, G.A., The Optical Performance of the Human Eye — VIII, 51

GABOR, D., Light and Information — I, 109
GAMO, H., Matrix Treatment of Partial Coherence — III, 187
GHATAK, A., K. THYAGARAJAN, Graded Index Optical Waveguides: A Review — XVIII, 1
GHATAK, A.K., see Sodha, M.S. — XIII, 169
GIACOBINO, E., B. CAGNAC, Doppler-Free Multiphoton Spectroscopy — XVII, 85
GIACOBINO, E., see Reynaud, S. — XXX, 1
GIANNESSI, L., see Dattoli, G. — XXXI, 321
GINZBURG, V.L., see Agranovich, V.M. — IX, 235
GINZBURG, V.L., Radiation by Uniformly Moving Sources. Vavilov–Cherenkov effect, Doppler effect in a medium, transition radiation and associated phenomena — XXXII, 267
GIOVANELLI, R.G., Diffusion Through Non-Uniform Media — II, 109
GLASER, I., Information Processing with Spatially Incoherent Light — XXIV, 389
GNIADEK, K., J. PETYKIEWICZ, Applications of Optical Methods in the Diffraction Theory of Elastic Waves — IX, 281
GOODMAN, J.W., Synthetic-Aperture Optics — VIII, 1
GOZANI, J., see Charnotskii, M.I. — XXXII, 203
GRAHAM, R., The Phase Transition Concept and Coherence in Atomic Emission — XII, 233
GREDESKUL, S.A., see Freilikher, V.D. — XXX, 137

HACHE, F., see Flytzanis, C. — XXIX, 321
HALL, D.G., Optical Waveguide Diffraction Gratings: Coupling between Guided Modes — XXIX, 1
HARIHARAN, P., Colour Holography — XX, 263
HARIHARAN, P., Interferometry with Lasers — XXIV, 103
HARWIT, M., J.A. DECKER JR, Modulation Techniques in Spectrometry — XII, 101
HASEGAWA, A., see Kodama, Y. — XXX, 205
HEIDMANN, A., see Reynaud, S. — XXX, 1
HELSTROM, C.W., Quantum Detection Theory — X, 289
HERRIOT, D.R., Some Applications of Lasers to Interferometry — VI, 171
HUANG, T.S., Bandwidth Compression of Optical Images — X, 1

IMOTO, N., see Yamamoto, Y. — XXVIII, 87

JACOBSSON, R., Light Reflection from Films of Continuously Varying Refractive Index — V, 247
JACQUINOT, P., B. ROIZEN-DOSSIER, Apodisation — III, 29
JAMROZ, W., B.P. STOICHEFF, Generation of Tunable Coherent Vacuum-Ultraviolet Radiation — XX, 325
JONES, D.G.C., see Allen, L. — IX, 179

KASTLER, A., *see* Cohen-Tannoudji, C.   V, 1
KHOO, I.C., Nonlinear Optics of Liquid Crystals   XXVI, 105
KIELICH, S., Multi-Photon Scattering Molecular Spectroscopy   XX, 155
KINOSITA, K., Surface Deterioration of Optical Glasses   IV, 85
KITAGAWA, M., *see* Yamamoto, Y.   XXVIII, 87
KLEIN, M.C., *see* Flytzanis, C.   XXIX, 321
KLYATSKIN, V.I., The Imbedding Method in Statistical Boundary-Value Wave Problems   XXXIII, 1
KODAMA, Y., A. HASEGAWA, Theoretical Foundation of Optical-Soliton Concept in Fibers   XXX, 205
KOPPELMAN, G., Multiple-Beam Interference and Natural Modes in Open Resonators   VII, 1
KOTTLER, F., The Elements of Radiative Transfer   III, 1
KOTTLER, F., Diffraction at a Black Screen, Part I: Kirchhoff's Theory   IV, 281
KOTTLER, F., Diffraction at a Black Screen, Part II: Electromagnetic Theory   VI, 331
KRAVTSOV, YU.A., Rays and Caustics as Physical Objects   XXVI, 227
KRAVTSOV, YU.A., *see* Barabanenkov, Yu.N.   XXIX, 65
KUBOTA, H., Interference Color   I, 211

LABEYRIE, A., High-Resolution Techniques in Optical Astronomy   XIV, 47
LEAN, E.G., Interaction of Light and Acoustic Surface Waves   XI, 123
LEE, W.-H., Computer-Generated Holograms: Techniques and Applications   XVI, 119
LEITH, E.N., J. UPATNIEKS, Recent Advances in Holography   VI, 1
LETOKHOV, V.S., Laser Selective Photophysics and Photochemistry   XVI, 1
LEVI, L., Vision in Communication   VIII, 343
LIPSON, H., C.A. TAYLOR, X-Ray Crystal-Structure Determination as a Branch of Physical Optics   V, 287
LUGIATO, L.A., Theory of Optical Bistability   XXI, 69
LUKŠ, A., *see* Peřinová, V.   XXXIII, 129

MACHIDA, S., *see* Yamamoto, Y.   XXVIII, 87
MAINFRAY, G., C. MANUS, Nonlinear Processes in Atoms and in Weakly Relativistic Plasmas   XXXII, 313
MALACARA, D., Optical and Electronic Processing of Medical Images   XXII, 1
MALACARA, D., *see* Vlad, V.I.   XXXIII, 261
MALLICK, S., *see* Françon, M.   VI, 71
MANDEL, L., Fluctuations of Light Beams   II, 181
MANDEL, L., The Case For and Against Semiclassical Radiation Theory   XIII, 27
MANDEL, P., *see* Abraham, N.B.   XXV, 1
MANUS, C., *see* Mainfray, G.   XXXII, 313
MARCHAND, E.W., Gradient Index Lenses   XI, 305
MARTIN, P.J., R.P. NETTERFIELD, Optical Films Produced by Ion-Based Techniques   XXIII, 113
MASALOV, A.V., Spectral and Temporal Fluctuations of Broad-Band Laser Radiation   XXII, 145
MAYSTRE, D., Rigorous Vector Theories of Diffraction Gratings   XXI, 1
MEESSEN, A., *see* Rouard, P.   XV, 77
MEHTA, C.L., Theory of Photoelectron Counting   VIII, 373
MEYSTRE, P., Cavity Quantum Optics and the Quantum Measurement Process.   XXX, 261
MIHALACHE, D., M. BERTOLOTTI, C. SIBILIA, Nonlinear Wave Propagation in Planar Structures   XXVII, 227
MIKAELIAN, A.L., M.L. TER-MIKAELIAN, Quasi-Classical Theory of Laser Radiation   VII, 231
MIKAELIAN, A.L., Self-Focusing Media with Variable Index of Refraction   XVII, 279
MILLS, D.L., K.R. SUBBASWAMY, Surface and Size Effects on the Light Scattering Spectra of Solids   XIX, 45

MILONNI, P.W., B. SUNDARAM, Atoms in Strong Fields: Photoionization and Chaos   XXXI,   1
MIYAMOTO, K., Wave Optics and Geometrical Optics in Optical Design   I,   31
MOLLOW, B.R., Theory of Intensity Dependent Resonance Light Scattering and
    Resonance Fluorescence   XIX,   1
MURATA, K., Instruments for the Measuring of Optical Transfer Functions   V, 199
MUSSET, A., A. THELEN, Multilayer Antireflection Coatings   VIII, 201

NARDUCCI, L.M., see Abraham, N.B.   XXV,   1
NETTERFIELD, R.P., see Martin, P.J.   XXIII, 113
NISHIHARA, H., T. SUHARA, Micro Fresnel Lenses   XXIV,   1

OHTSU, M., T. TAKO, Coherence in Semiconductor Lasers   XXV, 191
OKOSHI, T., Projection-Type Holography   XV, 139
OOUE, S., The Photographic Image   VII, 299
OSTROVSKAYA, G.V., YU.I. OSTROVSKY, Holographic Methods of Plasma Diagnostics   XXII, 197
OSTROVSKY, YU.I., see Ostrovskaya, G.V.   XXII, 197
OSTROVSKY, YU.I., V.P. SHCHEPINOV, Correlation Holographic and Speckle Interferom-
    etry   XXX,   87
OUGHSTUN, K.E., Unstable Resonator Modes   XXIV, 165
OZ-VOGT, J., see Beran, M.J.   XXXIII, 319
OZRIN, V.D., see Barabanenkov, Yu.N.   XXIX, 65

PAL, B.P., Guided-Wave Optics on Silicon: Physics, Technology and Status   XXXII,   1
PATORSKI, K., The Self-Imaging Phenomenon and Its Applications   XXVII,   1
PAUL, H., see Brunner, W.   XV,   1
PEGIS, R.J., The Modern Development of Hamiltonian Optics   I,   1
PEGIS, R.J., see Delano, E.   VII, 67
PEŘINA, J., Photocount Statistics of Radiation Propagating Through Random and
    Nonlinear Media   XVIII, 127
PEŘINOVÁ, V., A. LUKŠ, Quantum Statistics of Dissipative Nonlinear Oscillators   XXXIII, 129
PERSHAN, P.S., Non-Linear Optics   V,   83
PETYKIEWICZ, J., see Gniadek, K.   IX, 281
PICHT, J., The Wave of a Moving Classical Electron   V, 351
POPOV, E., Light Diffraction by Relief Gratings: A Macroscopic and Microscopic
    View   XXXI, 139
PORTER, R.P., Generalized Holography with Application to Inverse Scattering and
    Inverse Source Problems   XXVII, 315
PSALTIS, D., see Casasent, D.   XVI, 289
PSALTIS, D., Y. QIAO, Adaptive Multilayer Optical Networks   XXXI, 227

QIAO, Y., see Psaltis, D.   XXXI, 227

RAYMER, M.G., I.A. WALMSLEY, The Quantum Coherence Properties of Stimulated
    Raman Scattering   XXVIII, 181
RENIERI, A., see Dattoli, G.   XXXI, 321
REYNAUD, S., A. HEIDMANN, E. GIACOBINO, C. FABRE, Quantum Fluctuations in Optical
    Systems   XXX,   1
RICARD, D., see Flytzanis, C.   XXIX, 321
RISEBERG, L.A., M.J. WEBER, Relaxation Phenomena in Rare-Earth Luminescence   XIV, 89
RIEKEN, H., Statistical Properties of Laser Light   VIII, 239
RODDIER, F., The Effects of Atmospheric Turbulence in Optical Astronomy   XIX, 281

ROIZEN-DOSSIER, B., *see* Jacquinot, P.	III, 29
RONCHI, L., *see* Wang Shaomin	XXV, 279
ROSENBLUM, W.M., J.L. CHRISTENSEN, Objective and Subjective Spherical Aberration Measurements of the Human Eye	XIII, 69
ROTHBERG, L., Dephasing-Induced Coherent Phenomena	XXIV, 39
ROUARD, P., P. BOUSQUET, Optical Constants of Thin Films	IV, 145
ROUARD, P., A. MEESSEN, Optical Properties of Thin Metal Films	XV, 77
ROUSSIGNOL, PH., *see* Flytzanis, C.	XXIX, 321
RUBINOWICZ, A., The Miyamoto–Wolf Diffraction Wave	IV, 199
RUDOLPH, D., *see* Schmahl, G.	XIV, 195
SAICHEV, A.I., *see* Barabanenkov, Yu.N.	XXIX, 65
SAÏSSE, M., *see* Courtès, G.	XX, 1
SAITO, S., *see* Yamamoto, Y.	XXVIII, 87
SAKAI, H., *see* Vanasse, G.A.	VI, 259
SALEH, B.E.A., *see* Teich, M.C.	XXVI, 1
SCHEERMESSER, T., *see* Bryngdahl, O.	XXXIII, 389
SCHIEVE, W.C., *see* Englund, J.C.	XXI, 355
SCHMAHL, G., D. RUDOLPH, Holographic Diffraction Gratings	XIV, 195
SCHUBERT, M., B. WILHELMI, The Mutual Dependence Between Coherence Properties of Light and Nonlinear Optical Processes	XVII, 163
SCHULZ, G., J. SCHWIDER, Interferometric Testing of Smooth Surfaces	XIII, 93
SCHULZ, G., Aspheric Surfaces	XXV, 349
SCHWIDER, J., *see* Schulz, G.	XIII, 93
SCHWIDER, J., Advanced Evaluation Techniques in Interferometry	XXVIII, 271
SCULLY, M.O., K.G. WHITNEY, Tools of Theoretical Quantum Optics	X, 89
SENITZKY, I.R., Semiclassical Radiation Theory Within a Quantum-Mechanical Framework	XVI, 413
SHCHEPINOV, V.P., *see* Ostrovsky, Yu.I.	XXX, 87
SIBILIA, C., *see* Mihalache, D.	XXVII, 227
SIMPSON, J.R., *see* Dutta, N.K.	XXXI, 189
SIPE, J.E., *see* Van Kranendonk, J.	XV, 245
SIPE, J.E., *see* De Sterke, C.M.	XXXIII, 203
SITTIG, E.K., Elastooptic Light Modulation and Deflection	X, 229
SLUSHER, R.E., Self-Induced Transparency	XII, 53
SMITH, A.W., *see* Armstrong, J.A.	VI, 211
SMITH, D.Y., D.L. DEXTER, Optical Absorption Strength of Defects in Insulators	X, 165
SMITH, R.W., The Use of Image Tubes as Shutters	X, 45
SNAPP, R.R., *see* Englund, J.C.	XXI, 355
SODHA, M.S., A.K. GHATAK, V.K. TRIPATHI, Self-Focusing of Laser Beams in Plasmas and Semiconductors	XIII, 169
SOROKO, L.M., Axicons and Meso-Optical Imaging Devices	XXVII, 109
SPREEUW, R.J.C., J.P. WOERDMAN, Optical Atoms	XXXI, 263
STEEL, W.H., Two-Beam Interferometry	V, 145
STOICHEFF, B.P., *see* Jamroz, W.	XX, 325
STROHBEHN, J.W., Optical Propagation Through the Turbulent Atmosphere	IX, 73
STROKE, G.W., Ruling, Testing and Use of Optical Gratings for High-Resolution Spectroscopy	II, 1
SUBBASWAMY, K.R., *see* Mills, D.L.	XIX, 45
SUHARA, T., *see* Nishihara, H.	XXIV, 1
SUNDARAM, B., *see* Milonni, P.W.	XXXI, 1

SVELTO, O., Self-Focusing, Self-Trapping, and Self-Phase Modulation of Laser Beams	XII, 1
SWEENEY, D.W., see Ceglio, N.M.	XXI, 287
SWINNEY, H.L., see Cummins, H.Z.	VIII, 133
TAKO, T., see Ohtsu, M.	XXV, 191
TANAKA, K., Paraxial Theory in Optical Design in Terms of Gaussian Brackets	XXIII, 63
TANGO, W.J., R.Q. TWISS, Michelson Stellar Interferometry	XVII, 239
TATARSKII, V.I., V.U. ZAVOROTNYI, Strong Fluctuations in Light Propagation in a Randomly Inhomogeneous Medium	XVIII, 204
TATARSKII, V.I., see Charnotskii, M.I.	XXXII, 203
TAYLOR, C.A., see Lipson, H.	V, 287
TEICH, M.C., B.E.A. SALEH, Photon Bunching and Antibunching	XXVI, 1
TER-MIKAELIAN, M.L., see Mikaelian, A.L.	VII, 231
THELEN, A., see Musset, A.	VIII, 201
THOMPSON, B.J., Image Formation with Partially Coherent Light	VII, 169
THYAGARAJAN, K., see Ghatak, A.	XVIII, 1
TONOMURA, A., Electron Holography	XXIII, 183
TORRE, A., see Dattoli, G.	XXXI, 321
TRIPATHI, V.K., see Sodha, M.S.	XIII, 169
TSUJIUCHI, J., Correction of Optical Images by Compensation of Aberrations and by Spatial Frequency Filtering	II, 131
TWISS, R.Q., see Tango, W.J.	XVII, 239
UPATNIEKS, J., see Leith, E.N.	VI, 1
UPSTILL, C., see Berry, M.V.	XVIII, 257
USHIODA, S., Light Scattering Spectroscopy of Surface Electromagnetic Waves in Solids	XIX, 139
VAMPOUILLE, M., see Froehly, C.	XX, 63
VAN DE GRIND, W.A., see Bouman, M.A.	XXII, 77
VAN HEEL, A.C.S., Modern Alignment Devices	I, 289
VAN KRANENDONK, J., J.E. SIPE, Foundations of the Macroscopic Electromagnetic Theory of Dielectric Media	XV, 245
VANASSE, G.A., H. SAKAI, Fourier Spectroscopy	VI, 259
VERNIER, P.J., Photoemission	XIV, 245
VLAD, V.I., D. MALACARA, Direct Spatial Reconstruction of Optical Phase from Phase-Modulated Images	XXXIII, 261
WALMSLEY, I.A., see Raymer, M.G.	XXVIII, 181
WANG SHAOMIN, L. RONCHI, Principles and Design of Optical Arrays	XXV, 279
WEBER, M.J., see Riseberg, L.A.	XIV, 89
WEIGELT, G., Triple-Correlation Imaging in Optical Astronomy	XXIX, 293
WELFORD, W.T., Aberration Theory of Gratings and Grating Mountings	IV, 241
WELFORD, W.T., Aplanatism and Isoplanatism	XIII, 267
WELFORD, W.T., see Bassett, I.M.	XXVII, 161
WHITNEY, K.G., see Scully, M.O.	X, 89
WILHELMI, B., see Schubert, M.	XVII, 163
WINSTON, R., see Bassett, I.M.	XXVII, 161
WOERDMAN, J.P., see Spreeuw, R.J.C.	XXXI, 263
WOLTER, H., On Basic Analogies and Principal Differences between Optical and Electronic Information	I, 155

WYNNE, C.G., Field Correctors for Astronomical Telescopes   X, 137
WYROWSKI, F., see Bryngdahl, O.   XXVIII,   1
WYROWSKI, F., see Bryngdahl, O.   XXXIII, 389

YAMAGUCHI, I., Fringe Formations in Deformation and Vibration Measurements using
   Laser Light   XXII, 271
YAMAJI, K., Design of Zoom Lenses   VI, 105
YAMAMOTO, T., Coherence Theory of Source-Size Compensation in Interference
   Microscopy   VIII, 295
YAMAMOTO, Y., S. MACHIDA, S. SAITO, N. IMOTO, T. YANAGAWA, M. KITAGAWA,
   G. BJÖRK, Quantum Mechanical Limit in Optical Precision Measurement and
   Communication   XXVIII, 87
YANAGAWA, T., see Yamamoto, Y.   XXVIII, 87
YAROSLAVSKY, L.P., The Theory of Optimal Methods for Localization of Objects in
   Pictures   XXXII, 145
YOSHINAGA, H., Recent Developments in Far Infrared Spectroscopic Techniques   XI, 77
YU, F.T.S., Principles of Optical Processing with Partially Coherent Light   XXIII, 221
YU, F.T.S., Optical Neural Networks: Architecture, Design and Models   XXXII, 61

ZAVOROTNY, V.U., see Charnotskii, M.I.   XXXII, 203
ZAVOROTNYI, V.U., see Tatarskii, V.I.   XVIII, 204
ZUIDEMA, P., see Bouman, M.A.   XXII, 77